Computer vision technology in the food and beverage industries

Related titles:

Robotics and automation in the food industry
(ISBN 978-1-84569-801-0)
The cost of electronics has significantly reduced in recent years and investment in robotics and automation in the food sector is increasing. These technologies have great potential to improve safety, quality and profitability by optimising process monitoring and control, and reducing manual labour. This book reviews current and emerging robotics and automation technologies and their applications in different industry sectors.

Rapid and on-line instrumentation for food quality assurance
(ISBN 978-1-85573-674-0)
With its high volumes of production, the food industry has an urgent need for instrumentation that can be used online and which produces rapid measurements. As a result, there has been a wealth of research in this field, which is summarised in this important collection. The first part of the book looks at techniques for the rapid detection of contaminants such as pesticides, veterinary drug residues and foreign bodies. The second part of the book considers methods for identifying ingredients, including additives, and measuring product quality.

Food process modelling
(ISBN 978-1-85573-565-1)
Food process modelling provides an authoritative review of one of the most exciting and influential developments in the food industry. The modelling of food processes allows analysts not only to understand such processes more clearly but also to control them more closely and make predictions about them. Modelling thus aids the search for greater and more consistent food quality. Written by a distinguished international team of experts, *Food process modelling* covers both the range of modelling techniques and their practical applications across the food chain.

Details of these books and a complete list of titles from Woodhead Publishing can be obtained by:

- visiting our web site at www.woodheadpublishing.com
- contacting Customer Services (e-mail: sales@woodheadpublishing.com; fax: +44 (0) 1223 832819; tel.: +44 (0) 1223 499140 ext. 130; address: Woodhead Publishing Limited, 80, High Street, Sawston, Cambridge CB22 3HJ, UK)
- in North America, contacting our US office (e-mail: usmarketing@woodheadpublishing.com; tel.: (215) 928 9112; address: Woodhead Publishing, 1518 Walnut Street, Suite 1100, Philadelphia, PA 19102-3406, USA)

If you would like e-versions of our content, please visit our online platform: www.woodheadpublishingonline.com. Please recommend it to your librarian so that everyone in your institution can benefit from the wealth of content on the site.

Woodhead Publishing Series in Food Science, Technology and Nutrition:
Number 238

Computer vision technology in the food and beverage industries

**Edited by
Da-Wen Sun**

Oxford   Cambridge   Philadelphia   New Delhi

Published by Woodhead Publishing Limited,
80 High Street, Sawston, Cambridge CB22 3HJ, UK
www.woodheadpublishing.com
www.woodheadpublishingonline.com

Woodhead Publishing, 1518 Walnut Street, Suite 1100, Philadelphia, PA 19102-3406, USA

Woodhead Publishing India Private Limited, G-2, Vardaan House,
7/28 Ansari Road, Daryaganj, New Delhi – 110002, India
www.woodheadpublishingindia.com

First published 2012, Woodhead Publishing Limited
© Woodhead Publishing Limited, 2012; except Chapter 1 © Elsevier, 2007 and Chapter 17 © Canadian Grain Commission, Government of Canada, 2012
The authors have asserted their moral rights.

British Library Cataloguing in Publication Data
A catalogue record for this book is available from the British Library.

Library of Congress Control Number: 2012942829

ISBN 978-0-85709-036-2 (print)
ISBN 978-0-85709-577-0 (online)
ISSN 2042-8049 Woodhead Publishing Series in Food Science, Technology and Nutrition (print)
ISSN 2042-8057 Woodhead Publishing Series in Food Science, Technology and Nutrition (online)

The publisher's policy is to use permanent paper from mills that operate a sustainable forestry policy, and which has been manufactured from pulp which is processed using acid-free and elemental chlorine-free practices. Furthermore, the publisher ensures that the text paper and cover board used have met acceptable environmental accreditation standards.

Typeset by Newgen Publishing and Data Services, India
Printed by TJ International Ltd, Padstow, Cornwall, UK

Contents

Contributor contact details .... *xi*

About the editor .... *xv*

Woodhead Publishing Series in Food Science, Technology and Nutrition .... *xviii*

Part I An introduction to computer vision in the food and beverage industries **1**

1 Computer vision and infrared techniques for image acquisition in the food and beverage industries **3**

M. Z. Abdullah, Universiti Sains Malaysia, Malaysia

1.1 Introduction 3

1.2 The electromagnetic spectrum 5

1.3 Image acquisition systems 7

1.4 Conclusions 24

1.5 References 24

1.6 Appendix: nomenclature and abbreviations 26

2 Hyperspectral and multispectral imaging in the food and beverage industries **27**

J. Qin, United States Department of Agriculture, USA

2.1 Introduction 27

2.2 Spectral image acquisition methods 28

2.3 Construction of spectral imaging systems 30

2.4 Calibration of spectral imaging systems 41

2.5 Spectral images and analysis techniques 47

2.6 Applications for food and beverage products 54

2.7 Conclusions 58

2.8   Further information..   58
2.9   References...   59

**3   Tomographic techniques for computer vision in the
       food and beverage industries..   64**
       M. Z. Abdullah, Universiti Sains Malaysia, Malaysia
3.1   Introduction...   64
3.2   Nuclear tomography..   65
3.3   Electrical impedance...   69
3.4   Image reconstruction..   86
3.5   Applications..   89
3.6   Conclusions..   93
3.7   References...   93
3.8   Appendix: nomenclature and abbreviations........................   95

**4   Image processing techniques for computer vision in the
       food and beverage industries..   97**
       N. A. Valous and D.-W. Sun, University College Dublin, Ireland
4.1   Introduction...   97
4.2   Digital image analysis techniques......................................   99
4.3   Classification..   113
4.4   Relevance, impact and trends for the
       food and beverage industry...   119
4.5   Conclusions..   122
4.6   References...   123

**Part II   Computer vision applications in food and beverage
            processing operations/technologies.......................   131**

5   Computer vision in food processing: an overview...........   133
       R. Lind and A. Murhed, SICK IVP AB, Sweden
5.1   Introduction to computer vision...   133
5.2   Technology selection..   136
5.3   Selection of image analysis methods.................................   139
5.4   Application examples..   143
5.5   Conclusion..   148
5.6   References...   148

**6   Computer vision for automatic sorting in the food
       industry..   150**
       E. R. Davies, Royal Holloway, University of London, UK
6.1   Introduction...   150
6.2   Basic techniques and their application...............................   152
6.3   Advanced techniques and their application........................   160
6.4   Alternative image modalities..   172

6.5   Special real-time hardware for food sorting 173
6.6   Recent advances in computer vision for food sorting.................. 175
6.7   Future trends 176
6.8   Conclusion 177
6.9   Sources of further information and advice.................................. 177
6.10   Acknowledgements... 178
6.11   References.. 178

**7 Computer vision for foreign body detection and
removal in the food industry** **181**
*N. Toyofuku and R. P. Haff, Plant Mycotoxin Research
Unit, USDA ARS WRRC, USA*
7.1   Introduction... 181
7.2   Optical inspection 183
7.3   Fundamentals of X-ray inspection... 188
7.4   X-ray inspection of food products 195
7.5   Conclusions... 200
7.6   References.. 200

**8 Automated cutting in the food industry using computer
vision**.. **206**
*W. D. R. Daley, Georgia Institute of Technology, USA and
O. Arif, King Abdullah University of Science and Technology,
Saudi Arabia*
8.1   Introduction... 206
8.2   Machine vision and computer vision 208
8.3   Feature selection, extraction and analysis................................... 211
8.4   Machine learning algorithms 212
8.5   Application examples: sensing for automated
cutting and handling.. 213
8.6   Future trends 228
8.7   Conclusions... 230
8.8   Acknowledgments.. 231
8.9   References.. 231

9 Image analysis of food microstructure **233**
J. C. Russ, North Carolina State University, USA
9.1   Introduction... 233
9.2   Quality control applications of digital imaging 234
9.3   Characterizing the internal structure 237
9.4   Volume, surface and length.. 238
9.5   Number and spatial distribution.. 243
9.6   Surfaces and fractal dimensions... 249
9.7   Conclusions... 250
9.8   References.. 251

Part III   Current and future applications of computer vision for quality control and processing of particular products **253**

10 Computer vision in the fresh and processed meat industries **255**
P. Jackman and D.-W. Sun, University College Dublin, Ireland
10.1   Introduction 255
10.2   Meat image features 256
10.3   Application and implementation 260
10.4   Application and implementation for lamb, pork and other processed meats 269
10.5   Future trends 271
10.6   Conclusions 271
10.7   References 272

11 Real-time ultrasound (RTU) imaging methods for quality control of meats **277**
S. R. Silva, CECAV, University of Trás-os-Montes e Alto Douro, Portugal and V. P. Cadavez, CIMO, ESA, Instituto Politécnico de Bragança, Portugal
11.1   Introduction 277
11.2   Historical background on ultrasound use for carcass composition and meat traits evaluation 278
11.3   Basic ultrasound imaging principles 282
11.4   Applications of real-time ultrasound (RTU) to predict carcass composition and meat traits in large animals 285
11.5   Applications of RTU to predict carcass composition and meat traits in small animals and fish 293
11.6   Using real-time ultrasonography to predict intramuscular fat (IMF) *in vivo* 303
11.7   Optimization of production system and market carcass characteristics 310
11.8   The future for RTU imaging in the meat industry 313
11.9   Conclusion 314
11.10 References 315

12 Computer vision in the poultry industry **330**
K. Chao, Henry A. Wallace Beltsville Agricultural Research Center, USA, B. Park, Richard B. Russell Research Center, USA and M. S. Kim, Henry A. Wallace Beltsville Agricultural Research Center, USA
12.1   Introduction 330
12.2   Poultry processing applications 331
12.3   Development of spectral imaging for poultry inspection 333
12.4   Case studies for online line-scan poultry safety inspection 336
12.5   Future trends 350

12.6   Conclusions.. 350
12.7   References.. 351

13 Computer vision in the fish industry **352**
J. R. Mathiassen, E. Misimi, S. O. Østvik and I. G. Aursand,
Department of Processing Technology, SINTEF Fisheries and
Aquaculture, Norway
13.1   Introduction.. 352
13.2   The need for computer vision in the fish industry 353
13.3   Automated sorting and grading.................................... 354
13.4   Automated processing.. 360
13.5   Process understanding and optimization..................... 367
13.6   Challenges in applying computer vision in the fish industry....... 373
13.7   Future trends 374
13.8   Further information.. 375
13.9   Conclusions.. 376
13.10 References... 376

14 Fruit, vegetable and nut quality evaluation and control
   using computer vision **379**
J. Blasco, IVIA – Instituto Valenciano de Investigaciones Agrarias,
Centro de Agroingeniería, Spain, N. Aleixos, I3BH, Universitat
Politècnica de València, Spain and S. Cubero and D. Lorente, IVIA –
Instituto Valenciano de Investigaciones Agrarias, Centro de
Agroingeniería, Spain
14.1   Introduction.. 379
14.2   Basics of machine vision systems for fruit,
       vegetable and nut quality evaluation and control 381
14.3   Applications of computer vision in the inspection
       of external features... 386
14.4   Real-time automatic inspection systems..................... 388
14.5   Future trends 392
14.6   Conclusions.. 394
14.7   Sources of further information.................................... 395
14.8   Acknowledgements.. 396
14.9   References.. 396

15 Grain quality evaluation by computer vision **400**
D. S. Jayas and C. B. Singh, University of Manitoba, Canada
15.1   Introduction.. 400
15.2   Colour imaging 402
15.3   Hyperspectral imaging.. 406
15.4   X-ray imaging 411
15.5   Thermal imaging... 415
15.6   Conclusions.. 418

15.7   Acknowledgements...   418
15.8   References..   419

16 Computer vision in the bakery industry **422**
C.-J. Du, University of Warwick, UK, Q. Cheng, University of
Reading, UK, and D.-W. Sun, University College Dublin, Ireland
16.1   Introduction...   422
16.2   Computer vision applications for analysing bread......................   423
16.3   Computer vision applications for analysing muffins..................   432
16.4   Computer vision applications for analysing biscuits.................   436
16.5   Computer vision applications for analysing pizza bases............   439
16.6   Computer vision applications for analysing
       other bakery products..   444
16.7   Future trends and further information....................................   445
16.8   Conclusions...   446
16.9   References..   447

17 Development of multispectral imaging systems for quality
   evaluation of cereal grains and grain products **451**
M. A. Shahin, D. W. Hatcher and S. J. Symons, Canadian Grain
Commission, Canada
17.1   Introduction...   452
17.2   Hyperspectral imaging..   456
17.3   Detection of mildew damage in wheat....................................   459
17.4   Detection of fusarium damage in wheat..................................   461
17.5   Sprout damage in wheat...   465
17.6   Determination of green immature kernels in cereal grains.........   469
17.7   Effect of mildew on the quality of end-products........................   471
17.8   Development of multispectral imaging systems........................   473
17.9   Conclusions...   477
17.10 Acknowledgements...   478
17.11 References..   478

Index..   *483*

Contributor contact details

(* = main contact)

Editor
Da-Wen Sun
FRCFT
University College Dublin
Belfield
Dublin 4
Ireland

E-mail: dawen.sun@ucd.ie

Chapters 1 and 3
M. Z. Abdullah
School of Electrical and Electronic
    Engineering
Engineering Campus
Universiti Sains Malaysia
14300 Penang
Malaysia

E-mail: mza@eng.usm.my

Chapter 2
Jianwei Qin
Environmental Microbial and Food
    Safety Laboratory
Henry A. Wallace Beltsville
    Agricultural Research Center
Agricultural Research Service
United States Department of
    Agriculture
USDA/ARS/EMFSL
Bldg. 303
BARC-East, 10300 Baltimore Ave.
Beltsville
MD 20705-2350
USA

E-mail: jianwei.qin@ars.usda.gov

Chapter 4
Nektarios A. Valous and Da-Wen Sun*
FRCFT
University College Dublin
Belfield
Dublin 4
Ireland

E-mail: dawen.sun@ucd.ie

Chapter 5

Rasmus Lind* and Anders Murhed
SICK IVP AB
Wallenbergs gata 4
583 30 Linköping
Sweden

E-mail: Rasmus.Lind@sickivp.se;
    Anders.Murhed@sickivp.se

Chapter 6

E. Roy Davies
Machine Vision Group
Department of Physics
Royal Holloway, University of
    London
Egham
Surrey
TW20 0EX
UK

E-mail: e.r.davies@rhul.ac.uk

Chapter 7

Natsuko Toyofuku* and Ron P. Haff
Plant Mycotoxin Research Unit
USDA ARS WRRC
800 Buchanan St
Albany, CA 94710
USA

E-mail: Natsuko.Toyofuku@ars.usda.
    gov; ron.haff@ars.usda.gov

Chapter 8

Wayne D. R. Daley*
GTRI/ATAS/FPTD
Georgia Institute of Technology
640 Strong Street
Atlanta, GA 30332-0823
USA

E-mail: wayne.daley@gtri.gatech.edu

Omar Arif
King Abdullah University of Science
    and Technology
Saudi Arabia

Chapter 9

John C. Russ
Professor Emeritus
College of Engineering
North Carolina State University
Raleigh, NC
USA

E-mail: John_Russ@NCSU.edu

Chapter 10

Patrick Jackman and Da-Wen Sun*
FRCFT
University College Dublin
National University of Ireland
Agriculture and Food Science Centre
Belfield
Dublin 4
Ireland

E-mail: dawen.sun@ucd.ie; patrick.
    jackman@manchester.ac.uk

Chapter 11

Severiano R. Silva*
CECAV Research Center
Department of Animal Science
University of Trás-os-Montes e Alto
  Douro
PO Box 1013
5001-811 Vila Real
Portugal

E-mail: ssilva@utad.pt

Vasco P. Cadavez
CIMO Research Center
Department of Animal Science
Escola Superior Agrária, Instituto
  Politécnico de Bragança
PO Box 1172
5301-855 Bragança
Portugal

E-mail: vcadavez@ipb.pt

Chapter 12

Kuanglin Chao*
USDA, ARS
Henry A. Wallace Beltsville
  Agricultural Research Center
Environmental Microbial and Food
  Safety Laboratory
10300 Baltimore Avenue
Building 303 BARC-East
Beltsville, MD 20705-2350
USA

E-mail: kevin.chao@ars.usda.gov

Bosoon Park
USDA, ARS
Richard B. Russell Research Center
PO Box 5677
Athens, GA 30602-5677
USA

Moon S. Kim
USDA, ARS
Henry A. Wallace Beltsville
  Agricultural Research Center
Environmental Microbial and Food
  Safety Laboratory
10300 Baltimore Avenue
Building 303 BARC-East
Beltsville, MD 20705-2350
USA

Chapter 13

J. R. Mathiassen*, E. Misimi, S. O.
  Østvik and I. G. Aursand
Department of Processing Technology
SINTEF Fisheries and Aquaculture
Brattørkaia 17C
NO-7010 Trondheim
Norway

E-mail: John.Reidar.Mathiassen@
  sintef.no; Ekrem.Misimi@sintef.no;
  Ida.G.Aursand@sintef.no

Chapter 14

José Blasco*
IVIA – Instituto Valenciano de
  Investigaciones Agrarias
Centro de Agroingeniería
Cra. Moncada-Náquera km 5
46113 Moncada (Valencia)
Spain

E-mail: blasco_josiva@gva.es

Nuria Aleixos
Instituto Interuniversitario de
  Investigación en Bioingeniería
  y Tecnología Orientada al Ser
  Humano
Universitat Politècnica de València
Camino de Vera s/n
46022 Valencia
Spain

E-mail: naleixos@dig.upv.es

Sergio Cubero and Delia Lorente
IVIA – Instituto Valenciano de
    Investigaciones Agrarias
Centro de Agroingeniería
Cra. Moncada-Náquera km 5
46113 Moncada (Valencia)
Spain

E-mail: lorente_del@gva.es; cubero_
    ser@gva.es

Chapter 15

Digvir S. Jayas* and Chandra B.
    Singh
Department of Biosystems
Engineering
University of Manitoba
Winnipeg
MB, R3T 5V6
Canada

E-mail: digvir.jayas@ad.umanitoba.
    ca; singhcb@gmail.com

Chapter 16

Cheng-Jin Du*
Warwick Systems Biology Centre
University of Warwick
Coventry
CV4 7AL
UK

E-mail: c.du@warwick.ac.uk

Qiaofen Cheng
Department of Food and Nutritional
Sciences
University of Reading
Whiteknights
Reading, RG6 6AP
UK

E-mail: q.cheng@reading.ac.uk

Da-Wen Sun
Food Refrigeration and Computerised
Food Technology
National University of Ireland
Dublin
(University College Dublin)
Agriculture and Food Science Centre
Belfield
Dublin 4
Ireland

E-mail: dawen.sun@ucd.ie

Chapter 17

Muhammad A. Shahin,* Dave
    W. Hatcher and Stephen J. Symons
Grain Research Laboratory
Canadian Grain Commission
1404-303 Main Street
Winnipeg
MB, R3C 3G8
Canada

E-mail: muhammad.shahin@
    grainscanada.gc.ca; Dave.hatcher@
    grainscanada.gc.ca; Stephen.
    symons@grainscanada.gc.ca

About the editor

Born in Southern China, Professor Da-Wen Sun is a world authority in food engineering research and education; he is a Member of the Royal Irish Academy which is the highest academic honour in Ireland, he is also a Member of Academia Europaea (The Academy of Europe), and a Fellow of International Academy of Food Science and Technology. His main research activities include: cooling, drying and refrigeration processes and systems; quality and safety of food products; bioprocess simulation and optimisation; and computer vision technology. His many scholarly works have become standard reference materials for researchers in the areas of computer vision, computational fluid dynamics modelling, vacuum cooling, etc. Results of his work have been published in some 600 papers including over 250 peer-reviewed journal papers (h-index = 36). He has also edited 13 authoritative books. According to Thomson Scientific's *Essential Science Indicators*[SM], based on data derived over a period of 10 years and 4 months (1 January 2000–30 April 2010) from the ISI Web of Science, a total of 2554 scientists are among the top 1% of the most cited scientists in the category of Agriculture Sciences, and Professor Sun tops the list with his ranking of 31.

He gained a first class BSc(Hons) and MSc in Mechanical Engineering, and a PhD in Chemical Engineering in China before working in various universities in Europe. He became the first Chinese national to be permanently employed in an Irish University when he was appointed College Lecturer at the National University of Ireland, Dublin (University College Dublin) in 1995, and was subsequently promoted in the shortest possible time to full professor. Dr Sun is now Professor of Food and Biosystems Engineering and Director of the Food Refrigeration and Computerised Food Technology Research Group at University College Dublin (UCD).

As a leading educator in food engineering, Professor Sun has significantly contributed to the field of food engineering. He has trained many PhD students, who have made their own contributions to the industry and academia. He has also given lectures on advances in food engineering on a regular basis in academic institutions internationally and delivered keynote speeches at international conferences. As a recognised authority in food engineering, he has been conferred adjunct/ visiting/consulting professorships from ten top universities in China. In recognition of his significant contribution to food engineering worldwide and for his outstanding leadership in the field, the International Commission of Agricultural and Biosystems Engineering (CIGR) awarded him the 'CIGR Merit Award' in 2000 and again in 2006; the Institution of Mechanical Engineers (IMechE) based in the UK named him 'Food Engineer of the Year 2004; and in 2008 he was awarded 'CIGR Recognition Award' in honour of his distinguished achievements (being in the top 1% of Agricultural Engineering scientists in the world). In 2007 he was presented with the only 'AFST(I) Fellow Award' in that year by the Association of Food Scientists and Technologists (India) and in 2010 he was presented with the 'CIGR Fellow Award', the title of Fellow is the highest honour in CIGR, and is conferred to individuals who have made sustained, outstanding contributions worldwide.

He is a Fellow of the Institution of Agricultural Engineers and a Fellow of Engineers Ireland (the Institution of Engineers of Ireland); he is also a Chartered Engineer. He has received numerous awards for teaching and research excellence, including the President's Research Fellowship, and has twice received the President's Research Award of University College Dublin. He is Editor-in-Chief of *Food and Bioprocess Technology – an International Journal* (Springer) (2010 Impact Factor = 3.576, ranked fourth among 126 ISI-listed food science and technology journals), Series Editor of the 'Contemporary Food Engineering' book series (CRC Press/Taylor & Francis), former Editor of *Journal of Food Engineering* (Elsevier), and Editorial Board Member for a number of international journals including *Journal of Food Process Engineering, Sensing and Instrumentation for Food Quality and Safety, Czech Journal of Food Sciences, Polish Journal of Food and Nutritional Sciences, etc.*

In May 2010 he was awarded membership of the Royal Irish Academy (RIA), which is the highest honour that can be attained by scholars and scientists working in Ireland, and at the 51st CIGR General Assembly held during the CIGR World Congress in Quebec City, Canada in June 2010, he was elected Incoming President of CIGR, and will become CIGR President in 2013–2014 – the term of his CIGR presidency is 6 years, with 2 years spent as Incoming President, President, and Past President respectively. In September 2011, he was elected to Academia Europaea (The Academy of Europe), which functions as the European Academy of Humanities, Letters and Sciences and is one of the most prestigious academies in the world, election to the Academia Europaea represents the highest academic distinction.

Woodhead Publishing Series in Food Science, Technology and Nutrition

1 Chilled foods: a comprehensive guide *Edited by C. Dennis and M. Stringer*
2 Yoghurt: science and technology *A. Y. Tamime and R. K. Robinson*
3 Food processing technology: principles and practice *P. J. Fellows*
4 Bender's dictionary of nutrition and food technology Sixth edition *D. A. Bender*
5 Determination of veterinary residues in food *Edited by N. T. Crosby*
6 Food contaminants: sources and surveillance *Edited by C. Creaser and R. Purchase*
7 Nitrates and nitrites in food and water *Edited by M. J. Hill*
8 Pesticide chemistry and bioscience: the food-environment challenge *Edited by G. T. Brooks and T. Roberts*
9 Pesticides: developments, impacts and controls *Edited by G. A. Best and A. D. Ruthven*
10 Dietary fibre: chemical and biological aspects *Edited by D. A. T. Southgate, K. W. Waldron, I. T. Johnson and G. R. Fenwick*
11 Vitamins and minerals in health and nutrition *M. Tolonen*
12 Technology of biscuits, crackers and cookies Second edition *D. Manley*
13 Instrumentation and sensors for the food industry *Edited by E. Kress-Rogers*
14 Food and cancer prevention: chemical and biological aspects *Edited by K. W. Waldron, I. T. Johnson and G. R. Fenwick*
15 Food colloids: proteins, lipids and polysaccharides *Edited by E. Dickinson and B. Bergenstahl*
16 Food emulsions and foams *Edited by E. Dickinson*
17 Maillard reactions in chemistry, food and health *Edited by T. P. Labuza, V. Monnier, J. Baynes and J. O'Brien*
18 The Maillard reaction in foods and medicine *Edited by J. O'Brien, H. E. Nursten, M. J. Crabbe and J. M. Ames*
19 Encapsulation and controlled release *Edited by D. R. Karsa and R. A. Stephenson*
20 Flavours and fragrances *Edited by A. D. Swift*
21 Feta and related cheeses *Edited by A. Y. Tamime and R. K. Robinson*
22 Biochemistry of milk products *Edited by A. T. Andrews and J. R. Varley*
23 Physical properties of foods and food processing systems *M. J. Lewis*

24 Food irradiation: a reference guide *V. M. Wilkinson and G. Gould*
25 Kent's technology of cereals: an introduction for students of food science and agriculture Fourth edition *N. L. Kent and A. D. Evers*
26 Biosensors for food analysis *Edited by A. O. Scott*
27 Separation processes in the food and biotechnology industries: principles and applications *Edited by A. S. Grandison and M. J. Lewis*
28 Handbook of indices of food quality and authenticity *R. S. Singhal, P. K. Kulkarni and D. V. Rege*
29 Principles and practices for the safe processing of foods *D. A. Shapton and N. F. Shapton*
30 Biscuit, cookie and cracker manufacturing manuals Volume 1: ingredients *D. Manley*
31 Biscuit, cookie and cracker manufacturing manuals Volume 2: biscuit doughs *D. Manley*
32 Biscuit, cookie and cracker manufacturing manuals Volume 3: biscuit dough piece forming *D. Manley*
33 Biscuit, cookie and cracker manufacturing manuals Volume 4: baking and cooling of biscuits *D. Manley*
34 Biscuit, cookie and cracker manufacturing manuals Volume 5: secondary processing in biscuit manufacturing *D. Manley*
35 Biscuit, cookie and cracker manufacturing manuals Volume 6: biscuit packaging and storage *D. Manley*
36 Practical dehydration Second edition *M. Greensmith*
37 Lawrie's meat science Sixth edition *R. A. Lawrie*
38 Yoghurt: science and technology Second edition *A. Y. Tamime and R. K. Robinson*
39 New ingredients in food processing: biochemistry and agriculture *G. Linden and D. Lorient*
40 Benders' dictionary of nutrition and food technology Seventh edition *D. A. Bender and A. E. Bender*
41 Technology of biscuits, crackers and cookies Third edition *D. Manley*
42 Food processing technology: principles and practice Second edition *P. J. Fellows*
43 Managing frozen foods *Edited by C. J. Kennedy*
44 Handbook of hydrocolloids *Edited by G. O. Phillips and P. A. Williams*
45 Food labelling *Edited by J. R. Blanchfield*
46 Cereal biotechnology *Edited by P. C. Morris and J. H. Bryce*
47 Food intolerance and the food industry *Edited by T. Dean*
48 The stability and shelf-life of food *Edited by D. Kilcast and P. Subramaniam*
49 Functional foods: concept to product *Edited by G. R. Gibson and C. M. Williams*
50 Chilled foods: a comprehensive guide Second edition *Edited by M. Stringer and C. Dennis*
51 HACCP in the meat industry *Edited by M. Brown*
52 Biscuit, cracker and cookie recipes for the food industry *D. Manley*
53 Cereals processing technology *Edited by G. Owens*
54 Baking problems solved *S. P. Cauvain and L. S. Young*
55 Thermal technologies in food processing *Edited by P. Richardson*
56 Frying: improving quality *Edited by J. B. Rossell*
57 Food chemical safety Volume 1: contaminants *Edited by D. Watson*
58 Making the most of HACCP: learning from others' experience *Edited by T. Mayes and S. Mortimore*
59 Food process modelling *Edited by L. M. M. Tijskens, M. L. A. T. M. Hertog and B. M. Nicolaï*
60 EU food law: a practical guide *Edited by K. Goodburn*
61 Extrusion cooking: technologies and applications *Edited by R. Guy*

62 **Auditing in the food industry: from safety and quality to environmental and other audits** *Edited by M. Dillon and C. Griffith*
63 **Handbook of herbs and spices Volume 1** *Edited by K. V. Peter*
64 **Food product development: maximising success** *M. Earle, R. Earle and A. Anderson*
65 **Instrumentation and sensors for the food industry Second edition** *Edited by E. Kress-Rogers and C. J. B. Brimelow*
66 **Food chemical safety Volume 2: additives** *Edited by D. Watson*
67 **Fruit and vegetable biotechnology** *Edited by V. Valpuesta*
68 **Foodborne pathogens: hazards, risk analysis and control** *Edited by C. de W. Blackburn and P. J. McClure*
69 **Meat refrigeration** *S. J. James and C. James*
70 **Lockhart and Wiseman's crop husbandry Eighth edition** *H. J. S. Finch, A. M. Samuel and G. P. F. Lane*
71 **Safety and quality issues in fish processing** *Edited by H. A. Bremner*
72 **Minimal processing technologies in the food industries** *Edited by T. Ohlsson and N. Bengtsson*
73 **Fruit and vegetable processing: improving quality** *Edited by W. Jongen*
74 **The nutrition handbook for food processors** *Edited by C. J. K. Henry and C. Chapman*
75 **Colour in food: improving quality** *Edited by D MacDougall*
76 **Meat processing: improving quality** *Edited by J. P. Kerry, J. F. Kerry and D. A. Ledward*
77 **Microbiological risk assessment in food processing** *Edited by M. Brown and M. Stringer*
78 **Performance functional foods** *Edited by D. Watson*
79 **Functional dairy products Volume 1** *Edited by T. Mattila-Sandholm and M. Saarela*
80 **Taints and off-flavours in foods** *Edited by B. Baigrie*
81 **Yeasts in food** *Edited by T. Boekhout and V. Robert*
82 **Phytochemical functional foods** *Edited by I. T. Johnson and G. Williamson*
83 **Novel food packaging techniques** *Edited by R. Ahvenainen*
84 **Detecting pathogens in food** *Edited by T. A. McMeekin*
85 **Natural antimicrobials for the minimal processing of foods** *Edited by S. Roller*
86 **Texture in food Volume 1: semi-solid foods** *Edited by B. M. McKenna*
87 **Dairy processing: improving quality** *Edited by G. Smit*
88 **Hygiene in food processing: principles and practice** *Edited by H. L. M. Lelieveld, M. A. Mostert, B. White and J. Holah*
89 **Rapid and on-line instrumentation for food quality assurance** *Edited by I. Tothill*
90 **Sausage manufacture: principles and practice** *E. Essien*
91 **Environmentally-friendly food processing** *Edited by B. Mattsson and U. Sonesson*
92 **Bread making: improving quality** *Edited by S. P. Cauvain*
93 **Food preservation techniques** *Edited by P. Zeuthen and L. Bøgh-Sørensen*
94 **Food authenticity and traceability** *Edited by M. Lees*
95 **Analytical methods for food additives** *R. Wood, L. Foster, A. Damant and P. Key*
96 **Handbook of herbs and spices Volume 2** *Edited by K. V. Peter*
97 **Texture in food Volume 2: solid foods** *Edited by D. Kilcast*
98 **Proteins in food processing** *Edited by R. Yada*
99 **Detecting foreign bodies in food** *Edited by M. Edwards*
100 **Understanding and measuring the shelf-life of food** *Edited by R. Steele*
101 **Poultry meat processing and quality** *Edited by G. Mead*
102 **Functional foods, ageing and degenerative disease** *Edited by C. Remacle and B. Reusens*

103 Mycotoxins in food: detection and control *Edited by N. Magan and M. Olsen*

104 Improving the thermal processing of foods *Edited by P. Richardson*

105 Pesticide, veterinary and other residues in food *Edited by D. Watson*

106 Starch in food: structure, functions and applications *Edited by A.-C. Eliasson*

107 Functional foods, cardiovascular disease and diabetes *Edited by A. Arnoldi*

108 Brewing: science and practice *D. E. Briggs, P. A. Brookes, R. Stevens and C. A. Boulton*

109 Using cereal science and technology for the benefit of consumers: proceedings of the 12th International ICC Cereal and Bread Congress, 24 – 26th May, 2004, Harrogate, UK *Edited by S. P. Cauvain, L. S. Young and S. Salmon*

110 Improving the safety of fresh meat *Edited by J. Sofos*

111 Understanding pathogen behaviour: virulence, stress response and resistance *Edited by M. Griffiths*

112 The microwave processing of foods *Edited by H. Schubert and M. Regier*

113 Food safety control in the poultry industry *Edited by G. Mead*

114 Improving the safety of fresh fruit and vegetables *Edited by W. Jongen*

115 Food, diet and obesity *Edited by D. Mela*

116 Handbook of hygiene control in the food industry *Edited by H. L. M. Lelieveld, M. A. Mostert and J. Holah*

117 Detecting allergens in food *Edited by S. Koppelman and S. Hefle*

118 Improving the fat content of foods *Edited by C. Williams and J. Buttriss*

119 Improving traceability in food processing and distribution *Edited by I. Smith and A. Furness*

120 Flavour in food *Edited by A. Voilley and P. Etievant*

121 The Chorleywood bread process *S. P. Cauvain and L. S. Young*

122 Food spoilage microorganisms *Edited by C. de W. Blackburn*

123 Emerging foodborne pathogens *Edited by Y. Motarjemi and M. Adams*

124 Benders' dictionary of nutrition and food technology Eighth edition *D. A. Bender*

125 Optimising sweet taste in foods *Edited by W. J. Spillane*

126 Brewing: new technologies *Edited by C. Bamforth*

127 Handbook of herbs and spices Volume 3 *Edited by K. V. Peter*

128 Lawrie's meat science Seventh edition *R. A. Lawrie in collaboration with D. A. Ledward*

129 Modifying lipids for use in food *Edited by F. Gunstone*

130 Meat products handbook: practical science and technology *G. Feiner*

131 Food consumption and disease risk: consumer-pathogen interactions *Edited by M. Potter*

132 Acrylamide and other hazardous compounds in heat-treated foods *Edited by K. Skog and J. Alexander*

133 Managing allergens in food *Edited by C. Mills, H. Wichers and K. Hoffman-Sommergruber*

134 Microbiological analysis of red meat, poultry and eggs *Edited by G. Mead*

135 Maximising the value of marine by-products *Edited by F. Shahidi*

136 Chemical migration and food contact materials *Edited by K. Barnes, R. Sinclair and D. Watson*

137 Understanding consumers of food products *Edited by L. Frewer and H. van Trijp*

138 Reducing salt in foods: practical strategies *Edited by D. Kilcast and F. Angus*

139 Modelling microorganisms in food *Edited by S. Brul, S. Van Gerwen and M. Zwietering*

140 Tamime and Robinson's Yoghurt: science and technology Third edition *A. Y. Tamime and R. K. Robinson*

141 Handbook of waste management and co-product recovery in food processing Volume 1 *Edited by K. W. Waldron*

142 **Improving the flavour of cheese** *Edited by B. Weimer*
143 **Novel food ingredients for weight control** *Edited by C. J. K. Henry*
144 **Consumer-led food product development** *Edited by H. MacFie*
145 **Functional dairy products Volume 2** *Edited by M. Saarela*
146 **Modifying flavour in food** *Edited by A. J. Taylor and J. Hort*
147 **Cheese problems solved** *Edited by P. L. H. McSweeney*
148 **Handbook of organic food safety and quality** *Edited by J. Cooper, C. Leifert and U. Niggli*
149 **Understanding and controlling the microstructure of complex foods** *Edited by D. J. McClements*
150 **Novel enzyme technology for food applications** *Edited by R. Rastall*
151 **Food preservation by pulsed electric fields: from research to application** *Edited by H. L. M. Lelieveld and S. W. H. de Haan*
152 **Technology of functional cereal products** *Edited by B. R. Hamaker*
153 **Case studies in food product development** *Edited by M. Earle and R. Earle*
154 **Delivery and controlled release of bioactives in foods and nutraceuticals** *Edited by N. Garti*
155 **Fruit and vegetable flavour: recent advances and future prospects** *Edited by B. Brückner and S. G. Wyllie*
156 **Food fortification and supplementation: technological, safety and regulatory aspects** *Edited by P. Berry Ottaway*
157 **Improving the health-promoting properties of fruit and vegetable products** *Edited by F. A. Tomás-Barberán and M. I. Gil*
158 **Improving seafood products for the consumer** *Edited by T. Børresen*
159 **In-pack processed foods: improving quality** *Edited by P. Richardson*
160 **Handbook of water and energy management in food processing** *Edited by J. Klemeš, R.. Smith and J.-K. Kim*
161 **Environmentally compatible food packaging** *Edited by E. Chiellini*
162 **Improving farmed fish quality and safety** *Edited by Ø. Lie*
163 **Carbohydrate-active enzymes** *Edited by K.-H. Park*
164 **Chilled foods: a comprehensive guide Third edition** *Edited by M. Brown*
165 **Food for the ageing population** *Edited by M. M. Raats, C. P. G. M. de Groot and W. A Van Staveren*
166 **Improving the sensory and nutritional quality of fresh meat** *Edited by J. P. Kerry and D. A. Ledward*
167 **Shellfish safety and quality** *Edited by S. E. Shumway and G. E. Rodrick*
168 **Functional and speciality beverage technology** *Edited by P. Paquin*
169 **Functional foods: principles and technology** *M. Guo*
170 **Endocrine-disrupting chemicals in food** *Edited by I. Shaw*
171 **Meals in science and practice: interdisciplinary research and business applications** *Edited by H. L. Meiselman*
172 **Food constituents and oral health: current status and future prospects** *Edited by M. Wilson*
173 **Handbook of hydrocolloids Second edition** *Edited by G. O. Phillips and P. A. Williams*
174 **Food processing technology: principles and practice Third edition** *P. J. Fellows*
175 **Science and technology of enrobed and filled chocolate, confectionery and bakery products** *Edited by G. Talbot*
176 **Foodborne pathogens: hazards, risk analysis and control Second edition** *Edited by C. de W. Blackburn and P. J. McClure*
177 **Designing functional foods: measuring and controlling food structure breakdown and absorption** *Edited by D. J. McClements and E. A. Decker*
178 **New technologies in aquaculture: improving production efficiency, quality and environmental management** *Edited by G. Burnell and G. Allan*

179 **More baking problems solved** *S. P. Cauvain and L. S. Young*
180 **Soft drink and fruit juice problems solved** *P. Ashurst and R. Hargitt*
181 **Biofilms in the food and beverage industries** *Edited by P. M. Fratamico, B. A. Annous and N. W. Gunther*
182 **Dairy-derived ingredients: food and neutraceutical uses** *Edited by M. Corredig*
183 **Handbook of waste management and co-product recovery in food processing Volume 2** *Edited by K. W. Waldron*
184 **Innovations in food labelling** *Edited by J. Albert*
185 **Delivering performance in food supply chains** *Edited by C. Mena and G. Stevens*
186 **Chemical deterioration and physical instability of food and beverages** *Edited by L. H. Skibsted, J. Risbo and M. L. Andersen*
187 **Managing wine quality Volume 1: viticulture and wine quality** *Edited by A. G. Reynolds*
188 **Improving the safety and quality of milk Volume 1: milk production and processing** *Edited by M. Griffiths*
189 **Improving the safety and quality of milk Volume 2: improving quality in milk products** *Edited by M. Griffiths*
190 **Cereal grains: assessing and managing quality** *Edited by C. Wrigley and I. Batey*
191 **Sensory analysis for food and beverage quality control: a practical guide** *Edited by D. Kilcast*
192 **Managing wine quality Volume 2: oenology and wine quality** *Edited by A. G. Reynolds*
193 **Winemaking problems solved** *Edited by C. E. Butzke*
194 **Environmental assessment and management in the food industry** *Edited by U. Sonesson, J. Berlin and F. Ziegler*
195 **Consumer-driven innovation in food and personal care products** *Edited by S. R. Jaeger and H. MacFie*
196 **Tracing pathogens in the food chain** *Edited by S. Brul, P.M. Fratamico and T.A. McMeekin*
197 **Case studies in novel food processing technologies: innovations in processing, packaging, and predictive modelling** *Edited by C. J. Doona, K. Kustin and F. E. Feeherry*
198 **Freeze-drying of pharmaceutical and food products** *T.-C. Hua, B.-L. Liu and H. Zhang*
199 **Oxidation in foods and beverages and antioxidant applications Volume 1: understanding mechanisms of oxidation and antioxidant activity** *Edited by E. A. Decker, R. J. Elias and D. J. McClements*
200 **Oxidation in foods and beverages and antioxidant applications Volume 2: management in different industry sectors** *Edited by E. A. Decker, R. J. Elias and D. J. McClements*
201 **Protective cultures, antimicrobial metabolites and bacteriophages for food and beverage biopreservation** *Edited by C. Lacroix*
202 **Separation, extraction and concentration processes in the food, beverage and nutraceutical industries** *Edited by S. S. H. Rizvi*
203 **Determining mycotoxins and mycotoxigenic fungi in food and feed** *Edited by S. De Saeger*
204 **Developing children's food products** *Edited by D. Kilcast and F. Angus*
205 **Functional foods: concept to product Second edition** *Edited by M. Saarela*
206 **Postharvest biology and technology of tropical and subtropical fruits Volume 1: fundamental issues** *Edited by E. M. Yahia*
207 **Postharvest biology and technology of tropical and subtropical fruits Volume 2: açai to citrus** *Edited by E. M. Yahia*
208 **Postharvest biology and technology of tropical and subtropical fruits Volume 3: cocona to mango** *Edited by E. M. Yahia*

209   **Postharvest biology and technology of tropical and subtropical fruits Volume 4: mangosteen to white sapote** *Edited by E. M. Yahia*

210   **Food and beverage stability and shelf life** *Edited by D. Kilcast and P. Subramaniam*

211   **Processed Meats: improving safety, nutrition and quality** *Edited by J. P. Kerry and J. F. Kerry*

212   **Food chain integrity: a holistic approach to food traceability, safety, quality and authenticity** *Edited by J. Hoorfar, K. Jordan, F. Butler and R. Prugger*

213   **Improving the safety and quality of eggs and egg products Volume 1** *Edited by Y. Nys, M. Bain and F. Van Immerseel*

214   **Improving the safety and quality of eggs and egg products Volume 2** *Edited by F. Van Immerseel, Y. Nys and M. Bain*

215   **Animal feed contamination: effects on livestock and food safety** *Edited by J. Fink-Gremmels*

216   **Hygienic design of food factories** *Edited by J. Holah and H. L. M. Lelieveld*

217   **Manley's technology of biscuits, crackers and cookies Fourth edition** *Edited by D. Manley*

218   **Nanotechnology in the food, beverage and nutraceutical industries** *Edited by Q. Huang*

219   **Rice quality: a guide to rice properties and analysis** *K. R. Bhattacharya*

220   **Advances in meat, poultry and seafood packaging** *Edited by J. P. Kerry*

221   **Reducing saturated fats in foods** *Edited by G. Talbot*

222   **Handbook of food proteins** *Edited by G. O. Phillips and P. A. Williams*

223   **Lifetime nutritional influences on cognition, behaviour and psychiatric illness** *Edited by D. Benton*

224   **Food machinery for the production of cereal foods, snack foods and confectionery** *L.-M. Cheng*

225   **Alcoholic beverages: sensory evaluation and consumer research** *Edited by J. Piggott*

226   **Extrusion problems solved: food, pet food and feed** *M. N. Riaz and G. J. Rokey*

227   **Handbook of herbs and spices Second edition Volume 1** *Edited by K. V. Peter*

228   **Handbook of herbs and spices Second edition Volume 2** *Edited by K. V. Peter*

229   **Breadmaking: improving quality Second edition** *Edited by S. P. Cauvain*

230   **Emerging food packaging technologies: principles and practice** *Edited by K. L. Yam and D. S. Lee*

231   **Infectious disease in aquaculture: prevention and control** *Edited by B. Austin*

232   **Diet, immunity and inflammation** *Edited by P. C. Calder and P. Yaqoob*

233   **Natural food additives, ingredients and flavourings** *Edited by D. Baines and R. Seal*

234   **Microbial decontamination in the food industry: novel methods and applications** *Edited by A. Demirci and M. Ngadi*

235   **Chemical contaminants and residues in foods** *Edited by D. Schrenk*

236   **Robotics and automation in the food industry: current and future technologies** *Edited by D. G. Caldwell*

237   **Fibre-rich and wholegrain foods: improving quality** *Edited by J. A. Delcour and K. Poutanen*

238   **Computer vision technology in the food and beverage industries** *Edited by D.-W. Sun*

239   **Encapsulation technologies and delivery systems for food ingredients and nutraceuticals** *Edited by N. Garti and D. J. McClements*

240   **Case studies in food safety and authenticity** *Edited by J. Hoorfar*

Part I

An introduction to computer vision in the food and beverage industries

1

Computer vision and infrared techniques for image acquisition in the food and beverage industries*

M. Z. Abdullah, Universiti Sains Malaysia, Malaysia

Abstract: Computer vision systems are ideally suited for routine inspection and quality assurance tasks which are common in the food and beverages industries. Backed by powerful machine intelligence and state-of-the-art electronic technologies, machine vision provides a mechanism by which the human thinking process is simulated artificially. Depending on the nature of application and the sensitivity needed to perform the inspection, an image can be acquired at different wavelengths, extending from visible to invisible electromagnetic spectrum. In this chapter the basic electronics needed to acquire an image at three different wavelengths are described. The spectrums covered include the visible, ultrasonic and infrared spectrums. The limitations and drawbacks of each system are also briefly discussed.

Key words: electromagnetic spectrum, image acquisition system, computer vision, illumination system, ultrasound, thermographic imaging, charge-coupled device (CCD) camera.

1.1   Introduction

In making a physical assessment of agricultural materials and foodstuffs, *images* are undoubtedly the preferred method for representing concepts to the human brain. Many of the quality factors affecting foodstuffs can be determined by visual inspection and image analysis. Such inspections determine market price and to some extent the 'best-if-used-before-date'. Traditionally, quality inspection is performed by trained *human inspectors* who approach the problem of quality assessment in two ways: seeing and feeling. In addition to being costly, this method is highly variable and decisions are not always consistent between inspectors or

---

*Adapted from *Computer vision technology for food quality evaluation*, ed. Da-Wen Sun, Chapter 1, Image acquisition systems, pp. 3–35. Copyright (2007), with permission from Elsevier.

from day to day. This is, however, changing with the advent of *electronic imaging* systems and with rapid decline in costs of computers, peripherals and other digital devices. Moreover, the inspection of foodstuffs for various quality factors is a very repetitive task, which is also very subjective in nature. In this type of environment *machine vision* systems are ideally suited for routine inspection and quality assurance tasks. Backed by powerful artificial intelligence systems and the state-of-the-art electronic technologies, machine vision provides a mechanism in which the human thinking process is simulated artificially. To-date machine vision has extensively been applied to solve various food engineering problems, ranging from simple quality evaluation of food products to complicated robot guidance applications (Abdullah *et al.*, 2000; Pearson, 1996; Tao *et al.*, 1995). Despite the general utility of machine vision images as a first-line inspection tool, their capabilities for more in-depth investigation are fundamentally limited. This is due to the fact that images produced by vision camera are formed using a narrow band of radiation, extending from 10^{-4} m to 10^{-7} m in wavelength. Due to this, scientists and engineers have invented camera systems that allow patterns of energy from virtually any part of the electromagnetic spectrum to be visualized. Camera systems such as *computed tomography* (CT), *magnetic resonance imaging* (MRI), *nuclear magnetic resonance* (NMR), *single photon emission computed tomography* (SPECT) and *positron emission tomography* (PET) operate at shorter wavelengths ranging from 10^{-8} m to 10^{-13} m. The details are covered in Chapter 3. On the opposite side of the electromagnetic spectrum, there are *infrared* (IR) and *radio* cameras which enable visualization to be performed at wavelengths greater than 10^3 m and 10^6 m, respectively. All these imaging modalities rely on acquisition hardware featuring an array or ring of detectors which measure the strength of some form of *radiation*, either due to reflection or after the signal has passed transversely through the object. Perhaps one thing that these camera systems have in common is the requirement to perform *digital image processing* of the resulting signals using modern computing power. While digital image processing is usually assumed to be the process of converting radiant energy in three-dimensional (3-D) world into a two-dimensional (2-D) radiant array of numbers, this is certainly not so when the detected energy is outside the visible part of the spectrum. The reason is that the technology used to acquire the imaging signals are quite different depending on the camera modalities. The aim of this chapter is, therefore, to give a brief review of the present state-of-the-art of *image acquisition* technologies which have found many applications in the food industry.

Section 1.2 summarizes the *electromagnetic spectrum* which is useful in image formation. Section 1.3 includes a summary of the principle of operation of the machine vision technology, followed by *illumination* and electronics requirements. Other imaging modalities, particularly the acquisition technologies operating at the non-visible range are also briefly discussed. In particular, technologies based on ultrasound and IR are addressed, followed by some of their successful applications in food engineering found in literatures. Section 1.4, which is the final conclusions section, addresses likely future developments in this exciting field of *electronic imaging*.

1.2   The electromagnetic spectrum

As discussed earlier, images are derived from the electromagnetic radiation in both visible and non-visible range. Radiation energy travels in space at the speed of light, in the form of sinusoidal waves with known wavelengths. Arranged from shorter to longer wavelengths, the electromagnetic spectrum provides information on frequency as well as *energy distributions* of the electromagnetic radiation. Figure 1.1 gives the electromagnetic spectrum of all *electromagnetic waves*.

Referring to Fig. 1.1, the *gamma rays* with *wavelengths* less than 0.1 nm constitute the shortest wavelengths of the electromagnetic spectrum. Traditionally, gamma radiation is important for medical and astronomical imaging, leading to the development of various types of anatomical imaging modalities such as the CT, MRI, SPECT and PET. In CT the radiation is projected into the target from diametrically opposed source, while with others it originates from the target – by simulated emission in the case of MRI and through the use of radio-pharmaceuticals in SPECT and PET. On the other hand, the longest waves are radiowaves, which have wavelengths of many kilometres. The well-known *ground-probing radar* (GPR) and other *microwave*-based imaging modalities operate in this frequency range. Located in the middle of the electromagnetic spectrum is the visible range, consisting of a narrow portion of the spectrum, from 400 nm (blue) to 700 nm (red). The popular *charge coupled device* (CCD) camera operates in this spectrum range. IR light lies between the visible and microwave portions of the electromagnetic

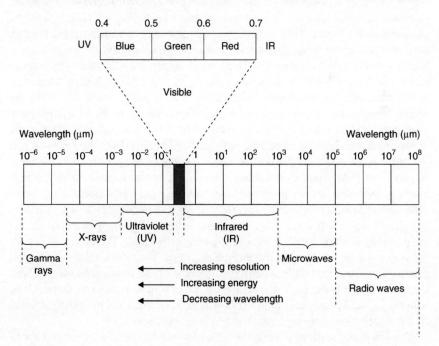

Fig. 1.1   The electromagnetic spectrum comprising the visible and non-visible range.

band. Just like the visible light, IR has wavelengths that range from near (shorter) IR to far (longer) IR. The latter belongs to the thermally sensitive region which makes it useful in imaging applications that rely on heat signature. One example of such an imaging device is the indium–galium–arsenide (InGaAs)-based *near infrared* (NIR) camera which gives optimum response from the 900 nm to 1700 nm band (Doebelin, 1996). *Ultraviolet* (UV) has shorter wavelength than visible light. Similar to IR, the UV part of the spectrum can be divided into three regions: *near ultraviolet* (NUV) (300 nm), *far ultraviolet* (FUV) (30 nm) and *extreme ultraviolet* (EUV) (3 nm). NUV is closest to visible band. In contrast, EUV is closest to the X-ray region, and therefore is the most energetic of the three types. FUV, meanwhile, lies between the near and extreme UV regions. It is the least explored of the three regions. To date there exist many types of CCD camera that provide sensitivity down near UV wavelength range. The sensitivity of such a camera usually peaks at around 369 nm while offering coverage down to 300 nm.

Mathematically, wavelength (λ), *frequency* (f) and *energy* (E) are related by *Planck's* equation which is given by:

$$E = h\frac{c}{\lambda}$$

[1.1]

where h is the Planck's constant (6.626076×10^{-34} Js) and c is the speed of light (2.998×10^{8} m/s). Consequently, energy increases as wavelength decreases. Therefore, gamma rays which have the shortest wavelengths have the highest energy of all the electromagnetic waves. This explains why gamma rays can easily travel through most objects without being affected. In contrast, radiowaves have the longest wavelength and hence the lowest energy. Therefore, their penetrative power is an order of magnitude lower (by hundreds, at least) compared to gamma or X-rays. Moreover, both gamma and X-rays travel in a straight line and the paths are not affected by the object through which these signals propagate. This is known as the *hard field* effect. Conversely, radiowaves do not travel in straight lines and their paths depend strongly on the medium of propagation. This is the *soft field* effect. Both the hard and soft field effects have a direct implication on the quality of images produced by these signals. Soft field effect causes many undesirable artefacts, most notably, image blurring. Therefore, images produced by gamma rays generally appear much better compared to images produced by radiowaves. Another important attribute, which is wavelength dependent, is the *image resolution*. In theory, the image spatial resolution is essentially limited to half of the interrogating wavelength. Therefore, the *spatial resolution* also increases as the wavelength decreases. Thus, the resolution of typical gamma rays is less than 0.05 nm, enabling this type of *electromagnetic wave* to 'see' extremely small objects such as water molecules. In summary, these attributes, along with the physical properties of the sensor materials, establish the fundamental limits on the capability of imaging modalities and their applications.

The following sections explain the technology of image acquisition and applications for all the imaging modalities discussed, focusing on the visible modality or computer vision system – since this device has been used extensively for solving various food engineering problems. Moreover, given the progress in computer technology, computer vision hardware is now relatively inexpensive and easy to use. To date, some personal computers offer capability for a basic vision system, by including a camera and its interface within the system. However, there are specialized systems for vision, offering performance in more than one aspect. Naturally, as with any specialized equipment, such systems can be expensive.

1.3   Image acquisition systems

In general, images are formed by incident light in the *visible spectrum* falling on a partially reflective, partially absorptive surface, with the scattered *photons* being gathered up in the camera lens and converted to electrical signals either by vacuum tube or CCD. In practice this is only one of many ways in which images can be generated. Generally, *thermal*, *ultrasonic*, X-rays, radiowaves and other techniques can all generate image. This section examines the methods and procedures in which images are generated for computer vision applications.

1.3.1   Computer vision
Hardware configuration of computer-based machine vision system is relatively standard. Typically, a vision system consists of the *illumination device* to illuminate the sample under test; the solid-state CCD array camera to acquire an image; the *frame grabber* to provide scanning of the A/D (analogue-to-digital) conversion of scan lines into picture elements or pixels digitized in an N row by M column image; personal computer or microprocessor system to provide disk storage of images and computational capability with vendor-supplied software and specific application programs; and a high-resolution colour monitor which aids in visualizing images and the effects of various image analysis routines. Figure 1.2 shows a typical set-up. The set-up shown in Fig. 1.2 is where the investigator

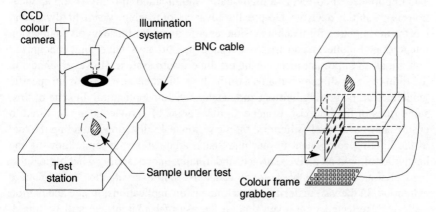

Fig. 1.2   Essential elements of a typical computer vision system.

Published by Woodhead Publishing Limited, 2012

needs to start, in terms of experimenting with machine vision applications. All essential components are commercially available and the price for the elementary system can be as low as $2000.

The set-up shown in Fig. 1.2 is an example of a computer vision system which can be found in many food laboratories, mainly for research and imaging applications. In this case the objective is to ultimately free human inspectors from undertaking tedious, laborious, time-consuming and repetitive inspection tasks, allowing them to focus on more demanding and skilful jobs. Computer vision technology not only provides a high level of flexibility and repeatability at relative low cost, but also, more importantly, it permits fairly high plant throughput without compromising accuracy. The food industry continues to be among the fastest-growing segments of machine vision application, and it ranks among the top ten industries that use machine vision systems (Gunasekaran, 1996). Currently, several commercial vendors offer automatic vision-based quality evaluations for the food industry.

Even though machine vision systems have become increasingly simple to use, the applications themselves can still be extremely complicated. A developer needs to know precisely what needs to be achieved in order to ensure successful implementation of a machine vision application. Key characteristics include not only the specific part dimensions and part tolerances, but also the level of measurement precision required and the speed of the production line. Virtually all manufacturing processes will produce some degree of variability and, while, the best machine vision technology is robust enough to compensate automatically for minor differences over time, the applications themselves need to take major changes into account. Additional complexity arises for companies with complex *lighting* and *optical* strategies, or unusual materials handling logistics. For these reasons, it is essential to understand the characteristics of the part and sub-assemblies of machine system, as well as the specifications of the production line itself.

The illumination

The importance of correct and high-quality illumination in many vision applications is absolutely decisive. Despite the advances of machine vision hardware and electronics, lighting for machine vision remains an art for those involved in vision integration. Engineers and machine vision practitioners have long recognized lighting as an important piece of the machine vision system. However, choosing the right lighting strategy remains a difficult problem because there is no specific guideline for integrating lighting and machine vision application. In spite of this, some rules of thumb exist. In general, three areas of knowledge are required to ensure successful level of lighting for the vision task: first, understanding the role of the lighting component in machine vision applications; second, knowing the behaviour of light on a given surface; and, finally, understanding what basic lighting techniques are available that will cause the light to create the desired feature extraction. In the vast majority of machine vision applications, image acquisition deals with reflected light, even though the use of backlight can still be found. Therefore, the most important aspect of lighting is to understand what happens when light hits the surface – more specifically, to know how to control the reflection so that the image is of reasonably good quality.

Another major area of concern is the choice of illuminant as this is instrumental in the capability of any machine vision to accurately represent image. This is due to the fact that the sensor response of a standard imaging device is given by a *spectral integration* process (Matas *et al.*, 1995). Mathematically,

$$p_k^x = \int_{\lambda_1}^{\lambda_2} \rho_k(\lambda) L(\lambda) d\lambda \qquad [1.2]$$

where p_k^x is the *response* of the *k*th sensor at location *x* of the sensor array, $\rho_k(\lambda)$ is the *responsivity function* of the *k*th sensor and $L(\lambda)$ is the light reflected from the surface that is projected on pixel *x*. For CCD camera the *stimulus* $L(\lambda)$ is the product of the *spectral power distribution* $S(\lambda)$ of the light that illuminates the object and the *spectral reflectance* $C(\lambda)$ of the camera itself – that is,

$$L(\lambda) = S(\lambda) C(\lambda) \qquad [1.3]$$

Hence, two different illuminants $S_1(\lambda)$ and $S_2(\lambda)$ may yield different stimulus using the same camera. Therefore, the illuminant is an important factor that one must take into account when considering machine vision integration. Frequently, a knowledge selection of an illuminant is necessary for specific vision application.

Traditionally, the two most common illuminants are *fluorescent* and *incandescent* bulbs, even though other light sources such as the *light emitting diode* (LED) and electroluminescence are also useful. Figure 1.3 shows the spectral distributions of three different light sources: the sun, an incandescent bulb and standard cool white fluorescent light. Referring to this figure, the only difference between daylight and electric light is the amount of energy emitted at each wavelength.

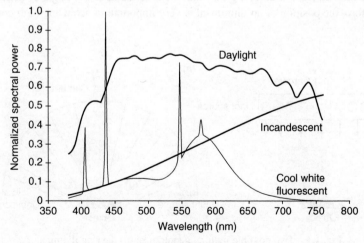

Fig. 1.3 Comparison of relative spectral energy distribution between daylight, incandescent and cool white fluorescent light (Stiles and Wyszecki, 2000).

Even though the light energy itself is fundamentally the same, however, the more optimum light will have more intensity than the other source. When the light is not as intense as it should be, three possible damaging effects occur. First, there may not be sufficient signal-to-noise ratio at the camera. Second, the electrical noise tends to increase as the light gets dimmer and less intense. Third, and most importantly, a less intense light will cause a significant loss in the camera depth of field. Additionally, effects from ambient light are more likely to occur under poor lighting conditions.

Referring to Fig. 1.3 again, it can be seen that the incandescent source has a fairly normal distribution over the visible spectrum while the fluorescent has sharp peaks in some regions. This means that objects under an incandescent source produce image with a much lower signal-to-noise ratio. This is not acceptable in some cases, especially those that are concerned with colour image processing (Daley *et al.*, 1993). In contrast, fluorescent bulbs are inherently more efficient and produce more intense illumination at specific wavelengths. Moreover, fluorescent light provides a more even, *uniform dispersion* of light from the emitting surface and, hence, does not require the use of diffusing optics to disseminate the light source over the field of view, as in the incandescent bulbs. For these reasons, a fluorescent bulb, particularly the cool white type, is a popular choice for many machine vision practitioners (Abdullah *et al.*, 2001, and 2005; Pedreschi *et al.*, 2006; Tao *et al.*, 1995). However, care must be taken when using fluorescent light since this source is normally AC driven. The 50 Hz fluorescent bulb usually introduces artefacts in the image resulting from the over-sampling of the analogue-to-digital converter. In order to reduce flickering, high-frequency florescent bulbs, operating at frequencies in the range of a few tens of kilohertz, are preferred over the low-frequency ones.

Apart from the illuminant, the surface geometry is also important in the illumination design. The key factor is to determine whether the surface is *specular* or *diffuse*. Light striking the diffused surface is scattered because of the multitude of surfaces angles. In comparison, light striking a glossy surface is reflected at the angle of incidence. Therefore, the position of an illuminant is very important in achieving high contrast

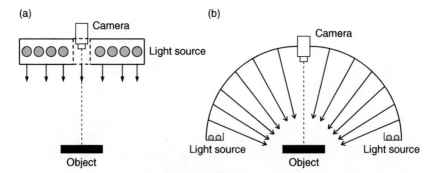

Fig. 1.4   Two possible lighting geometries: (a) the ring illuminator and (b) the diffuse illuminator.

in an image. The two most common geometries for the illuminators are (1) the *ring illuminator* and (2) the *diffuse illuminator*. Figure 1.4 illustrates these geometries.

The ring illuminator is the simplest geometry of the illumination and it is usually intended for general purpose applications, especially for imaging flat surfaces. The diffuse illuminator, meanwhile, delivers virtually 180 degrees of diffuse illumination and is used for imaging challenging, reflective objects. Since most food products are basically 3-D objects, the diffuse illuminator is well suited for this imaging application. However, the ring illuminator can solve some lighting problems in food engineering. For instance, a ring light, together with a 90 kHz ultra-high-frequency fluorescent bulb, was found effective in colour and shape grading of star fruits (Abdullah *et al.*, 2005). In an attempt to produce uniform lighting, Paulsen (1990) mounted a ring light into the cylindrical diffuse lighting chamber. Such a set-up is extremely useful for visual inspection of grains and oilseed, with success rates almost reaching 100%. In spite of the general utility of ring light, however, the majority of machine vision applications are based on diffuse illumination. Heinemann *et al.* (1994) employed this type of illumination system for shape grading of mushrooms. The same illuminator was investigated by Steinmetz *et al.* (1996) for quality grading of melons. They have all reported a successful application of machine vision, with grading accuracy exceeding 95%. There are many other applications involving diffuse illumination and computer vision integration. Batchelor (1985) reviewed some of the important factors to consider when designing a good illumination system.

The electronics
Capturing the image electronically is the first step in digital image processing. Two key elements are responsible for this: the *camera* and the *frame grabber*. First, the camera converts photons to electrical signals and, second, the frame grabber digitizes these signals into stream of data or a bitmap image. There are many types of camera, ranging from the older pick-up tubes, such as the *vidicons*, to the most recent solid-state imaging devices, such as *complementary metal oxide silicon* (CMOS) cameras. The latter is now the dominant camera technology, revolutionizing imaging science with the invention of CCD devices in 1970. As CCD cameras have less noise, higher sensitivity and greater dynamic range, the CCD has also become the device of choice for a wide variety of food engineering applications.

In general, CCD sensors comprise of a *photo sensitive diode* and a *capacitor* connected in parallel. There are two different modes in which the sensor can be operated: (i) the *passive* and (ii) the *active* modes. Figure 1.5 shows the details of the schematics. Referring to Fig. 1.5, the *photodiode* converts light into electrical charges which are then stored in the capacitor. The charges are proportional to the light intensity. In passive mode, these charges are transferred to a bus line when 'select' signal is activated. In active mode, charges are first amplified before being transferred to a bus line, thus compensating for the limited fill factor of the photodiode. An additional 'reset' signal allows the capacitor to be discharged when the image is rescanned.

Depending on sensing applications, CCD imagers come in various architectures. The simplest form is the *linear CCD* scanner shown schematically in Fig. 1.6a. This architecture is used mostly in office scanner machines. It consists of a single row of photodiodes, which capture the photons. The sensors are lined up adjacent to a CCD shift register, which does the readout. The picture or document to be scanned is moved, one line at a time, across the scanner by mechanical or optical means. Meanwhile Fig. 1.6b and 1.6c show 2-D CCD area arrays which are mostly associated with modern digital cameras. The circuit in Fig. 1.6b portrays the *interline CCD* architecture while 1.6c shows the architecture of *a frame-transfer imager.*

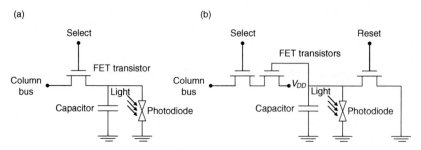

Fig. 1.5   Sensor operation in (a) passive mode and (b) active mode.

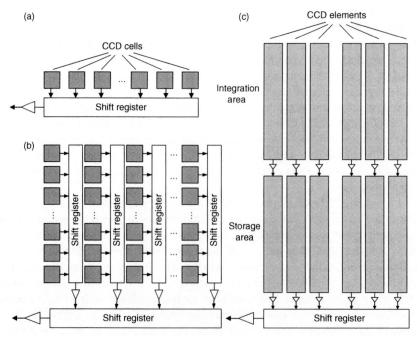

Fig. 1.6   Three possible CCD architectures: (a) linear, (b) interline and (c) frame-transfer.

Published by Woodhead Publishing Limited, 2012

Basically, the interline CCD comprises a stack of vertical linear scanners connected by an additional, horizontal shift register that collects and passes on the charge readout from linear scanners, row by row. In the case of a frame-transfer architecture, the CCD elements whose entire surface are covered by photo sensitive device form the photo sensing area. It can be seen from Fig. 1.6c that the frame-transfer comprises integration and storage areas, forming the integration and storage frames, respectively. The storage frame array captures an image and transfers the charge to the adjacent storage frame array. In this way, the integration array can capture a new image while storage array reads out the previous image. Both interline and frame-transfer architectures are suitable for capturing motion images, while the linear scanner is best suited for scanning still pictures. Full-frame CCD cameras with 4 million pixels and a *frame rate* of more than 30 frames per second (fps) are now commercially available. Modern CCD cameras come with analogue, digital or both outputs. The analogue signals are conformant with European CCIR (Comite Consultatif International des Radiocommunication) or US RS170 video standards. In spite of reduced dynamic range, the analogue cameras work well for slower applications (<20 MHz). For high-speed applications, digital cameras are preferred. These cameras have internal digitization circuitry and usually produce a parallel digital output. Typically, data are output on a 8-32 bit wide parallel bus, clocking rates of up to 40 MHz and utilizing RS422 or RS644 international video standards. For the purpose of camera control and PC interfacing, analogue cameras require an analogue frame grabber. Likewise, digital cameras require a digital frame grabber. Generally, a frame grabber comprises signal conditioning elements, A/D converters, look-up table, image buffer and peripheral component interface (PCI) bus. Figure 1.7 illustrates some of the basic elements of a typical frame grabber card.

However, it must be borne in mind that the internal working circuitries of modern and state-of-the-art frame grabbers are more complex than the one shown in Fig. 1.7. Nevertheless, the basic elements remain relatively the same from one manufacturer to another. The latest digital frame grabbers may feature sophisticated on-board camera control modules and on-the-fly data resequencing capabilities, which are intended for high-speed applications. In addition to PCI bus, some frame grabbers feature PC104 capability and firewire plug-and-play style.

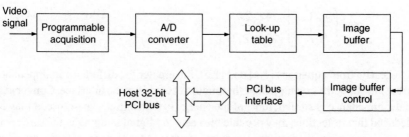

Fig. 1.7   General structure of frame grabber card showing some important elements.

Published by Woodhead Publishing Limited, 2012

In summary, three key considerations in deciding on a frame grabber are: (i) choice of camera, (ii) speed requirements and (iii) computer choice.

1.3.2   Ultrasound

In addition to computer vision, there is growing interest in using *ultrasound* in food quality evaluation. One reason for this is that changes in acoustic properties can be related to *density* changes in the food product (McClements, 1995). Furthermore, ultrasound has the ability to differentiate between the propagation velocity within various media and the differences in *acoustic impedance* between different regions within a given volume. In the past ultrasound has been used for measuring the moisture content of food products (Steele, 1974); predicting the intramuscular fat content of bovine products (Morlein *et al.*, 2005); and studying and evaluating the turgidity and hydration of orange peel (Camarena and Martínez-Mora, 2005). In addition to these, one of the most widespread and also most promising *ultrasonic* applications is the utilization of ultrasound for composition measurement. Recent studies have shown that ultrasonic velocity measurements can accurately be used to predict fat, water, protein and other chemical compositions of meat-based products (Simal *et al.*, 2003).

The information in ultrasound originates from the *reflection* or *transmission* of *sound waves* emitted by an external source. *Refraction, absorption* and *scattering* also play a role, but mainly as factors that degrade the ultrasonic measurements. A typical source/detector unit is based on a *piezoelectric crystal* resonating between 1 and 10 MHz. The basic physical parameters of importance are the frequency of the wave and the acoustical impedance of the object through which the sound wave travels. The acoustical impedance Z itself is a function of the speed of sound v and the density ρ of the object, and the relationship $Z = \rho v$ holds. When an ultrasonic plane wave propagates in a medium with $Z = Z_1$ and is incident normally at an interface with $Z = Z_2$, the *intensity reflection* and *transmission coefficients* (α_r and α_t) are given in terms of the incident (p_i), reflected (p_r) and transmitted (p_t) sound pressures (Wells, 1969). Mathematically,

$$\alpha_r = \left(\frac{p_r}{p_i}\right)^2 = \left[\frac{(Z_2 - Z_1)}{(Z_2 + Z_1)}\right]^2 \qquad [1.4]$$

$$\alpha_t = \left(\frac{p_t}{p_i}\right)^2 = \left[\frac{4Z_1 Z_2}{(Z_2 + Z_1)^2}\right] \qquad [1.5]$$

Clearly, from equations [1.4] and [1.5], the greater the difference in impedance at the interface, the greater will be the amount of energy at the interface. Conversely, if the impedance is similar, most of the energy is transmitted. Therefore, it may be deduced that reflections may provide the basis for digital imaging in the same way that a computer vision system generates images. McClements (1995) publishes useful data on impedance and interaction processes for a wide range of human tissues, including materials of interest in food technology. For instance, the density of human muscle is typically 1000 kg/m³; the speed of sound is therefore 1500 m/s. In this case α_r at a muscle-fat interface is approximately 1% and this value increases sharply to 99.9% at a skin–air interface. Therefore, this technique generally requires a coupling medium between test sample and transducer surface. *Couplants* like water, gel and oil are routinely used for certain inspection situations. However, it must be stressed that the use of a couplant may not always be suitable for some specialized applications, especially when the material property might change, or when contamination damage would result. Gan *et al.* (2005) have attempted to use a non-contact-based ultrasound inspection technique for evaluating food properties. They have used a pair of capacitive devices in air to deliver high ultrasonic energy, with application of *pulse compression* technique for signal recovery and analysis. The use of such a processing technique is necessary to recover signals buried in noise, as a result of large acoustic impedance and *mismatch* between air and the sample. They have experimented with this technique in monitoring palm oil crystallization, employing a *chirp signal* with a centre frequency of 700 kHz and a bandwidth of 600 kHz. From the results, they deduced that both contact and non-contact measurements are well correlated.

The majority of the studies found in the literature rely on the measurement of ultrasonic velocity because this is the simplest and most reliable ultrasonic measurement. It utilizes an impulse, or a sequence of discrete impulses, transmitted into the medium, and the resulting interactions provide raw data for imaging. Here the transduction would be sensed in two ways: (i) from the same viewpoint by the partial reflection of the ultrasonic energy, or (ii) from an opposing viewpoint by transmission, where the energy is partially attenuated. For either view, the *time-of-flight* (TOF) and hence the velocity may be

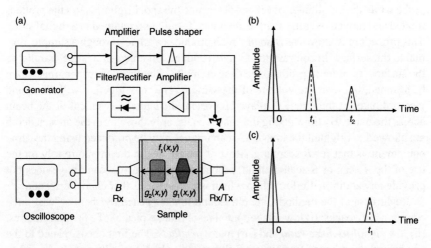

Fig. 1.8   Ultrasonic measuring system showing (a) essential elements, (b) reflection and (c) transmission measurements.

estimated from both *reflection* and *transmission* sensor data. Figure 1.8 demonstrates these general concepts.

In Fig. 1.8a, *A* represents the composite ultrasonic transducer acting as both the *transmitter* (Tx) and the *receiver* (Rx) probe; *B* represents the ultrasonic receiver probe. The interposing area between transducer probes consists of a homogeneous fluid of density distribution $f_1(x,y)$. Within the sample are two inhomogeneities having densities of $g_1(x,y)$ and $g_2(x,y)$ in which $g_2(x,y) > g_1(x,y) > f_1(x,y)$. The interactions between ultrasound and the sample can be explained as follows. Probe *A* transmits the ultrasonic wave which travels in straight line until it reaches the $f_1(x,y)$ and $g_1(x,y)$ interface, causing reflection which is detected by the same probe that now acts as a receiver. The amplified signals are fed into the *y*-plates of the oscilloscope and a timebase is provided, synchronized to the transmitter pulse. Some of the energy, however, continues to travel until it reaches the $f_1(x,y)$ and $g_2(x,y)$ interface, where some energy is again reflected and hence detected by *A*. In similar fashion, some of the remaining energy continues to travel until it reaches probe *B* where it is again detected and measured. Consequently, probe *A* forms a graph detailing the echo signal in which the height corresponds to the size of the inhomongeneity and the timebase provides its range or depth. Such a pattern is known as the *A-scan* and it is shown in Fig. 1.8b. Meanwhile, the graph in Fig. 1.8c shows the attenuated transmitted energy as observed by probe *B*. Both graphs show that information relating to amplitude of both the transmitted and reflected pulses can be measured and this can also be used for imaging. As shown in Fig. 1.8, the signals are usually rectified and filtered to present a simple 1-D picture, and the timebase can be delayed to allow for a couplant gap.

To provide a full 2-D image, the ultrasonic probe must be scanned over the surface of the sample. In so doing, the Tx/Rx probe is connected via mechanical linkages to position transducers which measure its *x* and *y* co-ordinates and its orientation. In this case, the output signals determine the origin and direction of the probe while the amplitudes of echoes determine the spot brightness. As the probe is rotated and moved over the sample, an image is built and retained in a digital store. This procedure is known as *B-scan* which produces a 'slice' through a sample, normal to the surface. In contrast, the *C-scan* produces an image of a 'slice' parallel to the surface. In order to produce the *C*-scan image, the ultrasonic probe must again be scanned, but over the volume of the sample. The time together with the *x* and *y* co-ordinates of the image displayed represent the lateral movement of the beam across the plane. By time-gating the echo signals, only those from the chosen depth are allowed to brighten the image. *C*-scan images may be produced using the same equipment as that for *B*-scanning. Most of the studies in the literature rely on the use of the *A*-scan or *B*-scan method, probably because the *C*-scan image does not provide any additional information which is useful for further characterization.

Regardless of the methods, the ultrasound images generally share at least three common drawbacks: (i) low image spatial resolution, typically of a few millimetres, (ii) low *signal-to-noise ratio* and (iii) many *artefacts*. The first one is related to the wavelength, and hence frequency of ultrasound, which typically ranges from 2 to 10 MHz. In order to improve the resolution, some ultrasound devices operate at frequencies higher than this range. However, this must be exercised with care since the *skin effect* increases with increasing frequency. Both factors have to be balanced against each other. The second and third drawbacks are due to the coherent nature of the sound wave and physics of reflection. Any *coherent pulse* will interfere with its reflected, refracted and transmitted components, giving rise to speckle, similar to the speckle observed in laser light (Fishbane *et al.*, 1996). On the other hand, reflection occurs when the surface has a normal component parallel to the direction of the incident wave. Interfaces between materials that are parallel to the wave will not reflect the wave and therefore not be seen in ultrasound images. Those parallel interfaces form a hole in the ultrasound images. Despite these drawbacks, the technology is safe and relatively inexpensive. Current research tends to eliminate artefacts, improve image contrast and simplify the presentation of data. To date many efforts are directed towards 3-D data acquisitions and image representations.

1.3.3   Infrared

Presumably, when both computer vision and ultrasound systems failed to produce the desired images, food engineers and technologists could resort to the use of much longer wavelength for image acquisition. The IR range lies in the region of 700–1000 nm, and the technique responsible for generating such images is called *thermographic* photography. Thermographic imaging is based on the simple fact that all objects emit a certain amount of *thermal* radiation as a function of their temperature. Generally, the higher temperature the object is at, the more IR radiation it emits. A specially built camera, known as the IR camera, can detect this radiation in a similar way to an ordinary camera detecting visible light. However, unlike computer vision, thermal imaging does not require an illumination source for spectral reflectance, which can be affected by the varied surface colour of a target or by the illumination set-up.

Thermographic signatures of food are very different for different materials and, hence, IR imaging has found applications and many other uses in the food industry, such as identification of foreign bodies in food products (Ginesu *et al.*, 2004). Moreover, many major physiological properties of foodstuffs (firmness, soluble solids content and acidity) appear highly correlated with IR signals, implying that image analysis of IR thermography is suitable for quality evaluation and shelf-life determination of a number of fruit and vegetable products (Gómez *et al.*, 2005). Therefore, thermal imaging offers a potential alternative technology for *non-destructive* and *non-contact* image sensing applications. Good thermographic images can be obtained by leaving the object at rest below the IR camera, applying a heat pulse produced by flash light and monitoring the decreasing temperature as a function of time. Because of different *thermal capacities* or *heat conductivities*, the objects will cool down at different speeds. Therefore, thermal conductivity of an object can be measured by the decreasing temperature calculated from a sequence of IR images. Using these relatively straightforward procedures, experiments were performed on objects with different thermal properties, aiming to simulate *foreign body* contamination in real experiments (Ginesu

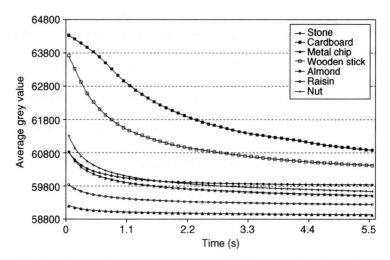

Fig. 1.9   Decreasing temperature curves of different materials plotted as a
function of time (Ginesu *et al.*, 2004).

et al., 2004). Both the long (500 fps) and short (80 fps) sequence modes were used
to record the images, enabling the radiation patterns of objects with low and high
thermal capacities to be monitored, respectively. Temperature data were presented
in terms of average grey levels computed from 10×10 image pixels in the neighbourhood of each object. Figure 1.9 shows the results.

It can be seen, from Fig. 1.9, that cardboard pieces and wooden sticks behave
quite differently from other materials, as they appear much hotter at the beginning but decrease in temperature rather quickly. This is due to the fact that these
materials are dry and light, whereas foods contain a large quantity of water that
heats up more slowly and reaches a lower temperature, thus maintaining the heat
for a longer time and cooling down slowly. By plotting and analysing the absolute
differences between radiation curves of different materials, it is possible to distinguish between food and foreign objects.

Theoretically, the existence of such unique thermal signatures of different
materials, as shown in Fig. 1.9, is due to the concept of a *black body*, defined as
an object which does not reflect any radiation. Planck's law describes the radiation emission from a black body as (Gaossorgues, 1994):

$$R(\lambda, \theta) = \frac{2\pi h c^2 \lambda^{-5}}{\exp\left(\dfrac{hc}{\lambda \sigma \theta}\right) - 1} \qquad [1.6]$$

where $h = 6.626076 \times 10^{-34}$ Js is Planck's constant, $\sigma = 1.38054 \times 10^{-23}$ JK^{-1} is
the Stefan-Boltzman's constant, $c = 2.998 \times 10^8$ ms^{-1} is the speed of light, θ is the
absolute temperature in degrees kelvin and λ is again the wavelength. Usually,
objects are not black bodies. Consequently, the above law does not apply without
certain corrections. Non-black bodies absorb a fraction A, reflect a fraction R and

transmit a fraction T. These fractions are selective, depending on the wavelength and on the angle of incident radiation. By introducing the spectral emissivity $\varepsilon(\lambda)$ to balance the absorbance, it can be found:

$$A(\lambda) = \varepsilon(\lambda) \qquad [1.7]$$

and

$$\varepsilon(\lambda) + R(\lambda) + T(\lambda) = 1 \qquad [1.8]$$

Using these corrections, equation [1.6] can be simplified, yielding:

$$R(\lambda, \theta) = \varepsilon(\lambda) R_{blackbody}(\lambda, \theta) \qquad [1.9]$$

This means that the emission coefficient $\varepsilon(\lambda)$ relates the ideal radiation of a black body with real non-black bodies. In summary, an ideal black body is a material that is a perfect *emitter* of heat energy and, therefore, has the emissivity value equal to unity. In contrast, a material with zero emissivity would be considered as a perfect thermal mirror. However, most real bodies, including food objects, show wavelength-dependent emissivities. Since emissivity varies with material, this parameter is the important factor in thermographic image formation. For accurate measurement of temperature, the *emissivity* should be provided to the camera manually for its inclusion in temperature calculation. The function which describes thermographic image, $f(x,y)$, can be expressed as follows:

$$f(x,y) = f\left[\theta(x,y), \varepsilon(x,y)\right] \qquad [1.10]$$

where x and y are the co-ordinates of individual image pixels, $\theta(x,y)$ is the temperature of the target at image co-ordinates (x,y) and $\varepsilon(x,y)$ is the emissivity of the sample also at co-ordinates (x,y). From the computer vision viewpoint, thermographic images are a function of two variables: temperature and emissivity. Contrast in thermographic images may be the result of either different temperatures of different objects on the scene or different emissivities of different objects with same temperature. It can also be the combination of both temperature and emissivity variations.

As mentioned previously, IR or thermographic cameras operate in wavelengths as long as 14,000 nm or 14 μm. The IR sensor array is equivalent to CCD in the ordinary camera; the sensors with resolution of 160 × 120 pixels or higher are widely available, and their response time is sufficient to provide live thermographic video at 25 frames per second. However, and unlike sensors used in conventional imaging systems, the processes of image formation and acquisition in thermographic cameras are quite complex. Broadly speaking, thermographic cameras can be divided into two types: those with cooled IR image detectors and those without cooled detectors. These are discussed in the following section.

Cooled IR detectors
Cooled IR detectors are typically contained in a vacuum-sealed case and cryogenically cooled. This greatly increases their sensitivity since their body temperatures

are much lower than that of the objects from which they are meant to detect radiation. Typically, cooling temperatures range from –163°C to –265°C, with –193°C being the most common. Similar to common digital cameras that detect and convert light to electrical charge, the IR detectors detect and convert *thermal radiation* to electrical signals. In the case of IR cameras, *cooling* is needed in order to suppress thermally emitted *dark currents*. A further advantage of cooling is suppression of noise from ambient radiation emitted by the apparatus. Materials used for IR detection include liquid-helium-cooled *bolometers*, photon-counting superconducting tunnel junction arrays and a wide range of cheaper narrow gap semiconductor devices. The mercury–cadmium–telluride (*HgCdTe*), indium–antimonide (*InSb*) and *InGaAs* are the most common types of semiconductor IR detectors, with newer composition such as mercury–manganese–telluride (*HgMnTe*) and mercury–zinc–telluride (*HgZnTe*) developing. However, HgCdTe and its extension remain among the most common IR detectors. Principles of the operation of a HgCdTe-based detector are illustrated in Fig. 1.10.

In Fig. 1.10, the sensor is represented by a detector *diode* which is mechanically bonded to a silicon (Si) multiplexer for readout operation. An electrical connection is required between each pixel and the rest of the circuitry. This is formed by the heat and pressure bonding of an indium bump or solder bond. Row and column shift registers allow sequential access to each pixel. Similar to other semiconductor devices, this type of sensor is constructed using modern fabrication technology such as the vapour deposition epitaxy (Campbell, 2001). In this method, the diode is grown by depositing CdTe on sapphire, followed by liquid epitaxy growth of HgCdTe. A complete HgCeTe IR detector system usually comprises a small printed circuit board (PCB), completed with digital signal processor chip (DSP) and an optical system responsible for focusing the scene onto the plane of array. At present a large 2-D array comprising of 2048 × 2048 pixels, with each pixel of size 18 μm assembled on 40 × 40 mm device, and complete IR camera systems are commercially available. They operate in the bands from 3–5 μm or 8–12 μm and need cooling at −196°C.

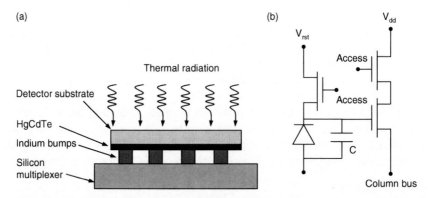

Fig. 1.10   Hybrid focal plane architecture for HgCdTe-based IR detector showing (a) cell structure and (b) equivalent circuit.

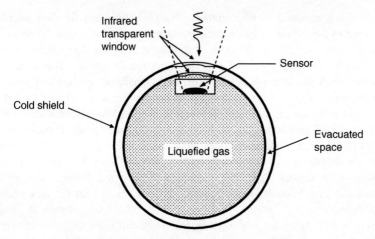

Fig. 1.11   Schematic diagram of a typical Dewar.

There are different ways to cool the detectors – mainly *liquefied gas, cryogenic engine, gas expansion* and *thermoelectric* effect. The most common method is cryogenic cooling, employing liquefied gas stored in a vacuum called a *Dewar*. This device is named after Sir James Dewar, a Scottish scientist who successfully liquefied hydrogen for the first time in 1892. Figure 1.11 shows the construction of a typical Dewar, highlighting all the important elements. Typically, the sensor is mounted directly on the cold surface with a cold shield and IR transparent window. Usually a protective coating such as *zinc sulphide* is applied onto the surface of HgCeTe in order to increase its life span. The most commonly used and cheapest liquefied gas is *liquid nitrogen*, providing a sustainable cold temperature at −196°C without regular filling.

Another common method to achieve cooling is through the use of the *Joule-Thompson* gas expansion method. High-pressure gas such as nitrogen or *argon* produces droplets of liquid nitrogen at −187°C after quick expansion. Compared to Dewar, this method is, however, noisy and cumbersome. When refilling is not practical, such as application in remote areas, a cooling method using a closed Stirling cycle can then be employed. This machine cools through repetitive compression and expansion cycles of a gas piston. Therefore, this method is also cumbersome compared to Dewar. Another more practical approach of cooling is the thermoelectric elements based on *Peltier-Thompson effects* (Fraden, 1997). This method utilizes a junction of dissimilar metals carrying current; temperature rises or falls, depending on the direction of current. One direction of current results in the Peltier effect whereas the opposite direction produces the Thompson effect by the same law of physics. Unfortunately, Peltier elements are unattractive for temperatures below −73°C due to high current consumption. In spite of this drawback, the thermoelectric cooling has no moving parts, and is quiet and reliable. Therefore, it is also widely used in most IR cameras.

Uncooled IR detectors
As the name applies, *uncooled* thermal cameras use sensors that operate at room temperature. Uncooled IR sensors work by the change of resistance, voltage or

current when exposed to IR radiation. These changes are then measured and compared to the values at the operating temperature of the sensor. Unlike cooled detectors, uncooled IR cameras can be stabilized at an *ambient temperature* and, thus, they do not require bulky, expensive cryogenic coolers. This makes such IR cameras smaller and less costly. Their main disadvantages are lower sensitivity and longer response time. However, these problems are almost solved with the advent of surface micro-machining technology. Most uncooled detectors are based on *pyroelectric* materials or *microbolometer* technology. Pyroelectricity is the ability of some materials to generate an electrical potential when they are heated or cooled. It was first discovered in minerals such as quartz, tourmaline and other ionic crystals. The first generation of uncooled thermal cameras looks very similar to the conventional *cathode ray tube*, except for the face plate and target material. The schematic is shown in Fig. 1.12. As IR signals impinge on the pyroelectric plate, the surface temperature of this plate changes. In turn this induces the charge which accumulates on pyroelectric material. The electron beam scans this material and two things may happen depending on whether a charge is absent or present. In the absence of charge – that is, no radiation – the electron beam is deflected towards the mesh by the action of *x* and *y* deflection plates. In contrast, the electron beam is focused onto the spot in the presence of charge, thus causing current to flow into an amplifier circuit. In this way a video signal is built as the electron beam is scanned over the entire surface of the pyroelectric plate. Since the accumulation of charge only occurs when the temperature of the pyroelectric material changes, this pyroelectric tube is only suitable for imaging a dynamic scene. This effect will benefit certain applications, such as monitoring a drying process, where only the fast changes of temperature are recorded (Fito *et al.*, 2004).

With the advent of semiconductor technology, it is now possible to produce pyroelectric solid-state arrays with resolution reaching 320 × 240 pixels. This

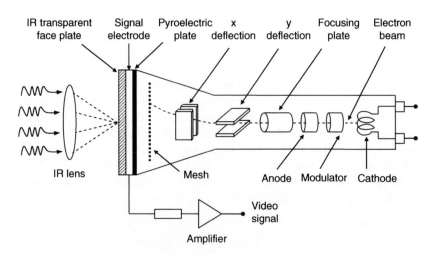

Fig. 1.12   Schematic diagram of the first-generation pyroelectric tube.

Published by Woodhead Publishing Limited, 2012

type of camera offers high detectivity but produces image at relatively low speed (typically 1 Hz). Furthermore, absolute temperature measurement often requires individual calibration of each element, which significantly slows down the image acquisition time. However, the main advantage lies with its ability to produce images without the necessity of cooling. This makes it suitable for a wide range of non-destructive applications, especially in industry.

Another kind of IR camera is based on microbolometer technology. Theoretically, a microbolometer is a *monolithic sensor* capable of detecting IR radiation through the direct or indirect heating of a low-mass film that has temperature-dependent physical properties. Popular materials include thermistors with high temperature coefficients of resistance such as the *vanadium oxide* (VO_x), silicon devices such as *Schottky barrier* diodes and transistors, and thermoelectrics such as the silicon p-n junctions. One example of the bolometer-type uncooled infrared focal plane array (IRFPA), having a 320 × 240 pixel array and operating at 60 Hz frame rate, has been experimented with for use in industry (Oda *et al.*, 2003). Figure 1.13 shows the schematic structure of each bolometer pixel. The pixel is divided into two parts: (i) a silicon readout integrated circuit (ROIC) in the lower part and (ii) a suspended microbridge structure in the upper part. The two parts are separated by a cavity. The microbridge structure is composed of a diaphragm and supported by two beams, thereby thermally isolating the former from the latter heat sink.

Fabrication of microbolometers such as the one shown in Fig. 1.13 use microelectromechanical techniques, originally developed at Bell Labs for air-bridge isolation integrated circuits. It has been carefully engineered so that part of the IR radiation is absorbed by the silicon passivation layers in the diaphragm and partly transmitted. The transmitted radiation is perfectly returned by the reflecting layer and is again absorbed by the passivation layers. In this way, more than 80% of the incident IR radiation is absorbed. The absorbed radiation heats the diaphragm and changes the bolometer resistance. Supplying a bias current enables the resistance change to be converted to voltage and detected by ROIC. The analogue signal voltage of the ROIC is digitized by analogue-to-digital converter of the

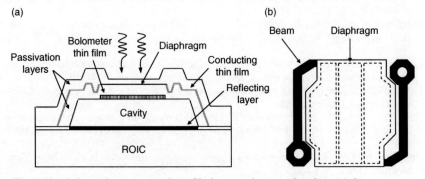

Fig. 1.13   Schematic representation of bolometer detector showing: (a) the cross-sectional view and (b) the plan view of each bolometer pixel.

Published by Woodhead Publishing Limited, 2012

receiving circuits. These data are, first, corrected for non-uniformity in bolometer responsitivity and, second, adjusted for video output. The pixel size of such a detector is $37 \times 37\mu m$ and the fill factor is about 72%.

1.4   Conclusions

Presently, there are various technologies that can be used to acquire digital images for food and beverages studies. A CCD camera system is one of the most commonly used image acquisition systems because of its reliability, economy and speed. Since a CCD camera is reflective technology, its application is limited to food objects which can reflect light when radiated upon by an illuminator. When food materials are transparent to light, then a CCD camera will not work. In this case other types of imaging technologies such as ultrasound and IR should be considered. The former is suitable when food objects are made up of materials which can alter the behaviour of ultrasonic waves, either through reflection, diffraction or deflection. Meanwhile, the latter is suitable when thermal capacities or heat conductivities of food materials vary with temperature. The future direction in this area is the fusion of different imaging technologies, providing a variety of information and making the inspection, monitoring and control processes more reliable.

1.5   References

Abdullah, M. Z., Abdul-Aziz, S. and Dos-Mohamed, A. M. (2000) Quality inspection of bakery products using a color-based machine vision system. *Journal of Food Quality*, **23**, 39–50.

Abdullah, M. Z., Fathinul-Syahir, A. S. and Mohd-Azemi, B. M. N. (2005) Automated inspection system for colour and shape grading of starfruit (*Averrhoa carambola* L.) using machine vision sensor. *Transactions of the Institute of Measurement and Control*, **27(2)**, 65–87.

Abdullah, M. Z., Guan, L. C. and Mohd-Azemi, B. M. N. (2001) Stepwise discriminant analysis for colour grading of oil palm using machine vision system. *Transactions of IChemE*, Part C, **57**, 223–31.

Batchelor, B. G. (1985) Lighting and viewing techniques. In Batchelor, B. G., Hill, D. A. and Hodgson, D. C. (eds), *Automated Visual Inspection*. Bedford, UK: IFS, 103–79.

Camarena, F. and Martínez-Mora, J. A. (2005) Potential of ultrasound to evaluate turgidity and hydration of the orange peel. *Journal of Food Engineering*, **75(4)**, 503–7.

Campbell, S. A. (2001) *The Science and Engineering of Microelectronic Fabrication*. New York: Oxford University Press.

Daley, W., Carey, R. and Thompson, C. (1993) Poultry grading inspection using colour imaging. *SPIE Proceedings Machine Vision Applications in Industrial Inspection*, **1907**, 124.

Doebelin, E. O. (1996) *Measurement Systems: Application and Design*. New York: McGraw Hill.

Fishbane, P. M., Gasiorowiczs, S. and Thornton, S. T. (1996) *Physics for Scientists and Engineers*. New Jersey: Prentice-Hall.

Fito, P. J., Ortolá, M. D., De los Reyes, R., Fito, P. and De los Reyes, E. (2004) Control of citrus surface drying by image analysis of infrared thermography. *Journal of Food Engineering*, **61(3)**, 287–90.

Fraden, J. (1997) *Handbook of Modern Sensors: Physics, Designs and Applications*. New York: American Institute of Physics Press .

Gan, T. H., Pallav, P. and Hutchins, D. A. (2005) Non-contact ultrasonic quality measurements of food products. *Journal of Food Engineering*, **77(2)**, 239–47.

Gaossorgues, G. (1994) *Infrared Thermography*. London: Chapman and Hall.

Ginesu, G., Guisto, D. G., Märgner, V. and Meinlschmidt, P. (2004) Detection of foreign bodies in food by thermal image processing. *IEEE Transactions of Industrial Electronics*, **51(2)**, 480–90.

Gómez, A. H., He, Y. and Pereira, A. G. (2005) Non-destructive measurement of acidity, soluble solids and firmness of Satsuma mandarin using Vis/NIR-spectroscopy techniques. *Journal of Food Engineering*, **77(2)**, 313–19.

Gunasekaran, S. (1996) Computer vision technology for food quality assurance. *Trends in Food Science and Technology*, 7, 245–56.

Heinemann, P. H., Hughes, R., Morrow, C. T., Sommer, III, H. J., Beelman, R. B. and Wuest, P. J. (1994) Grading of mushrooms using machine vision system. *Transactions of the ASAE*, **37(5)**, 1671–7.

Matas, J., Marik, R. and Kittler, J. (1995) Colour-based object recognition under spectrally non-uniform illumination. *Image and Vision Computing*, **13(9)**, 663–9.

McClements, D. J. (1995) Advances in the application of ultrasound in food analysis and processing. *Trends in Food Science and Technology*, 6, 293–9.

Morlein, D., Rosner, F., Brand, S., Jenderka, K. V., and Wicke, M. (2005) Non-destructive estimation of the intramuscular fat content of the *longissimus* muscle of pigs by means of spectral analysis of ultrasound echo signals. *Meat Science*, 69, 187–99.

Oda, N., Tanaka, Y., Sasaki, T., Ajisawa, A., Kawahara, A. and Kurashina, S. (2003) Performance of 320 × 240 bolometer-type uncooled infrared detector. *NEC Research and Development*, **44(2)**, 170–4.

Paulsen, M. (1990) Using machine vision to inspect oilseeds. *INFORM*, **1(1)**, 50–5.

Pearson, T. (1996) Machine vision system for automated detection of stained pistachio nuts. *Lebensmittel Wissenschaft und Technologie*, **29(3)**, 203–9.

Pedreschi, F., León, J., Mery, D. and Moyano, P. (2006) Development of a computer vision system to measure the color of potato chips. *Food Research International*, 39, 1092–8.

Simal, S., Benedito, J., Clemente, G., Femenia, A. and Rosell?, C. (2003) Ultrasonic determination of the composition of a meat-based product. *Journal of Food Engineering*, **58(3)**, 253–7.

Steele, D. J. (1974) Ultrasonics to measure the moisture content of food products. *British Journal Non-destructive Testing*, **16**, 169–73.

Steinmetz, V., Crochon, M., Bellon-Maurel, V., Garcia-Fernandez, J. L., Barreiro-Elorza, P. and Vestreken, L. (1996) Sensors for fruit firmness assessment: comparison and fusion. *Journal of Agricultural Engineering Research*, **64(1)**, 15–28.

Tao, Y., Heinemann, P. H., Varghese, Z., Morrow, C. T. and Sommer, III, H. J. (1995) Machine vision for color inspection of potatoes and apples. *Transactions of the ASAE*, **38(5)**, 1555–61.

Wells, P. N. T. (1969) *Physical Principles of Ultrasonic Diagnosis*. New York: Academic Press.

1.6   Appendix: nomenclature and abbreviations

1.6.1   Nomenclature
θ	absolute temperature, K
ρ	density
σ	Stefan-Boltzman's constant, 1.38054×10^{-23} JK^{-1}
λ	wavelength, m
c	speed of light, 2.998×10^8 ms^{-1}
E	energy, J
f	frequency, Hz
h	Planck's constant, 6.626076×10^{-34} Js
v	speed of sound, ms^{-1}
Z	acoustical impedance, Ω

1.6.2   Abbreviations
A/D	analogue-to-digital converter
CAT	computer-assisted tomography
CCD	charge-coupled device
CCIR	Comite Consultatif International des Radiocommunication
CdTe	cadmium telluride
CMOS	complementary metal oxide silicon
CT	computerized tomography
D/A	digital to analogue converter
DSP	digital signal processor
EUV	extreme ultraviolet
FET	field effect transistor
fps	frames per second
FUV	far ultraviolet
GPR	ground-probing radar
HgCdTe	mercury–cadmium–telluride
HgMnTe	mercury–manganese–telluride
InGaAs	indium–gallium–arsenide
InSb	indium–antimonide
IR	infrared
IRFPA	infrared focal plane array
LED	light emitting diode
MRI	magnetic resonance imaging
NIR	near infrared
NMR	nuclear magnetic resonance
NUV	near ultraviolet
PC	personal computer
PCI	peripheral component interface
PET	positron emission tomography
ROIC	readout integrated circuit
Rx	receiver
Si	silicon
SPECT	single photon emission computed tomography
TOF	time-of-flight
Tx	transmitter
UV	ultraviolet
VO_x	vanadium oxide

2

Hyperspectral and multispectral imaging in the food and beverage industries

J. Qin, United States Department of Agriculture, USA

Abstract: This chapter presents hyperspectral and multispectral imaging technologies in the food and agricultural area. It puts emphasis on the introduction and demonstration of spectral imaging techniques for practical uses. The main topics include spectral image acquisition methods, essential components for building spectral imaging systems (e.g., light sources, wavelength dispersive devices and area detectors), methods for calibrating spectral imaging systems (e.g., spectral and spatial calibrations) and techniques for analyzing spectral images (e.g., data preprocessing, dimension reduction and band selection). The spectral imaging applications for evaluating food and beverage products are also presented.

Key words: hyperspectral, multispectral, machine vision, food quality and safety, nondestructive evaluation.

2.1   Introduction

Traditional optical sensing techniques, such as imaging and spectroscopy, have limitations for acquisition of adequate information for nondestructive evaluation of food and agricultural products. In recent years, spectral imaging (i.e., hyperspectral and multispectral) has emerged as a better tool for quality and safety inspection of various agricultural commodities. Spectral imaging techniques combine conventional imaging and spectroscopy techniques; it is possible to obtain both spatial and spectral information from the target, which is essentially useful for evaluating individual food items. The technique has drawn tremendous interest from both academic and industrial areas, and it has been developed rapidly during the past decade.

This chapter focuses on hyperspectral and multispectral imaging technologies for food and agricultural applications. There is an introduction to spectral image acquisition methods in Section 2.2, with basic concepts and ground rules for the rest of the chapter. The emphasis of the chapter is put on the introduction to spectral imaging techniques for practical uses, including essential components for building a spectral imaging system (Section 2.3), methods for calibrating a spectral imaging system (Section 2.4) and techniques for analyzing spectral images (Section 2.5). Applications of hyperspectral and multispectral imaging for food and beverage products are summarized in Section 2.6. Conclusions are presented in Section 2.7, summarizing the chapter and addressing future development. Finally, further information, including useful books and websites, are given.

2.2   Spectral image acquisition methods

Spectral images are three-dimensional (3-D) in nature, with two spatial dimensions and one spectral dimension. Based on the continuity of the data stored in the wavelength domain, spectral imaging can be divided into two main techniques: hyperspectral imaging and multispectral imaging. The hyperspectral technique acquires images with abundant (tens or hundreds) and continuous wave (CW) bands, while the multispectral technique acquires images with few (generally less than ten) and discrete wavebands. A full spectrum can be extracted from each pixel in hyperspectral images. Multispectral images produce a set of isolated data points for each pixel due to the separate wavebands stored in the dataset.

2.2.1   Hyperspectral imaging

Hyperspectral imaging is intended to collect images with high spatial and spectral resolutions for fundamental research. The process usually involves a relatively long time for image acquisition under laboratory conditions and relatively complicated procedures for offline image analysis. Generally, there are three approaches for acquiring 3-D hyperspectral cubes [hypercubes (x, y, λ)]. They are point-scan, line-scan and area-scan methods, as illustrated in Fig. 2.1. In the point-scan method (also known as the whiskbroom method), a single point is

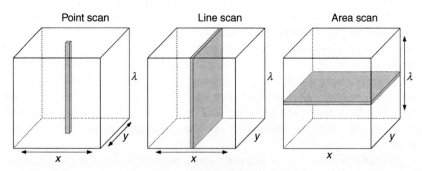

Fig. 2.1   Methods for acquiring 3-D hyperspectral cubes with both spatial (x and y) and spectral (λ) information. Arrows represent scan directions, and grey areas represent data acquired at a time.

scanned along two spatial dimensions (x and y) by moving either the sample or the detector. A spectrophotometer equipped with a point detector is used to acquire a single spectrum for each pixel in the scene. Hyperspectral image data are accumulated pixel by pixel. Two-axis motorized positioning tables are usually needed to move the sample to finish the scan.

The line-scan method (also known as the pushbroom method) is an extension of the point-scan method. Instead of scanning one point each time, this method simultaneously acquires a slit of spatial information as well as full spectral information for each spatial point in the field-of-view (FOV). A special 2-D image (y, λ) with one spatial dimension (y) and one spectral dimension (λ) is taken. A complete hypercube is obtained as the slit is scanned in the direction of motion (x). Hyperspectral systems based on imaging spectrographs with either fixed or moving slits work in the line-scan mode.

Both point-scan and line-scan methods are spatial-scan methods. The area-scan method (also known as band sequential [BSQ] method), on the other hand, is a spectral-scan method. This approach acquires a 2-D single-band grayscale image (x, y) with full spatial information. A hypercube containing a stack of single-band images is built up as the scan is performed in the spectral domain. No relative movement between the sample and the detector is required for this method. Imaging systems using filters (e.g., filter wheels and electronically tunable filters) belong to the area-scan method.

2.2.2   Multispectral imaging

Multispectral imaging aims to acquire spatial and spectral information that is directly useful for real-time applications in the field (e.g., fruit packing houses and food processing plants). The process generally involves fast image acquisition and simple algorithms for image processing and decision-making. Reducing the total volume of the data required in collecting both spatial and spectral domains is the key for building multispectral imaging systems. In practice, this means acquiring images with relatively low spatial resolutions at few important wavelengths. Hyperspectral images are usually used as fundamental datasets to determine optimal wavebands that can be used by a multispectral imaging solution for a particular application.

The aforementioned point-scan method is practically not feasible for fast image acquisition because of its time-consuming scan along two spatial dimensions. The other two methods (i.e., line-scan and area-scan), however, can be adjusted to satisfy the requirements of multispectral image acquisition. As illustrated in Fig. 2.2, both line-scan and area-scan methods can be implemented to collect wavelengths much less than those in hyperspectral imaging. For the line-scan method, this can be achieved by specifying the positions of all the useful tracks (corresponding to the selected important wavelengths) along the spectral dimension of the charge-coupled device (CCD) detector. It is usually executed using the random-track readout mode of the camera. Only the data from the selected tracks are acquired, which reduces the amount of the data for each line-scan image (y, λ) and

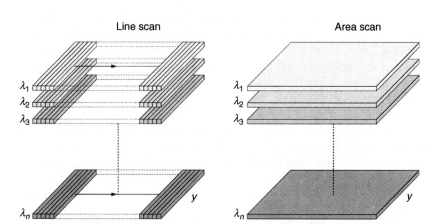

Fig. 2.2   Methods for acquiring multispectral images with few and discrete wavelengths ($\lambda_1-\lambda_n$, n is generally less than 10).

consequently shortens the acquisition time. The bandwidth of the selected tracks can be adjusted through the pixel binning along the spectral dimension.

On the other hand, the area-scan method for multispectral imaging collects single-band images at selected wavelengths simultaneously. Light from the spatial scene is usually divided into several parts by certain optical separation devices (e.g., a beam-splitter). The divided scenes will go through preset bandpass filters (corresponding to the selected important wavelengths) separately. Narrowband images are then formed on several cameras or on one camera with a large CCD sensor. By avoiding the scan in the spectral domain used in hyperspectral imaging, this method can also save the image acquisition time for multispectral imaging applications.

Besides the spectral image acquisition methods discussed above, a new approach called single-shot method has been recently developed. This method is intended to record both spatial and spectral information on an area detector with one exposure. No scan in either spatial or spectral domains is needed for acquiring a hypercube, making it attractive for real-time applications. This method is still in the early stages and not fully developed. Only a few implementations relying on complicated fore-optics design and computationally intensive postprocessing for image reconstructions are currently available (Bodkin *et al.*, 2008), with limitations for ranges and resolutions for spatial and spectral dimensions.

2.3   Construction of spectral imaging systems

A spectral imaging system generally consists of a light source, a wavelength dispersive device and an area detector (Fig. 2.3). The components for building spectral imaging systems are presented in the following sections.

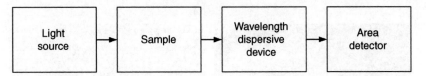

Fig. 2.3   General configuration for a spectral imaging system.

2.3.1   Light sources

Light is the information carrier for imaging systems. Numerous types of light sources are available for various spectral imaging applications. In general, these light sources can be classified into two categories: illumination sources and excitation sources. Broadband light sources are generally used as the illumination sources for reflectance and transmittance imaging systems. The spectral constitution of the incident light is not changed after light-sample interactions. The measurement is performed based on the intensity changes at different wavelengths. Narrowband (or monochromatic) light sources are the common excitation sources. When excited by a high-intensity monochromatic light, some biological materials (e.g., animal and plant tissues) emit low-intensity light in a broad wavelength range. The energy change (or frequency shift) can cause fluorescence emission or Raman scattering, which carries information about the target that can be used for various inspection purposes.

Illumination sources
Halogen lights are the most common broadband illumination source. Light is generated through incandescent emission when high temperature is applied on the tungsten filament that is housed in a quartz glass envelope filled with halogen gas. Quartz tungsten halogen (QTH) lamps generate a smooth spectrum in the wavelength range from visible to infrared. The QTH lamps can be used to directly illuminate the target or be placed in a lamp housing, where light is delivered via an optical fiber. A halogen fiberoptic illuminator is shown in Fig. 2.4a. It generates light by a 150-watt halogen lamp and delivers broadband light through fiberoptic guides for different illumination purposes, such as line light for line-scan systems and ring light for area-scan systems. The halogen sources have been intensively used in hyperspectral/multispectral reflectance measurements for food surface inspections (Kim *et al.*, 2001; Park *et al.*, 2002; Lu, 2003). High-intensity QTH lights have also been used in transmittance measurements for detecting inside of agricultural commodities (Qin and Lu, 2005; Ariana and Lu, 2008; Yoon *et al.*, 2008).

Light-emitting diode (LED) technology has advanced rapidly owing to the demands for cheap, powerful, robust and reliable sources. LEDs are solid-state sources that emit light when electricity is applied to a semiconductor. Depending on the materials of the p-n junction inside the LEDs, they can generate narrowband light at different wavelengths (colors). High-intensity broadband LEDs have recently been developed by mixing red, blue and green monochromatic lights. As a new type of light source, LEDs have many advantages over traditional lighting,

(a)   (b)

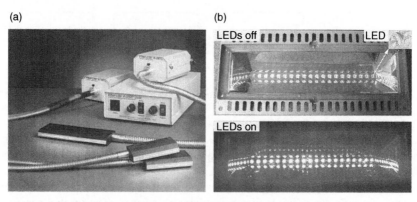

Fig. 2.4   Light sources for spectral imaging systems: (a) a halogen fiberoptic light as an illumination source for reflectance measurement (photo courtesy of Dolan-Jenner Industries, Boxborough, MA, USA) and (b) a high-power blue LED line light as an excitation source for fluorescence measurement.

such as long lifetime, low power consumption, low heat generation, small size, fast response, robustness and non-sensitivity to vibration. They can be assembled in different arrangements (e.g., spot, line and ring lights) to satisfy different lighting requirements. Broadband LEDs have started to be used in the area of food quality and safety inspection (Lawrence *et al.*, 2007; Chao *et al.*, 2008). The development of LED technology is still ongoing, and the use of LEDs as new light sources for spectral imaging applications is likely to expand in the near future.

Excitation sources

Lasers are powerful monochromatic sources that are widely used for excitation purposes. Light from lasers is generated through stimulated emission, which usually occurs inside a resonant optical cavity filled with a gain medium (e.g., gas, dye solution, semiconductor or crystal). They can operate in CW mode or pulse mode in terms of temporal continuity of the output. Lasers are the ideal excitation sources for fluorescence and Raman measurements owing to their unique features, such as highly concentrated energy, perfect directionality and real monochromatic emission. Various lasers have been used in hyperspectral fluorescence and Raman imaging systems for inspection of food and agricultural commodities (Kim *et al.*, 2003; Noh and Lu, 2007; Qin *et al.*, 2010). Narrowband LED lights can also be used as excitation sources (Qin *et al.*, 2011). Figure 2.4b shows a high-powered blue LED light. Twenty 3 W LEDs (SAILUX Semiconductor Lighting, Gwangju, South Korea) are mounted in a line on top of the back mirror. The unit generates light with a spectral peak at 400 nm and a full width at half maximum (FWHM) bandwidth of 20 nm. It can be used to excite biological samples for fluorescence measurement. Other types of light sources such as ultraviolet (UV) fluorescent lamps (Kim *et al.*, 2001), high-pressure arc lamps (e.g., xenon) and low-pressure metal vapor lamps (e.g., mercury) can also serve as excitation sources.

2.3.2   Wavelength dispersive devices

Wavelength dispersive devices are the core component of spectral imaging systems. Their function is to disperse broadband light into different wavelengths and project the dispersed light to the area detectors. Many optical and electro-optical instruments can be used for this purpose. The common dispersive devices are introduced in this section.

Imaging spectrographs

An imaging spectrograph is an optical device that disperses broadband light into different wavelengths from different spatial regions of a target. It differs from the traditional spectrograph in that it can also carry spatial information. The imaging spectrograph operates in line-scan mode, and it is the important component for line-scan spectral imaging systems. Most imaging spectrographs are built based on diffraction gratings, which include two major types: transmission gratings and reflection gratings.

Figure 2.5a shows the operating principle of a transmission grating-based imaging spectrograph. Specifically, it is built using a prism-grating-prism (PGP) component. An incoming light from the entrance slit is collimated by the front lens. The collimated beam is dispersed at the PGP component, where light

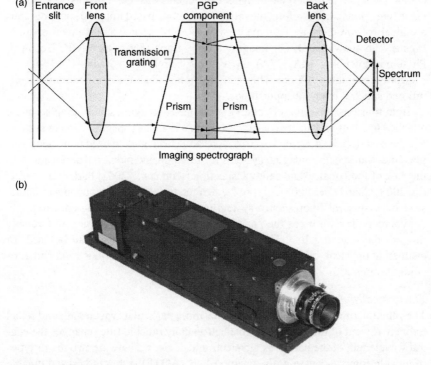

Fig. 2.5   PGP imaging spectrograph: (a) operating principle and (b) an ImSpector imaging spectrograph produced by Specim (photo courtesy of Specim, Oulu, Finland).

propagation direction is dependent on its wavelength. The central wavelength passes symmetrically through the PGP, while the shorter and longer wavelengths are dispersed up and down relative to the central axis. As a result, the light from the scanning line is dispersed into different wavelengths. The dispersed light is then projected onto an area detector through the back lens, creating a special 2-D image: one dimension represents spatial and the other spectral. A PGP imaging spectrograph (ImSpector Fast10, Specim, Oulu, Finland) is shown in Fig. 2.5b. It covers a wavelength range of 400–1000 nm with a spectral resolution of 15 nm. The Fast10 spectrograph allows high light throughput and fast readout rate by compressing spectral dispersion on the detector, making it suitable for high-speed spectral image acquisition (up to 1500 lines per second).

Reflection gratings have been recently adopted to build imaging spectrographs. A reflection grating-based imaging spectrograph is illustrated in Fig. 2.6a. It is constructed based on an Offner configuration. The basic structure includes a pair of concentric spherical mirrors coupled with a convex reflection grating. The lower mirror guides light from the entrance slit to the reflection grating, where the beam is dispersed into different wavelengths. The upper mirror then reflects the dispersed light to the detector, where a continuous spectrum is formed. The Offner spectrograph is built entirely of reflective optics. The reflection grating is not limited by the transmission properties of the grating substrate, and the reflective components (e.g., mirrors) have higher efficiencies than the transmission components (e.g., prisms). Therefore, the reflection grating-based imaging spectrographs are ideal for low-light measurements such as fluorescence and Raman imaging. Figure 2.6b shows an Offner imaging spectrograph (Hyperspec NIR, Headwall Photonics, Fitchburg, MA, USA). It covers a spectral region of 900–1700 nm with a spectral resolution of 5 nm. This spectrograph is ideal for high-resolution near-infrared spectral imaging applications.

Both transmission- and reflection grating-based imaging spectrographs can be attached to a lens and an area detector to form a line-scan spectral camera system. For the past decade, imaging spectrographs have been widely used to develop various line-scan spectral imaging systems for research and industrial applications in the area of food quality and safety inspection (Kim et al., 2001; Park et al., 2002; Lu, 2003; Chao et al., 2008). Figure 2.7 demonstrates main components of a line-scan hyperspectral fluorescence system using an Offner imaging spectrograph. A hypercube is built up as the positioning table moves the samples transversely through the scanning line. The acquired hyperspectral images can be used for evaluating physical, chemical and biological properties of various food and agricultural products.

Electronically tunable filters
The function of a bandpass filter is to transmit a particular wavelength and reject light energy out of the bandpass. An electronically tunable filter changes the central wavelength of the bandpass via electronic devices. There are two major types of tunable filters: acousto-optic tunable filters (AOTF) and liquid crystal tunable filters (LCTF).

An AOTF is a solid-state device that isolates a single wavelength from a broadband light based on light–sound interactions in a crystal. The operating principle of the AOTF and an AOTF camera adapter (VA210, Brimrose, Sparks, MD, USA) are shown in Fig. 2.8. The acoustic transducer generates high-frequency acoustic waves through the crystal, which changes the refractive index of the crystal. The variations of the refractive index make the crystal to act like a transmission diffraction grating. Light is diffracted into two first-order beams with orthogonal polarizations. The zero-order beam and the undesired diffracted beams are blocked by the beam stop. The AOTF only diffracts light at one particular wavelength at a time. The isolated wavelength is a function of the frequency of the acoustic waves. Thus, the passing wavelength can be controlled by varying the frequency of the RF source.

(a)

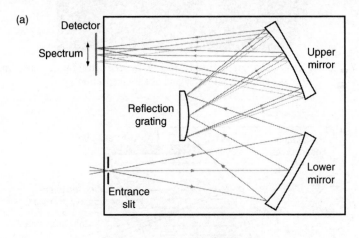

(b)

Fig. 2.6   Offner imaging spectrograph: (a) operating principle and (b) a Hyperspec imaging spectrograph produced by Headwall Photonics (photo courtesy of Headwall Photonics, Fitchburg, MA, USA).

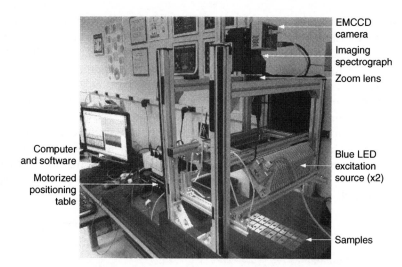

Fig. 2.7   An imaging spectrograph-based line-scan hyperspectral fluorescence imaging system for food safety and quality inspection.

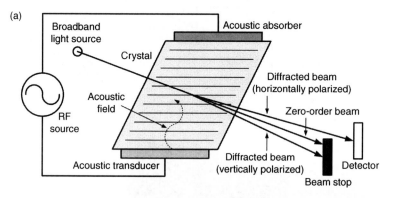

Fig. 2.8   Acousto-optic tunable filter (AOTF): (a) operating principle and (b) an AOTF camera adapter produced by Brimrose (photo courtesy of Brimrose Corporation, Sparks, MD, USA).

An LCTF is a solid-state device utilizing electronically controlled liquid crystal cells to transmit light with a specific wavelength. The LCTF is constructed by a series of optical stacks, each consisting of a retarder and a liquid crystal layer between two polarizers. A single filter stage and example LCTFs (VariSpec, Cambridge Research and Instrumentation, Woburn, MA, USA) are shown in Fig. 2.9. The incident light is polarized through the polarizer and then separated into two rays by the retarder. The separated rays emerge with a phase delay that is dependent upon the wavelength. The polarizer behind the retarder only transmits wavelengths in phase to the next stage. Each stage transmits light as a sinusoidal function of the wavelength. The transmitted light adds constructively in the desired bandpass and destructively in the other spectral regions. All the stages function together to transmit a single wavelength. An electric field is applied to each liquid crystal layer to cause retardance changes. The LCTF controller, thus, can shift the narrow bandpass region throughout a broad wavelength range.

Wavelength switching of the electronically tunable filters is much faster than that of the mechanical devices such as filter wheels. Other advantages for building spectral imaging systems include high optical throughput, narrow bandwidth, broad spectral range, accessibility of random wavelength and flexible controllability and programmability

(a)

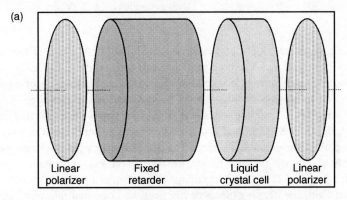

| Linear polarizer | Fixed retarder | Liquid crystal cell | Linear polarizer |

(b)

Fig. 2.9   Liquid crystal tunable filter (LCTF): (a) single filter stage and (b) VariSpec LCTFs produced by Cambridge Research and Instrumentation (CRi) (photo courtesy of CRi, Woburn, MA, USA).

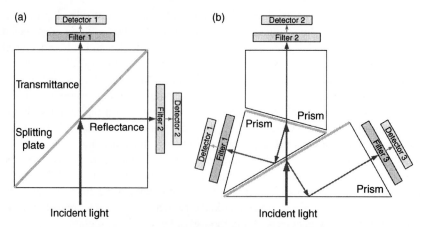

Fig. 2.10   Beam-splitting devices for multispectral imaging: (a) a two-channel plate beam-splitter and (b) a three-channel prism beam-splitter.

(Morris *et al.*, 1994). The AOTFs and LCTFs have been used to develop various area-scan spectral imaging systems for many agricultural applications (Peng and Lu, 2006; Safren *et al.*, 2007; Zhang *et al.*, 2007; Gómez-Sanchis *et al.*, 2008a).

Beam-splitting devices

Both imaging spectrographs and electronically tunable filters can collect continuous (hyperspectral) and discrete (multispectral) spectral data. The image acquisition generally needs to scan in spatial or spectral domain. Beam-splitting devices, on the other hand, can simultaneously acquire narrowband images at selected wavelengths. Their function is to divide light into two or more parts. The useful wavelengths are predetermined, and the corresponding bandpass filters are placed on each path of the separated beams. The divided scenes pass through the filters separately, and the single-band images are formed on several cameras or one camera with a large CCD sensor. Depending on the wavelength constitution of each separated light, the beam-splitting devices can be classified into two categories: color splitting and neutral splitting. The color-splitting devices guide particular wavebands (e.g., visible or infrared) to each output, while the neutral splitting devices guide particular portions of the total light energy to each output.

Figure 2.10 shows two example neutral beam-splitting devices. Figure 2.10a is a two-channel plate beam-splitter. Light with a 45° angel of incidence is divided into two portions after hitting the plate painted with separation coatings. Different ratios between reflectance and transmittance are available for different applications (e.g., 30R/70T, 50R/50T and 70R/30T). If the splitting plate is replaced by a cold mirror or a hot mirror, the unit becomes a color-splitting device since a cold mirror reflects visible and transmits infrared and a hot mirror does the opposite. Figure 2.10b is a three-channel prism beam-splitter. The unit consists of three prism components that are cemented together by two neutral films. The incoming light is separated into three parts after interacting with the neutral films and the hypotenuses of the prisms. After passing through three preset filters, light will form three narrowband images on three detectors for the same scene.

Since beam-splitting devices can acquire multiple images instantly, they can be used to build multispectral imaging systems. Two-band or three-band systems may be sufficient for certain applications using simple detection algorithms (e.g., band ratio). More image channels can be achieved by using more complicated separation prisms (e.g., four-channel prisms) or combining color and neutral splitting devices, such as specialty mirrors (e.g., cold and hot mirrors) and two-channel beam-splitters (Kise *et al.*, 2010). When all the separated light beams are aligned to the same propagation direction, all the filters can be arranged in parallel and one camera can collect all the narrowband images on different regions of the CCD sensor (Chao *et al.*, 2001).The multispectral imaging systems based on beam-splitting devices have found many agricultural applications, especially for real-time and online inspection tasks (Aleixos *et al.*, 2002; Kleynen *et al.*, 2005; Kise *et al.*, 2007; Lu and Peng, 2007; Park *et al.*, 2007a).

2.3.3   Area detectors
In a typical spectral imaging system, light signals from the source interact with the target and then pass through the wavelength dispersive device. The light carrying the sample information is eventually collected by an area detector. The function of the area detector is to acquire the spatial distribution of the light intensity by converting radiation energy into electrical signals. Currently CCD cameras are the mainstream devices used in spectral imaging systems.

Silicon CCD cameras
The CCD sensor is composed of many (usually millions) small photodiodes (called pixels) that are made of light-sensitive materials, such as silicon (Si) or indium gallium arsenide (InGaAs). Each photodiode acts like an individual spot detector that converts incident photons to electrons, generating an electrical signal proportional to total light exposure. All the electrical signals are shifted out of the detector and then digitalized to form the images. Spectral response of the CCD sensor, which is quantified by quantum efficiency (QE), is primarily governed by the substrate materials used to make the photodiodes. Owing to the natural sensitivity to the visible light, silicon is intensively used as sensor material for making the cameras working in visible and short-wavelength near-infrared region. A silicon CCD camera (Luca-R, Andor Technology, Belfast, Northern Ireland, UK) and its QE curve are shown in Fig. 2.11a. The spectral response of the silicon sensors is generally a bell-shaped curve with QE values declining towards both ultraviolet and near-infrared regions.

InGaAs CCD cameras
For the near-infrared region, InGaAs, which is made of an alloy of indium arsenide (InAs) and gallium arsenide (GaAs), is the common substrate material of the image sensors. An InGaAs detector (SU640KTS-1.7RT, Sensors Unlimited, Princeton, NJ, USA) and its QE curve are shown in Fig. 2.11b. InGaAs has fairly

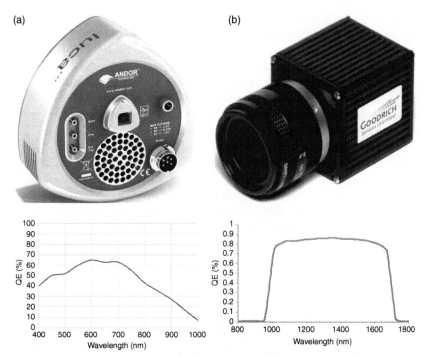

Fig. 2.11   Area detectors for spectral imaging systems: (a) a silicon EMCCD detector and its quantum efficiency (QE) (photo courtesy of Andor Technology, Belfast, Northern Ireland, UK) and (b) an InGaAs photodiode detector and its quantum efficiency (photo courtesy of Sensors Unlimited, Princeton, NJ, USA).

flat and high QE in the near-infrared region. Standard InGaAs (53% InAs and 47% GaAs) sensors cover the spectral region of 900–1700 nm. Extended wavelength range (e.g., 1100–2600 nm) can be achieved by changing the percentages of InAs and GaAs used for making the sensors (Sensors Unlimited, 2006). The silicon and InGaAs CCD cameras have been widely used in various spectral imaging systems using the spectral information in visible and near-infrared regions (Kim *et al.*, 2001; Park *et al.*, 2002; Lu, 2003; Nicolaï *et al.*, 2006).

High-performance CCD cameras
Regular CCD cameras can acquire high-quality images when there is sufficient light and no short exposure is required (e.g., reflectance and transmittance). Low-light imaging applications (e.g., fluorescence and Raman) and fast spectral image acquisition usually need high-performance cameras such as electron multiplying CCD (EMCCD) and intensified CCD (ICCD) cameras. An EMCCD differs from a regular CCD by adding a unique electron multiplication register to the end of the normal readout register. This built-in register multiplies the weak charge signals before any readout noise is imposed by the output amplifier, achieving real gain for the useful signals. The camera shown in Fig. 2.11a is an EMCCD camera. The EMCCD cameras have started to find applications for inspection of food and

agricultural products (Chao *et al.*, 2008; Kim *et al.*, 2011; Qin *et al.*, 2011). ICCD is another type of high-performance image sensor that can detect weak optical signals. The ICCD utilizes an image intensifier tube to apply the gain to the incident light before it reaches the sensor. The amplified light signals are then coupled to the CCD. Hence, the EMCCD is based on electronic amplification, while the ICCD is based on optical amplification. The ICCD cameras can realize fast gate times (in a nanosecond or picosecond), making them suitable for acquiring short-duration signals, such as time-dependent fluorescence emissions (Kim *et al.*, 2003).

2.4   Calibration of spectral imaging systems

Spectral imaging systems obtain rich raw information (e.g., spatial, spectral and light intensity) from the target. Appropriate calibration is an essential step for acquiring meaningful image data. The commonly used calibration methods for the spectral imaging systems are introduced in the following sections.

2.4.1   Spectral calibration

Spectral calibration for hyperspectral/multispectral imaging systems aims to define the wavelengths for the pixels along the spectral dimensions of the image data. The calibration results can be used to determine the range and the resolution of the spectral information. The area-scan spectral imaging systems using electronically tunable filters (e.g., AOTFs and LCTFs) collect single-band images at a series of known wavelengths. The passing wavelengths are determined by their electronic controllers. Therefore, the wavelength calibration is usually not necessary for the area-scan imaging systems.

The line-scan spectral imaging systems using imaging spectrographs acquire images with unknown wavelengths. Spectral calibration is needed to map the pixel indices along the spectral dimension of the detector to the wavelengths. The calibration can be performed using spectrally established light sources, such as spectral calibration lamps and lasers. The spectral calibration lamps are the most common calibration sources. They generate narrow spectral lines from the excitation of rare gases and metal vapors, which are used as reference wavelengths for the calibration. Various calibration lamps (e.g., argon, krypton, neon, xenon, mercury, mercury-argon, mercury-neon and mercury-xenon) are available for the wavelength range from ultraviolet to near-infrared, and they have various forms (e.g., pencil-style lamps and battery-powered lamps) to meet different calibration requirements. Figure 2.12 shows an example of spectral calibration for a line-scan hyperspectral imaging system. It was performed using two pencil-style calibration lamps (xenon and mercury-argon lamps, Newport, Irvine, CA, USA), which have several good peaks in the spectral region of 400–1000 nm. Two images on the top are the line-scan images from the two lamps. Two spectral profiles are extracted along the vertical axis (spectral dimension) of the images. The pixel positions and the corresponding wavelengths are identified, and their

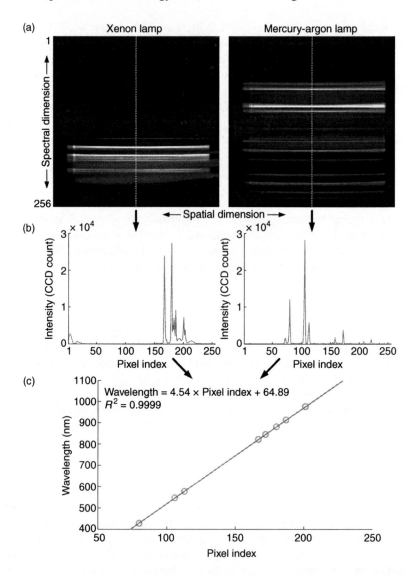

Fig. 2.12 Spectral calibration for a line-scan hyperspectral imaging system using calibration lamps: (a) line-scan images, (b) spectral profiles and (c) a linear regression model.

relationship is established using a linear regression function. The linear model can then be used to determine all the wavelengths along the spectral dimension. Nonlinear regression models are also used for the spectral calibration (Park *et al.*, 2002; Chao *et al.*, 2008).

Besides the absolute spectral calibration discussed above, relative spectral calibration is sometimes used under certain circumstances. For example, a Raman spectrum is generally presented as a shift in energy from that of the excitation

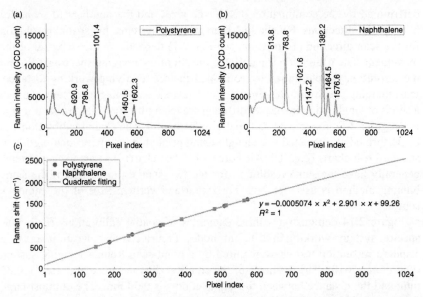

Fig. 2.13 Spectral calibration for a Raman imaging system using Raman shift standards. Raman spectrum of (a) polystyrene, (b) naphthalene and (c) a quadratic regression model.

source (e.g., laser). Raman shift, which is traditionally expressed as wavenumber in cm^{-1}, is essentially a relative unit. For spectral calibration of Raman imaging systems, it is more useful to use wavenumbers as references instead of wavelengths used in the absolute spectral calibration. Usually an excitation source with fixed spectral output and chemicals with known relative wavenumber shifts are used to calibrate Raman spectral imaging systems. A guide to Raman shift standards has been established by the American Society for Testing and Materials (ASTM) International (ASTM Standards, 2007). This guide provides Raman shift wavenumbers of eight chemicals, which cover a wide wavenumber range (85–3327 cm^{-1}). Figure 2.13 shows an example of spectral calibration for a Raman imaging system using a 785 nm excitation laser and two Raman shift standards (polystyrene and naphthalene) (Qin *et al.*, 2010). After identifying the pixel positions and the corresponding wavenumbers of the Raman peaks, a quadratic regression model is established to determine all the wavenumbers along the spectral dimension.

2.4.2 Spatial calibration

Spatial calibration determines the range and the resolution for the spatial information stored in the image data. The calibration results are useful for adjusting the FOV and estimating the spatial detection limit. Different methods can be used to calibrate spectral imaging systems working in different acquisition modes. The spatial resolution of the point-scan imaging systems is determined by the step sizes used for the two scan directions (x and y). The spatial range is determined by the combination of the step sizes and the numbers of scans for both spatial directions. For the line-scan imaging systems, the spatial resolution for the scan direction (x direction, see Fig. 2.1) depends on the step size of the movement. The y direction is parallel to the slit of the imaging spectrograph, and the resolution is determined by combined factors involving working distance, lens, imaging spectrograph, camera, etc. The area-scan imaging systems collect a series of single-band images at different wavelengths. Each narrowband image is a regular 2-D image with full spatial information. The spatial calibration can be performed at a selected wavelength using printed targets with square grids or standard test charts (e.g., US Air Force 1951 test chart). The area-scan systems generally have the same resolution for the two spatial dimensions if the same binning method is used for both horizontal and vertical axes of the camera's sensor.

Figure 2.14 demonstrates three examples of spatial calibrations for hyperspectral systems working in different modes. Figure 2.14a shows an image of a standard resolution test chart acquired by a point-scan Raman imaging system (Qin *et al.*, 2010). The diameter of the smallest dots in the central area is 0.25 mm, and the distance between the adjacent dots is 0.50 mm. The outmost large dots are positioned within a 50 mm square. A step size of 0.1 mm was used to scan both x and y directions. The 0.25 mm dots can be clearly discerned owing to the small step sizes used to scan the chart. Figure 2.14b is a 256 × 256 line-scan image of a piece of white paper printed with thin parallel lines 2 mm apart. The image was obtained by a hyperspectral system designed for light-scattering measurement for fruits and vegetables carried by a motor-controlled stage (Qin and Lu, 2008). The spatial axis of the imaging spectrograph is aligned to the horizontal dimension of the CCD sensor. Thus, the horizontal dimension represents spatial and the vertical dimension spectral. The step size used for image acquisition is 1.0 mm. Hence, the spatial resolution for the x direction (see Fig. 2.1) is 1.0 mm/pixel. The resolution for the y direction can be determined by dividing the distance by the number of pixels in this range. There are 150 pixels within 30 mm distance, thus the spatial resolution can be calculated as 30 mm/150 pixels = 0.2 mm/pixel. The spatial range along the horizontal dimension is 0.2 mm/pixel × 256 pixels = 51.2 mm. Figure 2.14c shows an area-scan image of a piece of white paper printed with 10 mm square grids, which was collected by a LCTF hyperspectral system. Similarly, the spatial resolution can be calculated as 90 mm/225 pixels = 0.4 mm/pixel. This system may not be able to differentiate the smallest dots in Fig. 2.14a since its resolution is lower than the diameter of these dots (0.25 mm).

2.4.3   Image registration

For the spectral imaging systems using beam-splitting devices, light from the spatial scene generally travels along different paths and interacts with different optical components before reaching the CCD detectors. Image misalignment usually occurs for the acquired narrowband images due to various factors associated

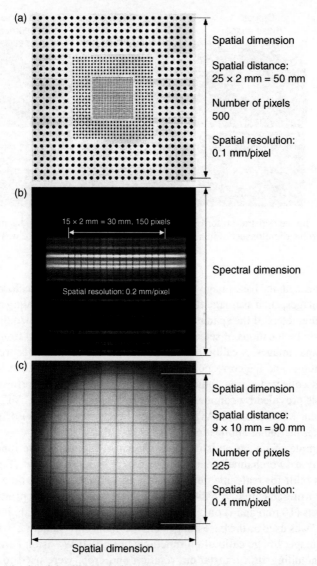

Fig. 2.14 Spatial calibrations for (a) point-scan, (b) line-scan and (c) area-scan hyperspectral imaging systems.

with the imperfect components in the system, such as lens distortions, positional tolerance of the CCD sensors, different sensor types and differences among the mechanical parts (e.g., lens mounts and connectors). The misaligned images will cause problems in spectral image analysis. Therefore, image registration is needed to align two or more images of the same scene. Typically, one single-band image (called base image) is selected as the reference, to which images at other bands (called input images) are compared. The goal of image registration is to align the

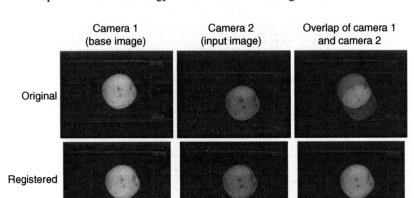

Fig. 2.15   Image registration for two narrowband images acquired by a beam-splitter-based two-band multispectral imaging system developed for real-time citrus surface disease detection.

input images with the base image by applying necessary spatial transformations to the input images, such as image flip, translation, rotation, crop, scaling and shearing. The parameters of the spatial transformations can be determined through the mapping for the locations of selected points (called control points) from a pair of base and input images. A calibration template is usually used in this process. The transformations and the corresponding parameters are then saved for transforming images collected in the future. Image registration can be performed using related tools provided by commercial software packages (e.g., LabVIEW Vision Development Module and MATLAB Image Processing Toolbox) or self-made programs for particular applications (Kise *et al.*, 2010).

An example of image registration is shown in Fig. 2.15. The images were obtained by a two-channel beam-splitter-based imaging system (Fig. 2.10a), which was built for real-time detection of citrus surface disease on a commercial fruit-sorting machine. A calibration template was created by printing a grid of black dots (10 mm apart) on a piece of white paper. The template image from 'Camera 1' was used as the base image, and the one from 'Camera 2' was used as the input image. Image calibration functions in LabVIEW Vision Development Module, including flip, translation, rotation and crop, were used to align the two images. The parameters determined from the template images were used to register the images from the citrus samples. The two original images in Fig. 2.15 were acquired by two cameras at two bands in short-wavelength near-infrared region from a citrus sample moving at a speed of five fruits per second. The overlap of the original images from 'Camera 1' and 'Camera 2' clearly reveals the misalignment of the two images. After image registration, the double image disappeared and only one fruit is observed in the overlay of the registered images. The image size was reduced from 640 × 480 (original images) to 608 × 415 (registered images) due to the crop operations in the image spatial transformation.

2.4.4   Other calibrations

Besides the calibration methods discussed above, there are other types of calibrations that can be performed for spectral imaging systems to satisfy the requirements of different applications. Radiometric calibration, for example, is required when the absolute spectral radiance of the sample needs to be determined. An integrating sphere typically serves as a radiance standard for the radiometric calibration. Various spectral imaging applications in the area of food quality and safety inspection can also create particular calibration needs. For example, intensity values of the spectral images from spherical fruit samples are not uniform due to the curvature effects, which cause problems in image processing and classification. To tackle this problem, Qin and Lu (2008) developed a method to correct the spatial profiles extracted from line-scan scattering images of apples acquired by an imaging spectrograph-based hyperspectral system. Gómez-Sanchis *et al.* (2008b) developed a method to correct the area-scan images of citrus samples collected by a LCTF-based hyperspectral system. Various methods and procedures for calibration, characterization and enhancement have been developed to fulfill the potential of the hyperspectral/multispectral imaging techniques (Lawrence *et al.*, 2003; Polder *et al.*, 2003; Gebhart *et al.*, 2007; Qin and Lu, 2007; Kim *et al.*, 2011). Accompanied with the introduction of new measurement concepts and instruments, improved and new calibration and correction approaches are expected in the future.

2.5   Spectral images and analysis techniques

Three-dimensional spectral images provide a large amount of spatial and spectral information. Spectral image analysis, which involves both chemometrics and image processing techniques, is a crucial step to obtain useful information for hyperspectral/multispectral imaging applications. The spectral image data and common analysis methods are introduced in the following sections.

2.5.1   Spectral image data

In general, 3-D spectral image data acquired by point-scan, line-scan and area-scan methods are stored in the formats of band interleaved by pixel (BIP), band interleaved by line (BIL), and band sequential (BSQ), respectively. BIP format stores the first pixel of all the bands, followed by all other pixels of all the bands in a sequential order. BIL format stores the first line of the first band, followed by all other bands of the first line. Subsequent lines are interleaved in a similar manner. BSQ format stores the spatial image of the first band, followed by the images of all other bands. The three formats have different advantages for interactive analysis and image processing. The BIP and BSQ formats offer optimal performances for spectral and spatial accesses, respectively. The BIL format gives a compromise in performance between spatial and spectral analysis. The three formats can be converted to each other without altering the spatial and spectral contents

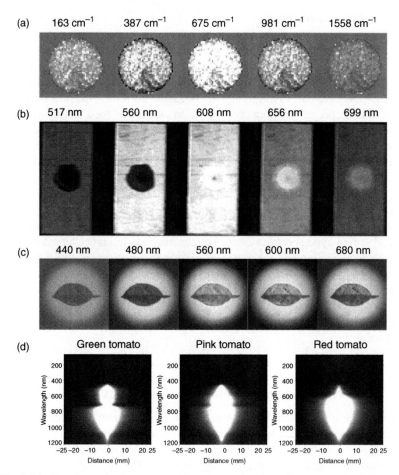

Fig. 2.16   Representative hyperspectral images acquired by different methods: (a) point-scan Raman images of a milk–melamine mixture, (b) line-scan fluorescence images of chicken blood on a stainless steel sheet, (c) area-scan reflectance images of a diseased leaf using an LCTF and (d) line-scan scattering images of tomatoes at different ripeness stages.

stored in the images. The spectral images are usually saved in two files: a binary file containing all the image data and an associated text header file containing information needed to read the image data, such as image dimensions, image format and data type. When the image acquisition is conducted at a few wavelengths in multispectral imaging applications, such as the two-band images demonstrated in Fig. 2.15, images can be saved in separate files for offline analysis using standard image formats (e.g., TIFF, BMP and PNG) or directly processed in real time without saving.

Figure 2.16 shows some representative hyperspectral images obtained from different acquisition methods. The images shown in Fig. 2.16a, b and c were collected using point-scan, line-scan and area-scan methods, respectively. The spatial scenes

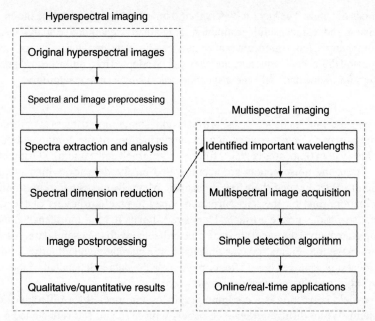

Fig. 2.17   Flowchart of main steps for hyperspectral and multispectral image analysis.

at different bands can be obtained from the hyperspectral images, regardless of the methods used for image acquisition. Different types of spectrum (i.e., Raman, fluorescence and reflectance) can be extracted at each pixel in the images. The dimensions of the hyperspectral Raman image in Fig. 2.16a are 50 × 50 × 1024. This hypercube can be regarded as 1024 single-band Raman images (each with 50 × 50 pixels), or 2500 Raman spectra (each with 1024 spectral data points). The images were acquired by a 16-bit CCD camera (Qin *et al.*, 2010); thus, each pixel occupies two bytes. The size of the image data file can be calculated as 50 × 50 × 1024 × 2 bytes = 5.12 megabytes. The file size of tens or hundreds of megabytes is common for hyperspectral images with large spatial and spectral dimensions. The images shown in Fig. 2.16d are light-scattering images, which were acquired by scanning one line to the samples [2-D image (y, λ), see Fig. 2.1]. Other than covering the whole sample surface, the scattering images aim to reveal the light-scattering patterns illuminated by a point source, which can provide useful information for evaluating internal qualities of food and agricultural products (Qin and Lu, 2008). As shown in Fig. 2.16d, the distinct narrowing in the scattering area around 675 nm for the green tomato is due to strong light absorption by chlorophyll in the fruit, which in turn reduces the scattering distance. The scattering bottleneck is not observed for the red tomato because its chlorophyll content is low or nonexistent.

2.5.2   Spectral image analysis

Main procedures for analyzing hyperspectral and multispectral images are summarized in Fig. 2.17. Hyperspectral images contain redundant information across tens or hundreds of bands. The key for hyperspectral image analysis is to reduce the spectral dimension and extract useful information for qualitative/quantitative analysis. A subset containing a few significant wavebands can be identified during the process of the spectral dimension reduction, and they can be adopted by a multispectral imaging solution for online and real-time inspection tasks using a simple algorithm.

Data preprocessing
Raw spectral images have noises, artifacts and, sometimes, useless signals due to the measurement environments and the imperfections of the components in the imaging system (e.g., source, lens, filter, spectrograph and camera). During image acquisition, the noise counts are accumulated on the detector, which increases the pixel values beyond their true intensities. Many factors, such as non-uniform illumination, dust on the lens surface or pixel-to-pixel variations of the CCD, can cause various image artifacts. Undesired signals (e.g., fluorescence emission during Raman measurement) may also be collected with the useful data. All these factors make the original images unsuitable for qualitative/quantitative analysis. Preprocessing removes these effects and makes the data independent of the imaging systems and the measurement conditions.

Flat-field correction is a common preprocessing method for reflectance measurement. White diffuse reflectance panels with high and flat reflectance over a broad spectral region (e.g., 250–2500 nm) are usually used as standards. The correction can be conducted using the following equation:

$$Rs(\lambda) = \frac{Is(\lambda) - Id(\lambda)}{Ir(\lambda) - Id(\lambda)} \times Rr(\lambda) \qquad [2.1]$$

where *Rs* is the relative reflectance image of the sample, *Is* is the intensity image of the sample, *Ir* is the reference image of the white panel, *Id* is the dark current image acquired with the light source off and the lens covered, *Rr* is the reflectance factor of the white panel and λ is the wavelength. In practice, a constant reflectance factor (*Rr*) of 100% for all the wavelengths is usually used for simplification, although the actual reflectance values of the white panel are slightly lower and there are also small variations over a certain wavelength range. Most food and agricultural products have lower reflectance than the white panel, thus the relative reflectance values calculated by equation [2.1] are in the range of 0–100%. They can be multiplied by a constant factor (e.g., 10 000) to have a large dynamic range and to reduce the rounding errors. The relative (or percentage) reflectance instead of the absolute intensity data are usually used for further data analysis. Besides the flat-field correction, other preprocessing methods, such as spectral smoothing, normalization, baseline correction, image masking and spatial filtering, can be used in spatial and spectral domains to deal with different types of spectral images.

Spectral dimension reduction and optimal band selection
Since hyperspectral images can be viewed as numerous spatially organized spectra, many chemometric methods and multivariate analysis techniques, such as

spectral matching methods, principal component analysis (PCA), partial least squares (PLS), artificial neural networks (ANNs), support vector machines (SVMs), linear discriminant analysis (LDA) and correlation analysis (CA), can be used for spectral analysis, dimension reduction and wavelength selection. To facilitate spectral analysis, spectra can be extracted from regions of interest (ROIs) selected at particular image areas. The whole hyperspectral image can also be unfolded and reshaped to form a 2-D spectral data matrix, on which multivariate analysis methods can be performed directly. The results can generally be folded back to the image format.

In many hyperspectral applications for inspecting food and agricultural products, it is required to identify target pixels on the sample surfaces. Spectral matching algorithms are usually used to perform statistical comparisons between known reference spectra in a spectral library and unknown spectra extracted from the hyperspectral images. Various spectral similarity measures have been developed for target detection and spectral classification. Frequent choices for such spectral metrics include spectral angle mapper (SAM), spectral correlation mapper (SCM), Euclidean distance (ED) and spectral information divergence (SID) (Chang, 2000; van der Meer, 2006). SAM, SCM, ED and SID calculate angle, correlation, distance and divergence between two spectra (or vectors), respectively. The smaller the values of these similarity metrics are, the smaller the differences between two spectra. In general, a reference spectrum is established first, and it is used to calculate a selected similarity metric for each hyperspectral pixel. As a result, a 2-D rule image with same spatial dimensions with the hyperspectral image is generated, which can be used for image classification and target detection. The spectral matching methods have been used for spectral dimension reduction and target identification in various hyperspectral applications (Park *et al.*, 2007b; Qin *et al.*, 2009).

Figure 2.18 demonstrates an example of using SID mapping to reduce the spectral dimension of the hyperspectral images. The images were collected from reflectance measurement of grapefruit samples in the wavelength range of 450–930 nm (92 bands) for detecting a particular disease (i.e., canker) on the fruit peel (Qin *et al.*, 2009). ROIs of canker were first selected from the images of the reference samples. Mean spectrum, which was calculated from the spectra extracted from all the canker ROIs, was used as the reference spectrum of canker. SID mapping was then performed to the hyperspectral images of the test samples to obtain the rule images. Finally, a simple thresholding method was applied to the rule images to separate canker lesions from the fruit peel and other surface diseases.

The spectral matching example in Fig. 2.18 reduces 92 bands to one band by using all the spectral information, which, however, does not provide important wavelengths. PCA, on the other hand, is a useful tool that can not only reduce the spectral dimension but also identify important bands. The basic idea of PCA is to find far fewer components than the original variables (wavelengths) through orthogonal transformation to maximize representation of the original data. The redundant data can, thus, be largely reduced by observing few scores (weighted sums of the original variables) without significantly losing useful information. Loadings of PCA (weighting coefficients) usually can be used to identify important variables that are

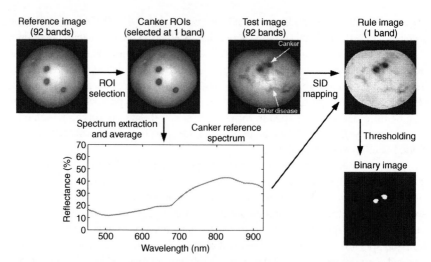

Fig. 2.18 Spectral information divergence (SID) mapping for reducing the spectral dimension of the hyperspectral reflectance images of a grapefruit sample and identifying a disease (canker) on the fruit peel.

responsible for specific features appearing in the corresponding scores. When PCA is used for hyperspectral images, the 3-D data are usually unfolded first so that each single-band image becomes a vector. The 3-D image data are consequently reshaped to form a 2-D matrix, on which PCA can be performed in the same manner as the regular spectral data. After PCA, each score vector for the selected principal components (PCs) is folded back to form a 2-D score image with the same dimensions of the single-band image. PCA and other similar methods – for example, independent component analysis (ICA) and minimum noise fraction (MNF) transform – have been widely used for analyzing hyperspectral images and determining important wavelengths (Kim *et al.*, 2002; Park *et al.*, 2002; Lu, 2003; Zhu *et al.*, 2007).

Figure 2.19 shows an example of using PCA to reduce the spectral dimension and identify the important bands. The hyperspectral reflectance images were acquired from grapefruit samples in the spectral region of 400–900 nm (99 bands) for canker detection (Qin *et al.*, 2008). PCA was performed to the hyperspectral images using all 99 wavelengths. The first four score images demonstrated different patterns for different parts of the fruit. The score images of the fifth and subsequent PCs did not provide meaningful information. The PC-3 score images showed great potential for canker detection and they were used for further image classification. Four important bands were identified at two local maxima and two local minima on the PC-3 loading curve, and they have the potential to be used in a multispectral imaging solution for canker detection.

Besides the methods discussed above, many other multivariate analysis techniques can be used to reduce spectral dimension and identify important wavelengths. Examples include CA (Lee *et al.*, 2008), ANN (Bajwa *et al.*, 2004), genetic algorithm (GA) (Xing *et al.*, 2008), sequential forward selection (SFS) method (Nakariyakul and Casasent, 2008), etc. After the spectral dimension reduction, image postprocessing

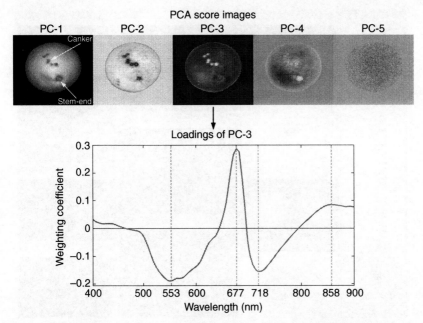

Fig. 2.19 Principal component analysis (PCA) for reducing the spectral dimension of the hyperspectral reflectance images of a grapefruit sample and identifying important wavelengths for surface disease (canker) detection.

operations (e.g., thresholding, morphological filtering) are usually followed to produce the final results (e.g., identification, classification and mapping) for qualitative/quantitative analysis. Simple algorithms (e.g., band ratio) can be developed based on the selected significant bands for multispectral imaging applications.

2.5.3   Spectral image analysis software

Various software packages are available to facilitate spectral image analysis. One of the most popular packages used in the community is the Environment for Visualizing Images (ENVI) (ITT Visual Information Solutions, Boulder, CO, USA). ENVI is a powerful tool for hyperspectral/multispectral image analysis, and it provides numerous functions for data transformation, filtering, classification, mapping, visualization, etc. Figure 2.20 shows a snapshot of the ENVI's user interface when used for analyzing hyperspectral reflectance images of diseased citrus samples. As shown in Fig. 2.20, the single-band images can be displayed at selected bands. The spatial and the spectral profiles can be extracted and plotted at the selected pixel. Various image-processing operations (e.g., masking, band ratio, thresholding and morphological filtering) can be used to generate the final classification result. Other similar packages include HyperCube (US Army Geospatial Center, Alexandria, VA, USA), Hyperspec (Headwall Photonics, Fitchburg, MA, USA), PLS_Toolbox and MIA_Toolbox (Eigenvector Research, Wenatchee, WA, USA). Besides the existing software, spectral image-processing routines can also

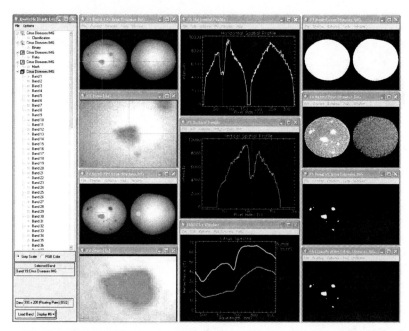

Fig. 2.20   ENVI software for hyperspectral image analysis of citrus surface diseases.

be developed using computer programming languages, such as MATLAB, C/
C++, Visual Basic and LabVIEW. The self-made programs are usually more flexible than the general software packages for particular applications.

2.6   Applications for food and beverage products

Conventional spectroscopy techniques, such as reflectance, transmittance, fluorescence and Raman, provide unique spectra that carry physical, chemical and biological information of the samples. Such fingerprints form the basis for qualitative
and quantitative evaluations of food and agricultural products. As an emerging
sensing technique, hyperspectral imaging equips conventional spectroscopy methods with the capability of spatial information acquisition, which greatly enhances
their detection abilities and expands the scope of the applications. During the past
decade, hyperspectral imaging has been intensively explored for analysing physical, chemical and biological properties of a broad range of food and agricultural
products. Table 2.1 summarizes the main hyperspectral imaging applications in
the area of quality and safety evaluation of food and beverage products. As shown
in Table 2.1, food surface inspection (e.g., defects on fruits, contaminants on meat
and pathogens on grain kernels) is a major application area. Reflectance and fluorescence spectral information in the visible and near-infrared region is widely used
to inspect external features of different types of products, including fruits, nuts,
vegetables, meat, grains and beverages. Imaging spectrograph-based line-scan

Table 2.1 Hyperspectral imaging applications for food and beverage products

Class	Product	Application	Image acquisition method	Wavelength (nm)	Reference
Fruits and nuts	Apple	Bruise detection	Line-scan reflectance	900–1700	Lu (2003)
	Apple	Bitter pit detection	Line-scan reflectance	954–1350	Nicolaï et al. (2006)
	Apple	Firmness evaluation	Area-scan scattering	650–1000	Peng and Lu (2006)
	Apple	Defect and feces detection	Line-scan reflectance/ fluorescence	400–1000	Kim et al. (2007)
	Banana	Quality and maturity evaluation	Line-scan reflectance	400–1000	Rajkumar et al. (2011)
	Cantaloupe	Feces detection	Line-scan fluorescence	425–774	Vargas et al. (2005)
	Cherry	Pit detection	Line-scan transmittance	450–1000	Qin and Lu (2005)
	Citrus	Rottenness detection	Area-scan reflectance	460–1020	Gómez-Sanchis et al. (2008a)
	Citrus	Canker detection	Line-scan reflectance	450–930	Qin et al. (2009)
	Peach	Firmness evaluation	Line-scan scattering	500–1000	Lu and Peng (2006)
	Strawberry	Quality evaluation	Line-scan reflectance	400–1000	ElMasry et al. (2007)
	Walnut	Shell and meat differentiation	Line-scan fluorescence	425–775	Zhu et al. (2007)
Vegetables	Cucumber	Chilling injury detection	Line-scan reflectance	447–951	Liu et al. (2005)
	Cucumber	Quality evaluation	Line-scan reflectance/ transmittance	400–1000	Ariana and Lu (2008)
	Mushroom	Bruise detection	Line-scan reflectance	400–1000	Gowen et al. (2008)
	Onion	Sour skin disease detection	Area-scan reflectance	950–1650	Wang et al. (2010)
	Potato	Cooking time prediction	Line-scan reflectance	400–1000	Nguyen Do Trong et al. (2011)
	Tomato	Ripeness evaluation	Line-scan reflectance	396–736	Polder et al. (2002)

(Continued)

Table 2.1 Continued

Class	Product	Application	Image acquisition method	Wavelength (nm)	Reference
Meat	Beef	Tenderness evaluation	Line-scan reflectance	400–1000	Naganathan et al. (2008)
	Beef	Microbial spoilage detection	Line-scan scattering	400–1100	Peng et al. (2011)
	Chicken	Skin tumor detection	Line-scan reflectance	420–850	Chao et al. (2002)
	Chicken	Feces and ingesta detection	Line-scan reflectance	400–900	Park et al. (2002)
	Chicken	Bone fragment detection	Line-scan reflectance/transmittance	364–1024	Yoon et al. (2008)
	Fish	Moisture and fat evaluation	Line-scan reflectance	460–1040	ElMasry and Wold (2008)
	Fish	Ridge detection	Line-scan reflectance	400–1000	Sivertsen et al. (2009)
	Lamb	Lamb type discrimination	Line-scan reflectance	900–1700	Kamruzzaman et al. (2011)
Grains	Pork	Quality evaluation	Line-scan reflectance	430–1000	Qiao et al. (2007)
	Corn kernel	Constituent evaluation	Area-scan transmittance	750–1090	Cogdill et al. (2004)
	Corn kernel	Oil and oleic acid evaluation	Area-scan reflectance	950–1700	Weinstock et al. (2006)
	Corn kernel	Aflatoxin detection	Line-scan fluorescence	400–600	Yao et al. (2010)
	Wheat kernel	Fungus detection	Area-scan reflectance	1000–1600	Zhang et al. (2007)
	Wheat kernel	Sprout damage detection	Line-scan reflectance	1000–2500	Xing et al. (2009)
	Wheat kernel	Fusarium damage detection	Line-scan reflectance	400–1700	Delwiche et al. (2011)
Beverages	Milk	Fat content evaluation	Line-scan scattering	530–900	Qin and Lu (2007)
	Milk	Melamine detection	Point-scan Raman	102–2538 cm^{-1}	Qin et al. (2010)
	Tea	Quality classification	Line-scan reflectance	408–1117	Zhao et al. (2009)

systems are used as major hyperspectral imaging tools, while area-scan systems using electronically tunable filters are also utilized in many applications. When it comes to inspection of internal attributes (e.g., pits in cherries, internal defects in cucumbers and bone fragments in chicken), transmittance measurement is usually performed using high-intensity light sources. Light-scattering technique provides another approach for internal quality evaluation (e.g., apple firmness measurement). Also, combining different image acquisition methods (e.g., reflectance/ transmittance and reflectance/fluorescence) can give the systems more inspection capacities than those using a single imaging mode. Such image acquisition combinations, along with data fusion techniques, are likely to expand in the future for building multitask food inspection systems.

Owing to the fast hardware and software development, hyperspectral imaging has evolved from a research platform into a useful tool for many practical applications. Generally, significant wavelengths are determined first for a particular application, and they are then used in a multispectral system for fast image acquisition. Table 2.2 summarizes applications of multispectral imaging for food and beverage products. Area-scan imaging systems using various beam-splitting devices (e.g., two-band, three-band and four-band) are used for food surface inspection. Bandpass filters with the nearest central wavelengths to the identified important

Table 2.2 Multispectral imaging applications for food and beverage products

Product	Application	Image acquisition method	Wavelength (nm)	Reference
Apple	Feces detection	Area-scan fluorescence	450, 550, 680, 730	Kim et al. (2003)
Apple	Surface defect detection	Area-scan reflectance	740, 950	Bennedsen et al. (2005)
Apple	Surface defect detection	Area-scan reflectance	450, 500, 750, 800	Kleynen et al. (2005)
Apple	Firmness evaluation	Area-scan scattering	680, 880, 905, 940	Lu and Peng (2007)
Apple	Defect and feces detection	Line-scan reflectance/ fluorescence	530, 665, 750, 800	Kim et al. (2008)
Chicken	Heart disease detection	Area-scan reflectance	495, 535, 585, 605	Chao et al. (2001)
Chicken	Feces detection	Area-scan reflectance	520, 560	Kise et al. (2007)
Chicken	Feces detection	Area-scan reflectance	517, 565, 802	Park et al. (2007a)
Chicken	Wholesomeness inspection	Line-scan reflectance	580, 620	Chao et al. (2008)
Peach	Maturity evaluation	Area-scan reflectance	450, 675, 800	Lleó et al. (2009)
Tea	Category sorting	Area-scan reflectance	580, 680, 800	Wu et al. (2008)
Tomato	Maturity evaluation	Area-scan reflectance	530, 595, 630, 850	Hahn (2002)

bands are used in these area-scan systems to obtain the narrowband images. The filters can be physically changed to different central wavelengths and bandwidths for different applications. On the other hand, line-scan imaging systems that can work in both hyperspectral and multispectral modes have recently been developed (Chao *et al.*, 2008; Kim *et al.*, 2008). The multispectral imaging mode allows flexible selection for the number of bands, the central wavelengths and the bandwidths through software control. With the help of the low-light detection ability of the EMCCD cameras, such systems are able to scan hundreds of lines per second using an exposure time at the millisecond level, making them especially suitable for real-time inspection of fast-moving food items on the processing lines. The fast line-scan spectral imaging technique represents a new direction in the area of online food inspection, and it has great potential to be adopted by food processing industries in the future.

2.7   Conclusions

This chapter has presented hyperspectral and multispectral imaging technologies in the area of quality and safety evaluation of food and agricultural products. Building a working spectral imaging system involves many theoretical and practical aspects. This chapter gave an overview of the factors needed to be considered for using hyperspectral/multispectral imaging in practice, including spectral image acquisition methods, components for building a spectral imaging system, methods for calibrating a spectral imaging system and techniques for analyzing spectral images. The commonly used practices in the area and the applications for inspecting food and beverage products were reviewed to reflect the current status of the spectral imaging technique. Driven by both academic and industrial forces in the food and agricultural area, the spectral imaging technique has been developed rapidly during the past decade. Line-scan spectral imaging systems that can acquire hundreds of lines per second have been developed for online food inspection, and they have great potential to become the standard facilities for routine uses in a variety of food processing plants. Imaging spectrographs that can scan more than 1000 lines are already available in the market. New hardware design concepts will be continuously introduced to produce various improved and novel components for building high-performance systems. The fast-growing computing capacity of the computers will facilitate handling large data files and processing spectral images in real time. The advances in both spectral imaging instruments and spectral image analysis techniques will propel the development of hyperspectral/multispectral imaging technologies in the future.

2.8   Further information

2.8.1   Books for further reading
Burns, D. A. and Ciurczak, E. W. (ed.) (2008) *Handbook of Near-infrared Analysis*, 3rd edn. Boca Raton: CRC Press.

Grahn, H. F. and Geladi, P. (ed.) (2007) *Techniques and Applications of Hyperspectral Image Analysis*. Chichester: John Wiley & Sons.

Richards, J. A. and Jia, X. (2006) *Remote Sensing Digital Image Analysis: An Introduction*, 4th edn. Berlin: Springer-Verlag.

Sun, D. (ed.) (2010) *Hyperspectral Imaging for Food Quality Analysis and Control*. San Diego: Academic Press, Elsevier.

2.8.2   Free software for hyperspectral image analysis
HyperCube
http://www.agc.army.mil/hypercube/
MatlabHyperspectral Toolbox
http://matlabhyperspec.sourceforge.net/

2.9   References

Aleixos, N., Blasco, J., Navarrón, F. and Moltó, E. (2002) Multispectral inspection of citrus in real-time using machine vision and digital signal processors. *Computers and Electronics in Agriculture*, **33**, 121–37.

Ariana, D. P. and Lu, R. (2008) Quality evaluation of pickling cucumbers using hyperspectral reflectance and transmittance imaging: Part I. Development of a prototype. *Sensing and Instrumentation for Food Quality and Safety*, **2**, 144–51.

ASTM Standards (2007) E1840-96: Standard Guide for Raman shift standards for spectrometer calibration. West Conshohocken, PA: ASTM.

Bajwa, S. G., Bajcsy, P., Groves, P. and Tian, L. (2004) Hyperspectral image data mining for band selection in agricultural applications. *Transactions of the American Society of Agricultural Engineers*, **47**, 895–907.

Bennedsen, B. S., Peterson, D. L. and Tabb, A. (2005) Identifying defects in images of rotating apples. *Computers and Electronics in Agriculture*, **48**, 92–102.

Bodkin, A., Sheinis, A. I. and Norton, A. (2008) *Hyperspectral Imaging Systems*, US Patent Application Publication, Pub. NO.: US 2008/0088840 A1.

Chang, C. I. (2000) An information theoretic-based approach to spectral variability, similarity and discriminability for hyperspectral image analysis. *IEEE Transactions on Information Theory*, **46**, 1927–32.

Chao, K., Chen, Y. R., Hruschka, W. R. and Park, B. (2001) Chicken heart disease characterization by multi-spectral imaging. *Applied Engineering in Agriculture*, **17**, 99–106.

Chao, K., Mehl, P. M. and Chen, Y. R. (2002) Use of hyper- and multi-spectral imaging for detection of chicken skin tumors. *Applied Engineering in Agriculture*, **18**, 113–19.

Chao, K., Yang, C. C., Kim, M. S. and Chan, D. (2008) High throughput spectral imaging system for wholesomeness inspection of chicken. *Applied Engineering in Agriculture*, **24**, 475–85.

Cogdill, R. P., Hurburgh, C. R. and Rippke, G. R. (2004) Single-kernel maize analysis by near-infrared hyperspectral imaging. *Transactions of the American Society of Agricultural Engineers*, **47**, 311–20.

Delwiche, S. R., Kim, M. S. and Dong, Y. (2011) Fusarium damage assessment in wheat kernels by Vis/NIRhyperspectral imaging. *Sensing and Instrumentation for Food Quality and Safety*, **5**, 63–71.

ElMasry, G. and Wold, J. P. (2008) High-speed assessment of fat and water content distribution in fish fillets using online imaging spectroscopy. *Journal of Agricultural and Food Chemistry*, **56**, 7672–7.

ElMasry, G., Wang, N., ElSayed, A. and Ngadi, M. (2007) Hyperspectral imaging for non-destructive determination of some quality attributes for strawberry. *Journal of Food Engineering*, **81**, 98–107.

Gebhart, S. C., Thompson, R. C. and Mahadevan-Jansen, A. (2007) Liquid-crystal tunable filter spectral imaging for brain tumor demarcation. *Applied Optics*, **46**, 1896–910.

Gómez-Sanchis, J., Gómez-Chova, L., Aleixos, N., Camps-Valls, G., Montesinos-Herrero, C., Moltó, E. and Blasco, J. (2008a) Hyperspectral system for early detection of rottenness caused by Penicillium digitatum in mandarins. *Journal of Food Engineering*, **89**, 80–6.

Gómez-Sanchis, J., Moltó, E., Camps-Valls, G., Gómez-Chova, L., Aleixos, N. and Blasco, J. (2008b) Automatic correction of the effects of the light source on spherical objects. An application to the analysis of hyperspectral images of citrus fruits. *Journal of Food Engineering*, **85**, 191–200.

Gowen, A. A., O'Donnell, C. P., Taghizadeh, M., Cullen, P. J., Frias, J. M. and Downey, G. (2008) Hyperspectral imaging combined with principal component analysis for bruise damage detection on white mushrooms (Agaricusbisporus). *Journal of Chemometrics*, **22**, 259–67.

Hahn, F. (2002) Multi-spectral prediction of unripe tomatoes. *Biosystems Engineering*, **81**, 147–55.

Kamruzzaman, M., ElMasry, G., Sun, D. and Allen, P. (2011) Application of NIR hyperspectral imaging for discrimination of lamb muscles. *Journal of Food Engineering*, **104**, 332–40.

Kim, M. S., Chen, Y. R. and Mehl, P. M. (2001) Hyperspectral reflectance and fluorescence imaging system for food quality and safety. *Transactions of the American Society of Agricultural Engineers*, **44**, 721–9.

Kim, M. S., Lefcourt, A. M., Chao, K., Chen, Y. R., Kim, I. and Chan, D. E. (2002) Multispectral detection of fecal contamination on apples based on hyperspectral imagery. Part I. Application of visible and near-infrared reflectance imaging. *Transactions of the American Society of Agricultural Engineers*, **45**, 2027–37.

Kim, M. S., Lefcourt, A. M. and Chen, Y. R. (2003) Multispectral laser-induced fluorescence imaging system for large biological samples. *Applied Optics*, **42**, 3927–34.

Kim, M. S., Chen, Y. R., Cho, B. K., Chao, K., Yang, C. C., Lefcourt, A. M. and Chan, D. (2007) Hyperspectral reflectance and fluorescence line-scan imaging for online defect and fecal contamination inspection of apples. *Sensing and Instrumentation for Food Quality and Safety*, **1**, 151–9.

Kim, M. S., Lee, K., Chao, K., Lefcourt, A. M., Jun, W. and Chan, D. (2008) Multispectral line-scan imaging system for simultaneous fluorescence and reflectance measurements of apples: multitask apple inspection system. *Sensing and Instrumentation for Food Quality and Safety*, **2**, 123–9.

Kim, M. S., Chao, K., Chan, D. E., Jun, W., Lefcourt, A. M., Delwiche, S. R., Kang, S. and Lee, K. (2011) Line-scan hyperspectral imaging platform for agro-food safety and quality evaluation: System enhancement and characterization. *Transactions of the American Society of Agricultural and Biological Engineers*, **54**, 703–11.

Kise, M., Park, B., Lawrence, K. C. and Windham, W. R. (2007) Design and calibration of a dual-band imaging system. *Sensing and Instrumentation for Food Quality and Safety*, **1**, 113–21.

Kise, M., Park, B., Heitschmidt, G. W., Lawrence, K. C. and Windham, W. R. (2010) Multispectral imaging system with interchangeable filter design. *Computers and Electronics in Agriculture*, **72**, 61–8.

Kleynen, O., Leemans, V. and Destain, M. F. (2005) Development of a multi-spectral vision system for the detection of defects on apples. *Journal of Food Engineering*, **69**, 41–9.

Lawrence, K. C., Park, B., Windham, W. R. and Mao, C. (2003) Calibration of a push-broom hyperspectral imaging system for agricultural inspection. *Transactions of the American Society of Agricultural Engineers*, **46**, 513–21.

Lawrence, K. C., Park, B., Heitschmidt, G. W., Windham, W. R. and Thai, C. N. (2007) Evaluation of LED and tungsten-halogen lighting for fecal contaminant detection. *Applied Engineering in Agriculture*, **23**, 811–18.

Lee, K., Kang, S., Delwiche, S. R., Kim, M. S. and Noh, S. (2008) Correlation analysis of hyperspectral imagery for multispectral wavelength selection for detection of defects on apples. *Sensing and Instrumentation for Food Quality and Safety*, **2**, 90–6.

Liu, Y., Chen, Y. R., Wang, C., Chan, D. E. and Kim, M. S. (2005) Development of a simple algorithm for the detection of chilling injury in cucumbers from visible/near-infrared hyperspectral imaging, *Applied Spectroscopy*, **59**, 78–85.

Lleo, L., Barreiro, P., Ruiz-Altisent, M. and Herrero, A. (2009) Multispectral images of peach related to firmness and maturity at harvest. *Journal of Food Engineering*, **93**, 229–35.

Lu, R. (2003) Detection of bruises on apples using near-infrared hyperspectral imaging. *Transactions of the American Society of Agricultural Engineers*, **46**, 523–30.

Lu, R. and Peng, Y. (2006) Hyperspectral scattering for assessing peach fruit firmness. *Biosystems Engineering*, **93**, 161–71.

Lu, R. and Peng, Y. (2007) Development of a multispectral imaging prototype for real-time detection of apple fruit firmness. *Optical Engineering*, **46**, 123201.

Morris, H. R., Hoyt, C. C. and Treado, P. J. (1994) Imaging spectrometers for fluorescence and Raman microscopy – acousto-optic and liquid-crystal tunable filters. *Applied Spectroscopy*, **48**, 857–66.

Naganathan, G. K., Grimes, L. M., Subbiah, J., Calkins, C. R., Samal, A. and Meyer, G. E. (2008) Visible/near-infrared hyperspectral imaging for beef tenderness prediction. *Computers and Electronics in Agriculture*, **64**, 225–33.

Nakariyakul, S. and Casasent, D. P. (2008) Hyperspectral waveband selection for contaminant detection on poultry carcasses. *Optical Engineering*, **47**, 087202.

Nguyen Do Trong, N., Tsuta, M., Nicolaï, B. M., Baerdemaeker, J. De and Saeys, W. (2011) Prediction of optimal cooking time for boiled potatoes by hyperspectral imaging. *Journal of Food Engineering*, **105**, 617–24.

Nicolaï, B. M., Lotze, E., Peirs, A., Scheerlinck, N. and Theron, K. I. (2006) Non-destructive measurement of bitter pit in apple fruit using NIR hyperspectral imaging. *Postharvest Biology and Technology*, **40**, 1–6.

Noh, H. K. and Lu, R. (2007) Hyperspectral laser-induced fluorescence imaging for assessing apple fruit quality. *Postharvest Biology and Technology*, **43**, 193–201.

Park, B., Lawrence, K. C., Windham, W. R. and Buhr, R. J. (2002) Hyperspectral imaging for detecting fecal and ingesta contaminants on poultry carcasses. *Transactions of the American Society of Agricultural Engineers*, **45**, 2017–26.

Park, B., Kise, M., Lawrence, K. C., Windham, W. R., Smith, D. P. and Thai, C. N. (2007a) Real-time multispectral imaging system for online poultry fecal inspection using unified modeling language. *Sensing and Instrumentation for Food Quality and Safety*, **1**, 45–54.

Park, B., Windham, W. R., Lawrence, K. C. and Smith, D. P. (2007b) Contaminant classification of poultry hyperspectral imagery using a spectral angle mapper algorithm. *Biosystems Engineering*, **96**, 323–33.

Peng, Y. and Lu, R. (2006) An LCTF-based multispectral imaging system for estimation of apple fruit firmness: Part I. Acquisition and characterization of scattering images. *Transactions of the American Society of Agricultural Engineers*, **49**, 259–67.

Peng, Y., Zhang, J., Wang, W., Li, Y., Wu, J., Huang, H., Gao, X. and Jiang, W. (2011) Potential prediction of the microbial spoilage of beef using spatially resolved hyperspectral scattering profiles, *Journal of Food Engineering*, **102**, 163–9.

Polder, G., van der Heijden, G. W. A. M. and Young, I. T. (2002) Spectral image analysis for measuring ripeness of tomatoes. *Transactions of the American Society of Agricultural Engineers*, **45**, 1155–61.

Polder, G., van der Heijden, G. W. A. M., Keizer, L. C. P. and Young, I. T. (2003) Calibration and characterisation of imaging spectrographs. *Journal of Near Infrared Spectroscopy*, **11**, 193–210.

Qiao, J., Ngadi, M. O., Wang, N., Gariepy, C. and Prasher, S. O. (2007) Pork quality and marbling level assessment using a hyperspectral imaging system. *Journal of Food Engineering*, **83**, 10–16.

Qin, J. and Lu, R. (2005) Detection of pits in tart cherries by hyperspectral transmission imaging. *Transactions of the American Society of Agricultural Engineers*, **48**, 1963–70.

Qin, J. and Lu, R. (2007) Measurement of the absorption and scattering properties of turbid liquid foods using hyperspectral imaging. *Applied Spectroscopy*, **61**, 388–96.

Qin, J. and Lu, R. (2008) Measurement of the optical properties of fruits and vegetables using spatially resolved hyperspectral diffuse reflectance imaging technique. *Postharvest Biology and Technology*, **49**, 355–65.

Qin, J., Burks, T. F., Kim, M. S., Chao, K. and Ritenour, M. A. (2008) Citrus canker detection using hyperspectral reflectance imaging and PCA-based image classification method. *Sensing and Instrumentation for Food Quality and Safety*, **2**, 168–77.

Qin, J., Burks, T. F., Ritenour, M. A. and Bonn, W. G. (2009) Detection of citrus canker using hyperspectral reflectance imaging with spectral information divergence. *Journal of Food Engineering*, **93**, 183–91.

Qin, J., Chao, K. and Kim, M. S. (2010) Raman chemical imaging system for food safety and quality inspection. *Transactions of the American Society of Agricultural and Biological Engineers*, **53**, 1873–82.

Qin, J., Chao, K., Kim, M. S., Kang, S., Cho, B. K. and Jun, W. (2011) Detection of organic residues on poultry processing equipment surfaces by LED-induced fluorescence imaging. *Applied Engineering in Agriculture*, **27**, 153–61.

Rajkumar, P., Wang, N., ElMasry, G., Raghavan, G. S. V. and Gariepy, Y. (2011) Studies on banana fruit quality and maturity stages using hyperspectral imaging. *Journal of Food Engineering*, doi: 10.1016/j.jfoodeng.2011.05.002.

Safren, O., Alchanatis, V., Ostrovsky, V. and Levi, O. (2007) Detection of green apples in hyperspectral images of apple-tree foliage using machine vision. *Transactions of the American Society of Agricultural and Biological Engineers*, **50**, 2303–13.

Sensors Unlimited (2006) What Is InGaAs? (Application Note). Princeton, NJ: Sensors Unlimited, Inc.

Sivertsen, A. H., Chu, C. K., Wang, L. C., Godtliebsen, F., Heia, K. and Nilsen, H. (2009) Ridge detection with application to automatic fish fillet inspection. *Journal of Food Engineering*, **90**, 317–24.

van der Meer, F. (2006) The effectiveness of spectral similarity measures for the analysis of hyperspectral imagery. *International Journal of Applied Earth Observation*, **8**, 3–17.

Vargas, A. M., Kim, M. S., Tao, Y., Lefcourt, A. M., Chen, Y. R., Luo, Y., Song, Y. and Buchanan, R. (2005) Detection of fecal contamination on cantaloupes using hyperspectral fluorescence imagery. *Journal of Food Science*, **70**, E471–E476.

Wang, W., Li, C., Gitaitis, R., Tollner, E. W., Rains, G. and Yoon, S. C. (2010) Near-infrared hyperspectral reflectance imaging for early detection of sour skin disease in vidalia sweet onions. *ASABE Annual International Meeting*, ASABE Paper No. 1009106.

Weinstock, B. A., Janni, J., Hagen, L. and Wright, S. (2006) Prediction of oil and oleic acid concentrations in individual corn (Zea mays L.) kernels using near-infrared reflectance hyperspectral imaging and multivariate analysis. *Applied Spectroscopy*, **60**, 9–16.

Wu, D., Yang, H., Chen, X., He, Y. and Li, X. (2008) Application of image texture for the sorting of tea categories using multi-spectral imaging technique and support vector machine. *Journal of Food Engineering*, **88**, 474–83.

Xing, J., Guyer, D., Ariana, D. and Lu, R. (2008) Determining optimal wavebands using genetic algorithm for detection of internal insect infestation in tart cherry. *Sensing and Instrumentation for Food Quality and Safety*, **2**, 161–7.

Xing, J., Huang, P., Symons, S., Shahin, M. and Hatcher, D. (2009) Using a short wavelength infrared (SWIR) hyperspectral imaging system to predict alpha amylase activity in individual Canadian western wheat kernels. *Sensing and Instrumentation for Food Quality and Safety*, **3**, 211–18.

Yao, H., Hruska, Z., Kincaid, R., Brown, R., Cleveland, T. and Bhatnagar, D. (2010) Correlation and classification of single kernel fluorescence hyperspectral data with aflatoxin concentration in corn kernels inoculated with Aspergillusflavus spores. *Food Additives & Contaminants, Part A*, **27**, 701–9.

Yoon, S. C., Lawrence, K. C., Smith, D. P., Park, B. and Windham, W. R. (2008) Bone fragment detection in chicken breast fillets using transmittance image enhancement. *Transactions of the American Society of Agricultural and Biological Engineers*, **51**, 331–39.

Zhang, H., Paliwal, J., Jayas, D. S. and White, N. D. G. (2007) Classification of fungal infected wheat kernels using near-infrared reflectance hyperspectral imaging and support vector machine. *Transactions of the American Society of Agricultural and Biological Engineers*, **50**, 1779–85.

Zhao, J., Chen, Q., Cai, J. and Ouyang, Q. (2009) Automated tea quality classification by hyperspectral imaging. *Applied Optics*, **48**, 3557–64.

Zhu, B., Jiang, L., Jin, F., Qin, L., Vogel, A. and Tao, Y. (2007)'Walnut shell and meat differentiation using fluorescence hyperspectral imagery with ICA-kNN optimal wavelength selection. *Sensing and Instrumentation for Food Quality and Safety*, **1**, 123–31.

3

Tomographic techniques for computer vision in the food and beverage industries

M. Z. Abdullah, Universiti Sains Malaysia, Malaysia

Abstract: The term 'tomography' refers to the general class of devices and procedures for producing two-dimensional cross-sectional images of a three-dimensional object. Therefore, tomographic imaging is a technique which is capable of revealing internal structures through the use of ionizing or non-ionizing penetrating waves in a non-invasive and non-destructive manner. Ionizing sources include X-ray, while electrical techniques and microwaves belong to the class of non-ionizing radiation. With the safety advantage of non-ionizing radiation, the use of soft-field sensors for food and beverages imaging has been a subject of interest for several decades, realizing the potential and challenges, and exploring several techniques. This chapter discusses different types of sensors and electronics needed to acquire tomographic signals, with emphasis given to emerging modalities like electrical impedance systems and ultra wide band sensing.

Key words: computerized tomography, magnetic resonance imaging, electrical resistance tomography, electrical capacitance tomography, ultra wide band imaging, image reconstruction, sensitivity analysis.

3.1   Introduction

While a computer vision system is useful for surface-type inspection, in many specialized investigations, food technologists and scientists frequently need to 'see' the internal view of the sample. It should now be recognized that a clear image displaying the object internal view cannot be formed with conventional imaging instruments because wave motions are continuous in space and time. Wave motion brought to a focus within a neighbourhood of a particular point necessarily converges before and diverges after, thereby inherently contaminating the values registered outside the neighbourhood. Therefore, the image of the surface of the body formed by conventional methods can be clear but the image depicting the internal structure of the sample must be contaminated. Therefore, the terms 'computer-assisted *tomography*' (CAT) and 'computerized tomography' (CT) have emerged following the development of a CT machine in 1972 by Nobel Prize winner Geoffrey Hounsfield at EMI, which has revolutionized clinical radiology. Nevertheless, food tomography is a relatively new subject since such an application requires expensive and costly outlays. The price of a typical medical CT scanner ranges from a few hundred to tens of millions of pounds. With no comparable increases in reimbursement, the purchase of such a system for use other than medical diagnostics could not be easily justified. However, some interesting applications involving the use of tomography for food applications have started to emerge recently. The first part of this chapter focuses on the conventional tomographic modalities, involving ionizing radiation like X-ray and nuclear magnetic resonance (NMR). In the literatures such systems are referred to as hard-field sensing. Meanwhile, the second part of this chapter is devoted to new and emerging techniques, such as electrical impedance systems and microwave imaging. These are referred to as soft-field sensing techniques.

3.2   Nuclear tomography

As the name applies, *nuclear* tomography involves the use of nuclear energy for imaging the two-dimensional (2-D) spatial distribution of physical characteristics of the object, from series of one-dimensional (1-D) projections. All nuclear imaging modalities rely upon acquisition hardware featuring a ring detector which measures the strength of radiation produced by the system. There are two general classes of sources of radiation, typified by the degree of control exerted over them by the experimenter. First, there are interior sources inside the body, which are usually beyond the direct control of the experimenter. Second, there are exterior sources outside the body which are usually completely under the control of experimenters. The first method is also known as remote sensing and the second method is termed remote probing. Figure 3.1 highlights the difference. For instance, in CT where radiation is projected into the object falls under first category, while stimulated emission in the case of magnetic resonance imaging (MRI) and through the use of radio-pharmaceuticals in single photon emission computed tomography (SPECT) and positron emission tomography (PET) fall under the second category.

Regardless of the scanning geometry, tomographic imaging shares one common feature: that is, the requirement to perform complex mathematical analysis of the resulting signals using modern digital computing. There are many good reviews on this subject. Interested readers are referred to publications by Brooks and Di Chiro (1975, 1976) and Kak (1979). Here, a brief description on various tomographic modalities is provided, focusing on the advancement of the technology since its inception more than 30 years ago.

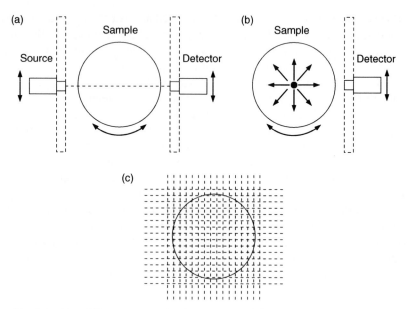

Fig. 3.1   Two different geometries for tomographic imaging: (a) remote probing, (b) remote sensing and (c) typical scanning pattern showing two orthogonal projections.

3.2.1   Computed tomography (CT)

As shown in Fig. 3.1, essentially, CT involves scanning the source and detector sideways to produce one projection data. This procedure is repeated at many viewing angles until the required set of all projection data is obtained. Image reconstruction from projection remains one of the important tasks in CT; this can be performed using a variety of methods. The history of these reconstruction techniques began in 1917 with the publication of a paper by the Austrian mathematician J. Radon, in which he proved that a 2-D or three-dimensional (3-D) object can be reconstructed uniquely from the infinite set of all its projections (Herman, 1980). To date there are hundreds of publications on CT imaging. A good summary on this subject is given in a book by Kak and Slaney (1988).

When the first CT machines were introduced in 1972, the spatial resolution achievable was three line pairs per millimetre, on a grid of 80 × 80 per projection. The time taken to perform each projection scan was approximately five minutes. In contrast, a modern machine achieves 15 line pairs per millimetre, on a grid of 1024 × 1024 per projection, with a scan time per projection down to less than a second. The projection thickness typically ranges from 1 to 10 mm and the density discrimination achievable is better than 1%. These machines use an X-ray source which rotates in a circular path around the sample. A *collimator* is employed in order to produce a sharp pencil-beam X-ray, which is measured using detectors comprising of a static ring of several hundreds of *scintillators*. These have been constructed from xenon ionization chambers, but a more compact solution is

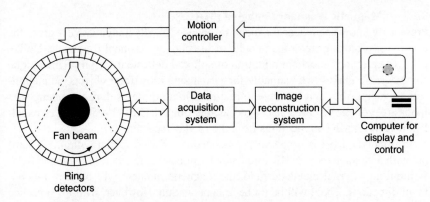

Fig. 3.2   Modern CT usually employs fan beam geometry in order to
reduce the data capturing speed.

offered by solid-state systems, where a scintillation crystal is closely coupled to a
photodiode. This source-detector combination measures parallel projections, one
sample at a time, by stepping linearly across the object. After each projection, the
gantry rotates to a new position and these procedures are repeated until all data are
gathered at sufficient viewing angles.

The latest generation CT machines employ a *fan beam* arrangement as opposed
to the *parallel-beam* geometry. In this way, the size of the beam can be enlarged
to cover the object field of view. Consequently, the gantry needs only to rotate,
thus speeding up the acquisition time. Employing a stationary ring comprising
of, typically, 1000 detectors, the data acquisition time of the modern CT scanners
is generally less than 0.1 s. Figure 3.2 illustrates the essential elements of such
systems.

Since CT is based upon the attenuation of X-rays, its primary strength is the
imaging of calcified objects such as bone and denser tissues. This limits its applications in food technology since food objects are mostly composed of soft or
semi-fluid objects. It is for this reason that CT imaging was initially limited to
medical applications. However, in the 30 years since its inception, capabilities and
applications have been expanded as a result of the advancement of technology and
software development. While medical disorders are still common reasons for CT
imaging, many other scientific fields such as geology, forestry, archaeology and
food science have found CT imaging to be the definitive tool for diagnostic information. For instance, CT, combined with appropriate image analysis, has been
used to study the magnitude and salt gradients in dry-cured ham in meat water
phase (Vestergaard *et al.*, 2005). In studying the growth and development in animals, Kolstad (2001) used CT as a non-invasive technique for detailed mapping
of fat amounts and fat distribution in crossbreed Norwegian pigs. There are other
recent applications involving CT in agriculture and food tomography. Interested
readers are again directed to relevant publications (Sarigul *et al.*, 2003; Fu *et al.*,
2005; Babin *et al.*, 2006).

3.2.2   Magnetic resonance imaging (MRI)

Previously know as nuclear magnetic resonance (NMR) imaging, MRI gives the density of protons or hydrogen nuclei of the body at resonant frequency. Unlike CT, MRI provides excellent rendition of soft and delicate materials. This unique characteristic makes MRI suitable for visualizing most food objects. The applications range from non-invasive to real-time monitoring of dynamic changes as foods are processed, stored, packaged and distributed. Hills (1995) gives an excellent review on MRI applications from the food perspective.

In principle, MRI is based on the association of each spatial region in a sample with a characteristic NMR frequency by imposing an external magnetic field. Without an external magnetic field, the magnetic moment will point at random in all directions. There will be no net magnetization. However, with the presence of a large magnetic field, the hydrogen nuclei will preferentially align their spin in the direction along the magnetic field. This is the *Lamour effect* and the frequency with which the nucleus precesses around the axis is termed the 'Lamour frequency' (McCarthy, 1994). This effect implies a transfer of energy from the spin system to another system or lattice. The transfer of energy is characterized by an exponential relaxation law with time constants T_1 and T_2, which are also known as the spin-lattice excitation and spin-spin relaxation times, respectively (McCarthy, 1994). In commercial MRI, magnetic field ranges from 0.5 to 2.0 tesla (compared to earth's magnetic field of less than 60 μT), T_1 is typically of the order of 0.2–2 s and T_2 ranges from 10 to 100 ms. According to Planck's equation $E = hf$, for a field strength of 1.5 T, f corresponds to radiowaves with a frequency of 60 MHz. This is the resonant frequency of the system. Therefore, by applying a radio frequency (RF) field at resonant frequency, the magnetic moments of the spinning nuclei lose equilibrium and hence radiating signal, which is a function of the line integral of the magnetic resonance signatures in the object. This radiation reflects the distribution of frequencies and a Fourier transform of these signals provides an image of the spatial distribution of the magnetization (Rinck, 2001). The basic block diagram of a typical MRI data acquisition system (DAS) is shown in Fig. 3.3. In general, the MRI system comprises of a scanner which has bore diameter of a few tens of centimetres. The static *magnetic field* is generated by a *superconducting* coil. RF coils are used to transmit RF excitation into the material to be imaged. This excites a component of magnetization in the transverse plane which can be detected by an RF reception coil. The signals are transduced and conditioned prior to image reconstruction. Current MRI scanners generate images with sub-millimetre resolution of virtual slice through the sample. The thickness of the slices is also of the order of a millimetre. Contrast resolution between materials depends strongly on the strength of the magnetization, T_1, T_2 and movement of the nuclei during imaging sequences. The most striking artefacts appear when the magnetic field is disturbed by ferromagnetic objects. Other artefacts, such as ringing, are due to the image reconstruction algorithm and sensor dynamics.

Due to the fact that MRI provides a rapid, direct and most importantly non-invasive, non-destructive means for determination of not only the quantity of water present, but also the structure dynamic characteristics of water, this relatively new

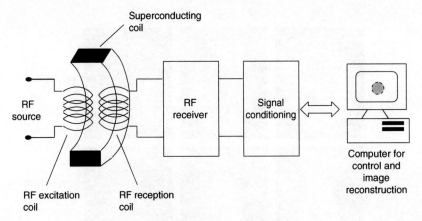

Fig. 3.3   Block diagram of a typical MRI system.

imaging technique has become useful for food engineering. There are numerous applications of MRI since water is the basic building block of many food materials. Figure 3.4 shows examples of MRI-captured images within corn kernels during the freezing process (Borompichaichartkul *et al.*, 2005). The brighter areas show locations where proton mobility is high, and thus water exists as a liquid. In this example, MRI provides useful information for characterizing the physical states of water in frozen corn.

Other interesting applications include real-time monitoring of ice gradients in a dough stick during freezing and thawing processes (Lucas *et al.*, 2005); mapping the temperature distribution patterns in food sauce during microwave-induced heating (Nott and Hall, 1999); and predicting sensory attributes related to the texture of cooked potatoes (Thybo *et al.*, 2004). This example list, which is far from exhaustive, serves to emphasize the potential of MRI for revolutionizing food science and engineering.

Similar to CT imagers, the major drawback is the current expense of an MRI machine, typically £500 000–1 million. Consequently, at present MRI machines are only useful as a research and development tool in food science. In order for MRI to be applied successfully on a commercial basis, the possible benefits must justify the expense. However, with the rapidly decreasing cost of electronic components, combined with the ever-increasing need for innovation in the food industry, it should not be too long before a commercial and affordable MRI machine is developed for food quality control.

3.3   Electrical impedance

A characteristic of common foodstuffs is a chemical composition including a high water content. For instance the water content of apples, apple purée and apple juices ranges from 85% to 90% in most cases (Ashurst, 1995). Other important

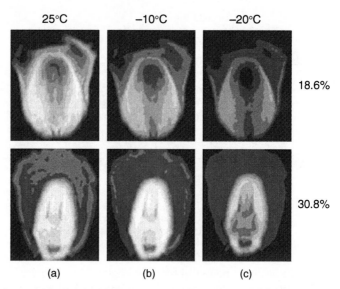

Fig. 3.4 Examples of MRI images showing the distribution of water and its freezing behaviour in different areas within the corn kernels: (a) images captured before freezing at different moisture contents; (b) and (c) images acquired after specified temperatures and moisture contents (Borompichaichartkul *et al.*, 2005).

constituents are mineral compounds and organic acids including components which are susceptible to ionic dissociation. Therefore, another characteristic trait of common foodstuffs is high electrolytic conductivity. This parameter can be characterized by electrical impedance, which is the property of a material to resist current flow. The model by Fricke, shown in Fig. 3.5, is usually used to model food or biological systems.

Parameters that are used to characterize electrolytic conductivity include impedance, resistance, admittance, conductance and capacitive reactance or capacitance. Among these parameters, impedance consisting of resistance and capacitance constitutes the most important component correlating with numerous quality factors in raw materials and foodstuffs. Generally, electrical impedance is a frequency-dependent parameter. At a wide current frequency range, the results of impedance measurements, in the literatures, are generally referred to as the resistivity or the inverse of conductivity. These electrical properties have been used extensively to study the chemical composition of liquid or semi-solid foodstuffs. For instance, the specific numerical characteristics of pasteurized apple purée and pulpy apple juices with different purée concentrations have been evaluated by resistivity and conductivity measurements (Żywica *et al.*, 2005). The study performed by these authors demonstrated that the values of resistivity and conductivity measured for the juice with 20% purée concentration, at a frequency of 100 Hz, reached 16.3 Ωm and 61.5 mS/m, respectively. In plant materials, these measurements have been used to determine the mechanical damage and biological changes in fruit

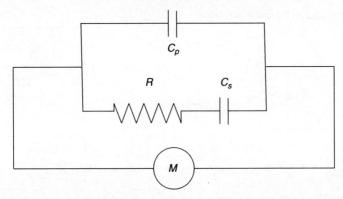

Fig. 3.5   Fricke's model of a food system. R is a resistance, C_p and C_s are equivalent parallel and series capacitances respectively, and M is the measuring equipment.

tissue during ripening and storage (Bauchot *et al.*, 2000); monitoring the sanitary state of ready-to-use vegetables (Orsi *et al.*, 1997). Measurements of conductivity were also found to be highly useful in monitoring the continuous processes of biomass production (Hauttmann and Müller, 2001), as well as the water content in drying and extrusion processes (Vlachos and Karapantios, 2000). In another study, electrical impedance analysis revealed that the connective tissue in a muscle was strongly comparable with meat permittivity (Foster and Schwan, 1989). Therefore, the mechanical and the structural properties of this food product can be studied via dielectric means. This includes permittivity measurements or the study of the interaction of an object with an alternating electric field. For instance, a strong correlation was observed between meat ageing estimated by electrical means with mechanical measurements performed using a standard benchmark compression test method (Damez *et al.*, 2008). Other studies have concluded that there is strong relationship between the electrical anisotropy of meat – that is, impedance varies according to whether current flows along or across muscle fibres, with muscle fibre mechanical resistant (Epstein and Foster, 1983; Lepetit *et al.*, 2002). In summary, electrical and dielectric properties could give some insight to the chemical composition, the structural and functional properties of biological products including foodstuffs. Unlike resistance and capacitance, however, inductance is not encountered in most food stuffs. Hence, this element does not play a very significant role in characterizing the electrical properties of food. Therefore, inductors are not included in the discussion.

The advent of instrumentation technologies and signal processing in recent years means that the impedance measurement can now be performed non-destructively in two or three dimensions. In this case, results are shown in the form of digital images instead of graphs and tables as in conventional techniques. Here, two of the most rapidly developing and exciting areas in impedance imaging are described. They are electrical resistance tomography (ERT) and electrical capacitance tomography (ECT), together with their successful applications in food

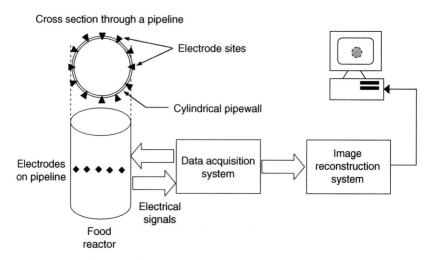

Fig. 3.6   Schematic block diagram of a typical ERT instrument.

imaging. Additionally, an emerging technology based on ultra wide band sensing is also included for completeness and thoroughness of discussion.

3.3.1   Electrical resistance tomography (ERT)

ERT belongs to the soft-field sensor systems, since the sensing field is altered by the density of distribution and physical properties of the object being imaged. This limits the resolution compared to hard-field sensing. Also known as electrical impedance imaging, ERT is a non-destructive technique that enables one to see the variation of electrical conductivity σ within the region. It uses low-frequency electrical current (usually less than 100 kHz) with tens of milliamperes in magnitude as a source of exploration. The ideal ERT would have sufficiently high resolution, in that it would enable accurate location where changes in the conductivity profile occur. By obtaining a sequence of images depicting the distribution of components, or concentration profiles of components across a given plane, it is possible to extract both qualitative and quantitative information of the media under investigation. This capability would be an asset in food imaging because one may sometimes infer, from absolute conductivity, additional physical information about a region such as the degree of non-homogeneity, total solid content, adulteration, separation, aeration and object-powder lump. This information can then be used, for example, to control and monitor the process or for design purposes. The instrumentation required to obtain such information is shown schematically in Fig. 3.6.

It is composed of three distinct parts: (1) the electrodes connected non-intrusively to the pipeline, (2) the DAS and (3) the image reconstruction system. Electrodes are mounted invasively but non-intrusively into the pipeline wall in order to make measurements of the distribution of electrical conductivity within

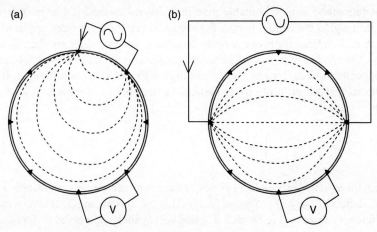

Fig. 3.7   ERT measuring strategies: (a) the adjacent method and (b) the opposite method. The dashed lines show the current paths.

an image plane. This is performed by injecting AC current at one pair of electrodes and measuring voltage at all other pair of electrodes. There a few means by which this can be implemented. Two of the most common strategies are (i) the adjacent and (ii) the opposite techniques. They are shown in Fig. 3.7.

Referring to Fig. 3.7, the adjacent (or neighbouring or four-electrode) method utilizes a linear array of four equally spaced electrodes, two of which carry applied current, while the other two measure the resulting voltages. This method minimizes the error caused by contact impedance at the electrode–electrolyte interfaces since it avoids measuring voltage on the current drive electrode. Measurements are performed by injecting AC current via one pair of electrodes and measuring the voltage at all other pairs. This procedure is repeated until all possible combinations have been current-driven. Allowing for reciprocity, the total number of unique or independent measurements obtained from n number of electrodes is determined by the following relationship:

$$N = \frac{n(n-3)}{2}$$
[3.1]

It can be seen from Fig. 3.7(a) that the current does not flow in straight line, rather it traces along the line of the least-resistive path. It can also be seen from Fig. 3.7(a) that the current density is generally much higher in regions near the boundary compared to those located at the centre. Consequently, it is expected that this method is more sensitive to changes of conductivity occurring near the boundary compared to those at the centre.

Meanwhile, the opposite method (sometimes referred to as the polar method), applies current at two electrodes which are located 180° apart. This method has been employed mainly to improve the current density and consequently the sensitivity distribution (Section 3.4) at the centre of the pipeline. Consequently, the opposite method is more suitable than the adjacent method if one is interested only in imaging the central part of the region in certain specialized applications. However, it suffers from one serious drawback: for the same number of electrodes, the number of independent current projections which can be applied using the opposite method is significantly less than with the adjacent technique. It can be shown that for an n electrode system, the number of independent measurements is given by:

$$N = \frac{n}{4}\left(\frac{3n}{2} - 1 \right)$$

[3.2]

Hence, for $n = 16$, equation [3.2] yields 92 unique measurements compared to 104 for the adjacent method (equation [3.1]). This has serious implications in image resolution, as discussed in Section 3.3, and will again be addressed in Section 3.4 in relation to the image reconstruction system.

Additionally, there are a number of limitations inherent to the ERT system which are a consequence of a trade-off between several parameters: the n electrodes placed around the periphery of the pipeline yield a finite number of unique measurements and the ability of the reconstruction algorithm to resolve small regions of different conductivity is dependent on N. However, if N is too large, the reconstruction algorithm imposes a high computational burden on the system and its real-time capability is degraded. The ability of the electronics to detect small changes in conductivity is dependent on the common-mode rejection ratio of the differential input amplifier, the stability of the AC current drive electronics and the associated demodulation stages. Obtaining an accurate and maximum amount of information describing the state of the conductivity distribution inside a pipeline is a responsibility of the DAS. Figure 3.8 shows the key sub-systems that constitute a typical DAS.

There are numerous publications on the electronics of the ERT DAS. Interested readers are referred to Brown and Seagar (1987), Dickin et al. (1992) and Cook et al. (1994) for more technical information. Even though these DASs were developed independently by the authors, the electronics, including that shown in Fig. 3.8, are fundamentally similar to that used in medical ERT applications (Brown and Seagar, 1987). The only difference is that the non-medical-based ERT systems are not required to operate within the same rigid band of current amplitudes and frequencies as in clinical applications. As shown in Fig. 3.8, the standard ERT data acquisition relied heavily on analogue electronics with all the signal processing performed in the analogue domain before digital conversion. This design works perfectly for imaging slow varying targets like detecting breast cancer in humans. However, it has the usual problems related with analogue circuitries, which ultimately limit the speed of its operation.

One of the most critical components of a current injection ERT system is the constant current generator. In most designs a voltage-to-current converter is employed. As the name applies, the output load current is made proportional to the input voltage. Two simple forms of voltage-to-current converter are (i) the

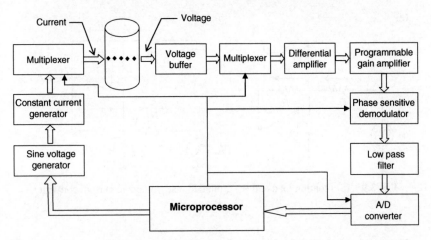

Fig. 3.8   Major components of the typical ERT DAS. Bigger arrow indicates data line and small arrow shows control line.

floating type and (ii) the grounded type. These are shown in the schematics in Fig. 3.9.

In a floating voltage-to-current converter, R_1 is not connected to the ground. As the input current to op-amp is zero, therefore,

$$i_l = i_1 = \frac{V_{in}}{R_l}$$ [3.3]

or

$$i_l \alpha V_{in}$$ [3.4]

Thus the load current is always proportional to the input voltage. Equation [3.3] holds regardless of the load. In EIT the load is resistive. Hence, the op-amp will draw current i_l whose magnitude is linearly dependent only on V_{in} or R_l. Such a circuit is also called the voltage controlled current source (VCCS). Even though the floating load-type voltage-to-current converter has many desirable properties, like stable load current and wide bandwidth, it cannot be used in a system with grounded load (which is prevalent in most applications). Positive feedback VCCS, shown in Fig. 3.9b, is a popular alternative. It is also known as the Howland current converter (from the name of the inventor). The circuit can be analysed by first determining V_l at the non-inverting input of op-amp and then establishing the relationship between V_l with i_l. Applying Kirchoff current law at node V_+ gives

$$V_+ = \frac{V_{in} + V_{out} - i_l R_2}{2}$$ [3.5]

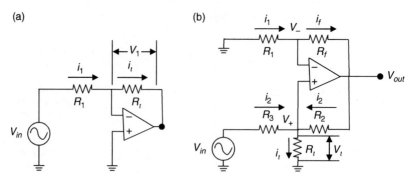

Fig. 3.9   (a) Floating load and (b) grounded load voltage to current converter.

The gain of the amplifier is given by

$$G = 1 + \frac{R_f}{R_1} \tag{3.6}$$

Using equations [3.5] and [3.6], the output voltage can be written as

$$V_{out} = V_1 = V_{in} + V_{out} - i_1 R_2 \tag{3.7}$$

Since $V_{out} = 0$, we have

$$i_1 = \frac{V_{in}}{R_2} \tag{3.8}$$

Equation [3.8] is similar to equation [3.3]. The VCCS in Fig. 3.9b forms the basis of modern EIT data collection systems which are now available commercially.

Another equally important component in ERT data collection electronics is the phase sensitive demodulator. The element is important since only the resistive component is needed in the image formation. However, the electrical impedance consists of two parts: (i) the real or resistive component and (ii) the imaginary or reactive component. In the presence of both components, the relationship between an applied alternating current and the resulting alternating voltage can be seen as both a phase shift and amplitude change. The measured voltage can be written as:

$$V_m = A\sin(\omega t - \phi) \tag{3.9}$$

where ϕ is the phase shift and A is the amplitude. Here, V_m can be intepreted as a sum of two sine wave voltages of equal frequency but different amplitudes. One

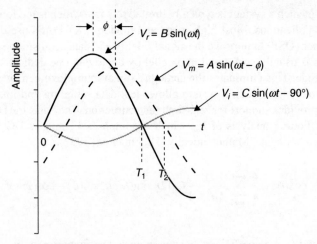

Fig. 3.10   Relationship between V_m, V_r and V_i.

is the sine wave V_r, which is in phase with the injected current, with amplitude related to the resistive component of the impedance. The other one is V_i, which is in quadrature (phase shift of 90°) with V_r. Here V_i is related to the reactive component of the impedance. The relationship between V_m, V_r and V_i is shown graphically in Fig. 3.10.

Algebraically, V_m can be expressed as follows:

$$V_m = V_r + V_i \tag{3.10}$$

Measuring V_m is eqivalent to integrating the signal from a period of time – that is, T_1 to T_2. Therefore,

$$V_m = \int_{T_1}^{T_2} V_r \, dt + \int_{T_1}^{T_2} V_i \, dt \tag{3.11}$$

Since only the real or resistive component needs to be measured, T_1 to T_2 can be selected so that the imaginary or capacitive component vanishes. Hence,

$$V_m = \int_{T_1}^{T_2} V_r \, dt \tag{3.12}$$

Electronically, this means switching the selected part of a period of V_m by means of an analogue switch. The element responsible for this is the phase sensitive detector. The integrated signal is low pass filtered to form an average or DC value – that is,

$$\bar{V}_m = \frac{1}{2\pi} \int_0^\pi B \sin(\omega t) d(\omega t) = \frac{B}{2\pi} [-\cos(\pi) + \cos(0)] = \frac{B}{\pi} \tag{3.13}$$

The speed of such a system is typically limited to 7 fps, which is too low for most industrial applications. Most high-speed and modern ERT DASs use digital signal processor (DSP) to improve the signal fidelity and data acquisition speed. The acquisition is usually performed in parallel by deploying one dedicated DSP for each electrode. This eliminates the analogue signal multiplexing, as well as analogue demodulation. Such a system allows fast data sampling and image frame rates, and provides support for real-time reconstruction. The basis for DSP measurement is Fourier analysis of V_m, which decomposes V_m into its DC value and harmonic components. Mathematically (Dickin *et al.*, 1992),

$$V_m = \frac{2}{\pi} C_m \cos(\phi) + \frac{4}{\pi} \sum_{n=1}^{\infty} \frac{1}{4n^2 - 1} C_n [2n \sin(2n\omega t) \sin(\phi) - \cos(2n\omega t) \cos(\phi)]$$

[3.14]

where C_m and C_n are coefficients, n is an integer. The first term on the right hand side of equation [3.14] is the DC, or resistive, component of V_m while the ones in the bracket are the harmonics. The idea is to obtain the first term only; this can be accomplished by, first, fast Fourier transform (FFT) and, second, low pass filtering. The electronic component which does this is the digital demodulator; the process is known as phase sensitive demodulation. All processing is done in the digital domain. This results in a signal-to-noise ratio much closer to theoretical value. An example of such a system is the high-performance ERT system utilizing the TMS320C6202/6713 processor and capable of capturing voltages at the speed of 914 dual fps with a root mean square error of less than 0.6% (Wang *et al.*, 2005). This system has the ability to provide a consistent, dynamic multi-dimensional image of the state of the overall homogeneity and inhomogeneity of the process, thus opening up a new opportunity for visualizing fast reactions in a pipeline.

3.3.2   Electrical capacitance tomography (ECT)

ECT provides a method of imaging the cross-section of the process pipelines or process reactors containing components with different permitivitties (ε). The first ECT system was the result of the work by Huang *et al.* (1989) at the University of Manchester, UK. Originally developed for visualizing multiphase flows (gas–liquid and gas–solids) in oil pipelines, ECT has been developed rapidly and used successfully in many applications, including pneumatic conveyors, fluidized beds and wet gas separators. ECT enables insight into the material distribution within closed vessels and consequently, into governing mechanisms in the process, both non-intrusively and non-invasively. Figure 3.11 shows a typical ECT system with eight-electrode configuration.

Referring to Fig. 3.11, ECT comprises of the primary sensor, the sensor electronics and capacitance measurement instrument and a computer for the reconstruction. The sensor generally consists of periphery mounted electrodes made from copper foil and a casing giving it mechanical stability and electromagnetic

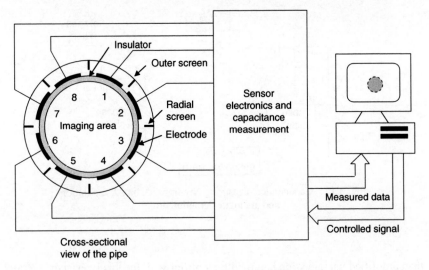

Fig. 3.11   Illustration of a typical ECT system.

stray immunity. During operation, an excitation signal in the form of AC voltage is applied to one of the electrodes; the remaining electrodes are kept at the ground potential, acting as detector electrodes. Subsequently, the voltage potential appearing on the remaining electrodes is measured, one at a time. For instance, the capacitance between electrode pairs 1-2, 1-3, 1-4, ..., 1-8 is first measured. Then, the capacitance between electrode pairs 2-3, 2-4, 2-5, ..., 2-8 is measured. This process is repeated until all independent sets of linearly independent measurements are obtained. Assuming n electrode system, the total number of independent capacitance measurements is given by

$$N = \frac{n(n-1)}{2} \qquad\qquad [3.15]$$

The number of electrodes n reported in literature typically ranges from 6 to 32 (Yang, 2010). Correspondingly, the number of independent capacitance measurements N ranges from 15 to 496. This number is a critical parameter as it affects the quality of the reconstructed image. In general, the image resolution increases as n increases. However, a large n leads to a smaller surface area of the respective electrode, thus reducing the magnitude of the inter-electrode capacitance and lowering the signal-to-noise ratio. There are many possible ways of measuring the unknown capacitance. The one described here is based on sine-wave excitation and amplitude demodulation (Yang *et al.*, 1999). The circuit is shown in Fig. 3.12.

Referring to the circuit in Fig. 3.12, C_x is the unknown capacitance while C_1, and C_2 are stray capacitances. An excitation voltage V_{in} is appied to C_x which is

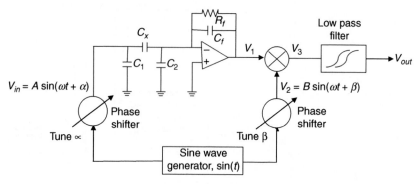

Fig. 3.12   Capacitance measuring circuit based on AC excitation and analogue demodulation.

then amplified using a wide bandwidth amplifier with feedback resistance R_f and feedback capacitance C_f. The output from the amplifier is given by

$$V_1 = -\frac{sC_xR_f}{1+sC_fR_f}V_{in} \qquad [3.16]$$

where $s = j\omega$. The circuit can be designed by choosing large feedback resistance and capacitance such that $sC_fR_f \gg 1$. Therefore,

$$V_1 \approx -\frac{C_x}{C_f}V_{in} \qquad [3.17]$$

or

$$V_1 = -\frac{C_x}{C_f}A\,\sin(\omega t + \alpha) \qquad [3.18]$$

The signal is then demodulated using analogue multiplier to give

$$V_2 = -\frac{C_x}{C_f}AB\,\sin(\omega t + \alpha)\sin(\omega t + \beta) \qquad [3.19]$$

From trigonometric identity

$$2\sin(x)\sin(y) = \cos(x - y) - \cos(x + y) \qquad [3.20]$$

Expanding equation [3.19] using equation [3.20] yields

$$V_2 = K\cos(\alpha - \beta) - \cos(2\omega t + \alpha + \beta)$$ [3.21]

where $K = -(C_x/(2C_f)AB)$. Tuning the two signals so that they are exactly in phase – that is, $\alpha = \beta$ – gives:

$$V_2 = K\{1 - \cos 2(\omega t + \alpha)\}$$ [3.22]

Clearly, from equation [3.22], the demodulated signal comprises of DC and the oscillating component. The latter can be removed by low pass filtering. Therefore,

$$\left|V_{out}\right| = \frac{C_x}{2C_f} AB$$ [3.23]

From equation [3.23] it can be seen that the signal after filtering represents the unknown capacitance. Choosing the excitation and reference voltages to be 10 $V_{p.p}$, $C_f = 10$ pF, then the sensitivity of the measuring system is 5V/pF. This is sufficient for most applications. The sensitivity can be further enhanced by incorporating the programmable gain amplifier into the circuit. It can also be seen from the above analysis that the stray capacitances C_1, and C_2 have no effect on the output voltage. Therefore, such a design is immune to stray capacitance, which is important especially when measuring very low capacitances. The design in Fig. 3.12 forms the basis of many modern and commercial ECT measuring instruments. The circuit in Fig. 3.12 is integrated with other electronic components, forming the ECT DAS. The schematic of such a system is shown in Fig. 3.13.

Referring to Fig. 3.13, the N capacitance measurements given by equation [3.23] are sent, one at a time, to a differential amplifier via a multiplexer. An offset signal is subtracted using a unity gain differential amplifier to balance the standing capacitance. The subtraction is necessary since the standing capacitance – that is, capacitance when the pipeline is filled up by high or low permittivity materials – is relatively large compared to the changes in capacitance measurements. Thus, the output from the differential amplifier represents the change in capacitance measurement. This signal is further amplified by adjustable gain amplifier, converted into a digital form and then sent to a computer for further processing.

3.3.3   Ultra wide band tomography (UWBT)

In a quest to develop a system which is capable of reconstructing an image with resolution comparable to ionizing radiations like X-ray, CT, gamma camera, MRI, etc., scientists and engineers began experimenting with UWBT in the early 2000s. The advantages of UWBT include its low cost, the use on non-ionizing radiation

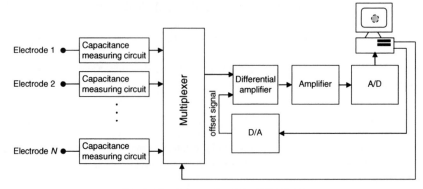

Fig. 3.13   Schematic of the ECT DAS.

and the ability to simultaneously image the electromagnetic inhomogeneities (i.e., dielectric permittivity as well as conductivity) of the media, which is not directly possible using other modalities. The biggest and perhaps the most important advantage of UWBT is the resolution of the image that it is capable of reconstructing. Theoretically, there is no known limit to the spatial resolution obtainable by UWBT, and image resolution as low as 1/30 of a wavelength has been reported for near-field imaging systems (Gilmore *et al.*, 2010). The challenge here is to develop a high-speed acquisition electronic and stable image reconstruction algorithm, in order to make UWBT a competitive and viable imaging modality. Presently, research in UWBT is actively being pursued in the laboratory by numerous research groups around the world, and oriented towards biomedical applications and medical diagnostics. Like ERT and ECT, it will be a matter of time before this advanced imaging technique reaches the food engineering laboratories.

The feasibility of applying UWBT for imaging food and agricultural products arises from the fact that the dielectric properties of these materials give an indication of the overall product's quality. There are two principal reasons why the dielectric properties are of interest to food technologists and scientists. First, it describes the behaviour of materials when subjected to high-frequency or microwave electric fields such as in dielectric heating applications. Second, it is useful for rapid determination of moisture content. Like ECT, UWBT produces an image which depicts the dielectric properties in 2-D or 3-D if measurements are taken from multiple planes. For practical use, the dielectric properties of usual interest are the dielectric constant ε' of the relative complex permittivity, $= \varepsilon' + \varepsilon''$. A brief discussion on the instrumentation system needed to perform this measurement is presented in the following paragraph.

The Federal Communications Committee (FCC) defines UWB as a system which uses electromagnetic signals having greater than 500 MHz bandwidth or 20% bandwidth around its centre frequency (FCC, 2002). One important element in any UWBT system is the antenna, which transmits and receives UWB signals.

(a)   (b)

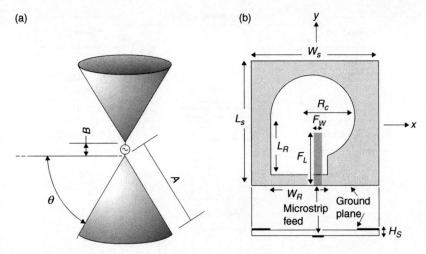

Fig. 3.14   Examples of UWB antenna: (a) biconical and (b) slotted types.

Apparently, most UWBT technologies are based on pulse sensing and, hence, there are requirements for designing UWB antennas which exhibit certain electrical and mechanical properties. Prominent among them include (1) wide band characteristics, (2) steady directional radiation patterns, (3) constant gain, (4) low cost, (5) compact and (6) relatively small size. There are few antenna geometries which fulfil these requirements. Two of the most popular geometries are shown in Fig. 3.14.

Referring to Fig. 3.14, the biconical antennas are known to provide very large impedance bandwidths when they are properly designed. As shown in Fig. 3.14a, there are three parameters useful in designing good biconical antennas. They are: (i) the cone length A, (ii) the gap between the cone B and (iii) the tapper angle θ. The combination of these parameters produces some of the antennas' important characteristics like the return loss, voltage standing wave ratio (VSWR) and gain. It has been reported that a biconical antenna with bandwidth 3.1 to 10.6 Ghz has been designed with $B = 1$ mm, $\theta = 35°$ and $A = 10.5$ mm (Amert and Whites, 2009). Another UWB antenna with omnidirectioal radiation pattern is presented by Tiang $et\ al.$ (2011) and shown geometrically in Fig. 3.14b. Known as p-shaped wide-slot antenna, this design offers several advantages in terms of its size, compactness, bandwidth and impulse response. By adjusting the slot size, this antenna configuration can be designed to have a stable radiation pattern and to operate at an intended impedance bandwidth. It has been shown both experimentally and in computer simulation that such an antenna can be operated at a bandwidth ranging from 4.9 to 11 GHz with the following design parameters: $R_c = 6$ mm, $L_R = 6.5$ mm, $W_R = 5.5$ mm, $F_L = 7.5$ mm, $F_W = 1.2$ mm, $L_s = 16$ mm, $W_s = 16$ mm and $H_s = 1.27$ mm. There are other types of UWB antennas but those shown in Fig. 3.14 are most suited for microwave imaging applications.

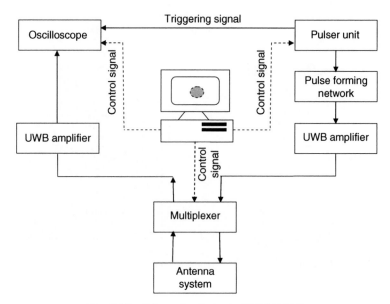

Fig. 3.15   Schematic of a typical UWB DAS.

The DAS for UWBT is fairly standard. All major elements are available commercially. As shown in Fig. 3.15, the heart of the system is the pulser system, which is capable of generating an impulse with picosecond rise and fall times. Signals produced by this device are fed into the pulse-forming network (PFN) which comprises of a number of capacitors and inductors. These components are arranged in a network so that the discharge pulses from the capacitors are spaced in time, resulting in a square or a trapezoidal current pulse with a relatively flat top. The resulting signals are further amplified using the UWB amplifier before they are radiated into the media by an antenna system via a multiplexer unit. Meanwhile, the high bandwidth oscilloscope is used as the main signal acquisition device. Triggering signals from the pulser unit are used to synchronize signal transmission with the digitizing oscilloscope. A personal computer is used to automate the data capturing process and perform image reconstruction.

Chen and Chew (2003) constructed an experimental UWB set-up similar to Fig. 3.15 using 4050B voltage step generator manufactured by Picosecond Pulse Lab (PSPL), a PSPL 4050RPH remote pulse head, two PSPL impulse forming networks, two UWB amplifiers, and 20 GHz 54120B digital oscilloscope manufactured by Hewlett-Packard. In their case, the PSPL 4050RPH generates a 10 V pulse with a rise time of 45 ps. A 50 ps impulse is obtained by passing the remote pulse head output through the PFN. Passing this impulse through a second PFN produces the desired monocycle pulse with repetition rate of about 100 kHz. Such a system produces UWB signal with bandwidth up to 2 GHz. It also has another advantage in that both measurement and image reconstruction could be performed in the time domain, which is the preferred method when dealing with the UWB

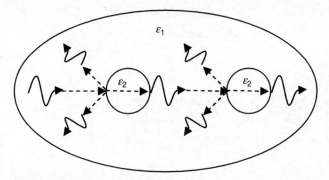

Fig. 3.16 Conversion of incident wavefield to reflected and transmitted components.

signal. However, this system has one major drawback. The maximum frequency that it can capture depends on the maximum bandwidth of the receiving unit – that is, the oscilloscope. As a rule of thumb, the scope's bandwidth must be at least five times higher than the highest frequency of the UWB. Hence, in order to measure at 10 GHz, an oscilloscope with 50 GHz bandwidth is needed in order to ensure minimum signal attenuation. Presently, there are oscilloscopes which have a bandwidth of more than 20 GHz but the sampling rate is not in real time. For instance, the latest Le Croy Corps' SDA100G digital oscilloscope offers unprecedented 100 GHz bandwidth. However, this oscilloscope has an acquisition rate of about 10 Mega samples per second which is still too slow for UWB capturing. It is for this reason that most researchers have resorted to the use of vector network analyser (VNA) for high-frequency UWB applications (Gilmore *et al.*, 2010; Zanoon and Abdullah, 2011). Unlike an oscilloscope, VNA performs measurement in the frequency domain and the medium is characterized in terms of network scattering parameters, or S parameters. These parameters unravel multiple scattering effects resulting from the propagation of the UWB waves in an inhomogeneous body. One can imagine the incident wave propagating in the media with permittivity ε_1, striking two objects with permittivity ε_2. As shown in Fig. 3.16, when the incident wave hits the first object, it generates two types of travelling waves: the reflected and transmitted components. The transmitted wave generates other reflected and transmitted waves when it hits the second object. This process is repeated when the transmitted component encounters another object with different permittivity. If the objects are made up of lossy materials, then a portion of the wave could be absorbed by the objects. VNA is an instrument which provides accurate assessment of the ratios of the reflected signal to incident signal, and the transmitted signal to incident signal.

This device utilizes synthesized frequency sources to provide a known test stimulus (usually sine wave) that can sweep across a range of frequencies or power levels. The source is usually a built-in phase-locked voltage-controlled oscillator. For a two-port VNA altogether there are four S parameters which can be measured. They are S_{11}, S_{12}, S_{21} and S_{22}, respectively corresponding to forward reflection coefficient, reverse transmission coefficient, forward transmission coefficient and

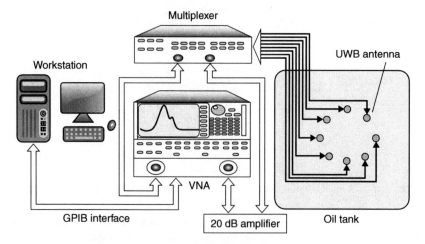

Fig. 3.17 Example of the UWB DAS with VNA.

reverse reflection coefficient. The numbering conventions for these parameters is that the first number following the S is the port where the signal emerges, and the second number is where the signal is applied. In most imaging applications either S_{12} or S_{21} are measured. These signals are usually inverse Fourier transform in order to produce the time domain equivalent. The latter version is used for image reconstruction. An example of VNA-based DAS is shown in Fig. 3.17.

3.4 Image reconstruction

In the electrical tomography systems described previously, sensors are placed around the periphery of a subject through which the measurements are performed. The reconstruction algorithm can be thought of simply as a series of procedures performed repeatedly on digitized measurement data in order to determine the distribution of regions of different electrical properties. In the case of EIT, boundary voltages are used to reconstruct the unknown conductivity or resistivity profiles. Meanwhile, the ECT system attempts to reconstruct an image depicting cross-sectional distribution of dielectric materials from the changes of capacitance measurements. In the same way, the goal of UWBT is to generate the permittivity distribution of the scaterrers from the measurement of the scattered fields outside the objects. In all situations there are a limited number of measurements obtained in response to the excitation applied to the sensors on the periphery. From these measurements, we must reconstruct a faithful and accurate representation of the electrical profiles inside the medium by solving an appropriate boundary value problem. This is referred to as the inverse problem, which often requires the linear approximation of the non-linear problem. A number of interesting reconstruction algorithms have been developed – the most popular among them is based on the optimization procedures, which is accepted as being one of the most theoretically correct reconstruction algorithms. This method requires the minimization of the cost function, which is defined as the sum of squared differences between the measured or desired values at a set of N locations, and the calculated values which are obtained by analytical or numerical methods. Mathematically,

$$E(\chi) = \sum_{i=1}^{N} \left\| \beta_i^d(\chi) - \beta_i^c(\chi) \right\|^2 \qquad [3.24]$$

where χ are source parameters describing the geometrical and electrical structure of the space under investigation – that is, conductivity or resistivity for EIT, dielectric constant for ECT and microwave permittivity for UWBT. Some of the most popular deterministic minimization methods include the steepest descent, non-linear conjugate gradient and regularized iterative scheme (Kelley, 2006). Generally, these methods require the Jacobian J with respect to the model parameters to be calculated. The calculation could be performed by the perturbation of measured variables expressed as:

$$J(\chi) = \Delta(\chi) \sum_{i=1}^{N} \{ \beta_i^d(\chi) - \beta_i^c(\chi) \Delta \beta_i^d(\chi) \} \qquad 3.25]$$

Given the Jacobian, the sensitivity of the system S can be derived by first assuming that the perturbation is very small and, second, expanding equation [3.25] via Taylor series. Simplifying and ignoring the higher order terms, the non-linear relationship between measured variables and model parameters can be linearized in matrix form as follows:

$$\Delta \beta = SG \qquad [3.26]$$

where, $\Delta \beta$ is an $n \times 1$ vector of the desired changes in measured variables, G is an $m \times 1$ vector of normalized image of distribution of model parameters and S is an $n \times m$ matrix known as the sensitivity matrix. In general, the sensitivity matrix describes the changes in the measured values in response to changes of the model parameters. It can be constructed by subdividing the imaging region into small pixels and determining the change in the measured values for each projection due to small change in the model parameters in each pixel. Examples of the sensitivity profiles obtained from UWBT system are shown in Fig. 3.18.

In general, sensitivity patterns of EIT and ECT systems are identical to Fig. 3.18 for a given frequency. It can be seen from this figure that the sensitivities are not uniform but change with pixel locations. In principle, the measured variable is changed everywhere in the imaging region in response to the changes of the model parameters in any given pixel. This feature distinguishes the electrical tomography from the hard-field sensings like X-ray

(a)

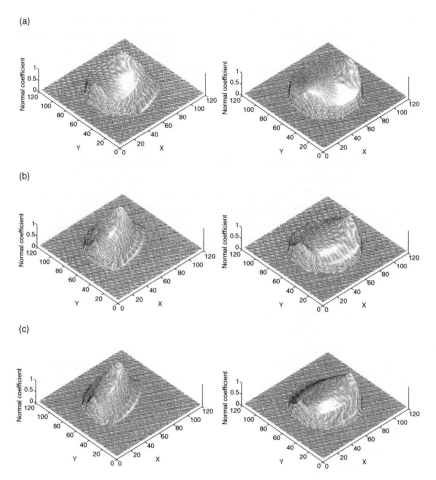

(b)

(c)

Fig. 3.18 Isometric view of the sensitivity distributions obtained from two different projections (left column, projection 1; right column, projection 2), and calculated for three different frequencies: (a) 1 GHz, (b) 3 GHz and (c) 5 GHz.

and MRI. In the latter cases, the sensitivity between two sensing probes is a straight line and approximately uniform, indicating only that part of the imaging region which lies along the path has an effect on the measurement. It is also for this reason that the image reconstruction problem in any electrical tomography system is highly non-linear. Reconstructing an image requires the inversion of the S matrix given in equation [3.26]. By virtue of the equivalence principle, the inverse problem is inherently ill-posed and the uniqueness of the solutions usually cannot be guaranteed. Despite this difficulty, there are numbers of tractable algorithms which have been developed to deal with ill-posed or ill-conditioned problems. The technical discussions are beyond the scope of this book and interested readers are referred to a paper by Yang *et al.* (1999), which describes various inversion techniques and image reconstruction algorithms.

3.5   Applications

Though the application of electrical imaging to engineering has been the subject of increasing international interest for several years, the same trend is not observed in food engineering. To date there are limited publications which centre on the use of ERT and ECT in food engineering applications. However, this is changing as the quality of images and electronics of these alternative devices continues to improve, together with the fact that they are cheaper and easy to maintain compared to other conventional ionizing instruments. Thus far, there are few interesting publications which describe the promising applications of the electrical tomography systems to food engineering processes.

One of the first attempts to investigate the feasibility of applying ERT for food imaging is illustrated by the work by Henningson *et al.* (2005), who studied velocity profiles of yoghurt and its rheological bahaviour in a pipe of industrial dimensions. A cross-correlation technique was used to transform the dual-plane conductivity maps into velocity profiles. Comparing simulated and experimental results, they discovered that ERT results have some noise, especially the velocity pixels near the wall of the pipe. In contrast, the centreline velocities are very well resolved with an error less than 7%. They concluded that ERT is a useful method for determination of the velocity profile of food; the information produced can be used in process conditioning in order to maximize loss of product. More recently Sharifi and Young (2011) have successfully applied ERT to the monitoring of milk storage and mixing tanks used in various stages of the milk processing industries. Employing the commercial 3-D ERT P2000 system manufactured by ITS, UK, they have qualitatively and quantitatively demonstrated that this imaging device is potentially useful in the milk processing industry in at least four different types of visualizations. They are (i) homogeneity and heterogeneity, (ii) adulteration, (iii) separation and (iv) aeration. Examples of ERT images produced from the authors' adulteration study are shown in Fig. 3.19. These images sequentially show the snap changes of milk conductivity after it was adulterated by addition of water to the surface of the tank. The water was prepared in such a way that it is more concentrated or conductive compared to homogeneous milk. The images were reconstructed at 25 s intervals and show the changes of the overall milk conductivity after water is added into the tank at 50 s (Fig. 3.19b). These examples also demonstrate how ERT can be used to visualize the dynamics of mixing in stirred tank reactor.

The first application demonstrating the potential use of ECT for monitoring and control tool in food industry involved fluidized bed dryers. Like other chemical processes, fluid bed drying is very common in food processing since most food or dairy products are required in agglomerate or granular form to achieve good instant properties. Drying of food products by fluidized bed is

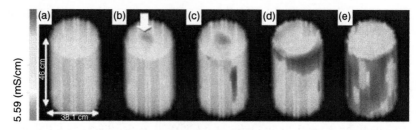

Fig. 3.19   Conductivity changes in skimmed milk after it was adulterated with addition of water to the surface of the tank. The images (a–e) were reconstructed every 25 s before and after adulteration. In this case the adulteration took place in frame (b) (Sharifi and Young, 2011).

accompanied by the reduction of moisture content from approximately 30% to 2% by weight, which in turn, results in the change of the dielectric properties of the bed hydrodynamics (Sacilik *et al.*, 2006). The changes in bed permittivity result in a change in bulk capacitance, which can be measured by means of ECT. A tomographic imaging technique applied to this unit operation can identify local changes in bed hydrodynamics associated with fines entrainment, segregation or non-uniform mixing, all of which may contribute to process efficiency and product degradation. Chaplin and Pugsley (2005) are among the earliest researchers to have pioneered the use of an ECT system as a monitoring or control tool of fluidized bed reactors. The components of the fluidized bed drying apparatus and ECT DAS used by these authors are presented in Fig. 3.20. In the experiments, the authors used the wet pharmaceutical granules which were bed packed at different loadings. The ECT electronics used was the eight-electrode, two-plane DAS manufactured by Process Tomography (Cheshire, UK). Capacitance data was collected as 28 unique measurements corresponding to all possible combinations from an eight-electrode sensor. Image reconstructions were performed using the pre-computed sensitivity matrices as discussed in Section 3.4. The dryings were achieved by passing the fluidizing air at superficial gas velocities, following similar operating conditions in the pharmaceutical industry for the drying of pharmaceuticals. In this case, the ECT system was used to monitor the dynamic behaviour of the packed granules while the bed was fluidized at a temperature of 110°C. Dynamic ECT data sets were collected at 10 ms intervals between frames and at a 100 Hz sampling rate. Examples of tomograms showing the dynamic of the fluidized bed reconstructed from 3.5 kg bed loading and a moisture content of 28% by weight are shown in Fig. 3.21. The inverse grey-scale value was used to display the tomograms, in which the grey-scale 0 and 1 represent the low and high permittivity regions, respectively. Clearly, from this figure, beginning from frame number 5 and ending at frame number 100, corresponding to 50 ms and 1000 ms drying times respectively, the measurement plane is dominated by a stable core of fluidizing air, while the dense bed of solids remains near the wall.

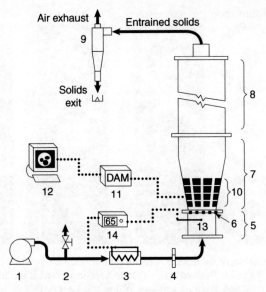

Fig. 3.20   Schematics of the ECT-based instrumentation of the fluidized bed dryer: (1) blower, (2) bypass, (3) heater, (4) orifice, (5) windbox, (6) distributor, (7) product bowl, (8) freeboard, (9) cyclone, (10) ECT sensors, (11) data acquisition module, (12) computer, (13) thermocouple and (14) controller (Chaplin and Pugsley, 2005).

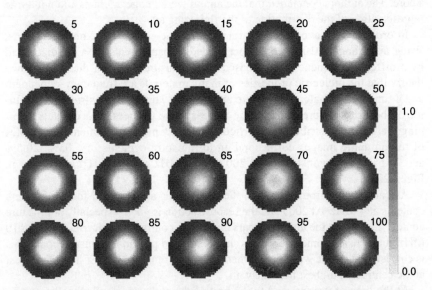

Fig. 3.21   Tomographic images showing the hydrodynamic of the fluidized bed drying at 3.5 kg bed loading, superficial gas velocity 2.5 m/s and moisture content 28% by weight (Chaplin and Pugsley, 2005).

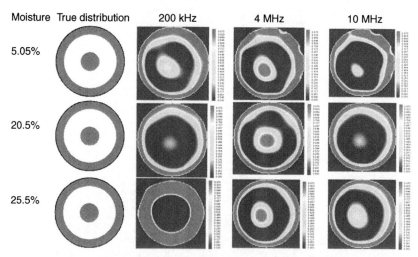

Fig. 3.22   Cross-sectional images depicting the annular-core distributions of wetted granular particles inside fluidized bed dryer, and reconstructed at different frequencies and moisture contents (Wang and Yang, 2011).

These results suggest the existence of some kind of centralized hydrodynamic behaviour, resulting in channelling of air through the centre of the bed. However, these results were not repeated on pharmaceutical granules with moisture content less than 18% by weight, as the air flow pattern appears like an annulus instead of a core. The authors concluded that the annuli were image artefacts and not a true representation of the distribution of the fluidizing air.

In order to improve the measurement accuracy and reduce noise artefacts, Wang and Yang (2010) recently performed measurements on a fluidized bed by a different frequency. They based their measurements on the well-known theory that the dielectric properties of many materials are frequency-dependent and a function of material moisture and density. Hence, reconstructing the fluidized bed at a different frequency increases the chances of accurately mapping the granule distributions, particularly at low moisture contents. Examples of tomographic images corresponding to moisture contents of 5.05%, 20.5% and 25.5%, and reconstructed at 200 kHz, 4 MHz and 10 MHz, are shown in Fig. 3.22.

As can be seen from Fig. 3.22, the best results are obtained for all moisture contents at 4 MHz. At 200 kHz, the reconstruction is good, especially at moisture content 25.5%, but the image appears blurred. In contrast, the blurring is less at 10 MHz but there are some artefacts in shape, size and position reconstructions. This is consistent with the well-known theory that the sensitivity of ECT is the lowest at the centre and highest near the wall.

At the time of writing, and to the best of the author's knowledge, there is no publication demonstrating the general feasibility of UWBT in visualizing food properties and material components. To date most of the developments are oriented towards biomedical applications, particularly early-stage breast cancer detection. Given the super-resolution capability of the microwave signals, together with rapid development in this subject, it is just a matter of time before UWBT finds its application in the food laboratory.

3.6   Conclusions

As discussed above, there are several powerful imaging modalities that are capable of visualizing food materials non-destructively, each having particular strengths and weaknesses. CT, because of its ionizing properties, is suitable for imaging hard and solid foods; MRI is good for soft materials; and EIT/UWBT is good for impedance and permittivity mapping. Some of these technologies are already available commercially, while some are still in the development stage. The latter includes ultra wide band sensing, which has attracted significant interest among leading researchers in the world because of its ability to reconstruct an image at super-resolution accuracy, reaching a resolution as good as one half to one third of a wavelength. This resolution offers the capability for UWBT to detect the location and size of millimetre-scaled targets in inhomogeneously complicated dielectric distributions (Zanoon and Abdullah, 2011). Also under development is a system that can combine results from various modalities in order to enhance and improve image quality further. With careful calibration, images from different modalities can be registered and superimposed, giving rise to what is presently known as 'multimodal imaging' or the 'sensor fusion technique'. With intense research being pursued in some of the world's leading laboratories, it will not be long before such emerging technologies reach food technologists and scientists.

3.7   References

Amert, A. K. and Whites, K. W. (2009) Miniturization of the biconical antenna for ultra wide-band applications. *IEEE Transactions on Antennas and Propagation*, **57(12)**, 3728–35.

Ashurst, P. R. (1995) *Production and Packaging of Noncarbonated Fruit Juices and fruit beverages*. London: Chapman & Hall.

Babin, P., Della Valle, G., Chiron, H., Cloetens, P., Hoszowska, J., Penot, P., Réguerre, A. L., Salva, L. and Dendieval, R. (2006) Fast x-ray tomography analysis of bubble growth and foam setting during bread making. *Journal of Cereal Science*, **43(3)**, 393–7.

Bauchot, A. D., Harker, F. R. and Arnold, A. M. (2000) The use of electrical impedance spectroscopy to assess the physiological condition of Kiwifruit. *Postharvest Biological Technology*, **18**, 9–18.

Borompichaichartkul, C., Moran, G., Srzednicki, G. and Price, W. S. (2005) Nuclear magnetic resonance (NMR) and magnetic resonance imaging (MRI) studies of corn at sub-zero temperatures. *Journal of Food Engineering*, **69(2)**, 199–205.

Brooks, R. A. and Di Chiro, G. (1975) Theory of image reconstruction in computed tomography. *Radiology*, **117**, 561–72.

Brooks, R. A. and Di Chiro, G. (1976) Principles of computer assisted tomography (CAT) in radiographic and radioisotope imaging. *Physics in Medical Biology*, **21(5)**, 689–732.

Brown, B. H. and Seagar, A. D. (1987) The Sheffield data collection system. *Clinical Physics and Physiology Measurement*, **8A**, 91–8 [N14].

Chaplin, G. and Pugsley, T. (2005) Application of electrical capacitance tomography to the fluidized bed drying of pharmaceutical granule. *Chemical Engineering Science*, **60**, 7022–33.

Chen, F. C. and Chew, W. C. (2003) Time-domain ultra wideband imaging radar experiment for verifying super-resolution in non-linear inverse scattering. *Journal of Electromagnetic Waves and Applications*, **17(9)**, 1243–60.

Cook, R., Saulnier, G. J., Gisser, G. D., Goble, J. C., Newell, J. C. and Isaacson, D. (1994) ACT3: A high-speed, high precision, electrical impedance tomography. *IEEE Transactions on Biomedical Engineering*, **41(8)**, 713–22.

Damez, J. L., Clerjon, S., Abouelkaram, S. and Lepetit, J. (2008) Electrical impedance probing of the muscle food anisotropy for meat ageing control. *Food Control*, **19**, 931–9.

Dickin, F. J., Wang, M., Abdullah, M. Z. and Zhao, X. J. (1992) *ICEMI Proceedings International Conference on Electronic Measurement and Instruments*, 38–47.

Epstein, B. R. and Foster, K. R. (1983) Anisotrophy in the dielectric properties of skeletal muscle. *Medical and Biological Engineering and Computing*, **21**, 51–5.

FCC (Federal Communications Commission) (2002) Notice of proposed rule making, Revision of part 15 of the commission's rules regarding ultra-wideband transmission systems, ET Docket, 98–153.

Foster, K. R. and Schwan, H. P. (1989) Dielectric properties of tissues and biological materials: A critical review. *Critical Reviews in Biomedical Engineering*, **17**, 24–104.

Fu, X., Milroy, G. E., Dutt, M., Bentham, A. C., Hancock, B. C. and Elliot, J. A. (2005) Quantitative analysis of packed and compacted granular systems by x-ray microtomography. *SPIE Proceedings Medical Imaging and Image Processing*, **5747**, 1955.

Gilmore, C., Mojabi, P., Zakaria, A., Pistorius, A. and LoVetri, J. (2010) On super-resolution with an experimental microwave tomography system. *IEEE Antennas and Wireless Propagation Letters*, **9**, 393–6.

Hauttmann, S. and Müller, J. (2001) In-situ biomass characterisation by impedance spectroscopy using a full-bridge circuit. *Bioprocess and Biosystems Engineering*, **24**, 137–41.

Henningsson, M., Ostergen, K. and Dejmek, P. (2005) Plug flow of yoghurt in piping as determined by cross-correlated dual-plane electrical resistance tomography. *Journal of Food Engineering*, **76(2)**, 163–168.

Herman, G. (1980) *Image Reconstruction from Projections*. New York: Academic Press.

Hills, B. (1995) Food processing: an MRI perspective. *Trends in Food Science & Technology*, **6**, 111–17.

Huang, S. M., Plaskowski, A. B., Xie, C. G. and Beck, M. S. (1989) Tomographic imaging of two-component flow using capacitance sensors. *Journal of Physics E: Scientific Instrument*, **22**, 173–7.

Kak, A. C. and Slaney, M. (1988) *Principles of Computerised Tomography Imaging*. New York: IEEE Press .

Kak, C. K. (1979) Computerised tomography with x-ray, emission and ultrasound sources. *Proceedings of IEEE*, **67(9)**, 1245–72.

Kelley, C. T. (2006) Iterative methods for optimization. *SIAM* , Philadelphia.

Kolstad, K. (2001) Fat deposition and distribution measured by computer tomography in three genetic groups of pigs. *Livestock Production Science*, **67**, 281–92.

Lepetit, J., Salé, P., Favier, R. and Dalle, R. (2002) Electrical impedance and tenderization in bovine meat. *Meat Science*, **49(2)**, 51–62.

Lucas, T., Greiner, A., Quellec, S., Le Bail, A. and Davanel, A. (2005) MRI quantification of ice gradients in dough during freezing or thawing processes. *Journal of Food Engineering*, **71(1)**, 98–108.

McCarthy, M. (1994) *Magnetic Resonance Imaging in Foods*. New York: Chapman & Hall .

Nott, K. P. and Hall, L. D. (1999) Advances in temperature validation of foods. *Trends in Food Science & Technology*, **10**, 366–74.

Orsi, C., Torriani, S., Battistitti, B. and Vescovo, M. (1997) Impedance measurements to assess microbial contamination of ready-to-use vegetable. *European Food Research and Technology*, **216(5)**, 248–50.

Rinck, P. A. (2001) *Magnetic Resonance in Medicine*. Berlin: Blackwell .

Sacilik, K., Tarimici, C. and Colak, A. (2006) Dielectric properties of flaxseeds as affected by moisture content and bulk density in the ratio frequency range. *Biosystems Engineering*, **93**, 153–60.

Sarigul, E., Abott, A. L. and Schmoldt, D. T. (2003) Rule driven defect detection in CT images of hardwood logs. *Computers and Electronics in Agriculture*, **41**, 101–19.

Sharifi, M. and Young, B. (2011) 3-Dimensional spatial monitoring of tanks for the milk processing industry using electrical resistance tomography. *Journal of Food Engineering*, **105(2)**, 312–19.

Tiang, S. S., Ain, M. F. and Abdullah, M. Z. (2011) Compact and wideband wide-slot antenna for microwave imaging system. *Proceedings of the IEEE International RF and Microwave Conference* (RFM 2011), Seremban, 12–14 December 2011.

Thybo, A. K., Szczpinski, P. M., Karlsoon, A. H., Donstrup, S., Stodkilde-Jorgesen, H. S. and Andersen, H. J. (2004) Prediction of sensory texture quality attributes of cooked potatoes by NMR-imaging (MRI) of raw potatoes in combination with different image analysis method. *Journal of Food Engineering*, **61(1)**, 91–100.

Vestergaard, C., Erbou, S. G., Thauland, T., Adler-Nisen, J. and Berg, P. (2005) Salt distribution in dry-cured ham measured by computed tomography and image analysis. *Meat Science*, **69**, 9–15.

Vlachos, N. A. and Karapantsios, T. D. (2000) Water content measurement of thin sheet starch product using a conductance technique. *Journal of Food Engineering*, **46**, 81–98.

Wang HG, Yang WQ (2010) Measurement of fluidized bed dryer by different frequency and different normalisation methods with electrical capacitance tomography. *Powder Technology*, **199**, 60–9.

Wang, M., Ma, Y., Holliday, N., Dai, Y., Williams, R. and Lucas, G. (2005) A high performance EIT system. *IEEE Sensors Journal*, **5(2)**, 289–98.

Yang, W. Q. (2010) Design of electrical capacitance tomography sensors. *Measurement Science and Technology*, **21(4)**, 42001–13.

Yang, W. Q., Spink, D. M., York, T. A. and McCann, M. (1999) An image reconstruction algorithm based on Landweber's iteration method for electrical-capacitance tomography. *Measurement Science Technology*, **10**, 1065–9.

Zanoon, T. F. and Abdullah, M. Z. (2011) Early stage breast cancer detection by means of time-domain ultra wide band sensing. *Measurement Science Technology*, **22(11)**, 114016.

Żywica, R., Pierzynowska-Korniak, G. and Wójcik, J. (2005) Application of food products electrical model parameters for evaluation of apple purée dilution. *Journal of Food Engineering*, **19**, 931–9.

3.8   Appendix: nomenclature and abbreviations

3.8.1   Nomenclature

c speed of light (2.998×10^8 ms^{-1})

C capacitance (farad)

C_p parallel capacitance (farad)

C_s series capacitance (farad)

E energy (J)

f frequency (Hz)

h Planck's constant (6.626076×10^{-34} Js)

i	current (A)
n	number of independent measurements
N	number of electrodes
R	resistance (Ω)
S	S parameters
S_{11}	forward reflection coefficient
S_{12}	reverse transmission coefficient
S_{21}	forward transmission coefficient
S_{22}	reverse reflection coefficient
V	voltage (V)
ϕ, α, β	phase (degree)
λ	wavelength (m)
σ	Stefan-Boltzman's constant (1.38054×10^{-23} JK^{-1})
ω	frequency (rad s^{-1})

3.8.2   Abbreviations

AC	alternating current
A/D	analogue to digital converter
CAT	computer-assisted tomography
CT	computerized tomography
DC	direct current
DAS	data acquisition system
DSP	digital signal processor
ECT	electrical capacitance tomography
EIT	electrical impedance tomography
ERT	electrical resistance tomography
FCC	Federal Communications Committee
FFT	fast Fourier transform
fps	frame per second
MRI	magnetic resonance imaging
NMR	nuclear magnetic resonance
PET	positron emission tomography
PFN	pulse forming network
PSPL	Picosecond Pulse Lab
RF	radio frequency
Rx	receiver
SPECT	single photon emission computed tomography
TOF	time of flight
Tx	transmitter
UWB	ultra wide band
UWBT	ultra wide band tomography
VCCS	voltage controlled current source
VNA	vector network analyser
VSWR	voltage standing wave ratio

4

Image processing techniques for computer vision in the food and beverage industries

N. A. Valous and D.-W. Sun, University College Dublin, Ireland

Abstract: The foremost underlying drivers for using image processing technologies are automation and improved rapid operations. Image analysis involves taking measurements of objects within an image, preferably automatically, and assigning them to groups or classes. The objective is to present conventional and more advanced digital image processing techniques that predominantly involve the image analysis class of operations: segmentation, feature selection and extraction, and classification. These approaches will be reviewed from the aspect of illustrating their role and impact in the food and beverage industry.

Key words: computer vision, image processing, segmentation, feature extraction, feature selection, classification.

4.1 Introduction

The explosive growth in both hardware platforms and software frameworks has led to many significant advances in imaging technology. In recent years, researchers have developed several non-contact methods for the assessment/ inspection of food and beverage products, overcoming most of the drawbacks of traditional methods such as human inspection. These methods are based on the automatic detection of various image features, which may correlate with attributes related to sensorial, chemical and physical properties. Computer vision coupled with image processing, which includes the capturing and analysing of images, facilitates the objective and rapid two- and three-dimensional (2-D and 3-D) assessment of visual characteristics, as well as characteristics that cannot be visually differentiated by human inspection – that is, structural and textural characteristics through the extraction of suitable features (Valous *et al.*, 2010a). Therefore, computer vision systems based on a variety of imaging sensors can be used by the food and beverage industry for inspection, control and grading (Jähne and Haußecker, 2000). The application potential of image processing techniques to the food and beverage industry has long been recognized. The food industry ranks among the top ten industries using image processing techniques, which have been proven successful for the objective and non-destructive evaluation of several products (Du and Sun, 2004).

Digital image processing has expanded and is further rapidly expanding from a few specialized applications into a standard scientific tool. Image processing techniques are now applied to virtually all the natural sciences and technical disciplines. Many real-world examples demonstrate that image processing enables complex phenomena to be investigated, which could not be adequately accessed with conventional measuring techniques. These may include (i) counting and gauging: a classic task for digital image processing is counting particles and measuring their size distribution or checking for defects; (ii) exploring the 3-D space: a large variety of range imaging and volumetric imaging techniques have been developed, therefore, image processing techniques are also applied to depth maps and volumetric images; (iii) exploring dynamic processes: this is possible by analysing image sequences on a variety of spatial and temporal scales and (iv) classification of objects observed in images (Jähne, 2005). Today, image processing systems are being designed for a variety of cases, incorporating specialized and dedicated processors for specific applications. Some applications require interactive and real-time processing, high-precision computation and dedicated display environments (Dhawan, 1990).

A complete digital image processing system is a collection of hardware and software tools that can acquire an image, using appropriate sensors to detect the radiation or field and capture the features of interest from the object in the best possible way. If the detected image is continuous, it will need to be digitized by an analogue-to-digital converter. The next step is to store the image, either temporarily using read/write memory devices known as random access memory or more permanently using magnetic media, optical media, or semiconductor technology. The final steps include the manipulation – that is, processing of the image using sets of programming functions – and the displaying of the image on a monitor. Image processing procedures can be grouped into five fundamental classes: enhancement, restoration, analysis, compression and synthesis (Dougherty, 2009). Table 4.1 summarizes the aforementioned digital image processing classes and examples of operations within them.

In this chapter we shall be interested predominantly in the image analysis class, which contains certain representative operations – that is, segmentation, feature extraction and selection – and object classification. Image analysis involves taking measurements of objects within an image, preferably automatically, and assigning them to groups or classes. Generally, the process begins with isolating the objects of interest from the rest (known as segmentation of

Table 4.1   Digital image processing classes, and examples of operations within them

Classes	Examples of operations
Image enhancement	Contrast enhancement, image averaging, frequency domain filtering.
Image restoration	Photometric correction, inverse filtering.
Image analysis	Segmentation, feature extraction and selection, object classification.
Image compression	Lossless and lossy compression.
Image synthesis	Tomographic imaging, 3D reconstruction.

Source: Dougherty (2009).

the image), measuring a number of features, such as size, shape and texture, and then classifying the objects into groups according to these features. This permits the categorization of a new object as either belonging to a particular group or not, depending on whether its features fall inside or outside the tolerance of that group, respectively. The aim of this chapter is to present conventional and more advanced digital image processing techniques that have been used in applications and predominantly involve the image analysis class of operations: segmentation, feature selection and extraction, and classification. These approaches are then reviewed from the aspect of illustrating their role and impact in the food and beverage industry. Although the techniques and methodologies to be reviewed are of relevance to the wider research community of digital image processing, they could also indicate future trends and suggest new routes and applications for the food and beverage industry.

4.2   Digital image analysis techniques

The following sections focus on the initial selection of an input, including assessment of various qualities determining the categorization process and the way in which this analysis is usefully put into practice.

4.2.1   Image segmentation

Description of techniques
Image segmentation is typically defined as an exhaustive partitioning of an input image into regions, each of which is considered to be homogeneous with respect to some image property of interest – for example intensity, colour or texture. This process is considered an essential component of any image analysis system, and problems associated with it have received considerable attention (Zheng and Sun, 2007; Bong and Rajeswari, 2011). In many applications of image processing, the grey levels of pixels belonging to an object are substantially different from the

grey levels of the pixels belonging to the background. Thresholding then becomes a simple but effective tool to separate objects from the background. The output of the thresholding operation is a binary image, above the threshold indicating the foreground objects, below the threshold corresponding to the background. Various factors, such as non-stationary and correlated noise, ambient illumination, busyness of grey levels within the object and its background, inadequate contrast, and object size not commensurate with the scene complicate the thresholding operation. In a non-destructive setting for inspection and quality control in the food and beverage industry, thresholding is often the first critical step in a series of processing operations such as morphological filtering, measurement and statistical assessment (Sezgin and Sankur, 2004). The most commonly used segmentation techniques can be classified into two broad categories: (i) region segmentation techniques that look for regions satisfying a given homogeneity criterion and (ii) edge-based segmentation techniques that look for edges between regions with different characteristics. Combined (hybrid) strategies have also been used in many applications (Sun and Du, 2004; Rogowska, 2009).

Region-based segmentation methods are, in general, less sensitive to noise. Examples of region-based segmentation methods include: region growing, split and merge, histogram thresholding, random field, watershed and clustering (Shaaban and Omar, 2009). Among existing techniques, thresholding is one of the most popular approaches in terms of simplicity, robustness and accuracy. Thresholding is a common region segmentation method; in this technique a threshold is selected, and an image is divided into groups of pixels having values less than the threshold and groups of pixels with values greater or equal to the threshold. There are several thresholding methods: global methods based on grey-level histograms, global methods based on local properties, local threshold selection and dynamic thresholding. Thresholding can also be classified into parametric and non-parametric approaches. In the parametric approach, grey-level distribution of an image is assumed to obey a given statistical model, and optimal parameter estimation for the model is sought by using a given histogram. The non-parametric one usually finds the optimal threshold by optimizing an objective function such as between-class variance and entropy (Li *et al.*, 2011). Region growing is another class of region segmentation algorithms that assign adjacent pixels or regions to the same segment if their image values are close enough, according to some pre-selected criterion of closeness. Region growing processes are more appropriate in some cases than clustering or thresholding approaches, which operate directly in colour space or in its three mono-dimensional representations, because they simultaneously take into account both colour distribution in the colour space and its repartition in the spatial domain (Tremeau and Borel, 1997). Clustering algorithms achieve region segmentation by partitioning the image into sets or clusters of pixels that have strong similarity in the feature space. In essence, they extract the region as a cluster, and are suited to the partitioning of larger parts (global segmentation). They are robust to noise, but are not suited to the extraction of small regions (Ryu and Miyanaga, 2004). Watershed segmentation is a region-based technique that utilizes image morphology. The morphological watershed algorithm can segment a binary image into different regions by treating its inverse distance map as a landscape and the local minima as markers. By labelling each segmented region with a unique index, different parts can be separated and identified. The performance of watershed segmentation depends on how the right markers are chosen. Spurious markers will lead to over-segmentation, which is the major drawback of the watershed algorithm (Sun and Luo, 2009). Watershed image segmentation is good for handling uniformed background and objects with blurry edges (Zhang *et al.*, 2010).

Edge-based segmentation algorithms try to find object boundaries and segment regions enclosed by the boundaries. These algorithms usually operate on edge magnitude or phase images produced by an edge operator suited to the expected characteristics of the image. For example, most gradient operators (e.g., Prewitt, Kirsch, Roberts, etc.) are based on the existence of an ideal step edge, and involve calculation of convolutions – for example weighted summations of the pixel intensities in local neighbourhoods. The weights can be listed as a numerical array in a form corresponding to the local image neighbourhood, also known as a mask, window, or kernel (Rogowska, 2009). Other edge-based segmentation techniques are graph searching and contour following. In graph searching each image pixel corresponds to a graph node, and each path in the graph corresponds to a possible edge in the image. A graph consists of a set of points called nodes and a set of links that determine how the nodes can be connected. Each node has a cost associated with it, which is usually calculated using the local edge magnitude, edge direction and *a priori* knowledge about the boundary shape or location. The cost of a path through the graph is the sum of costs of all nodes that are included in the path. By finding the optimal low-cost path in the graph, one can define the optimal border. The graph searching technique is very powerful, but it strongly depends on an application-specific cost function (Thedens *et al.*, 1995). In contour-based approaches, often the first step of edge detection is done locally. Subsequently, efforts are made to improve results by a global linking process that seeks to exploit curvilinear continuity. A criticism of this approach is that the edge/no edge decision is made prematurely. To detect extended contours of very low contrast, a very low threshold has to be set for the edge detector. This will cause random edge segments to be found everywhere in the image, making the task of the curvilinear linking process unnecessarily harder than if the raw contrast information was used (Malik *et al.*, 2001).

Application examples

Figure 4.1 demonstrates an example of a combined image segmentation approach that has some merit in the food industry, depicting: (a) a representative central section of a medium quality pork ham and (b) the same section but showing only segmented pores/defects and fat-connective tissue structures (Valous *et al.*, 2009). The identification and partition of these regions of interest (ROI) is based on the differences in colour between the structures and background matrix, using contrast enhancements and image filtering procedures, before applying typical histogram-based segmentation. An image with enhanced contrast has typically larger colour differences between ROIs, which helps to locate object boundaries and facilitates the partition of different structures in the image. In general, contrast enhancement amplifies these colour variations, thereby increasing feature visibility (Gauch, 1998). Figure 4.2 presents the hybrid approach that was used for the segmentation of pores/defects. The methodology was adaptable and provided a fine level of feature information.

Image pre-processing techniques help to improve the shape of the image histogram by making it more strongly bimodal. In the case of the fat-connective tissue segmentation more pre-processing steps are required due to the inhomogeneities in intensity and, particularly, due to the low contrast between the lean meat and fat-connective tissue. The hybrid approach consisted of finding the edges or boundaries of the pore objects as defined by the local pixel intensity gradient, and then selecting a global threshold to produce the binary image. Instead of global thresholds, local thresholds can be determined by splitting the image into sub-images and calculating thresholds for each sub-image or by examining the image intensities in the neighbourhood of each pixel. The identification and analysis of fatty regions and pores in hams is one of many important aspects that correlate well with consumer acceptance. Consumer desire for leaner meats has necessitated the production of meat products with less fat (Schilling et al., 2003). Porosity affects sensory properties and has a direct effect on other physical properties (Du and Sun, 2006). Moreover, the characterization of size distributions of pores is relevant to the ham industry because pore formation depends on the quality of the raw meat, pre-treatment and processing, which influences pore size, geometry and size distribution in the meat matrix; a well-structured matrix and a fine, uniform structure with numerous small pores would probably result in a more absorptive capacity and better retention of water compared to coarser structures with large pores.

Figure 4.3 demonstrates an example of a 2-D histogram variance thresholding technique on a raw beef joint, depicting: (a) the original RGB image of the beef joint and (b) its segmented version (Zheng et al., 2006). The use of 2-D histograms is justified by the fact that segmentation using thresholds selected from the 1-D histogram are not satisfactory when noise is present in the images (Brink, 1992). To obtain the

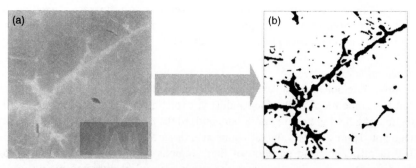

Fig. 4.1   Representative central section of a medium quality pork ham: (a) original RGB image and (b) binarized version of pores/defects and fat-connective tissue after segmentation.

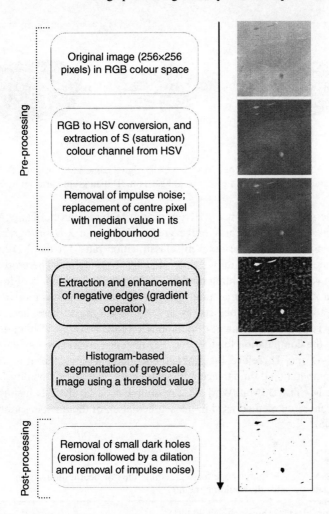

Fig. 4.2 Hybrid segmentation approach for pores/defects in a representative central section of a medium quality pork ham, depicted in flowchart mode.

2-D histogram of the image, the local average of a pixel, $f(x, y)$, is defined as the average intensity of its four neighbours denoted by $g(x, y)$ (Cheng *et al.*, 2000):

$$g(x,y) = \left[\frac{1}{4}[f(x,\ y+1) + f(x,\ y-1) + f(x+1,\ y) + f(x-1,\ y)] + 0.5\right] \quad [4.1]$$

A 2-D histogram is an array ($L \times L$) with the entries representing the number of occurrences of the pair [$f(x, y)$, $g(x, y)$]. A 2-D histogram can be viewed as a full Cartesian product of two sets X and Y, where X represents the grey levels and Y represents the local average grey levels: $X = Y = \{0, 1, 2, ..., L\text{-}1\}$. The pixels

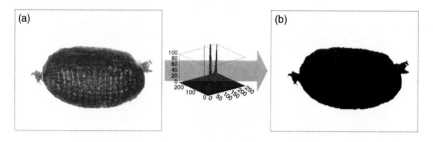

Fig. 4.3   A raw beef joint acquired with a CCD digital camera: (a) original RGB image and (b) segmented version of the beef joint based on 2D histogram variance thresholding.

having the same intensity but different spatial features can be distinguished in the second dimension (local average grey level) of the 2-D histogram. Zheng *et al.* (2006) found that the pixel misclassification rates for the 2-D histogram variance method were generally smaller than the 1-D method. Therefore, the 2-D histogram variance thresholding method can be used to segment beef joints from the background with higher accuracy. In general, thresholding is more preferable for images with only two objects, as in the case examined here, because of its simplicity and less requirement of computation load and time (Wu *et al.*, 1999). However, the computation load for the 2-D histogram variance method is larger than the 1-D one. Therefore, the problem encountered is whether it is worth sacrificing computational time to accomplish higher segmentation quality. Table 4.2 shows the MATLAB (MathWorks, USA) source code for the 2-D histogram variance thresholding method used in the segmentation of the raw beef joint (Zheng *et al.*, 2006).

Table 4.2   MATLAB source code for segmenting the raw beef joint using the 2-D histogram variance thresholding method

```
function [hist2d, t1, t2, bw] = threshold2d(I);
% bw: output image (binary), I: input image (8-bit grayscale)
% hist2d: 2D histogram, t1: threshold on x, t2: threshold on y
MI = medfilt2(I); % Median filtering
% 2D histogram
hist2d = []; hist2d (1:256,1:256) = 0; imagesize = size (I);
   for i = 1:imagesize (1)
      for j = 1:imagesize (2)
      hist2d (I(i, j)+1, MI (i, j)+1) = hist2d (I(i, j)+1, MI(i, j)+1) +1;
end
   end
```

Continued

Table 4.2   *Continued*

```
hist2d = hist2d/sum (sum(hist2d));
% Create the entry
entry1 = []; entry2 = [];
for i = 1:256
entry1 (:,i) = 1:1:256; entry2 (i,:) = 1:1:256;
  end
entry = entry1.*entry2;
% Find the optimal threshold
maxvar = 0;
  for i = 1:256
    for j = 1:256
      meana = sum (sum (hist2d (1:i, 1:j)));
      meanb = sum (sum (hist2d (i:256, j:256)));
      firsta = [ ]; firstb = [ ];
if(and (meana ~ = 0, meanb ~ = 0))
  firsta = sum (sum(hist2d(1:i,1:j).*entry(1:i,1:j)))/meana;
  firstb = sum (sum(hist2d(i:256,j:256).*entry(i:256,j:256)))/meanb;
  end
sumvar = meana*meanb*(firsta-firstb)*(firsta-firstb);
  if(sumvar>maxvar)
    maxvar = sumvar; t1 = i; t2 = j;
  end
  end
  end
bw = and(I>t1, MI>t2); % Image thresholding
```

4.2.2   Feature extraction and selection

Description of techniques

After identifying image objects by image segmentation, image features can be identified and computed. The features are characterized into three categories: colour, morphology, and texture. Morphological features are the physical features that describe the appearance of an object. Colour features are the most direct and simple features of all the image features, mainly due to the fact that are related to the object or scene (Chen and Wang, 2010). Textural properties of an object can be defined as the spatial distribution of the grey-level intensities. Depending upon the purpose of analysis, different image features have different levels of importance. The main difficulty in image processing is that it often produces a large number of image features, since it is not known in advance which features have significant impacts on the required parameters. Dealing with the handling of such a large number of parameters and modelling them can take huge computational time (Chatterjee and Bhattacherjee, 2011). Moreover, the redundant features may sometimes also mislead the model performance (Steppe *et al.*, 1996). Further, as the number of features grows, the number of training samples required for model development grows exponentially (Duda *et al.*, 2001). To improve performances, the sample dimensionality is reduced, thanks to feature extraction or selection schemes. Given an input set of features, dimensionality reduction can be achieved in two different ways. The first is to select the best subset of features of the input feature set. This process is termed 'feature selection'. The second approach is to create new features based on transformation from the original features to a lower dimensional space and is termed 'feature extraction'. This transformation may be a linear or nonlinear combination of the original features. The choice between the two depends on the application domain and the specific training data available (Pudil *et al.*, 2002).

Feature selection leads to savings in measurements cost, since some of the features are discarded and the selected features retain their original physical interpretation. In addition, the retained features may be important for understanding the physical process that generates the patterns. There are many potential benefits of variable and feature selection: facilitating data visualization and data understanding, reducing the measurement and storage requirements, reducing training and utilization times, defying the curse of dimensionality to improve prediction performance (Guyon and Elisseeff, 2003). Data samples can be either unlabelled or labelled, leading to the development of unsupervised or supervised feature selection techniques, respectively. Unsupervised feature selection measures the feature capacity by keeping the intrinsic data structure in order to evaluate its relevance. Supervised feature selection consists in evaluating feature relevance by measuring the correlation between the feature and class labels. Supervised feature selection requires sufficient labelled data samples in order to provide a discriminating feature space. However, the sample labelling process by the human user is fastidious and expensive. That is the reason why, in many real applications, there are huge unlabelled data and small labelled samples. To deal with this 'lack of labelled-sample problem', recent semi-supervised feature selection schemes have been developed (Kalakech *et al.*, 2011). On the other hand, transformed features generated by feature extraction may provide a better discriminative ability than the best subset of given features, but these new features may not have a clear physical meaning. Feature extraction is important since the curse of high dimensionality is usually a major cause of limitations of many practical technologies, and the large quantities of features may even degrade the performances of classifiers when the size of the training set is small compared to the number of features (Yang *et al.*, 2011). Unsupervised feature extraction methods try to generate representative features from raw data, thus helping any classifier to learn a more robust solution and achieving a better generalization performance (Teixera *et al.*, 2011).

The goal of feature extraction is to project the original data into a space so that data with different characteristics are grouped away from each other. Principal component analysis (PCA) and independent component analysis (ICA) are unsupervised learning techniques that have attracted considerable attention during the past decade for this kind of projection. Many variants of techniques were developed as an extension based on these two strategies (Greene *et al.*, 2010). Full spectral techniques for dimensionality reduction, like PCA and kernel PCA (kPCA), perform an eigendecomposition of a full matrix that captures the covariances between dimensions or the pairwise similarities between data points (in kPCA the feature space is constructed by means of a kernel function). In PCA, the data are projected into a subspace that minimizes the reconstruction error in the mean squared sense. The principal components in the kernel-based equivalent are computed in a feature space which results from the implicit nonlinear mapping of the source data. Thus, kPCA can take into account a wider class of higher-order statistics (Fauvel *et al.*, 2009). Kernel PCA computes the kernel matrix k of the data points x_i, and subsequently centres k which corresponds to subtracting the mean of the features in traditional PCA. The entries in the kernel matrix are defined by (Van der Maaten *et al.*, 2009):

$$k_{ij} = \kappa\left(x_i, x_j\right) \qquad [4.2]$$

where κ is a kernel function. To obtain the low-dimensional representation, data are projected onto the eigenvectors of the covariance matrix α_i. The result of the projection is given by:

$$y_i = \left(\sum_{j=1}^{n} \alpha_1^j \kappa(x_j, x_i), ..., \sum_{j=1}^{n} \alpha_d^j \kappa(x_j, x_i) \right) \qquad [4.3]$$

where α_1^j indicates the jth value in vector α_1, d are the principal eigenvectors and κ is the kernel function that was also used in the computation of the kernel matrix. Therefore, PCA and its variants reduce a large set of correlated variables to a smaller number of uncorrelated components. However, PCA can de-correlate only the second-order dependencies. As the second-order statistics provide only partial information on the statistics of images, it might become necessary to incorporate higher-order statistics as well. ICA is a relatively recent technique that can be considered as a generalization of PCA. Its aim is to exploit the higher-order statistical structure in the data. ICA can linearly transform the observed variables into such components that are not only uncorrelated but also as statistically independent from each other as possible (Chen *et al.*, 2008). In the basic ICA model, the observed mixture signals are expressed by a linear combination of the latent sources with a mixing matrix. The ICA solution is to find a de-mixing matrix that yields statistically independent signals (Hoyer and Hyvarinen, 2000). For the basic conceptual framework of the ICA algorithm it is assumed that m

measured variables, $\tilde{x} = [x_1, x_2, ..., x_m]^T$ can be expressed as linear combinations of n unknown latent source components $\tilde{s} = [s_1, s_2, ..., s_n]^T$ (Lu et al., 2011):

$$\tilde{x} = A\tilde{s} \qquad [4.4]$$

where A is the unknown mixing matrix of size $m \times m$. Here, we assume $m \geq n$ for matrix A to be full rank matrix. The vector $\tilde{s}$ is the latent source data that cannot be directly obtained from the observed mixture data $\tilde{x}$. ICA aims to estimate the latent source components $\tilde{s}$ and unknown mixing matrix A from $\tilde{x}$ with appropriate assumptions on the statistical properties of the source distribution. Thus, an ICA model is intended to find a de-mixing matrix W of size $m \times m$ so that (Lu et al., 2011):

$$\tilde{f} = W\tilde{x} \qquad [4.5]$$

where $\tilde{f} = [f_1, f_2, ..., f_n]^T$ is the independent component vector. The elements of f must be statistically independent and are called independent components (ICs). The ICs are used to estimate the source components s_j. The ICA modelling is formulated as an optimization problem by setting up the measure of the independence of ICs as an objective function followed by using some optimization techniques for solving the de-mixing matrix W. The matrix W can be determined using an unsupervised learning algorithm with the objective of maximizing the statistical independence of ICs. The ICs with non-Gaussian distributions imply the statistical independence (Tsai and Lai, 2008).

While feature selection can be applied to both supervised and unsupervised learning, the focus is on the former case where the class labels are known beforehand. The topic of feature selection for unsupervised learning is a more complex issue, and research into this field is recently getting more attention. In any case, from the perspective of classifier design, a feature selection algorithm is part of the classification rule, where feature selection is subsequently followed by a standard classifier that is applied to the selected subset of features (Hua et al., 2009). The objectives of feature selection are to avoid overfitting and improve model performance, to provide faster and more cost-effective models and to gain a deeper insight into the underlying processes that generated the data (Jain and Zongker, 1997). However, the advantages of feature selection techniques come at a certain price, as the search for a subset of relevant features introduces an additional layer of complexity in the modelling task. Feature selection techniques can be organized into three categories (Sayes et al., 2007):

- *Filter techniques* (e.g., chi-square, correlation-based feature selection, etc.): these assess the relevance of features by looking only at the intrinsic properties of the data. In most cases a feature relevance score is calculated, and low-scoring features are removed. Afterwards, this subset of features is presented as input to the classification algorithm.

- *Wrapper methods* (e.g., sequential forward selection, genetic algorithms, etc.): these embed the model hypothesis search within the feature subset search. A search procedure in the space of possible feature subsets is defined, and various subsets of features are generated and evaluated. The evaluation of a specific subset of features is obtained by training and testing a specific classification model, rendering this approach tailored to a specific classification algorithm.
- *Embedded techniques* (e.g., decision trees, the weight vector of support vector machines, etc.): the search for an optimal subset of features is built into the classifier construction, and can be seen as a search in the combined space of feature subsets and hypotheses. Just like wrapper approaches, embedded approaches are thus specific to a given learning algorithm.

Chi-square (χ^2) is a common filter-based feature selection method, in which the features are evaluated individually by measuring their chi-squared statistic with respect to the classes. Chi-square is based on comparing the obtained values of the frequency of a class because of the split, to the expected frequency of the class. The chi-squared statistic of a feature is defined as (Liu and Setiono, 1995):

$$\chi^2 = \sum_{i=1}^{I} \sum_{j=1}^{k} \frac{\left(A_{ij} - E_{ij}\right)^2}{E_{ij}} \qquad [4.6]$$

where k is the number of classes, I is the number of intervals, A_{ij} is the number of patterns of the jth class within the ith interval. The expected frequency of A_{ij} is $E_{ij} = R_i \times C_j / N$, where R_i is the number of patterns in the ith interval $\sum_{j=1}^{k} A_{ij}$, C_j is the number of patterns in the jth class $\sum_{j=1}^{k} A_{ij}$ and N is the total number of patterns $\sum_{i=1}^{I} R_j$. In general, the larger the χ^2 value, the more informative the corresponding feature is (Jin *et al.*, 2006). A fast supervised wrapper-based (Zheng and Zhang, 2008) dimensionality reduction method is the linear discriminant analysis (LDA) and its variants (non-parametric, orthonormal, generalized). LDA finds the optimal linear transformation, which minimizes the within-class distance and maximizes the between-class distance simultaneously (Webb, 2002). LDA produces an optimally discriminative projection for certain cases. LDA searches for the transformation by maximizing the ratio of the between-class distance to the within-class distance. Unlike PCA, which has little to do with the class information, LDA takes much consideration of the label information of the data and it is generally believed that LDA is able to enhance class separability (Zhao *et al.*, 2006). A popular embedded-based feature selection algorithm is the SVM-RFE (support vector machine – recursive feature elimination) (Guyon *et al.*, 2002). It is an iterative algorithm, consisting of two steps; feature weights, obtained by training a linear SVM on the training set, are used in a scoring function for ranking features, and the feature with minimum rank is removed from the data. In this way, a chain of feature subsets of decreasing size is obtained. SVM classifiers are trained on sets restricted to the feature subsets, and the classifier with best predictive performance is selected (Jong *et al.*, 2004).

Application examples

A pure quaternion has a zero real part and a full quaternion has a non-zero real part. Thus, a quaternion number is a complex number with real and imaginary parts, hence the term 'hypercomplex' (Sangwine, 1996) is used. An RGB colour image of size (m,n) may be converted to a quaternion matrix by placing the three colour components into the three quaternionic imaginary parts, leaving the real part zero so that the image function $IM_q(m,n)$ is given by the following representation (Moxey *et al.*, 2003):

$$IM_a(m,n) = IM_R(m,n)i + IM_G(m,n)j + IM_B(m,n)k \qquad [4.7]$$

There are several kinds of image features for recognition such as visual, algebraic, statistical moments and transform coefficients. Algebraic features represent intrinsic attributions of the image, ergo various transforms or decompositions can be used to extract them (Hong, 1991). The quaternionic singular value decomposition (SVD) is a technique to decompose a quaternion matrix into several component matrices, exposing useful properties of the original matrix. SVD has been exploited generally in what is known as reduced-rank signal processing where the idea is to extract the significant parts of a signal (Sangwine and Le Bihan, 2006). Specifically, in quaternionic SVD, for any arbitrary $m \times n$ quaternion matrix IM_q, there exist two quaternion unitary matrices U and V and a diagonal Σ so that the following factorization exists in the following form (Le Bihan and Mars, 2004):

$$IM_q = U\Sigma V^T, \text{ with } \Sigma = \begin{pmatrix} \Sigma_r & 0 \\ 0 & 0 \end{pmatrix} \qquad [4.8]$$

where U denotes an $m \times m$ unitary quaternion matrix, Σ is an $m \times n$ diagonal matrix with non-negative real numbers on the diagonal and V^T is the conjugate transpose of an $m \times n$ unitary quaternion matrix V. The diagonal entries of Σ, called singular values Σ_r, can be arranged in order of decreasing magnitude and the columns of U and V are called left and right quaternion singular vectors for IM_q, respectively. Since the quaternionic matrix designates a mathematical representation of a colour image, the computed singular value feature vectors are unique for the colour image and can be used for pattern recognition purposes.

Figure 4.4 depicts a typical double log spectrum of averaged singular values computed from images acquired from four cooked pork ham qualities (Valous *et al.*, 2010b). Singular values of larger magnitude encapsulate most of the colour and textural information (Ramakrishnan and Selvan, 2006). In this sense, the information carried by singular values is explicit with the bigger ones having larger information capacity, whereas the rest just bring about smaller variation terms. In general, since the pork ham images and its quaternionic SVD have a unique corresponding relationship, the extracted singular values can be regarded as robust features of these images, as they are measures of their energy. Singular values provide the energy information of the image as well as the knowledge of how the energy is distributed. These features have very good stability and are more or less not invariant to systemic distortions (scaling, lightness variations, etc.) but proportionally sensitive to them (Pan *et al.*, 2003). These properties are very useful for describing and recognizing images. Table 4.3 shows the MATLAB (MathWorks, USA) source code for computing the quaternionic SVD of a colour image (Valous *et al.*, 2010b). The code requires the open-source quaternion toolbox (Sangwine andLe Bihan, 2005); the toolbox allows computations with quaternion matrices in almost the same way as with matrices of complex numbers. An algorithm based on the transformation of a quaternion matrix to bidiagonal form, using the quaternionic Householder transformations (Sangwine andLe Bihan, 2005), is the default SVD algorithm that has been implemented in the toolbox.

Kernel PCA, with the aid of the open-source toolbox for dimensionality reduction (v0.7.2), can be used next, as an unsupervised feature extraction technique (Debruyne *et al.*, 2010), for defining a mapping from a typically higher dimensional data space of singular value features to a space of reduced dimension, resulting in an uncorrelated dataset. This toolbox contains MATLAB implementations of a lot of techniques for dimensionality reduction, intrinsic dimensionality estimators, and additional techniques for data generation, out-of-sample extension and prewhitening (Van der Maaten *et al.*, 2009). Table 4.4 shows the MATLAB (MathWorks, USA) source code for computing the kernel principal components (Van der Maaten *et al.*, 2009) and the cumulated eigenvalues and total variance.

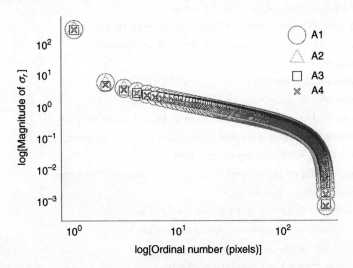

Fig. 4.4 Double log spectrum of averaged singular values computed from CCD images acquired from four cooked pork ham qualities, derived from silverside pork leg muscles (M. *biceps femoris*).

Table 4.3 MATLAB source code for computing the quaternionic singular value decomposition of a colour image, producing the singular value matrix and its Frobenius norm

```
% Requires the open-source quaternion toolbox
% Read the input colour RGB (3-channel) n x n x 3 pixels square image
I = imread ('input_colour_image.tif');
sz = size (I,2);
% Encode RGB image in the three imaginary parts of a quaternion
qI = convert (quaternion(I(:,:,1),I(:,:,2),I(:,:,3)),'double')./256;
% Compute the quaternionic singular value decomposition
[U S V] = svd (qI);
% S is the diagonal matrix of singular values with non-negative real numbers
for i = 1:sz
singular_values (i) = S(i, i);
end
% Returns the Frobenius norm of S
Frobenius = norm (S,'fro');
```

Table 4.4 MATLAB source code for computing the kernel principal components using the open-source toolbox for dimensionality reduction

```
% Requires the open-source toolbox for dimensionality reduction
% X: Input feature space of singular values
% no_dims: Number of kernel principal components (kPCs) to extract
no_dims = 20; % In this example, 20 kPCs will be extracted
ell = size (X,1);
K = gram (X,X,'gauss',1);
mapping.column_sums = sum(K)/ell; % column sums
mapping.total_sum = sum(mapping.column_sums) / ell; % total sum
J = ones(ell,1) * mapping.column_sums; % column sums (in matrix)
K = K – J – J'; K = K + mapping.total_sum;
% Compute first no_dims eigenvectors and store these in V, then store
% corresponding eigenvalues in L
K(isnan (K)) = 0; K(isinf (K)) = 0; [V,L] = eig (K);
% Sort eigenvalues and eigenvectors in descending order
[L, ind] = sort (diag(L), 'descend');
L = L (1:no_dims); V = V(:,ind (1:no_dims));
% Compute inverse of eigenvalues matrix L
invL = diag (1./ L);
```

```
% Compute square root of eigenvalues matrix L
sqrtL = diag (sqrt(L));
% Compute inverse of square root of eigenvalues matrix L
invsqrtL = diag (1./diag(sqrtL));
% Compute the new embedded points
mappedX = sqrtL * V';
% Set feature vectors in original format
mappedX = mappedX';
% Store information for out-of-sample extension
mapping.X = X; mapping.V = V; mapping.invsqrtL = invsqrtL;
% Cumulated eigenvalues and total variance
Eigenvalues = L; Total_variance = (L./sum(L))*100;
Cumul_eigenvalues = cumsum(L); Cumul_total_variance = cumsum(L./sum(L)*100)
```

4.3   Classification

The following sections investigate classification, once image segmentation and feature extraction and selection have taken place.

4.3.1   Description of techniques

In image pattern recognition, identifying, computing and extracting/selecting effective features is an important precursor step to successfully complete the task of classification. Classification analyses the numerical properties of various features and organizes data into categories (Du and Sun, 2007). Once these features have been found, the class of similar data can be labelled with a common label. Variations among different images and objects could be substantial, but yet only a small number of features are needed for classification (Zharkova and Jain, 2007). The separation of feature vectors into classes can be based upon so-called decision rules, established through the analysis of previous measurements during a training process. For the development of classifiers, two main aspects are considered: the basic assumptions that the classifier makes about the data (which results in a functional form of the classifier) and the optimization procedure to fit the model to the training data. It is possible to consider very complex classifiers, but without efficient methods to fit these classifiers to the data, they are not useful. Therefore, in many cases the functional form of the classifier is restricted by the optimization routines available (Duin and Tax, 2005). Pattern classification systems are not flawless and errors are expected due to several causes: (i) the features are inadequate or insufficient, (ii) the training samples used to design the classifier are not sufficiently representative, (iii) the classifier is not efficient enough in separating the classes and (iv) there is an intrinsic overlap of the classes that no classifier can resolve (Marques de Sá, 2001).

Machine learning algorithms are described as supervised or unsupervised. In supervised algorithms, the classes are predetermined and the task is to search for patterns and construct mathematical models. These models then are evaluated on the basis of their predictive capacity in relation to measures of variance in the data itself. Unsupervised algorithms seek out similarity between pieces of data in order to determine whether they can be characterized as forming a group; these groups are termed clusters (Theodoridis and Koutroumbas, 2009). In addition, numerous taxonomies for classification methods in pattern recognition have been presented in the literature (Alpaydin and Gürgen, 1998). The two most common approaches are statistical and neural classification. Statistical classifiers model the class-conditional densities and base their decisions on the posteriors which are computed using the class-conditional likelihoods and the priors. Likelihoods are assumed to either come from a given probability density family (e.g., normal), come from a mixture of such densities or be written in a completely non-parametric way (Mahmoud et al., 2004). Bayes decision theory then allows choosing the class that minimizes the decision risk. The parameters of the densities are estimated to maximize the likelihood of the given sample for that class. One major limitation of the statistical models is that they work well only when the underlying assumptions are satisfied. The efficiency of these methods depends to a large extent on the various assumptions or conditions under which the models are developed. Users must have a good knowledge of both data properties and model capabilities before the models can be successfully applied (Dehuri and Cho, 2010). A common unsupervised statistical classifier is the k-means algorithm. The technique finds a partition such that the squared error between the empirical mean of a cluster and the points in the cluster is minimized. The squared error between μ_k (mean of cluster c_k) and the points in cluster c_k is defined as (Jain, 2010):

$$J(c_k) = \sum_{x_i \in c_k} \left\| x_i - \mu_k \right\|^2$$

[4.9]

where $X = \{x_i\}$, $i = 1,...,n$ is the set of n d-dimensional points to be clustered into a set of K clusters, $C = \{c_k, k = 1,...,K\}$. The goal of k-means is to minimize the sum of the squared error over all K clusters (Jain, 2010):[4.10]

The algorithm starts with an initial partition with K clusters and assigns patterns to clusters so as to reduce the squared error. Since the squared error always decreases with an increase in the number of clusters K (with $J(C) = 0$ when $K = n$), it can be minimized only for a fixed number of clusters. The optimal number of clusters is not known a priori, and the clustering quality depends on the value of K specified by the user. The main steps of k-means algorithm are (Chicco et al., 2006):

- Step 1: select an initial partition with K clusters.
- Step 2: generate a new partition by assigning each pattern to its closest cluster centre.

- Step 3: compute new cluster centres.
- Repeat steps 2 and 3 until cluster membership stabilizes.

Statistical methods are in contrast with approaches where the discriminants are directly estimated. Neural classifiers are such approaches and their outputs can be converted directly to posteriors, eliminating the need to assume a statistical model (Kavzoglu and Mather, 2003). A multilayer network is a linear sum of nonlinear basis functions. In the neural network terminology, the nonlinear basis functions are called hidden units and the parameters are called connection weights. In a training process, given a training sample, the weights that minimize the difference between network outputs and required outputs are computed. Compared to most traditional classification approaches, neural classifiers are nonlinear, non-parametric and adaptive. They can theoretically approximate any fundamental relationship with arbitrary accuracy. They are ideally suitable for problems where observations are easy to obtain, but the data structure or underlying relationship is unknown (Lan et al., 2010). Table 4.5 summarizes the main statistical and neural pattern recognition techniques used for many practical problems in the area of computer vision (Meyer-Bäse, 2004).

Neural networks are online learning systems, intrinsically non-parametric and model-free, with the learning algorithms typically of the error correction type. On the other hand, most statistical classifiers can be interpreted to follow the Bayesian classification principle, either explicitly estimating the class densities and *a priori* probabilities, or the optimal discriminant functions directly by regression (Holmström et al., 1997). The performance of statistical and neural methods is usually discussed in connection with feature extraction/selection and classification. Efficient feature extraction is crucial for reliable classification and, if possible, these two subsystems should be matched optimally in the design of a complete pattern recognition system. Often, statistical methods are used as benchmarks for the performance evaluation of neural classifiers (Michalopoulou et al., 1995). The current understanding of biological neurons is that they communicate through pulses and employ the relative timing of the pulses to transmit information and perform computations. Spiking neural networks (SNNs), known as the third generation of artificial neural networks (ANNs), communicate by transmitting short transient spikes to other neurons, via weighted synaptic connections (Bohte et al., 2002). SNNs differ from traditional neural network models in that they utilize both timing and magnitude dynamics for the processing of neural information. SNNs are computationally more powerful than traditional threshold-based neuron models. This result has consequently encouraged interest in the area and their application to real-world classification problems (McGinley et al., 2010).

A more recent group of techniques, belonging to the fuzzy set methods, have achieved a high degree of success and popularity in many areas, such as control, multicriteria decision-making, approximate reasoning and pattern recognition (Keller et al., 1994). Fuzzy theory offers an efficient framework for handling the uncertainty encountered in image processing, and the diverse types of textures in computer vision problems. A fuzzy set consists of objects and their respective

Table 4.5   Important statistical and neural pattern recognition techniques used in computer vision

	Description
Statistical classifiers	
Maximum likelihood method	Estimates unknown parameters using a set of known feature vectors in each class.
Bayes method	Assumes a known *a priori* probability and minimizes the classification error probability. The class-conditional probability density functions describing the distribution of the feature vectors in each class must be known, otherwise they are estimated from the available training data.
Minimum distance classifier	Classifies an input vector based on its distance to the learned prototypes.
Entropy criteria	Classifies an input vector based on the minimization of the randomness of this vector.
Isodata or k-means or c-means clustering	The goal is to achieve a close partitioning of the data space. The number of classes and the initial values of the class centers need to be known *a priori*. Learning is an iterative process which adapts the class centers according to the training data.
Vector quantization	The input data are mapped onto a given number of code vectors (prototypes) which together form the code book.
Cluster swapping	It is applied mostly to a large data space, aiming to avoid a suboptimal class partitioning. The efficiency of a swap-based clustering algorithm depends on two factors: number of iterations (swaps) needed, and the time each iteration consumes.
Hierarchical clustering	Produces instead of a single clustering, a hierarchy of nested clustering. These algorithms have as many steps as the number of data vectors. At each step a new clustering is obtained based on the clustering produced at the previous step.
Neural classifiers	
Recurrent networks	They are nonlinear, fully interconnected systems and form an auto-associative memory. The neural network is trained via a storage prescription that forces stable states to correspond to local minima of a network 'energy' function. The Hopfield neural network is the most popular example of this type.
Multilayer feedforward neural networks	They implement a nonlinear mapping between an input and output space and are nonlinear function approximators, extracting higher-order statistics. The multilayer perceptron and the radial-basis neural network are the best known examples.
Local interaction-based neural networks	These networks are based on competitive learning; the output neurons of the network compete among themselves to be activated. The output that wins the competition is called the winning neuron. The local interaction is found in Kohonen maps, ART maps, and in the von der Malsburg model.

grades of membership in the set. The grade of membership of an object in the fuzzy set is given by a subjectively defined membership function. The value of the grade of membership of an object can range from 0 to 1, where the value of 1 denotes full membership, and the closer the value is to 0, the weaker the object's membership in the fuzzy set is (Friedman and Kandel, 1999). This theory provides an approximate and yet effective means for describing the characteristics of a system that is too complex or ill-defined to admit precise mathematical analysis. It is seen that the concept of fuzzy sets can be used at the feature level and at the classification level, for representing class membership of objects, and for providing an estimate (or representation) of missing information in terms of membership values. Fuzzy set theory has been extensively used in clustering problems where the task is to provide class labels to input data (partitioning of feature space) under unsupervised mode based on certain criterion. Applications of fuzzy pattern recognition and image processing have been reported in various domains (Mitra and Pal, 2005). The main difference between fuzzy and neural paradigms is that fuzzy set theory tries to mimic the human reasoning and thought process, whereas neural networks attempt to emulate the architecture and information representation scheme of the human brain (Meyer-Bäse, 2004). Integrating these two distinct paradigms by enhancing their individual capabilities can build a more intelligent processing system. This new processing paradigm of neuro-fuzzy computing – compared to standard neural networks or simple fuzzy classifiers – can be a more powerful computational paradigm (Mitrakis, *et al.*, 2008).

4.3.2   Application examples

As reviewed earlier, neural networks can be viewed as powerful, fault tolerant and reasonable alternatives to traditional classifiers (Kulkarni, 2001). Feedforward neural networks are especially attractive for supervised classification applications. On a more pragmatic level, numerous comparative studies have demonstrated that such networks are often, but not always, able to classify imagery more accurately than a variety of other widely used approaches (Yang *et al.*, 1999). The multilayer perceptron (MLP) network trained with a backpropagation or related learning algorithm has been frequently used for image classification, although in some cases the approach is not free from problems (Foody, 2004). MLP networks with one hidden layer and sigmoidal hidden layer activation functions are capable of approximating any decision boundary to arbitrary accuracy (Li *et al.*, 1999). MLPs are used due to their popularity and enhanced ability for generalization, which is related to the accurate prediction of data that are not part of the training dataset. Figure 4.5 shows a schematic of the MLP neural network architecture used in the construction of a supervised parsimonious classification model of binary lacunarity data from pork ham slice surface images (three classes), using a portion of salient features (Iqbal *et al.*, 2011).

To find the optimal ANN architecture for the classification, preliminary trial and error tests were carried out to determine the number of hidden neurons. The capacity of a feedforward neural network in learning the samples of the training

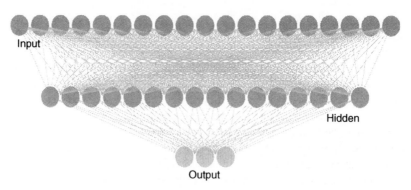

Fig. 4.5   Schematic of a multilayer perceptron classifier architecture of binary lacunarity data from pork ham slice surface images, consisting of an input layer of 19 salient features, one hidden layer of 16 nodes and an output layer of 3 categorical levels.

set is proportional to its complexity (Huang, 2003). Clearly, in some cases, more robust computational approaches on estimating the number of hidden layer nodes are better suited for the analysis (Teoh *et al.*, 2006). It is common practice to increase the number of nodes of the hidden layers or add more hidden layers to improve learning, but experiments showed that the speed of convergence is frequently made worse by the addition of extra nodes or more hidden layers (Li *et al.*, 1999). After a certain threshold of neurons has been reached, increasing their number beyond that threshold has a marginal effect on the resulting performance of the classifier (Bishop, 1995). The wider the hidden layer, the greater the capacity of the network to recognize patterns. This greater capacity has a drawback because the neural network can end up memorizing the training set instead of generalizing from it. To achieve generalization, the hidden layer should not be too wide. One rule of thumb is that it is not a good idea to use a hidden layer larger than the number of inputs (Berry and Linoff, 2004). In general, the number of hidden nodes determines the total number of weights in the network and thus, as a practical tip, there should not be more weights than the total number of training points in the dataset (Duda *et al.*, 2001). However, such a rule is generic and concerned with overfitting, in addition to not applying when regularization (one of the simplest forms of regularizer is called weight decay) is used (Long *et al.*, 2009); consequently more hidden nodes can be used. Weight decay was introduced into learning to improve the generalization capability of a model expressed as a linear combination of a set of given basis functions. Loosely speaking, the method penalizes large values of the coefficients (weights) of the linear combination. It can be modelled by adding to the empirical error functional, a term given by a squared norm (typically, the Euclidean norm) of the coefficient vector (Gnecco and Sanguineti, 2009). Weight decay training is advocated as an effective tool to learn from noisy data, and improves the performance achieved by standard backpropagation (Gupta and Lam, 1998). A network with a relatively smaller number of hidden nodes can represent only a limited variety of target functions (Neal,

1996). Hence, for a smaller training set, it is usually advisable to use regularization rather than to restrict the architecture to a small number of hidden nodes.

For the example of Figure 4.5, an adaptive supervised feedforward multilayer perceptron classifier, using a variant of the Quasi-Newton method; namely the BFGS/BP (Broyden Fletcher Goldfarb Shanno/Back Propagation) learning algorithm, was employed to obtain a suitable mapping from the input dataset. This algorithm performs better in terms of training speed and accuracy (Hui *et al.*, 1997) and requires more computation in each iteration, but has a faster convergence rate (Demuth *et al.*, 2009). The overall correct classification performance for the training, test and validation set were 86.7%, 86.7% and 85.0%, respectively.

4.4 Relevance, impact and trends for the food and beverage industry

We now consider precisely what all this means for the food and beverage industry, looking at current practical applications, possible implications and what the future looks likely to bring.

4.4.1   Introduction and relevance

Food and beverage products are produced in a large scale and consumed worldwide. The control of their safety/quality is essential to a company's survival in a competitive marketplace due to the effects of customer buying decisions and confidence (Gunasekaran, 1996). Within the manufacturing sector, the control of raw materials, together with in-process and endpoint inspection, play a significant role in the mass production of numerous food products. In this sense, digital image processing tools and methodologies are important non-destructive techniques that can help in expediting and optimizing product inspections, resulting in better quality products (Gill *et al.*, 2011). A big advantage for the industry is that computer vision is a robust and competitively priced sensing technique (Brosnan and Sun, 2004). In addition, the need to measure the physical properties of foods also arises from the increased regulatory action and heightened consumer concerns about food safety.

In recent years, the computer/machine vision market is estimated to have tripled in size, and projected future growth within existing and developing markets, over the next five years, has been estimated in double figures (Brosnan and Sun, 2002). Many food and beverage companies have yet to benefit from new opportunities offered through the introduction of imaging equipment and related techniques, either in established well-defined inspection tasks or newly evolving applications; however, there is increasing evidence of computer vision systems being adopted at the commercial level (Brosnan and Sun, 2002). In many cases, characterization and classification is still undertaken manually by skilled staff (Abdullah *et al.*, 2004). The variability associated with human assessment pertinent to automated inspections tasks, accentuates the need for objective measurement systems to provide reliable information throughout the production process

(Damez and Clerjon, 2008). This relative absence of automation may be attributed to existing methodologies, which are often unable to cope with a wide variety of products, yet a continued reduction in processor and memory costs would suggest that automated procedures have potential as cost-effective alternatives. The trend towards continuous automated production necessitates the measurement of properties and attributes, particularly in the area related to online or rapid at-line process control applications. The increased use of computer vision leads to a better understanding of the materials and the processes involved (Reh, 2008), which result in food products that are safer and improved in quality. The wider application of automated computer vision protocols would seem to offer a number of potential advantages, including improved safety, reduced labour costs, the elimination of subjective judgement and the creation of timely statistical product data (Smith, 2000).

4.4.2   Impact and practical examples

The sophistication of non-destructive methods has evolved rapidly with modern technologies. High-speed computers with high-capacity memory and high-speed data transfer can provide real-time evaluation and control in many applications. Similarly, high resolution sensors have become smaller, more sensitive and more robust (Mix, 2005). Computer vision is important to the modern manufacturing process, because the quality evaluation and grading requirements have moved beyond the limits of human inspection. In the processing plants, the manufactured items are often produced too quickly and with product tolerances too small to be analysed by human inspection protocols. Moreover, methods for assessing the quality of food and beverage products, either by sensory analysis or by instrumental techniques, are frequently destructive and time-consuming, and, therefore, they do not fit the conditions for a routine online analysis in an industrial environment. The industry is interested in rapid and non-destructive procedures. In this sense, digital image processing techniques are a non-destructive tool that has interesting applications for the overall characterization and quality inspection of food products (Chen and Sun, 1991). Computer vision is capable of extracting various features such as colour, image texture, shape and size, among others. These simple appearance features allow task-relevant analysis and interpretation with precision, objectivity and speed. In fact, these appearance properties may correlate well with many physical, chemical and sensorial indicators of food quality. These quality characteristics are also linked to features that can be explicitly measurable by non-destructive frameworks (Becker, 2002). Since quality characteristics are related to physico-chemical properties, many image processing methods have been developed, contributing significantly to industrial requirements for automated inspection and grading. As a practical example, Table 4.6 summarizes a subset of conventional and advanced image processing methods used in computer vision applications regarding the quality evaluation and control of food and beverage products, sorted by product type and publication year.

Table 4.6   Image processing methods used in computer vision applications for food and beverage product quality evaluation and control, sorted by product type and publication year

Products	Image processing methods	Reference
Mayonnaise	Computation of stereological microstructural parameters for estimating the droplet size, the interfacial surface area between the fat phase and water phase, and the size of the egg yolk aggregates.	Langton *et al.* (1999)
Apple	Region-based segmentation of surface defects and identification of stem-end and calyx areas using fractal features and artificial neural networks.	Li *et al.* (2002)
Tea	Computation of wavelet transform based image textural features for class sorting using multi-layer perceptron and learning vector quantization algorithms.	Borah *et al.* (2007)
Cheese	Correlation of colour measurements from CCD sensors with gravimetric measurements for automated inline monitoring of curd syneresis.	Everard *et al.* (2007)
Beef	Computation of colour and textural (wavelets) features to provide better information on beef eating quality, from correlations with quality measurements using partial least squares regression.	Jackman *et al.* (2008)
Pizza	Computation of effective shape and colour features for the multi-class categorization of pizza base, sauce spread and topping, using support vector machines.	Du and Sun (2008)
Wine	Classification based on features discretized by fuzzy c-means cluster analysis, using a backpropagation neural network.	Sun *et al.* (2008)
Bread	Colour and size measurements of crust and crumb using CCD sensors for analyzing physico-chemical effects occurring during the baking process.	Jusoh *et al.* (2009)
Grain	Orthogonal dimension measurements for the development of an automated computerized protocol.	Igathinathane *et al.* (2009)
Hazelnut	Kernel peeling evaluation using histogram thresholding and k-nearest neighbours clustering, for automated inspections.	Pallottino *et al.* (2010)

(Continued)

Table 4.6   (*Continued*)

Products	Image processing methods	Reference
Salmon	Correlation of colour measurements from CCD sensors with sensory panel evaluations for increased automation.	Quevedo *et al.* (2010)
Cooked ham	Quaternionic singular value decomposition of digital colour images for extraction of informative and uncorrelated singular values, as robust and stable features, used in neural network classification.	Valous *et al.* (2010b)
Chickpea	Computation of size, colour and morphological features for neural network classification, and correlations with trained inspectors.	Çakmak and Boyaci (2011)

4.4.3   Recent trends in accelerating image processing operations

An important aspect that can be considered in digital image processing is the trend towards higher resolutions and faster feature processing, in which the amount of data that need to be processed in a short amount of time or in real-time have increased dramatically. The key to cope with this issue is the concept of parallel processing, exploiting different forms of parallelism, which can be either data-level parallelism and/or instruction-level parallelism (Kehtarnavaz and Gamadia, 2006). A great deal of growth in the field of image processing is primarily due to the ever-increasing performance, due to optimizations in digital signal processors (DSP), field programmable gate arrays (FPGA) and multicore central processing units (CPU). Quite recently, an exciting new development in the field of high performance commodity computing is the emergence of the programmable graphics processing unit (GPU) (Babenko and Shah, 2008). A GPU can deliver peak performance in the order of 360 Gflops, more than seven times the performance of the fastest computer processor (Setoain *et al.*, 2008). Although the GPU architecture is not necessarily suitable for all kinds of parallel computations, the range of candidate applications is growing (Montrym and Moreton, 2005). Unfortunately, the level of effort and expertise required to maximize application performance on these kinds of systems is still quite high (Ryoo *et al.*, 2008), although software libraries (Allusse *et al.*, 2008) and commercial codes (Accelereyes, 2011) could facilitate the implementation of real-time capabilities in quality evaluation and control.

4.5   Conclusions

Owing to the imperfections of image acquisition systems, image pre-processing steps such as filtering and histogram manipulation are performed to remove noise, and to enhance contrast for the purpose of facilitating image segmentation. Although a large number of image segmentation techniques have been developed to date, no universal method can perform with ideal efficiency and accuracy across the wide spectrum of potential applications. Therefore, it is expected that a combination of several techniques will be needed in order to improve segmentation results and increase the adaptability of the methods. Likewise, there is no single best feature selection or extraction method. The proper choice depends on a number of issues, such as computational gains and particular implementation requirements. Given the proliferation of classification techniques, many of these have shown feasibility for the classification of food products. Unsurprisingly, it is not an easy task to select an optimal method that can be applied to different cases. It is impossible to advocate one technique as a general solution because each classification technique has its own strengths and weaknesses and is suitable for particular kinds of problems. As a result, one of the most interesting fields for further application is to combine several techniques, also known as classifier fusion. Indeed, there is a great interest in further researching and developing image analysis methods that have the potential to be robust, efficient and cost-competitive. The foremost underlying drivers for using image processing technologies are automation and improved rapid operations. In the food and beverage industry, quality evaluation and control still depend to a certain degree on manual inspection, which is tedious, laborious and costly, and is easily influenced by physiological factors inducing subjective and inconsistent evaluation results. Quality is a key factor for the modern food industry because high product quality is the basis for success in today's highly competitive market. To satisfy increased awareness, sophistication and greater expectation of consumers, it is necessary to improve quality control methodologies. Computer/machine vision, coupled with effective digital image processing techniques, is a proven technology that can provide useful information regarding product quality and the effects of the processing regime. Digital imaging technologies continue to change at a rapid pace, and the image processing techniques that have been developed can mature into industrial applications with the right integration framework.

4.6   References

Abdullah, M. Z., Guan, L. C., Lim, K. C., and Karim, A. A. (2004) The applications of computer vision system and tomographic radar imaging for assessing physical properties of food. *Journal of Food Engineering*, **61(1)**, 125–35.

Accelereyes (2011) Getting started guide. *Jacket GPU Computing with MATLAB*. <http://www.accelereyes.com/content/doc/GettingStartedGuide.pdf>. Accessed 19 May 2011, 1–98.

Allusse, Y., Horain, P., Agarwal, A. and Saipriyadarshan, C. (2008) GpuCV: a GPU-accelerated framework for image processing and computer vision. *Lecture Notes in Computer Science*, **5359**, 430–9.

Alpaydin, E. and Gürgen, F. (1998). Introduction to pattern recognition and classification in medical and astrophysical images. In C. T. Leondes (ed.), *Image Processing and Pattern Recognition*. San Diego: Academic Press, 61–87.

Babenko, P. and Shah, M. (2008) MinGPU: a minimum GPU library for computer vision. *Journal of Real-Time Image Processing*, **3(4)**, 255–68.

Becker, T. (2002) Defining meat quality. In J. Kerry, J. Kerry and D. Ledward (eds), *Meat Processing: Improving Quality*. Cambridge: Woodhead, 3–24.

Berry, M. J. A. and Linoff, G. S. (2004) *Data Mining Techniques for Marketing, Sales and Customer Relationship Management, 2nd edition*. Indiana: Wiley, 211–56.

Bishop, C. M. (1995) *Neural Networks for Pattern Recognition*. Oxford: Clarendon Press, 116–63.

Bong, C.-W. and Rajeswari, M. (2011) Multi-objective nature-inspired clustering and classification techniques for image segmentation. *Applied Soft Computing*, **11(4)**, 3271–82.

Bohte, S., Kok, J. and La Poutre, H. (2002) Error-backpropagation in temporally encoded networks of spiking neurons. *Neurocomputing*, **48(1–4)**, 17–37.

Borah, S., Hines, E. L. and Bhuyan, M. (2007) Wavelet transform based image texture analysis for size estimation applied to the sorting of tea granules. *Journal of Food Engineering*, **79(2)**, 629–39.

Brink, A. D. (1992) Thresholding of digital images using two-dimensional entropies. *Pattern Recognition*, **25(8)**, 803–8.

Brosnan, T. and Sun, D.-W. (2002) Inspection and grading of agricultural and food products by computer vision systems – a review. *Computers and Electronics in Agriculture*, **26(2–3)**, 193–213.

Brosnan, T. and Sun, D.-W. (2004) Improving quality inspection of food products by computer vision – a review. *Journal of Food Engineering*, **61(1)**, 3–16.

Chatterjee, S. and Bhattacherjee, A. (2011) Genetic algorithms for feature selection of image analysis-based quality monitoring model: an application to an iron mine. *Engineering Applications of Artificial Intelligence*, **24(5)**, 786–95.

Çakmak, Y. S. and Boyaci, I. H. (2011) Quality evaluation of chickpeas using an artificial neural network integrated computer vision system. *International Journal of Food Science & Technology*, **46(1)**, 194–200.

Chen, P. and Sun, Z. (1991) A review of non-destructive methods for quality evaluation and sorting of agricultural products. *Journal of Agricultural Engineering Research*, **49**, 85–98.

Chen, Y., Fu, Z. and Han, Y. (2008) Independent component analysis of Gabor features for texture classification. *Optical Engineering*, **47(12)**, 127003.

Chen, Z. and Wang, Q. (2010) Research of PCB image segmentation based on color features. *Proceedings of the 3rd International Conference on Advanced Computer Theory and Engineering*. Chengdu, China: 20–22 August, V2-543–V2-545.

Cheng, H. D., Chen, Y. H. and Jiang, X. H. (2000) Thresholding using two-dimensional histogram and fuzzy entropy principle. *IEEE Transactions on Image Processing*, **9(4)**, 732–5.

Chicco, G., Napoli, R. and Piglione, F. (2006) Comparisons among clustering techniques for electricity customer classification. *IEEE Transactions on Power Systems*, **21(2)**, 933–40.

Damez, J. L. and Clerjon, S. (2008) Meat quality assessment using biophysical methods related to meat structure. *Meat Science*, **80(1)**, 132–49.

Debruyne, M., Hubert, M. and Van Horebeek, J. (2010) Detecting influential observations in kernel PCA. *Computational Statistics and Data Analysis*, **54(12)**, 3007–19.

Dehuri, S. and Cho, S.-B. (2010) A hybrid genetic based functional link artificial neural network with a statistical comparison of classifiers over multiple datasets. *Neural Computing and Applications*, **19(2)**, 317–28.

Demuth, H., Beale, M. and Hagan, M. (2009) *Neural Network MATLAB Toolbox 6: User's Guide*. Natick: The MathWorks Inc.

Dhawan, P. A. (1990) A review on biomedical image processing and future trends. *Computer Methods and Programs in Biomedicine*, **31(3–4)**, 141–83.

Dougherty, G. (2009) *Digital Image Processing for Medical Applications*. Cambridge: Cambridge University Press, 3–15.

Du, C.-J. and Sun, D.-W. (2004) Recent developments in the applications of image processing techniques for food quality evaluation. *Trends in Food Science and Technology*, **15(5)**, 230–49.

Du, C.-J. and Sun, D.-W. (2006) Automatic measurement of pores and porosity in pork ham and their correlations with processing time water content and texture. *Meat Science*, **72(2)**, 294–302.

Du, C.-J. and Sun, D.-W. (2007) Object classification methods. In D.-W. Sun (ed.), *Computer Vision Technology for Food Quality Evaluation*. Burlington: Academic Press, 81–107.

Du, C.-J. and Sun, D.-W. (2008) Multi-classification of pizza using computer vision and support vector machine. *Journal of Food Engineering*, **86(2)**, 234–42.

Duda, R. O., Hart, P. E. and Stork, D. G. (2001) *Pattern classification, 2nd edition*. New York: John Wiley and Sons, 1–349.

Duin, R. P. W. and Tax, D. M. J. (2005) Statistical pattern recognition. In C. H. Chen and P. S. P. Wang (eds), *Handbook of Pattern Recognition and Computer Vision*. Singapore: World Scientific, 3–24.

Everard, C. D., O'Callaghan, D. J., Fagan, C. C., O'Donnell, C. P., Castillo, M. and Payne, F. A. (2007) Computer vision and color measurement techniques for inline monitoring of cheese curd syneresis. *Journal of Dairy Science*, **90(7)**, 3162–70.

Fauvel, M., Chanussot, J. and Benediktsson, A. (2009) Kernel principal component analysis for the classification of hyperspectral remote sensing data over urban areas. *EURASIP Journal on Advances in Signal Processing*, 14 pp., doi:10.1155/2009/783194.

Foody, G. M. (2004) Supervised image classification by MLP and RBF neural networks with and without an exhaustively defined set of classes. *International Journal of Remote Sensing*, **25(15)**, 3091–104.

Friedman, M. and Kandel, A. (1999) *Introduction to Pattern Recognition: Statistical, Structural, Neural and Fuzzy Logic Approaches*. London: Imperial College Press, 167–226.

Gauch, J. M. (1998) Noise removal and contrast enhancement. In S. J. Sangwine and R. E. N. Horne (eds), *The Colour Image Processing Handbook*. London: Chapman and Hall, 151–62.

Gill, G. S., Kumar, A. and Agarwal, R. (2011) Monitoring and grading of tea by computer vision – a review. *Journal of Food Engineering*, **106(1)**, 13–19.

Gnecco G. and Sanguineti, M. (2009) The weight-decay technique in learning from data: an optimization point of view. *Computational Management Science*, **6(1)**, 53–79.

Greene, W. N., Zhang, Y., Lu, T. T. and Chao, T.-H. (2010) Feature extraction and selection strategies for automated target recognition. *Proceedings of SPIE: The International Society for Optical Engineering*, 7703, 77030B.

Gunasekaran, S. (1996) Computer vision technology for food quality assurance. *Trends in Food Science & Technology*, **7(8)**, 245–56.

Gupta, A. and Lam, S. M. (1998) Weight decay backpropagation for noisy data. *Neural Networks*, **11(6)**, 1127–38.

Guyon, I., Weston, J., Barnhill S. and Vapnik, V. (2002) Gene selection for cancer classification using support vector machines. *Machine Learning*, **46(1–3)**, 389–422.

Guyon, I. and Elisseeff, A. (2003) An introduction to variable and feature selection. *Journal of Machine Learning Research*, **3(3)**, 1157–82.

Holmström, L., Koistinen, P. and Laaksonen, J. (1997) Neural and statistical classifiers – taxonomy and two case studies. *IEEE Transactions on Neural Networks*, **8(1)**, 5–17.

Hong, Z.-Q. (1991) Algebraic features extraction of image for recognition. *Pattern Recognition*, **24(3)**, 211–19.

Hoyer, P. O. and Hyvarinen, A. (2000) Independent component analysis applied to feature extraction from colour and stereo images. *Network: Computation in Neural Systems*, **11(3)**, 191–210.

Hua, J., Tembe, W. D. and Dougherty, E. R. (2009) Performance of feature selection methods in the classification of high-dimension data. *Pattern Recognition*, **42(3)**, 409–24.

Huang, G. (2003) Learning capability and storage capacity of two-hidden layer feedforward networks. *IEEE Transactions on Neural Networks*, **14(2)**, 274–81.

Hui, L. C. K., Lam, K.-Y. and Chea, C. W. (1997) Global optimization in neural network training. *Neural Computing and Applications*, **5(1)**, 58–64.

Igathinathane, C., Pordesimo, L. O. and Batchelor, W. D. (2009) Major orthogonal dimensions measurement of food grains by machine vision using ImageJ. *Food Research International*, **42(1)**, 76–84.

Iqbal, A., Valous, N. A., Sun, D.-W. and Allen, P. (2011) Parsimonious classification of binary lacunarity data computed from food surface images using kernel principal component analysis and artificial neural networks. *Meat Science*, **87(2)**, 107–14.

Jackman, P., Sun, D.-W., Du, C.-J., Allen, P. and Downey, G. (2008) Prediction of beef eating quality from colour, marbling and wavelet texture features. *Meat Science*, **80(4)**, 1273–81.

Jähne, B. (2005) *Digital Image Processing*. Berlin: Springer-Verlag, 3–30.

Jähne, B. and Haußecker, H. (2000) *Computer Vision and Applications: A Guide for Students and Practitioners*, San Diego: Academic Press, 1–10.

Jain, A. K. (2010) Data clustering: 50 years beyond *k*-means. *Pattern Recognition Letters*, **31(8)**, 651–66.

Jain, A. and Zongker, D. (1997) Feature selection: evaluation, application, and small sample performance. *IEEE Transactions on Pattern Analysis and Machine Intelligence*, **19(2)**, 153–8.

Jin, X., Xu, A., Bie, R. and Guo, P. (2006) Machine learning techniques and chi-square feature selection for cancer classification using SAGE gene expression profiles. In J. Li, Q. Yang and A.-H. Tan (eds), *Data Mining for Biomedical Applications*, PAKDD 2006 Workshop, BioDM 2006. Singapore: Lecture Notes in Computer Science, 3916, Springer, 106–15.

Jong, K., Marchiori, E., Sebag, M. and Van der Vaart, A. (2004) Feature selection in proteomic pattern data with support vector machines. *Proceedings of the IEEE Symposium on Computational Intelligence in Bioinformatics and Computational Biology*. La Jolla, USA: 7–8 October, 41–8.

Jusoh, Y. M. M., Chin, N. L., Yusof, Y. A. and Rahman, R. A. (2009) Bread crust thickness measurement using digital imaging and L*a*b* colour system. *Journal of Food Engineering*, **94(3–4)**, 366–71.

Kalakech, M., Biela, P., Macaire, L. and Hamad, D. (2011) Constraint scores for semi-supervised feature selection: a comparative study. *Pattern Recognition Letters*, **32(5)**, 656–65.

Kavzoglu, T. and Mather, P. M. (2003) The use of backpropagating artificial neural networks in land cover classification. *International Journal of Remote Sensing*, **24(23)**, 4907–38.

Kehtarnavaz, N. and Gamadia, M. (2006) *Real-Time Image and Video Processing: From Research to Reality*. San Rafael: Morgan & Claypool, pp. 1–55.

Keller, J. M., Gader, P., Tahani, H., Chiang, J.-H. and Mohamed, M. (1994) Advances in fuzzy integration for pattern recognition. *Fuzzy Sets and Systems*, **65(2–3)**, 273–83.

Kulkarni, A. D. (2001) *Computer Vision and Fuzzy-Neural Systems*. New Jersey: Prentice-Hall PTR, 227–280.

Langton, M., Jordansson, E., Altskär, A., Sørensen, C. and Hermansson, A.-M. (1999) Microstructure and image analysis of mayonnaises. *Food Hydrocolloids*, **13(2)**, 113–25.

Lan, J., Hu, M. Y., Patuwo, E. and Zhang, G. P. (2010) An investigation of neural network classifiers with unequal misclassification costs and group sizes. *Decision Support Systems*, **48(4)**, 581–91.

Le Bihan, N. and Mars, J. (2004) Singular value decomposition of quaternion matrices: a new tool for vector-sensor signal processing. *Signal Processing*, **84(7)**, 1177–99.

Li, Y., Rad, A. B. and Peng, W. (1999) An enhanced training algorithm for multilayer neural networks based on reference output of hidden layer. *Neural Computing and Applications*, **8(3)**, 218–25.

Li, Q., Wang, M. and Gu, W. (2002) Computer vision based system for apple surface defect detection. *Computers and Electronics in Agriculture*, **36(2–3)**, 215–23.

Li, Z., Liu, C., Liu, G., Yang, X. and Cheng, Y. (2011) Statistical thresholding method for infrared images. *Pattern Analysis & Applications*, **14(2)**, 109–26.

Liu, H. and Setiono, R. (1995) Chi2: feature selection and discretization of numeric attributes. *Proceedings of IEEE 7th International Conference on Tools with Artificial Intelligence*. Herndon, USA: 5–8 November, 338–391.

Long, N., Gianola, D., Rosa, G. J. M., Weigel, K. A. and Avendaño, S. (2009) Comparison of classification methods for detecting associations between SNPs and chick mortality. *Genetics Selection Evolution*, **41**, 18, doi: 10.1186/1297–9686-41-18.

Lu, C.-J., Shao, Y. E. and Li, P.-H. (2011) Mixture control chart patterns recognition using independent component analysis and support vector machine. *Neurocomputing*, **74(11)**, 1908–14.

Mahmoud, S., El-Melegy, M. T., and Farag, A. A. (2004). A comparative study of statistical and neural methods for remote-sensing image classification and decision fusion. *Proceedings of the 2004 International Conference on Image Processing* (vol. 5, 3347–50). Singapore: 24–27 October.

Malik, J., Belongie, S., Leung, T. and Shi, J. (2001) Contour and texture analysis for image segmentation. *International Journal of Computer Vision*, **43(1)**, 7–27.

Marques de Sá, J. P. (2001) *Pattern Recognition: Concepts, Methods and Applications*. Berlin: Springer, 1–20.

McGinley, B., O'Halloran, M., Conceição, R. C., Morgan, F., Glavin, M. and Jones, E. (2010) Spiking neural networks for breast cancer classification using radar target signatures. *Progress in Electromagnetics Research C*, **17**, 79–94.

Meyer-Bäse, A. (2004) *Pattern Recognition for Medical Imaging*. San Diego: Elsevier, 136–317.

Michalopoulou, Z.-H., Alexandrou, D. and de Moustier, C. (1995) Application of neural and statistical classifiers to the problem of seafloor characterization. *IEEE Journal of Oceanic Engineering*, **20(3)**, 190–7.

Mitra, S. and Pal, S. K. (2005) Fuzzy sets in pattern recognition and machine intelligence. *Fuzzy Sets and Systems*, **156(3)**, 381–6.

Mitrakis, N. E., Topaloglou, C. A., Alexandridis, T. K., Theocharis, J. B. and Zalidis, G. C. (2008) Decision fusion of GA self-organizing neuro-fuzzy multilayered classifiers for land cover classification using textural and spectral features. *IEEE Transactions on Geoscience and Remote Sensing*, **46(7)**, 2137–51.

Mix, P. E. (2005) *Introduction to Non-Destructive Testing – A Training Guide, 2nd edition*. New Jersey: Wiley Interscience, 1–14.

Montrym, J. and Moreton, H. (2005) The GeForce 6800. *IEEE Micro*, **25(2)**, 41–51.

Moxey, C. E., Sangwine, S. J. and Ell, T. A. (2003) Hypercomplex correlation techniques for vector images. *IEEE Transactions on Signal Processing*, **51(7)**, 1941–53.

Neal, R. M. (1996) *Bayesian Learning for Neural Networks*. New York: Springer, 29–54.

Pallottino, F., Menesatti, P., Costa, C., Paglia, G., De Salvador, F. R. and Lolletti, D. (2010) Image analysis techniques for automated hazelnut peeling determination. *Food and Bioprocess Technology*, **3(1)**, 155–9.

Pan, Q., Zhang, M.-G., Zhou, D.-L., Cheng, Y.-M. and Zhang, H.-C. (2003) Face recognition based on singular-value feature vectors. *Optical Engineering*, **42(8)**, 2368–74.

Pudil, P., Novovičová, J. and Somol, P. (2002) Feature selection toolbox software package. *Pattern Recognition Letters*, **35(12)**, 487–92.

Quevedo, R. A., Aguilera, J. M. and Pedreschi, F. (2010) Color of salmon fillets by computer vision and sensory panel. *Food and Bioprocess Technology*, **3(5)**, 637–43.

Ramakrishnan, S. and Selvan, S. (2006) Image texture classification using exponential curve fitting of wavelet domain singular values. *Proceedings of the IET International Conference on Visual Information Engineering.* Bangalore, India: 26–28 September, 505–10.

Reh, C. (2008) An overview of sensor technology in practice: the user's view. In J. Irudayaraj and C. Reh (eds), *Non-destructive Testing of Food Quality.* Ames: Wiley-Blackwell, 1–31.

Rogowska, J. (2009) Overview and fundamentals of medical image segmentation. In I. H. Bankman (ed.), *Handbook of Medical Image Processing and Analysis, 2nd edition.* Burlington: Academic Press, 73–90.

Ryoo, S., Rodrigues, C. I., Stone, S. S., Stratton, J. A., Ueng, S.-Z., Baghsorkhi, S. S. and Hwu, W. W. (2008) Program optimization carving for GPU computing. *Journal of Parallel and Distributed Computing*, **68(10)**, 1389–401.

Ryu, H. and Miyanaga, Y. (2004) A study of image segmentation based on a robust data clustering method. *Electronics and Communications in Japan (Part III: Fundamental Electronic Science)*, **87(7)**, 27–35.

Sangwine, S. J. (1996) Fourier transforms of colour images using quaternion or hypercomplex, numbers. *Electronic Letters*, **32(21)**, 1979–80.

Sangwine, S. J. and Le Bihan, N. (2005) Quaternion Toolbox for MATLAB. *Software library licensed under the GNU General Public License.* <http://qtfm.sourceforge.net>. Accessed 19 May 2011.

Sangwine, S. J. and Le Bihan, N. (2006) Quaternion singular value decomposition based on bidiagonalization to a real or complex matrix using quaternion Householder transformations. *Applied Mathematics and Computation*, **182(1)**, 727–38.

Sayes, Y., Inza, I. and Larranaga, P. (2007) A review of feature selection techniques in bioinformatics. *Bioinformatics*, **23(19)**, 2507–17.

Schilling, M. W., Mink, L. E., Gochenour, P. S., Marriott, N. G. and Alvarado, C. Z. (2003) Utilization of pork collagen for functionality improvement of boneless cured ham manufactured from pale, soft, and exudative pork. *Meat Science*, **65(1)**, 547–53.

Setoain, J., Prieto, M., Tenllado, C. and Tirado, F. (2008) GPU for parallel on-board hyperspectral image processing. *International Journal of High Performance Computing Applications*, **22(4)**, 424–37.

Sezgin, M. and Sankur, B. (2004) Survey over image thresholding techniques and quantitative performance evaluation. *Journal of Electronic Imaging*, **13(1)**, 146–65.

Shaaban, K. M. and Omar, N. M. (2009) Region-based deformable net for automatic color image segmentation. *Image and Vision Computing*, **27(2)**, 1504–14.

Smith, M. L. (2000) *Surface Inspection Techniques: Using the Integration of Innovative Machine Vision and Modelling Techniques.* London: John Wiley & Sons, 1–7.

Steppe, J. M., Bauer, K. W. and Rogers, S. K. (1996) Integrated feature and architecture selection. *IEEE Transactions on Neural Networks*, **7(4)**, 1007–14.

Sun, D.-W. and Du, C.J. (2004) Segmentation of complex food images by stick growing and merging algorithm. *Journal of Food Engineering*, **61(1)**, 17–26.

Sun, X., Tang, X. and Lei, Y. (2008) Continuous attribute discretization and application in Chinese wine classification using BP neural network. In *Proceedings of the 3rd International Conference on Intelligent System and Knowledge Engineering.* Xiamen, China: 17–18 November, 896–900.

Sun, H. Q. and Luo, Y. J. (2009) Adaptive watershed segmentation of binary particle image. *Journal of Microscopy*, **233(2)**, 326–30.

Teixera, A. R., Tomé, A. M. and Lang, E. W. (2011) Unsupervised feature extraction via kernel subspace techniques. *Neurocomputing*, **74(5)**, 820–30.

Teoh, E. J., Tan, K. C. and Xiang, C. (2006) Estimating the number of hidden neurons in a feed forward network using the singular value decomposition. *IEEE Transactions on Neural Networks*, **17(6)**, 1623–9.

Thedens, D. R., Skorton, D. J. and Fleagle, S. R. (1995) Methods of graph searching for border detection in image sequences with applications to cardiac magnetic resonance imaging. *IEEE Transactions on Medical Imaging*, **14(1)**, 42–55.

Theodoridis, S. and Koutroumbas, K. (2009) *Pattern Recognition, 4th edition*. Burlington: Academic Press, 1–12.

Tremeau, A. and Borel, N. (1997) A region growing and merging algorithm to color segmentation. *Pattern Recognition*, **30(7)**, 1191–203.

Tsai, D.-M. and Lai, S.-H. (2008) Defect detection in periodically patterned surfaces using independent component analysis. *Pattern Recognition*, **41(9)**, 2812–32.

Valous, N. A., Mendoza, F., Sun, D.-W. and Allen, P. (2009) Texture appearance characterization of pre-sliced pork ham images using fractal metrics: Fourier analysis dimension and lacunarity. *Food Research International*, **42(3)**, 353–62.

Valous, N. A., Mendoza, F. and Sun, D.-W. (2010a) Emerging non-contact imaging, spectroscopic and colorimetric technologies for quality evaluation and control of hams: a review. *Trends in Food Science & Technology*, **21(1)**, 26–43.

Valous, N. A., Mendoza, F. and Sun, D.-W. (2010b) Supervised neural network classification of pre-sliced cooked pork ham images using quaternionic singular values. *Meat Science*, **84(3)**, 422–30.

Van der Maaten, L. J. P., Postma, E. O. and Van den Herik, H. J. (2009) Dimensionality reduction: a comparative review. *Tilburg University Technical Report – TiCC-TR 2009-005*, 36 pp. <http://bit.ly/agSpru>. Accessed 19 May 2011.

Webb, A. R. (2002) *Statistical Pattern Recognition, 2nd edition*. Chichester: John Wiley & Sons, 123–68.

Wu, X.-J., Zhang, Y.-J. and Xia, L.-Z. (1999) A fast recurring two-dimensional entropic thresholding algorithm. *Pattern Recognition*, **32(12)**, 2055–61.

Yang, H., Van Der Meer, F., Bakker, W. and Tan, Z. J. (1999) A back-propagation neural network for mineralogical mapping from AVIRIS data. *International Journal of Remote Sensing*, **20(1)**, 97–110.

Yang, W., Sun, C., Du, H. S. and Yang, J. (2011) Feature extraction using Laplacian maximum margin criterion. *Neural Processing Letters*, **33(1)**, 99–110.

Zhang, M., Zhang, L. and Cheng, H. D. (2010) A neutrosophic approach to image segmentation based on watershed method. *Signal Processing*, **90(5)**, 1510–17.

Zheng, H. and Zhang, Y. (2008) Feature selection for high-dimensional data in astronomy. *Advances in Space Research*, **41(12)**, 1960–4.

Zhao, H., Sun, S., Jing, Z. and Yang, J. (2006) Local structure based supervised feature extraction. *Pattern Recognition*, **39(8)**, 1546–50.

Zharkova, V. V. and Jain, L. C. (2007) Introduction to pattern recognition and classification in medical and astrophysical images. In V. Zharkova and L. C. Jain (eds), *Artificial Intelligence in Recognition and Classification of Astrophysical and Medical Images*. Berlin: Springer, 1–18.

Zheng, C., Sun, D.-W. and Zheng, L. (2006) Segmentation of beef joint images using histogram thresholding. *Journal of Food Process Engineering*, **29(6)**, 574–91.

Zheng, C. and Sun, D.-W. (2007) Image segmentation techniques. In D.-W. Sun (ed.), *Computer Vision Technology for Food Quality Evaluation*. Burlington: Academic Press, 37–56.

Part II

Computer vision applications in food and beverage processing operations/ technologies

5

Computer vision in food processing: an overview

R. Lind and A. Murhed, SICK IVP AB, Sweden

Abstract: Computer vision is a technology that allows the automation of visual inspection and measurement tasks using digital cameras and image analysis techniques. This chapter provides an introduction to various aspects of the application of computer vision technology in the food industry. The motivation for using computer vision techniques and the advantages that they offer are discussed, along with an overview of the various types of technology available and their uses in this field.

Key words: computer vision, machine vision, image processing, quality inspection.

5.1 Introduction to computer vision

During the second half of the twentieth century, the food industry underwent a transformation from small-scale and low-tech production to modern mass production. Today, a large percentage of foods are produced by international groups operating in the global market. Like in any other industry sector, the desire for increased market share and higher profits lead to the need for increasingly efficient production methods. The trend towards higher levels of automation, quality control and process optimization is ongoing, particularly in countries where the cost of labor is high.

The purpose of this chapter is to provide an introductory overview of how computer vision technology is contributing to this trend. Included are an overview of the technology and applications in food processing, as well as some practical considerations of investing and installing computer vision systems.

5.1.1 What is computer vision?
Computer vision is a technology used to automate visual inspection and measurement tasks using digital cameras and image analysis techniques. The computer uses

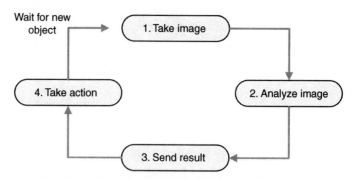

Fig. 5.1   A vision system typically operates in four-step cycles.

image data to perform pre-defined measurement tasks and to draw conclusions based on these data. A vision system typically operates in four-step cycles (Fig. 5.1):

1. Take an image
2. Analyze the image
3. Report result
4. Take action

Standard sensors or vision systems
About 20 years ago, when computer vision was relatively new, there was a clear distinction between standard sensors and a vision system. In simple terms, this distinction may be expressed as follows. A standard sensor measures a single physical property – for example, distance. A vision system, on the other hand, analyzes image array data to make more complex measurements and draw conclusions. An example of the latter in the food industry is the measurement of size and curvature of a cucumber to determine its quality class, which is something that can't be easily done with a standard sensor in the classical sense. Today, the categories have become less distinct and there is even some degree of overlap between them, as some 'standard sensors' are very sophisticated and some 'vision systems' are very simplified.

Computer vision versus humans
The fast, non-contact nature of vision technology makes it suitable for a wide range of applications in mass production industries. The highest level of performance is achieved when conditions are repeatable and standardized – for example, with regard to lighting, object presentation and type of inspection task. By contrast, the human eyes and brain of the experienced professional are still vastly superior when it comes to complex judgments.

For example, a portioning machine equipped with 3-D vision easily outperforms humans in determining where to cut a meat fillet to obtain pieces of equal volume. On the other hand, the butcher's experienced judgment is likely to be far more accurate in evaluating the overall meat quality.

5.1.2   Reasons for using computer vision

Computer vision is used to solve measurement and inspection tasks in situations when simpler technologies are insufficient. The main factors driving investment in computer vision technology in the food industry are its ability to:

1. *Improve quality.* If quality is improved, consumer complaints are reduced, brand perception is improved and food safety liability claims can be avoided. For example, hygiene standards of juice packages can be optimized by using computer vision technology to verify that the cap is tightened correctly.
2. *Increase throughput.* Throughput can be increased by automating portioning, sorting and packaging. For example, chicken nuggets can be portioned with a vision-guided water jet at full production speed.
3. *Reduce waste.* If faulty objects are automatically detected early in the process, these objects can be recycled or rejected as waste before any value is added. For example, the waste incurred by pouring chocolate on a defective wafer, which would then be rejected later in the process at a higher cost, can be avoided.
4. *Reduce give-away.* A computer vision system can prevent the overuse of expensive ingredients. For example, the sesame seeds on a hamburger bun are a major part of the ingredient cost. By counting them on each bun, the sprinkling process can be optimized in a control feedback loop.
5. *Reduce labor.* Operating expenses can be decreased, and monotonous manual tasks avoided. For example, a vision system can perform the high-speed, simple tasks, so that the operator can focus on tasks that require complex judgments.

5.1.3   Successful investment in computer vision

Because of the greater complexity of the vision system compared to standard sensors, the following financial aspects are important in making the decision to invest:

- What is the pay-back time?
- What will the total return on investment be in the end?
- What is the expected service life and maintenance cost?

Quantifying each aspect usually provides a response to the question of whether the investment is worthwhile or not. In the food industry, acceptable pay-back times typically range from a couple of weeks to a year.

Common expectations of computer vision systems, especially by novice buyers, are that they can do anything and, once installed, will run smoothly and be maintenance free. Practical experience tends to demonstrate that these expectations are inaccurate. An understanding of the limitations of the technology is, thus, just as important as an understanding of its capabilities. Although it takes expertise to judge each individual situation, the following aspects should be generally considered when investing in a vision system from an external provider:

- *Requirements and scope.* Are the application requirements documented and agreed? Are all the requirements in the scope necessary, or are they included simply because they are possible? If requirements are sorted in order of priority, cost and complexity can be reduced by deciding where to draw the line.
- *Testability and acceptance.* It is usually much easier to state a requirement than to test that it has been met. Are the requirements formulated such that they can be tested? It is in the interests of both buyer and seller to make an acceptance test specification a part of the business model, unless consultation is paid for at an hourly rate. This will involve slightly more work for both parties in the initial stages, but tends to pay off well later.
- *Degree of innovation and risk.* How novel are the inspection concepts that need to be developed, especially when viewed in the context of the competence of the integration resource? Risk handling should be considered, for example by implementing a limited prototype for proof-of-concept and by applying a project model with milestones.
- *Operating expenses.* Investing in computer vision often means more than just a capital expense. What competence and resources are needed to operate and maintain the vision system? These aspects should be included in the overall financial calculation and organization planning.

5.2   Technology selection

The following sections look at the criteria that need to be taken into consideration when making the choice between different branches of this technology, thus maximizing end results.

5.2.1   Vision system types

The setting up of a vision system varies widely in complexity, and can involve anything from a quick installation by an inexperienced user to a large project with PhD-level algorithm development and programming. The simplest vision systems, vision sensors, are self-contained cameras with built-in image triggering, lighting and embedded image analysis capabilities. Their configuration is simple and can be performed by any technically oriented person after a few hours of training.

In the mid-range of vision system complexity are the smart cameras. Like vision sensors, the analysis is embedded in the device. The difference is that there is a greater degree of flexibility in their hardware configuration and software programming. Typically, a period of training ranging from a few days to a few weeks is needed to become a proficient application developer.

The most flexible vision systems are the PC-based, meaning that the cameras only take images and the analysis is performed on a PC. There are well-packaged imaging libraries of ready-to-use functions that make the engineering task manageable without high algorithm competence. Expert application developers often prefer this type of system, since it allows full flexibility to create both customized system functionality and algorithms in demanding applications.

Most vision systems, including vision sensors, use peripheral equipment to perform their tasks – for example, a photoelectric switch for image triggering, a mechanism to reject faulty objects and a touch-panel operator interface for monitoring and control. Hence, the term 'vision system': it is seldom the case that a single product can complete the full task alone.

High-end vision systems sometimes include hyperspectral and multiscan capabilities, meaning that more image types are acquired simultaneously with the same hardware. For example, 3-D data can be used to locate objects and to verify their shape and volume, 2-D data to sort them according to contrast or color features, and hyperspectral data to identify ripeness or chemical contaminants. Compared to a system that comprises multiple cameras, a single multiscan system gives the possibility to perform sophisticated analysis at reduced hardware cost and within a compact installation (Åstrand, 1996; Hogan, 2010).

5.2.2   Imager types

Imagers for computer vision are usually of the CCD (charge-coupled device) or CMOS (complementary metal oxide semiconductor) type. Historically, there were clear advantages and disadvantages associated with each type, but today the differences are small and the precise details need only be known by specialists. An understanding of what type of data the different imagers can generate is, in fact, more important: for example, are the images produced in gray scale or color, line scan or 2-D snapshot, height profiles or height map 3-D? ... and so on. The purpose of the application is therefore key in the selection of the imager, as follows:

- Continuous inspection of product flow – for example, bulk volume of flour or beans – requires line-scanning imagers.
- Discrete objects with contrast features, such as labels on packages, are best captured by a 2-D gray-scale imager.
- Finding objects of a certain color among others requires a 2-D color imager. This is the case, for example, when quantifying the amount of green salad in a container of pre-processed food.
- Measuring height, volume and shape in low-contrast situations requires a 3-D imager with active lighting, such as an imager that is optimized for laser triangulation.
- Simultaneous acquisition of multiple features, such as 2-D color and 3-D, requires a multiscan imager type. For example, 3-D can be used to locate buns or cookies on a dirty conveyor (i.e., when contrast is lowered), and 2-D color can be used to measure the baking degree.

Fig. 5.2   Blue is a common choice of conveyor color to create contrast; this is effective until the surface becomes dirty.

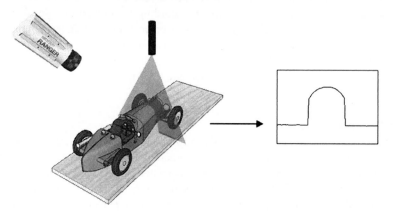

Fig. 5.3   A laser triangulation setup, where the view angle between camera and laser line creates a height profile corresponding to the cross-section of the object.

2-D vision can be used in situations when the contrast is good – that is, when the conveyor is clean and the product is of a different color. 3-D is superior when the conveyor may be dirty and when the products vary in height or color (Fig. 5.2).

In laser triangulation systems, a laser line projects onto the object surface and a single camera views it from the side. The camera's view angle makes the laser line represent the height profile of the object. Either each height profile is analyzed separately, or consecutive profiles are put together into a 3-D image while the object is moving. Since 3-D image acquisition requires movement of the product, the method is referred to as a scanning technology (Johannesson, 1995) (Fig. 5.3).

Stereo vision combines two or more 2-D images, taken from different positions, to create a 3-D image. Since no object or camera movement is needed, stereo vision creates 3-D images by snapshots, just like the single 'click' of a consumer camera. The main advantage of this technology over laser triangulation in industrial applications is its ability to take snapshot 3-D images. The main limitation is the lower height resolution of the images produced.

Spectroscopy is a method used to acquire and analyze individual colors, or wavelengths, separately. For example, it is very difficult for a normal color vision system to sort different plastics by their chemical constituents. Instead, many plastic materials reflect certain IR wavelengths to varying degrees. By analyzing the reflected intensity patterns by their respective wavelengths, the plastic type of each bottle in the recycling center can be classified and sorted by its unique 'fingerprint'.

5.3   Selection of image analysis methods

Once the image is taken and is available to be processed, image pre-processing and analysis is carried out. Pre-processing is a preparation of the image that facilitates a successful analysis. This section explains the most common techniques and algorithms used in computer vision pre-processing and analysis. Most vendors of vision system software provide this basic set of algorithms, or tools (Savage, 2011).

5.3.1   Calibration

Calibration is a term covering a number of operations – for example:

- *Image rectification.* Removal of lens and perspective distortion – that is, straightening out of the image (Fig. 5.4).
- *Unit conversion.* From pixels to physical values – for example, mm or inches.
- *Hand-eye calibration.* Alignment of camera and robot coordinate systems.
- *Color calibration.* Mapping of colors to a reference color space.

5.3.2   Filters

A filter manipulates the image in such a way as to enhance the features of interest and to suppress any unnecessary information, such as noise and clutter. The filter operates locally in the image, meaning that each pixel in the resulting image is a function of its neighboring pixels in the original image. The purpose of using filters is to improve the success rate and performance of the vision system (Sonka *et al.*, 2001, 68–9). For example, to count the sesame seeds on a hamburger, a

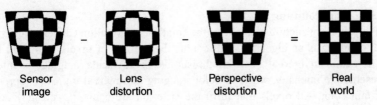

| Sensor image | Lens distortion | Perspective distortion | Real world |

Fig. 5.4   Image rectification through removal of lens and perspective distortion.

Fig. 5.5 Regions of interest: inspection of a logotype and reading of a bar code on an olive oil label is best carried out only in the regions where these features appear.

filter is first used to enhance small, embossed features and to suppress the relatively large-scale curvature of the bun's surface.

5.3.3   Region of interest (ROI)
An ROI, or region of interest, is a selected part of the image. It can be any shape, although rectangles and circles are most common. ROIs are used to decrease the execution time and to improve the success rate by limiting the analysis to the part of the image that contains the feature of interest. For example, inspection of a logotype and reading of a bar code on an olive oil label is best carried out only in the regions where these features appear. The algorithms can then concentrate on analyzing only the relevant parts (Fig. 5.5).

5.3.4   Pattern matching
Pattern matching is an algorithm that locates objects of known shape in the image. Once the object has been located, the algorithm can be further used to verify the shape, find defects in the pattern details, or to align ROIs for subsequent inspections of the object.

Pattern matching is the best algorithm for the location of complex patterns – for example, to check that the print on a cap is centered, allowing the rejection of any caps with misaligned prints (Fig. 5.6). Common matching algorithms are grayscale correlation and edge-based geometric matching (Sonka *et al.*, 2001, 190–3).

5.3.5   Pixel counting
A pixel, short for picture element, is the smallest information carrier in an image. Pixel counting is an algorithm that counts the number of pixels, inside an ROI, that lie within an interval, or within thresholds. In gray-scale imaging, the interval represents the intensity, while in color imaging the interval typically represents a certain color. All pixels that fulfil the criterion are counted, and they do not need to be connected. Pixel counting is the most basic of all image processing

(a)   (b)

Fig. 5.6   An example of pattern matching: (a) A correct, centered pattern on a cap and (b) a cap is judged as defective when the pattern is off center.

tools, and also one of the most widely used. It is especially useful for inspecting the presence of certain features. For example, all pixels with an orange hue in an image of a pre-processed food tray are counted and allow the amount of carrot to be calculated. Likewise, all green pixels represent the beans and so on.

5.3.6   Blob analysis
A blob is an area of connected pixels that fulfil certain criteria – for example, having a size and intensity within specified intervals. Unlike pattern matching, the shape is irrelevant in blob analysis. Blob analysis is used to find and count objects, and to make basic measurements of their shapes. The method is especially useful in food-picking applications, since organic objects tend to vary in both shape and size – for example, chicken breasts, shrimp, sausages and mushrooms (Fig 5.7).

5.3.7   Edge finding
An edge is a change in intensity (2-D) or height (3-D) data in an image. The task of an edge finder is to locate where the edge occurs in the image. Thanks to the relative simplicity of the algorithm, an edge finder can be a fast alternative to pattern matching when locating an object in the image. Edge finding is also used in calibrated systems to measure dimensions. For example, edge finders are used to count the number of slices in packs of bacon (Fig. 5.8).

5.3.8   Code reading
Code reading refers to seeing and decoding bar codes and 2-D codes – for example, EAN13 or DataMatrix codes. When code reading is the only purpose of the system, then a dedicated code-reading sensor is often the preferred choice. A vision system is needed if the code reading is to be combined with some other inspection or measurement in the same image.

Fig. 5.7   3-D image of chicken breasts.

(a)

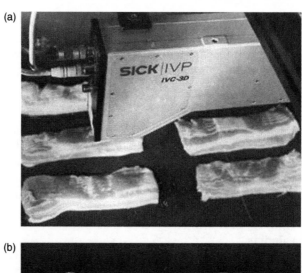

(b)

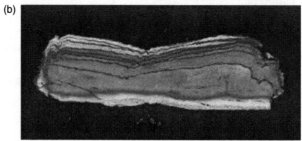

Fig. 5.8   (a) A 3-D smart camera scans bacon. (b) Edge finders are used to count the number of slices in each pack of bacon.

5.3.9   Optical character recognition (OCR) text reading

OCR (optical character recognition) is an algorithm that reads text. In food applications, OCR is most often used to read best before dates and batch numbers. The similar term OCV (optical character verification) refers to a specific type of OCR whereby the text string is known beforehand – for example, when a PLC (programmable logic controller) specifies a best before date to both the printer and the vision system that will verify the printer's result.

5.3.10   Special mechanical considerations

The food industry has special mechanical requirements that designers of vision systems need to take into account, particularly regarding working temperatures, choice of materials and mechanical shape. Temperatures during cleaning processes may cycle between cold and hot. The contracting and expanding air inside the components causes a pumping effect, allowing moisture to enter, which eventually causes condensation and corrosion problems. Examples of solutions to this type of problem are high IP ratings – for example, IP 67 or IP 69K – or active heating and cooling.

Food production processes commonly require the vision components to have food-grade stainless steel enclosures and shatter-free plastic windows. Such components are available on the market, but since the development of vision systems has historically been driven by other industries, such as electronics, the selection is still limited.

In wash-down environments, designed to high hygienic standards, it is important that food residues do not adhere to the equipment. For example, pieces of meat may stick to sharp, concave corners or grooves, which may allow bacteria to grow despite careful cleaning procedures. Therefore, to ensure effective cleaning, mechanical parts should be designed with rounded features and a minimal number of grooves.

5.4   Application examples

The following sections discuss some of the many and varied specialized applications of this technology to be found within the food and beverage industries.

5.4.1   Chocolate and cookie inspection

Smart cameras (operating in 2-D and 3-D) are frequently used in quality inspection of chocolates and cookies and the measurement of their dimensions (Fig. 5.9). Features such as length, width, diameter, height, volume and edge defects are measured. Emphasis is placed on improving quality and reducing waste and giveaway. Vision is sometimes also used for type sorting in situations where several product types are produced in the same line.

5.4.2   Quality grading of seafood

Foods are often graded according to shape and size. Large and well-shaped samples tend to be more attractive to consumers, which drives quality grading and

(a)

(b)

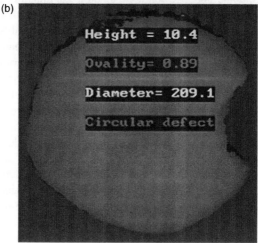

Fig. 5.9 (a) 3-D image of cookies. (b) 3-D image of defective cookie.
The height map is intensity coded: dark = low; bright = high.

price differentiation. For example, scallops are traditionally gauged and sorted manually by operators. Today, this labor-intensive and error-prone process can be replaced by a fully automatic 3-D vision system (Martínez *et al.*, 2011). Another example is oyster sorting, where high-speed 3-D vision can increase the through-put tenfold in quality grading lines (Fig. 5.10).

5.4.3   Freshness watchdog

The traditional 'best before' date is not always a reliable quality indicator for foods that require low storage temperature, such as meat and fish. A solution to this problem is the 'freshness watchdog', or time–temperature indicator. A special ink is printed on a label, which gradually changes color with time when exposed to high temperatures. The freshness of the product can be graded either manually by the consumer or automatically by a vision system. A reference color is printed

(a)

(b)

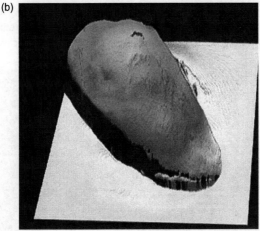

Fig. 5.10   (a) Bin of sorted oysters of the same grade. (b) 3-D image of an oyster.

on the same label as the time–temperature-sensitive ink to facilitate accurate decision making (*Food Marketing and Technology*, 2010b).

5.4.4   Foreign object detection

The harvesting of olives or nuts often generates unwanted debris among the produce. A vision system can identify such foreign objects – for example, if the bulk is transported along a conveyor with a free-fall passage. The sudden increase in speed separates the objects from one another and the vision system can activate an air nozzle to deflect the unwanted pieces from the main flow into a reject bin. X-ray vision is a commonly used technology in the detection of foreign objects inside foods – that is, when it is not possible to see the product as a whole.

A recent development in food quality inspection is three-way sorting, where the product is classified as accept, reject or recycle. Primary product is accepted, foreign objects are rejected and off-grade but still useful product is reworked or used for other purposes (*Food Marketing and Technology*, 2010a, 2011a, 2011b).

5.4.5   Bread shape and baking degree

Computer vision is used for a variety of different applications in bread-making – for example, shape and volume verification, baking degree measurement and automated cutting of patterns on the bread surface (Fig. 5.11). Multiscan technologies are applied to measure both the shape in 3-D and the baking degree in 2-D simultaneously, either by data fusion from several cameras or by a true multiscan camera that allows all necessary image acquisition to be carried out by the same piece of hardware.

5.4.6   Package fill level

The volume of product in containers can be measured by 3-D systems to optimize the fill level; this can apply to soup cans, margarine or ice cream, among many other products. A fill level that is too low can lead to customer complaints, while

(a)

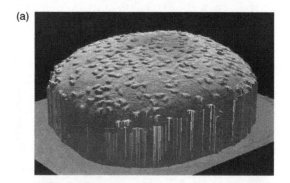

(b)

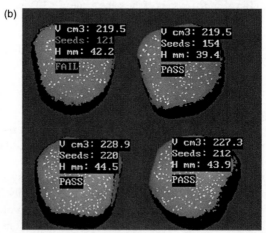

Fig. 5.11   (a) 3-D image of hamburger bun. (b) Inspection result based on size, shape and number of sesame seeds.

(a)

(b)

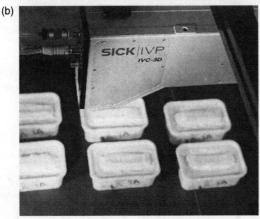

Fig. 5.12   (a) Fill level in soup can. (b) Fill level measurement of margarine packages.

an excessively high fill level may cause undesirable give-away costs and packaging problems (Fig. 5.12).

5.4.7   Meat slaughtering and portioning

Manual slaughtering and portioning of meat is labor intensive, and prolonged exposure to cold temperatures may lead to health concerns. The use of a computer vision system and automatic cutting machines to carry out these tasks not only removes the health risk but also substantially increases throughput. For example, a vision-guided robot used for removing intestines of pigs can process up to 500 carcasses per hour (Wilson, 2010, 2011).

5.4.8   Robotized sorting and picking

Robots are fast, reliable and can work continuously without frequent breaks. Through simultaneous quality inspection and tracking of the products on the

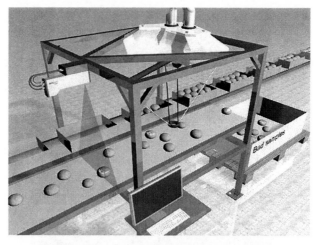

Fig. 5.13   Robotized sorting and bun picking.

conveyor, a vision-guided robot can scrap faulty items into a reject bin and sort good items into their final packages. Examples include vegetables, bread and meat products (Fig. 5.13).

5.5   Conclusion

The constant drive for greater efficiency in food production, as in all industries, is leading to ever-increasing use of automated processes such as computer vision systems. The technology offers a number of advantages, including improved quality control, increased throughput, reduced waste and reduced labor costs. Successful investment in computer vision technology has been shown to be dependent upon the selection of the correct system, and a thorough understanding of both its capabilities and its limitations: there are numerous types of system available to meet the multiple requirements of the food industry.

5.6   References

Åstrand, E. (1996) *Automatic Inspection of Sawn Wood*. Department of Electrical Engineering, Dissertation No. 424, Linköping, 54–65.

Food Marketing and Technology (2010a) The future of X-ray inspection systems in the food industry. **25(4)**, 23–5, Key No. 76399.

Food Marketing and Technology (2010b) Smart packages – with built-in freshness watchdogs. **25(1)**, 36–8, Key No. 75175.

Food Marketing and Technology (2011a) Breakthrough innovations: free fall sorter, now features shape recognition. **25(6)**, 23–5, Key No. 74717.

Food Marketing and Technology (2011b) Three-way sorting. **25(6)**, 30, Key No. 77265.

Hogan, H. (2010) Improving the picture of food production. *Photonics Spectra*, **44(11)**, 28–30.

Johannesson, M. (1995) *Sheet-of-light Range Imaging*. Linköping Studies in Science and Technology, Dissertation No. 399, Linköping, I-3–I-8.

Martínez, P., Alonso, M. and Álvarez, M. (2011) Scanning scallops in 3-D. *Vision Systems Design*, **16(6)**, 19–22.

Savage, L. (2011) Fully featured machine vision software drives manufacturing', *Photonics Spectra*, **45(11)**, 44–8.

Sonka, M., Hlavac, V. and Boyle, R. (2001) *Image Processing, Analysis, and Machine Vision*, 2nd Edition, Thomson Brooks/Cole.

Wilson, A. (2010) Pork process. *Vision Systems Design*, **15(3)**, 13–14.

Wilson, A. (2011) Vision-guided robots find a role in food production. *Vision Systems Design*, **16(10)**, 47–51.

6

Computer vision for automatic sorting in the food industry

E. R. Davies, Royal Holloway, University of London, UK

Abstract: Over the past decade, computer vision has been applied much more widely, uniformly and systematically for automatic sorting in the food industry. This has been ensured by continual developments in the constituent methodologies, namely image processing and pattern recognition. At the same time, advances in computer technology have permitted viable implementations to be achieved at lower cost. To some extent, progress is now being held up by the need for tailored development in each application; hence future algorithms will have to be made trainable to a greater extent than is currently possible. In addition, recent developments such as hyperspectral imaging should become accepted in a number of application areas – particularly for assessment of raw fruit and vegetable produce.

Key words: computer vision, image processing, pattern recognition, object location, food sorting.

6.1   Introduction

Computer vision has long been seen as the way forward to help many parts of industry – not to mention medicine, transport and a good number of other application areas – to monitor and even manage its varied operations (Davies, 2012). In particular, it has been applied to the inspection of products during manufacture, and in the case of food products it has also been important for inspecting the raw food materials arriving from farms and abattoirs (Davies, 2000). Above all, its use is seen as a way of maintaining control of quality, which is all the more exacting when the numbers of individual products arriving at a processing plant amount to millions per week, representing a typical flow of product of 20 items per second. Needless to say, for seeds or cereal grains, the corresponding figures are far greater, and it is more practical to measure flow rates in kilograms per second.

To apply computer vision, images of the various items first have to be obtained, using suitable cameras: these can be normal 'area' cameras, not dissimilar to the digital cameras found in the domestic market, but better adapted for continuous computer interfacing. However, with the steadily moving conveyor belt system found in many plants and factories, it is often beneficial to use a 'line-scan' camera to obtain in turn each vertical line of pixels for (effectively) an infinitely long horizontal image of the belt, from which normal 2-D images of the products and packages can be obtained.

Once suitable 2-D images have been acquired, they can be analysed to obtain the required visual information about the products. Much of the processing can be achieved by taking each image and converting it into a succession of other images that refine the data and steadily extract relevant details and measurements (Davies, 2012). Indeed, image processing is defined as the conversion of one image into another, in which the same objects occur at the same relative positions, but in successive images they may have noise removed, edges enhanced, edges detected, edges measured and then whole objects identified and assessed. Alternatively, grey-scale images may be thresholded to obtain binary images representing shapes, and these may then be 'thinned' and analysed further. These processes will be described in more detail in later sections of this chapter. Here we note that relatively simple image processing procedures are able to extract quite high-level information while working only in image space. However, if computer vision were limited to image processing, it would have severely restricted power, and it is necessary to add abstract pattern recognition processes to realise its full potential. Abstract pattern recognition is based on the idea that the computer should memorise both the characteristics of objects (such as their shapes) and their classifications, so that it can perform automatic interpretation whenever it sees such objects again. As we shall see, mere memorisation is virtually impossible because of the vast variations between different instances of any individual type of object, so special methods have to be invoked to perform abstract recognition (Webb, 2002). When this has been achieved, combining the capabilities of image processing and abstract pattern recognition provides the basis for viable computer vision systems.

Once suitable vision systems have been devised, they can be used for inspection, quality control, and also to oversee and control 'assembly' of more complex foods, such as pizzas and layer cakes. Certain products need X-ray inspection (e.g. to detect bones in chicken fillets), and in general it is necessary to select the image modalities that can provide the most appropriate information about products. Finally, there is a need for real-time processing so that any vision system can keep up with the flow rate on the product line: while it is commonplace for dedicated hardware accelerators to help with the computer processing, we are at last moving into an era where computers will be able to do all the necessary processing unaided.

The following section reviews basic techniques that arise from various image processing operations; Section 6.3 covers more advanced techniques, including pattern recognition methods: in both cases, subsections describe applications of the techniques to food sorting. Section 6.4 considers the need for alternative image modalities such as X-rays. Section 6.5 appraises the need for special hardware for real-time food sorting. In Section 6.6 the discussion is broadened to cover a much wider range of applications and recent advances in the food industry. Section 6.7 outlines likely future trends; the chapter ends with a conclusion and an appendix containing sources of further information.

6.2   Basic techniques and their application

Initially, basic principles and techniques are delineated, expanding on what can be achieved using computer vision technology and in what ways it can be used.

6.2.1   Thresholding and feature detection

A simple means of performing image processing is to take the pixel value at each pixel location and to place a modified value at the corresponding location in an output image space. For example, in the process known as thresholding, which is widely used for locating clearly defined objects in digital images, any pixel whose intensity is darker than a given threshold leads to a binary value of 1 in the output image space, while other pixels are given the value 0 (Fig. 6.1b). Such a process is called a 'pixel–pixel' operation. 'Window–pixel' operations are far more powerful and are achieved by examining the pixel values within a window around each

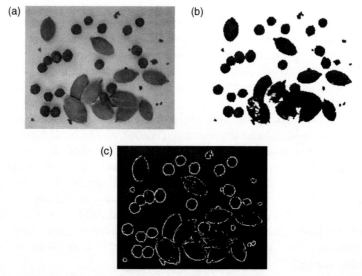

Fig. 6.1   Thresholding and edge detection. (a) Original seed image. (b) Thresholded version, showing that some seeds are not well presented. (c) Edge detected version, showing that edges are often broken and may be several pixels wide in some places.

(a)   (b)   (c)   (d)   (e)   (f)

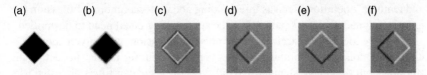

Fig 6.2   (a) Original figure of a square shape. (b)–(f) Effect of applying the five masks described in the text: (b) shows a slight blurring effect that is able to smooth out noise; (c) shows how a Laplacian mask gives a positive signal just inside an edge and a negative signal just outside it; (d) shows a positive edge signal on the right of the object and a negative signal on the left; (e) shows a positive edge signal on the top of the object and a negative signal at the bottom and (f) shows a similar effect to (d), but there is a stronger signal on the right-hand corner. In (c)–(e) the signals are shown on a grey pedestal so that positive and negative signals are easily identifiable.

pixel location and computing a value to be placed at the corresponding location in the output image space.

Among the most general and widely used window–pixel operations are convolutions, and for $(2k + 1) \times (2k + 1)$ windows, these are defined by the formula:

$$P'[i, j] = \sum_{m=-k}^{k} \sum_{n=-k}^{k} P[i - m, j - n] M[m, n] \qquad [6.1]$$

where P is the original image, M is the applied convolution mask and P' is the output image. Common convolutions of this sort are local averaging to help remove noise, enhancing small holes, enhancing vertical and horizontal edges and enhancing corners. Within 3×3 windows these respective operations can be achieved by applying the following convolution masks:

$$\frac{1}{9}\begin{bmatrix} 1 & 1 & 1 \\ 1 & 1 & 1 \\ 1 & 1 & 1 \end{bmatrix}, \frac{1}{8}\begin{bmatrix} -1 & -1 & -1 \\ -1 & 8 & -1 \\ -1 & -1 & -1 \end{bmatrix}, \frac{1}{3}\begin{bmatrix} -1 & 0 & 1 \\ -1 & 0 & 1 \\ -1 & 0 & 1 \end{bmatrix}, \frac{1}{3}\begin{bmatrix} 1 & 1 & 1 \\ 0 & 0 & 0 \\ -1 & -1 & -1 \end{bmatrix}, \frac{1}{20}\begin{bmatrix} -5 & 4 & 4 \\ -5 & -5 & 4 \\ -5 & 4 & 4 \end{bmatrix}$$

Basic properties of these masks are shown in Fig. 6.2. While the second mask tends to give a large signal just inside the boundary of an object (Fig. 6.2c), the signal is boosted at corners – and even more inside holes – where the many boundary contributions accumulate. At first sight the corner enhancement mask (Fig. 6.2f) gives a similar response to that of the horizontal edge mask (Fig. 6.2d); in fact, this is natural as the two 3×3 masks are quite similar; however, the right-hand corner does exhibit a greater signal in Fig. 6.2f than in Fig. 6.2d. The main fact to observe is that corners are more sizeable features than edges, and therefore require larger masks to identify them unambiguously (typically 7×7 pixels rather than 3×3).

While convolutions have wide utility and considerable power, equation [6.1] shows that they are always linear. In fact, nonlinear filters are needed to perform useful recognition functions. In particular, for actually detecting enhanced objects and features, operations such as thresholding and non-maximum suppression (i.e., ignoring signals that are smaller than the surrounding ones) need to be applied.

Next, we look more carefully at the noise suppression mask given above. This eliminates noise by smoothing it out over neighbouring pixels; in fact, it also smoothes out the image signal. To avoid this effect, median filters are often used. These involve taking the pixel values within the window, placing them in order of size, and taking the middle one in the resulting sequence. For a 3 × 3 window, the fifth of the nine values is taken as the output value; for example, if the nine values are 5, 3, 6, 2, 1, 18, 4, 6, 7, reordering them according to size gives the sequence 1, 2, 3, 4, 5, 6, 6, 7, 18; the middle one (the 'median') is 5, and is taken as the output value. Median filters are highly effective at eliminating noise without causing image blurring; in particular, they are especially good at removing 'impulse' noise, corresponding to sudden extreme values, such as 18 in the above sequence. In spite of these good qualities, this type of filter can introduce slight distortions by minutely shifting edges, and should therefore not be applied needlessly when accurate measurements are about to be made (Davies, 2012).

We now turn to the problem of edge detection. To achieve this we employ the outputs from the vertical and horizontal edge enhancement masks. In fact, these differentiate the image locally in the x and y directions, permitting the x and y components of intensity gradient, g_x, g_y, to be determined. As g_x and g_y indicate how rapidly the intensity is changing along the x and y axes, it is not surprising that this information can be used to find the magnitude and direction of maximum rate of change of intensity. In fact, the magnitude and direction can be calculated from the following equations:

$$g = \left[g_x^2 + g_y^2 \right]^{1/2}$$

[6.2]

$$\theta = \arctan\left(\frac{g_y}{g_x} \right)$$

[6.3]

The final process of edge detection is now achieved by thresholding g (Fig. 6.1c).

Next, we look more closely at the corner enhancement mask, which only enhances corners of a particular orientation. The fundamental reason for this is that the mask is a model corner, and moving a model corner over the whole image will only give a close match for corners of similar orientation. This means that detection of corners of arbitrary orientation will require eight masks of this type, each appropriately orientated. In general, this situation will apply when designing detectors for any arbitrary feature, though with larger features, larger and more intricate masks will be needed and greater numbers of masks with different orientations will have to be devised. In fact, designing general feature detectors illustrates the principle, well known in radar, that a 'matched' filter must be identical in form with the profile that it is to be used to detect, in order to achieve optimum signal-to-noise ratio. However, in image processing there is the caveat that this would give rise to sensitivity to the level of illumination of the background. Hence, matched filters used in image processing have to use masks whose coefficients sum to zero. This can readily be understood as follows: if the level of illumination rises by the same amount δ everywhere in the image, then the increase in response from the whole mask will be δ times the sum of the mask coefficients, and for a zero-sum mask the overall change in response will necessarily be zero. This applies for the last four of the convolution masks listed earlier. It is noted that the first mask is for suppressing noise and is not intended to emulate a matched filter.

To summarise, convolutions are highly important image processing operations as they provide optimal detection performance for a variety of image features, but to realise the detection itself, which results in the actual recognition of the features, their outputs need to be subject to nonlinear filters including thresholding and non-maximum suppression. In fact, non-maximum suppression is more important than it might be thought, because thresholding depends on the reliability with which global thresholds can be selected: in many cases, selection is difficult because it relies largely on there being a clear demarcation between true features and noise; in addition, while effective local thresholds can often be selected, global thresholds will vary with the level of illumination and the contrast, both of which are liable to vary significantly from one part of the image to another. Much effort has been devoted to these problems, not merely in the past, but also right up to the present day (Davies, 2007). This may appear to be an odd situation, as the eye can discern almost all types of feature and object without trouble. However, design of computer vision algorithms is often rendered extremely difficult by the fact that image data are incredibly variegated, with many aspects that are confusing for the computer interpretation system. It will serve to highlight this situation if we refer to one such problem here – that of secondary (reflective) lighting which changes the illumination of a given object when another object is placed nearby.

Application: locating insects in cereals
Thresholding and feature detection can find many applications in the agri-food industry; for example, they can be used to detect the location of insects in cereals. In cereal production, in spite of the undesirability of rodent droppings, and the poisonous nature of ergot (whose incidence has decreased markedly in recent years), insects pose an even greater threat to the cereals market: this is because insects have a breeding cycle that is measured in weeks; thus, the integrity of a store or shipment of grain can quickly be compromised. Hence, it is important to detect insect infestations at the earliest opportunity. The problem is exacerbated by the huge numbers of grains that are involved and the speeds with which they need to be checked: note, for example, that 30 tonne lorries may arrive at mills or shipping depots at intervals ranging from 3 to 20 min; while even a 3 kg sample will contain some 60 000 grains. Thus, devising effective inspection algorithms involves substantial computational problems (Davies *et al.*, 2002; Ridgway *et al.*, 2002).

In this case, thresholding was far from being a good starting point, because of shadows and dark colorations on the grains, together with rapeseeds and other artefacts. Hence, it was necessary to take account of the known shapes of insects, which included *Oryzaephilus surinamensis* (the saw-toothed grain beetle). However, bar detector algorithms turned out to be a promising line of investigation. In fact, the regions near the centres of the insects were readily detected by relatively large masks approximating to rings of pixels. The optimum situation was when the rings were half within and half outside the insects, in this case, an 'equal area' rule could be followed based on the matched filter concept. To achieve minimum computation, two masks were employed instead of the anticipated eight (see Section 6.2.1). This was found to be possible, first, by analogy with edge detection, which requires two masks, and, second, because line segments have 180° rotation symmetry, which means that a rigorous mathematical mapping to 360° rotations makes detection using two masks appropriate.

Tests of this procedure resulted in a false negative rate of around 1% against a total of 300 insects: the false negatives were due to cases where insects were viewed end-on or were partly hidden by the wheat grains. A small number of false positives arose from dark boundaries on some of the grains and the condition known as 'black-end-of-grain' was found that either could be eliminated if they were shown to lie within the region of a grain. To further increase discrimination relative to chaff, which can sometimes resemble insects, improvements were made to the bar detector algorithm. Instead of merely detecting the central part of the bar, the ends were also detected using a spot detector, which was quite close to being an end-of-bar matched filter; merging the main part of the bar with the ends then gave a very good approximation to the true size and shape of the insect and thus the capability for highly accurate recognition (Fig. 6.3).

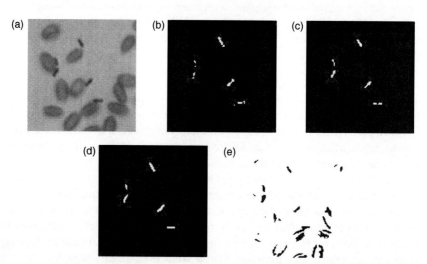

Fig 6.3 Insect location. (a) Original image. (b) Result of applying bar detector. (c) Result of applying end detector. (d) Combined result: 'bar-plus-end' detector. (e) Comparison with thresholded version of (a), showing the large number of false positives arising in that case. Reproduced from Davies *et al.* (2002) with permission from IOP Publishing Ltd.

6.2.2   Texture analysis

Object surfaces are said to have texture when they are rough, woven or composed of many small particles or strands. In such cases they acquire characteristic intensity patterns that have varying degrees of regularity and randomness. For the more regular patterns, periodicities will be measurable, but they may vary considerably over the surface; they will also exhibit strong directionality which may be manifest in several directions – as for the weave of a fabric. A pile of sand or seeds will have a much more random texture than a fabric, and will not be directional (Davies, 2012).

Because of the degree of randomness that is present in most textures, measuring texture patterns presents problems of statistics, so it becomes necessary to average over a sufficient area in order to arrive at an accurate assessment of the surface texture. This is important for the purpose of recognition, and for demarcation of object boundaries – and also for the location of any blemishes, defects or foreign bodies. Getting a unique characterisation of a texture can involve considerable computation, as each pixel must be accessed many times in order to measure textural coherence over different distances and directions. However, short-cuts may be acceptable in certain circumstances. For example, on a hypothetical 'bixit' line, only bixits should be present, so simple tests for (a) presence of a bixit, (b) presence of a non-bixit and (c) presence of a faulty bixit may be sufficient. Perhaps the simplest test is that for 'busy-ness', which may be measured as the number of edge points, or the number of tiny spots per unit area (spots can be detected using the second mask in Section 6.2.1) (Fig. 6.4).

6.2.3   Shape analysis

Shape analysis is one of the prime means by which humans can distinguish and recognise objects: it is no less important for computer recognition. Here we assume that any input images have been thresholded to produce binary images in which the objects appear as 1s in a background of 0s, as indicated in Section 6.2.

(a)
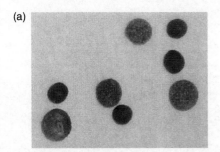
(b)
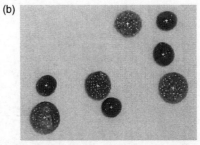

Fig. 6.4 Basic texture detection using the 'busy-ness' concept. (a) An original image of rape and charlock seeds. (b) Light highlights in the seeds have been located and are indicated by white dots. If more than about 20 white dots appear in the vicinity of the object centres (shown by white crosses), the corresponding seed is very reliably classified as a rape seed.

At this stage the computer will have to determine how many objects there are in an image and exactly what regions they cover. While all this is 'obvious' to a human examining the image, the computer has to determine it by 'connected components analysis'. This is not a trivial process for the following reason. When scanning over the image in a forward raster scan that is used on an analogue TV screen, first one object is met, then another, then the first one again, so labelling the pixels in sequence will not give objects unique sets of labels (Fig. 6.5). While it is often found that combinations of scans (e.g., forward raster scans followed by reverse raster scans) can help to arrive at a unique labelling, it is more reliable and more efficient to examine the first set of labels and to make a 'clash table' showing which labels co-exist. Careful iterative analysis of the clash table will then show how to achieve an ideal labelling. Once this has been carried out, the objects have necessarily been counted, and their areas, circumferences and linear measurements can be tabulated without difficulty. Clearly, connected components analysis makes it possible for many further size and shape measurements to be made and recorded.

If objects are to be sorted quickly, simple shape measurements that are invariant to size can be useful. Prime among these is the 'circularity' measure, $C = A/P^2$, where A is the area, P is the perimeter of a blob and C is largest for a circle and smallest for a thin stick. C is widely used for cell counting in biological samples, but is also useful for sorting small objects such as seeds. While not a sophisticated measure, it is useful for making a preliminary identification of the objects in an image, preparatory to a full assay of any doubtful cases, using more advanced and accurate tools. One such tool is that of moments, in which the first, second, third and higher moments of any complex shape are computed. As this can be done to a high degree of accuracy, it can leave little doubt as to the type of object that is

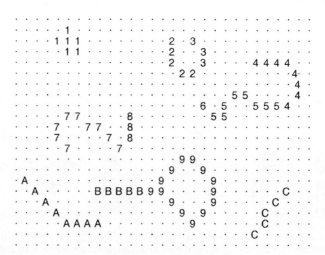

Fig. 6.5 Result of a basic algorithm for connected components analysis. In several cases objects have more than one label: analysis of label clashes is required in order to give a unique labelling.

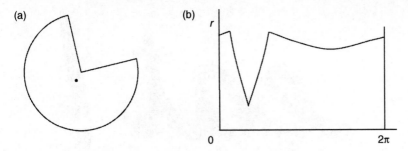

Fig. 6.6 Centroidal profile method for object shape analysis. (a) An original object. (b) Centroidal profile, which poses undue problems of interpretation.

being tested – so long as a large enough database of values is available. However, this method is somewhat computation-intensive, so boundary tracking methods that focus on object boundary points rather than all the points in the object shape are generally preferred.

The archetypical boundary tracking representation is the 'centroidal profile' method, which describes the object boundary in terms of a (r, θ) polar graph relative to the centroid location (Fig. 6.6). For a circle this is a simple graph consisting of a horizontal straight line at $r = a$, where a is the circle radius. For a square it is a set of four curves meeting at four cusps, which represent the four corners of the square; the orientation of the square is given by the angles θ at which the cusps occur. Clearly, such a graph will indicate the object's size and orientation, and will also permit any shape distortions to be detected and measured. Likewise it will permit the object to be recognised. The same situation applies for any of the objects that have been included in a suitable database. Unfortunately, this method is not very robust: for example, if two objects touch or overlap or if there is one object that is seriously misshapen, the centroid will be in an arbitrary location; as a result, the centroidal profile will be grossly distorted and will not be recognisable (Fig. 6.6). While there are alternative types of boundary profile, such as curvature–boundary distance (κ, s) curves, which are not reliant on accurate location of the centroid, these are clumsier to handle and less robust than the Hough transform approach, which we treat in Section 6.3.1.

Skeletonisation is a further approach to shape recognition. The idea is at its most natural when dealing with objects that are composed primarily of narrow strips and curves – particularly alphanumeric characters, asbestos fibres, branches in a tree and so on. In such cases, by eroding the boundaries of the objects repeatedly, the aim is to reach the innermost 'medial' lines, which in the case of hand-drawn characters would correspond to the paths followed by the pen (Fig. 6.7). However, the skeleton concept is very closely defined. It is meant to represent the shape of the connected object, and must never become disconnected during the erosion process. This makes algorithm design quite intricate and exacting. Once formed, the skeleton retains all the limbs and holes of the original shape; indeed, it is intended to be a topologically exact description, with all the junctions and line ends giving a useful, idealised width-independent description of the shape. It is a widely used technique,

Fig. 6.7   The skeleton concept. Here the skeleton is intended to show the path the pen travelled while drawing the shapes.

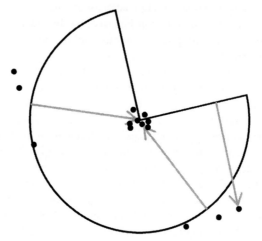

Fig. 6.8   Principle of the Hough transform. Here a partial circle is located robustly: irrelevant votes are scattered and do not interfere with object location.

but has the disadvantage that if noise in the original image leads to additional holes in the object shape, a misleading skeleton will result. However, if excessive noise can be eliminated, errors from this cause will be substantially reduced.

6.3   Advanced techniques and their application

The following sections look at how these techniques can be refined, thus making them more efficient, and how they can be usefully applied in varying industry environments.

6.3.1   Robust object location

In Section 6.2.1, we considered the efficient location of small features, typified by edge, hole and corner points, using the matched filter paradigm. Unfortunately, it is difficult to scale up this method to locate large objects, because of (a) variations in illumination across the objects, (b) variegated backgrounds causing different parts of their boundaries to suffer low or even inverted contrast and (c) the increased likelihood of partial obscuration or distortion. Hence, object location methods have to be designed to be much more robust.

To achieve a high level of robustness, a radical new approach is needed: it becomes necessary to *infer* the presence of objects from partial evidence. The Hough transform approach achieves this by a voting scheme in which only positive evidence for objects is counted. The simplest example of this is the detection of circular objects. All edge points in the image are made to vote for a circle centre by moving a distance equal to the supposed radius along the edge normal and casting a vote in a secondary image space called a parameter space (Fig. 6.8). If only p pixels are found on the circle boundary, p votes will still be cast at the circle centre, and a peak of this weight will be found at that position in the parameter space. Hence, the circle can be found robustly, in spite of the fact that much of its boundary may be misplaced (Fig. 6.9) (Davies, 1984). In addition, the problem of unknown circle radius may be solved in various ways – for example, by trying a sequence of radius values. The method can be extended to allow straight lines and ellipses to be located, or other shapes composed of these shapes. It can also be used to locate arbitrary shapes as long as these can be specified accurately. As indicated above, the power of the approach depends on accumulating positive evidence: erroneous evidence needs to be ignored as it would only serve to bias the results.

The method known as RANSAC works rather differently. Sets of data points are used to generate hypotheses and these are each tested to determine which gives greatest agreement with the image data. Despite its different strategy, it actually embodies a voting scheme, seeking the solution giving the greatest numbers of votes.

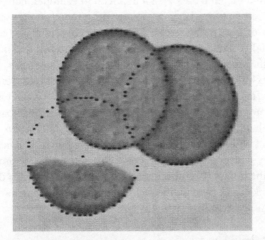

Fig. 6.9 Use of Hough transform for biscuit location. The broken biscuit and the partial occlusion of another do not prevent any of the biscuits from being located reliably. Reproduced from Davies (1984) with permission from IFS International Ltd.

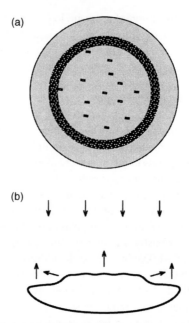

Fig. 6.10   Chocolate cakelets and mode of illumination. With this mode of illumination, a dark ring appears above the edge of jam positions. Reproduced from Davies (2001) with permission from Éditions Cépaduès, France.

Application: cake sorting

In the food industry, many modern foods have high added value relative to the original raw product – or even relative to the basic cooked product. This applies particularly to biscuits and cakes, and in some cases even small cakes are marvels of innovative manufacture. As an application example, here we consider the inspection of a certain type of cakelet, made with a round sponge base, to the centre of which a spot of jam has been added and the top of which is coated with chocolate (Davies, 2000). Chocolate is an expensive commodity, and hence it is important to control the amount of it that reaches the product: too much, and the cost will be too high; too little, and the layer of chocolate will be broken and the consumer will not be satisfied. In addition, the spot of jam must not spread too far away from the centre of the product, or else the characteristic dark ring that appears in the chocolate just above the edge of the jam will be absent, misshapen or distorted, and the final product will lose its pleasing symmetry (Fig. 6.10). Hence, while inspection is generally concerned predominantly with defects or contaminants, here the appearance of the product has also become important. While it is difficult to make a computer think like a human, the following computer-implementable rules can be laid down for acceptability with this type of cakelet:

1. The product should be round.
2. The diameter should lie within well-defined limits or the packing machine will be liable to jam.

3. The spot of jam should be central.
4. The spot of jam should be round and of the right diameter.
5. The chocolate should cover the whole of the top of the cakelet.
6. The chocolate should have a clear texture (imparted to it by the dipping grid).
7. The cakelet should be the right way up.
8. The cakelet should not be overlapped by, or even touching other cakelets, as this means that it will almost certainly be stuck to them.

Checking all these rules for each product at the required rate of some 20 products per second is computationally demanding. It is also a costly enterprise, because cameras, lighting systems and rejection machinery must be linked carefully to the product line. In addition, it is questionable where to place the inspection station. Placed at the end of the line, it can successfully monitor the overall quality of the product, but this will be too late to locate early faults such as lack of jam, so value added at a later stage (including expensive chocolate) will end up being wasted (Davies, 2000). So there is something to be gained by having an early inspection station as well as a final one, but this is seldom done for food products. However, the reverse situation applies for bread, where the raw cereals are more likely to be scrutinised than the final product.

The first stage of inspection is to locate the product; in this case the Hough transform is employed and works well. In particular, it is robust to products that are stuck together and to smudges of chocolate on the conveyor. Once the product has been located, a circular region of interest is defined around it. Then the degree of chocolate cover can be assessed highly accurately and the overall diameter can be measured. Next, the surface can be measured using an approach known as a radial intensity histogram, which contains information on the radial distribution of intensities in the texture of the chocolate (Davies, 1984). This permits the general appearance of the product to be assessed; it also permits the dark ring around the edge of the jam region to be measured for size and uniformity. With these approaches, all the rules listed above can be tested and the product flagged as acceptable or reject. It is important to note that it is the line manager's job to define the limits that have to be set on any one day: it is the machine's job to implement his wishes using the measurements made using the vision system.

Finally, some remarks about the lighting are in order. This had to be set up using four lights that were symmetrically placed around the camera, so as to provide reasonably uniform lighting over the product: great care over this was not necessary to the operation of the Hough transform, which is highly robust. However, reasonable care was still required in setting up the lighting (Davies, 2000), so as to minimise the number of separate measurements to be made and so that simple averaging procedures could be used to save computation.

6.3.2   Object location based on point features

A quite different approach called graph matching has frequently been employed when the objects being sought are characterised by sets of point features such as corners or holes – as in the case of hinges, brackets and certain types of biscuit. In such cases it is necessary to search for subsets of the observed point features that exactly match with subsets in the object template. The solution that is taken to be the most probable one is the largest subset that gives an exact match. Here, again, we see that this is a type of voting scheme in which the best solution is the one for which the greatest number of votes are recorded. Unfortunately, applying this idea rigorously incurs a computational load that is exponential in the number of features to be matched. This is because the definition of a subset is quite general and so all subsets of all sizes and compositions have to be tried. Exponential problems such as this have the property of requiring trivial amounts of computing for say three model features being sought in an image containing five features (in which case there are just ten possibilities to be checked), but the amount of computing escalates disproportionately when six model features have to be sought in an image containing ten features (in which case there are 210 possibilities to be checked), while for nine model features in an image containing 15 features there are a massive 5005 possibilities to be checked. Fortunately, for 2-D interpretation of biscuit images, this problem can be solved elegantly using a special form of the Hough transform, as shown below.

The reader is referred to Davies (2012) for further details and discussion of the three techniques (Hough transform, RANSAC and graph matching) outlined in Sections 6.3.1 and 6.3.2.

Application: biscuit sorting

Cream biscuits present a challenge for inspection because they consist of two wafer biscuits between which a layer of cream is sandwiched: exact alignment of the wafers is relatively unlikely, so from overhead the biscuit will not have an exact rectangular shape but one in which each side is composed from different parts of the upper and lower wafers. It would be better if the vision system could normalise itself on the upper wafer, so that any deviation between it and the ideal shape could be attributed to misalignment of the lower wafer and/or to cream oozing out from between the two wafers (Fig. 6.11). As for the cakelets already described, either of these two eventualities will lead to the product losing its symmetry and being less attractive to the consumer. Again, there are potential problems of packing machines becoming jammed.

In order to locate the upper wafer, location of the boundary will be hindered by either of the two deficiencies mentioned above, so it will be useful to locate it from its surface pattern. In this context it is useful to locate it from the 'docker' holes that it possesses (Davies, 2000, 2001). As any of these may be rendered invisible by excess cream, and as additional holes or points on the boundary can sometimes appear to be docker holes, all potential docker holes will have to be considered for matching to the ideal wafer template. While this can be carried out using the graph matching approach outlined in Section 6.3.2, a special version of the Hough transform was eventually used to locate the centre of the wafer and its orientation. The basic concept was to use each pair of features to predict the position of the centre of the wafer and to place a vote at that location in parameter space (Fig. 6.12). By this means wafer location was determined accurately and robustly and also more

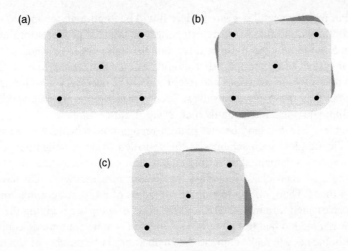

Fig. 6.11   Cream biscuit inspection. (a) Ideal appearance of cream biscuit. (b) Appearance when the two wafers are misaligned. (c) Appearance when cream is oozing out from between the wafers. Reproduced from Davies (2001) with permission from Éditions Cépaduès, France.

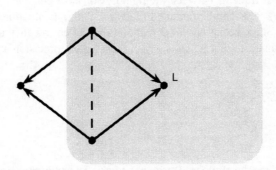

Fig. 6.12   Use of Hough transform for locating biscuit centre. For the two 'docker' holes located on the left of this biscuit, two potential centre locations are identified and votes are cast at these positions in parameter space. After all such votes have been cast, peaks show the location L of the true centre. Reproduced from Davies (2001) with permission from Éditions Cépaduès, France.

efficiently than by graph matching. In fact, there is a break-even point for this computation: this works out at about five docker holes, and for eight or more docker holes the Hough transform approach is hugely more rapid in operation.

6.3.3   Statistical pattern recognition

As indicated in Section 6.1, image processing alone will not permit reliable judgements to be made about classes of objects. However, it will permit the aggregation of data which can then be used in a complete computer vision system that is capable of making useful decisions. Here it will be important not only to classify and sort objects into one of many categories, or to place products into rejection bins, but also to know the 'false positive' and the 'false negative' rates, so that the classifier system can be optimised. It should be noted that there will be costs associated with erroneous decisions and it will usually be necessary to minimise these costs rather than the errors themselves. Such considerations are crucial when trying to eliminate poisonous moulds such as ergot, for example.

To achieve all this, an abstract pattern recognition scheme will be needed. One of the simplest such schemes is the so-called 'nearest neighbour' method, which may be implemented by setting up a 'training set', consisting of a database of instances of each class, including the feature measurements that have been made on them. Then, during testing, the features of a test pattern are compared with those of each training pattern, and the class of the pattern giving the closest match is assigned to that test pattern. This method is quite computation-intensive, because of the need to make comparisons with a very large number of stored training set patterns when testing. However, if a sufficiently large training set is used, the method has the advantage of a close to ideal (Bayes) error rate.

There are many methods giving better performance than the nearest neighbour method outlined above. However, whichever is used, it will be necessary to arrange the best tradeoff between the numbers of false positives (*FP*) and false negatives (*FN*). This can be regarded (Fawcett, 2006) as obtaining the best balance between *S*, the sensitivity of detection (also called 'recall'), and *D*, the discriminability against other types of object (also called 'precision'), where:

$$S = \frac{TP}{TP + FN} \qquad [6.4]$$

$$D = \frac{TP}{TP + FP} \qquad [6.5]$$

In each case *TP* is the number of 'true positives'. To obtain the correct balance, it is useful to optimise the '*F*-measure', for which the parameter γ is set appropriately, 0.5 often being a suitable value:

$$F = \frac{SD}{\gamma D + (1 - \gamma)S} \qquad [6.6]$$

To ensure setting up the system correctly, it is useful to plot the 'receiver operating characteristic' (ROC), which is a plot of the true positive versus the false positive rate (Fig. 6.13). However, it can be difficult to do this rigorously, if at all, when attempting tasks such as the optimum detection of rare defects.

6.3.4   Morphological analysis

While, intrinsically, morphology means much the same as shape analysis, it grew out of a different approach – that of application of simple window operations to modify

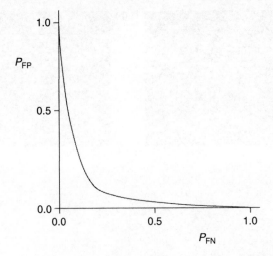

Fig. 6.13 Receiver operating characteristic (ROC). This graph of probability of false positives against false negatives is an example of an ROC curve. A typical working point is the position where $P_{FP} = P_{FN}$.

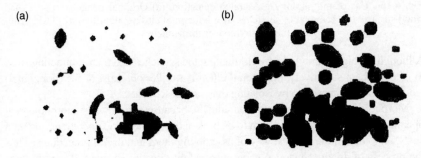

Fig. 6.14 Erosion and dilation. (a) Image of Fig. 6.1b eroded four times by a 3 × 3 erosion operator. (b) The same image dilated twice by a 3 × 3 dilation operator. Erosions of type (a) are useful for separating objects so that they can be counted; they also have the effect of removing tiny objects (in this case tiny flakes of extraneous vegetable matter). Dilations of type (b) can be useful for showing regions where objects appear at high density.

shapes so that they could be filtered to locate particular structures. The starting point was that of eroding objects, so that the outermost layer of pixels was eliminated, or dilating them, so that an additional boundary layer was added (Fig. 6.14): both methods could be applied isotropically or directionally – for example, parallel to the image x-axis. The purpose of these operations becomes clear if a set of vertical striations on a smooth surface is to be located. Eroding the image horizontally would eliminate the striations, and then dilating the image would return the image to its original state, but without the striations. Then, subtracting the final image from the original image would reveal the striations that had been eliminated. Such a method would, for example, reveal any scratches on a computer disc or polished metal surface. Figure 6.15 shows results in a similar but more exacting case – that of scratches on a wood surface.

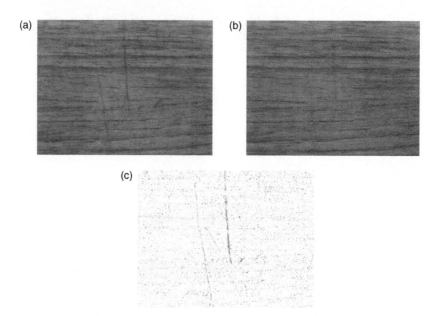

Fig. 6.15 Use of morphology for scratch detection. (a) Original image of a scratched wood surface. (b) Removal of scratches by a horizontal closing algorithm. (c) Using subtraction to recover the scratches.

Although the algorithm works by eliminating the scratches and then subtracting so as to recover them selectively, the overall effect is to filter out most of the (horizontal) wood grain markings, thereby isolating the vertical scratches.

Morphology is formulated mathematically, because carrying out morphological operations efficiently requires them to be broken down into basic operations and then analysed to determine the best way of assembling them into useful procedures. Thus the process of dilating image A using a (possibly directional) mask B is written as $A \oplus B$, while the process of eroding A using mask B is written as $A \ominus B$. In accordance with the types of processing indicated above, dilation followed by erosion is described as 'closing', represented by '•', and erosion followed by dilation is described as 'opening', represented by '∘'. These operations are written respectively as:

$$A \bullet B = (A \oplus B) \ominus B \qquad [6.7]$$

$$A \circ B = (A \ominus B) \oplus B \qquad [6.8]$$

With this notation, we can immediately write the process of identifying the striations described above as the following:

$$S = A - A \circ B \qquad [6.9]$$

While this approach shows how one sort of fault can be located, finding cracks in an egg might well require an operation of the following type:

$$T = A \bullet B - A \qquad [6.10]$$

This amounts to filling in the cracks and then comparing with the original image in order to accurately locate them.

Space does not permit a full exploration of the mathematics underlying morphology. However, we can write down two rules that give further insight into the rigour it imposes:

$$(A \oplus B) \oplus C = (A \oplus C) \oplus B \qquad\qquad [6.11]$$

$$(A \bullet B) \bullet B = A \bullet B \qquad\qquad [6.12]$$

These can be construed as 'it does not matter in what order two dilation operations are carried out' and 'once a hole has been filled in, it remains filled in, so repeated closing is wasteful and unnecessary'.

To make best use of morphology, the sizes and directionalities of the basic erosion and dilation operations need to be selected carefully. It should also be noted that grey-scale versions of these basic operations have been devised, some of these being as simple as maximum and minimum operations within pixel windows. Overall, morphology gives us an additional set of tools that are capable of efficiently locating a range of features in digital images, including particularly defects such as holes, concavities, cracks, spots, prominences and hairs – and also scratches, as mentioned earlier (Davies, 2012, ch. 7).

Application: locating contaminants in cereals
To illustrate the above principles, in this section we examine an important application of inspection – that of cereal grain inspection. There are several aspects that need to be assessed: (a) variety, (b) degree of contamination by other varieties, (c) quality, (d) presence of damaged or diseased grains and (e) presence of contaminants such as insects, rodent droppings and ergot – a poisonous naturally occurring mould. In the space available we will concentrate on the detection of some of the most important of these contaminants.

Many of the contaminants that can appear among wheat grains – such as insects, rodent droppings and ergot – are significantly darker than the rather grey wheat grains considered here. To the human eye this is what appears to make these contaminants recognisable; thus the computational task seems at first to be a rather trivial application of thresholding. However, tests show that this surmise is far from the truth – first, because of shadows between the grains and, second, because of dark patches on the grains themselves – while the presence of chaff, rape seeds and other natural substances, commonly known as extraneous vegetable matter (EVM), complicates the situation further.

First, we attend to the detection of rodent droppings and ergot, all of which are dark and have characteristic elongated shapes, while rape seeds are relatively small, dark and round. It turns out that detection is made more complicated because rodent droppings are often speckled, while ergot is shiny and often exhibits highlights, which appear as small light patches.

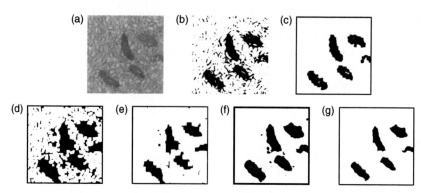

Fig. 6.16   Use of morphology to recover rat dropping shapes. (a) Original image of rat droppings among wheat grains. (b) Thresholded version of (a). (c) Result of erosion + dilation on (b). (d) Result of dilation + erosion on (b). (e) Result of erosion on (d). (f) Result of applying an 11 × 11 median filter on (b). (g) Result of erosion on (f). All erosions and dilations take place within a 7 × 7 window.

Although thresholding did not prove to be a complete solution to the problem, it presented a useful starting point, the main difficulty being the shadows between the grains. Thus, it seemed best to apply erosion operations to eliminate the shadows, and then to apply dilation operations to restore the shapes of the contaminants (Davies *et al.*, 1998). While successful in removing the shadows, this procedure was made worse by the problem of light patches due to speckle and highlights within the contaminant shapes (Fig. 6.16c). Thus, a closing rather than an opening type of operation seemed to be needed to 'consolidate' the contaminants before eliminating the shadows. However, tests showed that this procedure merely served to make the shadow problem worse (Fig. 6.16 d and e). What was needed to aid recognition of the contaminants was a way of (a) preserving the overall shapes, (b) eliminating the light patches within the shapes and (c) eliminating the shadows around the shapes – a highly exacting task. Eventually, it was found that applying median filters in unusually large windows was able to achieve all this (Fig. 6.16f). The reason why this was successful is that median filtering eliminates minority intensity values, whether dark or light, and thus tackles both (b) and (c): curiously it even eliminates both at the same time as it tries to find the central intensity value – and, in fact, achieves a best balance when minority dark and light pixels are present in equal measure. Nevertheless, some overall expansion of the contaminant shapes occur, because of the high density of shadows around the grains, but it was found that this could readily be offset by a final erosion operation operating in a small window (Fig. 6.16g). To ensure that ad hoc and thereby possibly deleterious processes were not introduced, the theory underlying this procedure was developed and the latter was shown to be a necessary and quantifiable inclusion in the overall algorithm.

Using this approach, good segmentation of the contaminants was achieved, *and* good preservation of their overall shapes, thereby making the subsequent recognition task more straightforward. In this case, final recognition has to be passed to an abstract pattern recognition system in order to discriminate between rat droppings, mouse droppings, ergot, large beetles, rape seeds and EVM. However, it was found that there was little need for use of a nearest neighbour or similar classifier, as a simple rule-based scheme was sufficiently accurate to deal with most of the possibilities: essentially, a rule-based classifier is organised as a tree, in which the computer starts at the root and makes successive decisions about which branch to take until it gets to the final solution in each case.

The main disadvantage of the above method is the need to apply a large median filter. In fact, there are many ways of speeding up median operations: in this case, by starting with thresholding, finding the median merely involves counting dark and light pixels within the chosen window, and this can be carried out very quickly. An interesting lesson is that, like dilation and erosion, median filtering is a morphological operation, though it is not often recorded as such. This may be due partly to its slow speed, making it less popular, though with modern computers, this is becoming a diminishing disadvantage.

In fact, large insects could also be found by using the above dark area approach. This could then be combined with the bar-plus-end detector approach described earlier, hence permitting detection of several types of insect ranging from the saw-toothed grain beetle to very large insects similar in size to mouse droppings. To achieve this, the areas obtained by the dark area detector and the bar-plus-end detector were combined, and used in a rule-based recognition system to classify the various objects detected; however, small insects were still flagged directly from the bar-plus-end detector output (Davies *et al.*, 2002). The architecture of the overall system is shown in Fig. 6.17.

Further work in this area has shown that under some lighting conditions ergot can be located reliably by thresholding. As indicated in Section 6.2.1, thresholding is subject to two major problems: one is the difficulty of finding a suitable global threshold and the other is the questionable validity of any global threshold under poor lighting conditions. However, the latter difficulty can often be solved by care in the setup of an inspection system, particularly by eliminating

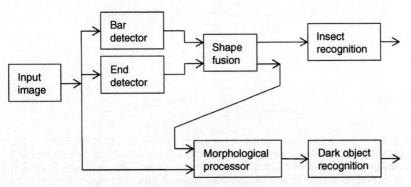

Fig. 6.17 Architecture of complete contaminant detection system. The dark object recognition system starts by carrying out connected components analysis and then identifies each component using rule-based recognition.

extraneous natural lighting, so finding a suitable threshold becomes the main difficulty. In part this is due to the fact that any contaminant forms a very small proportion of the area of an image, so its effect on the intensity histogram used to determine the thresholding level becomes minimal. However, a new 'global valley' approach, which is applicable to ergot detection, has shown how this may be achieved (Davies, 2007). The global valley approach involves searching for the valley in the intensity histogram that gives maximum response to a function f dependent on the depth d_l and d_r of the valley when measured respectively on its left and on its right: the most suitable formula for f was found to be the product of d_l and d_r. This makes it highly sensitive to tiny peaks near the left (dark) end of the intensity histogram, and thus provides a close to perfect threshold for discriminating ergot (Fig. 6.18).

6.4   Alternative image modalities

So far, the discussion has concentrated on the analysis of grey-scale and binary images. In fact, modern cameras normally have colour acquisition capabilities and

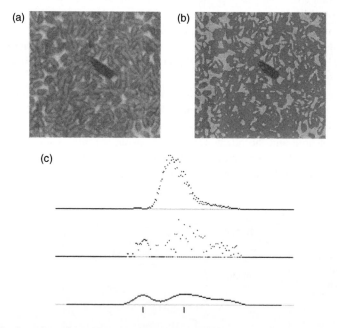

Fig. 6.18   Location of ergot among wheat grains. (a) Original image. (b) Result of applying two global thresholds, one locating the ergot and one identifying the background. (c) Analysis of the intensity histogram (top) using the 'global valley' approach (middle and bottom) to determine the best two threshold locations (two short vertical lines at the bottom). Note that the dark threshold (left) is determined surprisingly accurately considering the paucity of information at that position in the intensity histogram (top). Reproduced from Davies (2007) with permission from the Institution of Engineering and Technology (IET).

this can be immensely useful, though it tends to involve more complex processing techniques. Here red, green and blue (RGB) input is usual, but it is frequently useful to convert RGB images into other formats such as HSI (H is the 'hue', which is the main colour component, S is the 'saturation' or degree of colour vis-à-vis whiteness and I is the intensity) before proceeding with detailed image analysis. The particular advantage offered by HSI is that the effects of any random variations in illumination will be concentrated in the I channel and hardly echoed in the S and H channels. This means that when observing apples, for example, the H channel should indicate unambiguously whether green or red patches are being observed.

For some purposes, infrared imaging or thermal imaging can be useful. The latter applies to the detection of live objects such as insects or slugs, while the former can be useful for identifying the emission bands of certain molecules in vegetation or flesh. Ultraviolet light can also indicate traces from live matter – for example, fingerprints, rat urine, ergot and moulds that produce aflatoxin. While multispectral image data, covering visible light, infrared and the microwave regions, is frequently available from satellite data, this modality may be too complex for most food applications, but ways of combining the outputs of the various channels to obtain coherent information, such as using principal components analysis (PCA), are highly effective.

The other modality that is of immense value in the food industry is that of X-rays. Its penetrating powers are well known, but the prospect of radiation damage to the human body brings with it safety problems and the requirement for heavy screening. The result is dedicated machinery that is heavy, costly and non-portable. However, in recent years vast improvements have been made to semiconductor linear array detectors and at the same time dual-energy detection (DEXA) has been achieved and is in wide use. The value of DEXA technology is that by taking the ratios of the dual-energy detector outputs, the thickness of a specimen can largely be cancelled out. Hence, careful analysis can reveal more detailed structure in the X-ray images; thus detection of foreign bodies becomes much more reliable. Producing and processing X-ray images are rather specialised topics and cannot be covered in full here. For further details and guidance the reader is referred to Batchelor *et al.* (2004) and Kröger *et al.* (2006).

6.5   Special real-time hardware for food sorting

In this section we look back over the development of computer hardware for image processing with particular reference to product inspection and handling. From the time in the 1970s, when the possibilities for 100% inspection of products on product lines were first contemplated, it was found that available serial computers were nowhere near fast enough to meet real-time needs – the prime reasons being the large amounts of data in each image and the rate of processing that was entailed. Parallel processing using arrays of dedicated very large-scale integration (VLSI) logic chips was felt by many to be the key, though dedicated transputer farms and 'bit-slices' (very fast early types of microprocessor) were also fashionable. However, during the 1990s the digital signal processing (DSP) chip, originally posed as a solution to computing fast Fourier transformations (FFTs) for processing speech waveforms, emerged as the dominant approach, only to be largely usurped in the early 2000s by field programmable gate arrays (FPGAs). It was soon found that these random logic arrays could be vastly increased in power by including a few modest microprocessors on the same chip. This meant that any computer with which they were used could spend most of its highly flexible processing power doing the more complicated high-level vision tasks, while the FPGAs did the repetitive low-level tasks such as edge detection and morphology.

A decade later, PCs and other computers have become so much more powerful that in a number of applications there is no need for special dedicated hardware accelerators such as DSPs or even FPGAs with embedded microprocessors. This is advantageous because both types of device need to be programmed using special software, while a PC can be programmed in a simple high-level language such as C++, Java or Matlab. However, it should be added that applications requiring only standard unaided PCs (or stripped-down single-tasking PCs) are not yet very numerous, but as their speed and power increase we are at the exciting stage of envisaging more and more applications following this ideal. It should also be remarked that any PC chip needs a certain amount of random logic around it, so *some* FPGAs are also needed in any applied vision system. The result of all this is that, for many vision systems, the costs of vision processing are almost negligible vis-à-vis those of a complete inspection station. To take an important example, the cost of an X-ray setup on a food conveyor, including all the necessary screening, could easily be ~£100 000, compared with ~£10 000 or less for the PC vision processor. A further advantage of using an easily programmed PC as against a DSP or FPGA with embedded microprocessors is the reduced expertise and cost necessary to get the system working.

Overall, the problem to be tackled is that of understanding which inspection processes and algorithms can be implemented on a single PC and which need FPGA or other processing hardware as well. Added to this, there is great difficulty in knowing exactly what commercial vision systems can achieve. In fact, most of them have to be tailored to any new requirement, and prices can only be ascertained by extended discussions with the vendors and integrators: unlike the situation for other marketable devices and machines, prices are not available on the web; nor it is realistic to buy such a machine off the peg as the specifications of each application are so different. Indeed, the same system operating in the same application will be very dependent on the effectiveness of the lighting system, including the incidence of shadows, glints, ambient lighting, secondary illumination and so on. These factors mean that an article of this length can hardly scratch the surface in recommending how to approach the problem. However, a case study approach is invaluable in helping to understand this sort of situation, as we shall now see.

Looking again at the insect and non-insect contamination detection system outlined previously, the specification was that of producing a system that would inspect ~60 000 wheat grains in 3 min while costing only £5000. By the target date (around the year 2000), this was essentially achieved using a single PC (within a year advances in standard PC technology would in any case bring the speed fully up to specification). Thus, no extra processors were involved, and no special (e.g., FPGA, DSP) programming or handling skills were required – apart from C++ programming capability. However, if the task had been more sophisticated or much greater speed had been required, the unit costs would very quickly have increased substantially. As indicated earlier, it is generally difficult to estimate in advance the degree of sophistication that might be required – not least as lighting can dramatically alter the whole scenario: hence, it only remains to remark that much can be achieved already with single PC vision systems and that, over the coming decade, very much more will clearly be possible.

Interestingly, at the time of writing, a further battle is raging involving another way in which fast hardware can be married to PC technology – that is, via the new graphics processing units (GPUs) that have been developed for running the latest generation of computer games. GPUs have been identified as being able to dramatically speed up low-level vision functions in a way that is completely compatible with the software running on PCs and other workstations. They are exceptionally powerful, their main disadvantage being the amount of electrical power they consume: in special applications such as food sorting this will be unlikely to constitute a major problem, but in any case, in a few years, power levels would be expected to fall, and speed to increase further. Overall, there are definite signs that GPU technology could significantly influence the development of future real-time food sorting systems.

6.6   Recent advances in computer vision for food sorting

In recent years, the main driver of progress has been the need for greater control of quality, to more exacting standards, for a much wider range of foodstuffs. Whereas the challenges were already there ten years or so ago, computer power was insufficient and computer vision had not been developed far enough to permit wide implementation. However, the former difficulty has now largely been overcome, and any remaining lack of computer power can be made up by relatively cheap dedicated processors based, for example, on FPGA technology. Meanwhile, advances in algorithms and their gradual adaptation from one product to another have meant that the coverage of foodstuffs is much less patchy (Davies, 2009). For example, problems of defects on fruit and vegetables have been a matter of some importance for a good many years, with the result that surface defects, including spots, bruises, scab, scars, cracks, wrinkling, injury, mould, rot, discoloration and misshapes are all looked for much more consistently throughout the industry (Bennedsen et al., 2005; Blasco et al., 2007; Jahns et al., 2001; Riquelme et al., 2008; Xing et al., 2007). In fact, over the past decade or so, there has been a move away from concentration on highly controlled aspects of manufactured items such as cakes and biscuits to more careful scrutiny of the raw products as a whole. Whereas a decade ago publications centred largely on staple diet foods such as potatoes and apples, nowadays we see publications on quite specialised fruit and vegetables, including olives, pomegranates, dates and kiwifruit (Blasco *et al.*, 2009; Lee *et al.*, 2008; Moreda *et al.*, 2007; Riquelme *et al.*, 2008). For processed foods, some attention has been focused on cheeses, including those containing additional ingredients such as garlic, parsley and vegetables, which have to be checked for ingredient distribution (Jeliński *et al.*, 2007). Use of X-rays has become more widespread, a phenomenon that has been accelerated (e.g., Batchelor *et al.*, 2004; Kröger *et al.*, 2006) by the arrival of DEXA systems (see Section 6.4). At the same time, the use of faster computers has permitted greater attention to be paid to the analysis of colour (Blasco *et al.*, 2009; Riquelme *et al.*, 2008; Stajnko *et al.*, 2004). In addition, artificial neural networks (Mahesh *et al.*, 2008; Zhang *et al.*, 2005), fuzzy logic (Jahns *et al.*, 2001; Ureña *et al.*, 2001) and decision tree classifiers (Gómez *et al.*, 2007) have been successfully applied to a variety of foods, while morphology (Choudhary *et al.*, 2008; Jiang *et al.*, 2008) and thresholding (Chen and Qin, 2008; Davies, 2007) are probably the two most important means of discerning defects that have consistently moved forward in the last few years.

6.7   Future trends

Over the past decade or so we have seen considerable progress in the march to control quality, and its achievement for a great many different products and raw materials; as a result, coverage is now distinctly less patchy and also more soundly based and reliable. All this would not have been possible without considerable advances being made in the design of vision algorithms over this period. At the same time, computer power has advanced and its cost has been reduced so much that real-time implementation should be achievable for almost any envisaged process.

Nevertheless, while it may seem that the subject is now 'mature', it is still the case that the algorithms for each new application have to be designed carefully and individually, making the process quite manpower-intensive. What are currently needed are means to permit algorithms to be developed or adapted automatically. To some extent this can already be achieved by the use of trainable algorithms: for example, abstract pattern recognition algorithms are necessarily trainable as they have to be applied in a training plus testing regime. On the other hand, their training data normally have to be specially prepared and selected, and this is itself a manpower-intensive process. However, over time this side of the subject will progress much further, with the machine gathering its own training data, and this will have to encompass not only the classifier itself but also the image processing system that feeds it.

A separate area that has been progressing in recent years, and has gradually been catching on, is that of hyperspectral processing (Gómez-Sanchís *et al.*, 2008; Mahesh *et al.*, 2008; Qin and Lu, 2008). This involves feeding each input pixel not just with a single intensity or colour but with a whole spectrum that is output from a spectrum analyser. Such a system is complex, expensive and slow in operation. Basically, it works by using an area camera in place of a line-scan camera, the additional dimension being used for the spectral input. This type of system is slow not only to acquire images but also to process them, as a typical image 'hypercube' contains hundreds of megabytes of data. Indeed, this is an area where new types of algorithm will need to be developed to cope with the superabundance of data, but it has already been applied to food inspection – for example, for the inspection of citrus fruit (Gómez-Sanchís et al., 2008). It seems likely that over the coming decade this technique will be developed much further and used for everyday application in the food industry.

6.8   Conclusion

This chapter has aimed to give a background on available computer vision methodology as applied in the area of food processing: the basic principles are primarily those of image processing and abstract pattern recognition, which have to work in tandem in order for complete computer vision systems to be built. Other factors in complete designs of these systems are the image modalities to be used and appropriate methods of image acquisition. However, having computer hardware that is capable of the high speeds necessary for real-time operation is also a crucial factor.

6.9   Sources of further information and advice

The following books and reviews are particularly relevant to the present chapter:

Davies, E. R. (2000) *Image Processing for the Food Industry*. Singapore: World Scientific.
Davies, E. R. (2009) The application of machine vision to food and agriculture: a review. *Imaging Science*, 57, 197–217.
Davies, E. R. (2012) *Computer and Machine Vision: Theory, Algorithms, Practicalities*, 4th edition, Oxford, UK: Academic Press.
Edwards, M. (ed.) (2004) *Detecting Foreign Bodies in Food*. Cambridge, UK: Woodhead .
Graves, M. and Batchelor, B. G. (eds) (2003) *Machine Vision Techniques for Inspecting Natural Products*. London: Springer Verlag.
Mirmehdi, M., Xie, X. and Suri, J. (eds) (2008) *Handbook of Texture Analysis*. London: Imperial College Press.
Pinder, A. C. and Godfrey, G. (eds) (1993) *Food Process Monitoring Systems*. London: Blackie.
Webb, A. (2002) *Statistical Pattern Recognition*, 2nd edn. Chichester: Wiley.

In addition, the reader will find the journals and associations listed below to be of value.

Major food-related associations:
American Society of Agricultural and Biological Engineers (ASABE) – formerly the ASAE
Campden and Chorleywood Food Research Association (UK)
Home-Grown Cereals Authority (HGCA) (UK)
Leatherhead Food Research (UK)

Major machine vision and engineering associations:
British Machine Vision Association (BMVA) (UK)
European Machine Vision Association (EMVA)
Institute of Electrical and Electronic Engineers (IEEE) (USA)
Institution of Engineering and Technology (IET) (UK)
United Kingdom Industrial Vision Association (UKIVA)

Major food-related journals:
Biosystems Engineering
Cereal Chemistry
Computers and Electronics in Agriculture
Journal of Agricultural Engineering Research
Journal of Food Engineering
Journal of the Science of Food and Agriculture
Meat Science
Transactions of the ASABE
Trends in Food Science and Technology

Major computer vision journals:
IEEE Transactions on Pattern Analysis and Machine Intelligence
Imaging Science
Pattern Recognition
Pattern Recognition Letters
Real-time Image Processing – formerly *Real-time Imaging*

6.10   Acknowledgements

The author is pleased to acknowledge financial support from United Biscuits (UK) Ltd, Unilever UK Central Resources Ltd and the UK Science and Engineering Research Council (EPSRC) for financial support for the baked product work; and from the UK Home-Grown Cereals Authority (HGCA) for financial support for the cereal grain inspection work.

The author is grateful to Dr J. Chambers and Dr C. Ridgway of the Central Science Laboratory, MAFF (now Defra), York, UK, for useful discussions on the needs of the grain industry, and for providing the original images used to test the cereal grain inspection algorithms. He is also grateful to the HGCA Cereals R&D Committee for guidance on the relative importance of different types of contaminant. Finally, he would like to thank former RHUL colleagues Mark Bateman and David Mason for vital assistance with the implementation of key cereal inspection algorithms.

6.11   References

Batchelor, B. G., Davies, E. R. and Graves, M. (2004) Using X-rays to detect foreign bodies. Chapter 13 in Edwards, M. (ed.), *Detecting Foreign Bodies in Food*, Cambridge, UK: Woodhead, 226–64.

Bennedsen, B. S., Peterson, D. L. and Tabb, A. (2005) Identifying defects in images of rotating apples. *Computers and Electronics in Agriculture*, **48**, 92–102.

Blasco, J., Aleixos, N., Gómez, J. and Moltó, E. (2007) Citrus sorting by identification of the most common defects using multispectral computer vision. *Journal of Food Engineering*, **83**, 384–93.

Blasco, J., Cubero, S., Gómez-Sanchís, J., Mira, P. and Moltó, E. (2009) Development of a machine for the automatic sorting of pomegranate (Punica granatum) arils based on computer vision. *Journal of Food Engineering*, **90**, 27–34.

Chen, K. and Qin, C. (2008) Segmentation of beef marbling based on vision threshold. *Computers and Electronics in Agriculture*, **62**, 223–30.

Choudhary, R., Paliwal, J. and Jayas, D. S. (2008) Classification of cereal grains using wavelet, morphological, colour, and textural features of non-touching kernel images. *Biosystems Engineering*, **99**, 330–7.

Davies, E. R. (1984) Design of cost-effective systems for the inspection of certain food products during manufacture. In Pugh, A. (ed.), *Proceedings of the 4th Conference on Robot Vision and Sensory Controls*, London (9–11 October), 437–46.

Davies, E. R. (2000) *Image Processing for the Food Industry*. Singapore: World Scientific.

Davies, E. R. (2001) Some problems in food and cereals inspection and methods for their solution. *Proceedings of the 5th International Conference on Quality Control by Artificial Vision (QCAV 01)*, Le Creusot, France (21–23 May), 35–46.

Davies, E. R. (2007) A new transformation leading to efficient multi-level thresholding of digital images. *Proceedings of the IET International Conference on Visual Information Engineering*, Royal Statistical Society, London, 25–27 July, paper 26, 1–6.

Davies, E. R. (2009) The application of machine vision to food and agriculture: a review. *Imaging Science*, **57**, 197–217.

Davies, E. R. (2012) *Computer and Machine Vision: Theory, Algorithms, Practicalities*, 4th edition, Oxford, UK: Academic Press.

Davies, E. R., Bateman, M., Chambers, J. and Ridgway, C. (1998) Hybrid non-linear filters for locating speckled contaminants in grain. IEE Digest no 1998/284, Colloquium on Non-Linear Signal and Image Processing, IEE, 22 May, 12/1–5.

Davies, E. R., Chambers, J. and Ridgway, C. (2002) Combination linear feature detector for effective location of insects in grain images. *Measurement Science Technology*, **13(12)**, 2053–61.

Fawcett, T. (2006) An introduction to ROC analysis. *Pattern Recognition Letters*, **27**, 861–74.

Gómez, J., Blasco, J., Moltó, E. and Camps-Valls, G. (2007) Hyperspectral detection of citrus damage with a Mahalanobis kernel classifier. *Electronics Letters*, **43(20)**, 1082–4.

Gómez-Sanchís, J., Moltó, E., Camps-Valls, G., Gómez-Chova, L., Aleixos, N. and Blasco, J. (2008) Automatic correction of the effects of the light source on spherical objects. An application to the analysis of hyperspectral images of citrus fruits. *Journal of Food Engineering*, **85**, 191–200.

Jahns, G., Nielsen, H. M. and Paul, W. (2001) Measuring image analysis attributes and modelling fuzzy consumer aspects for tomato quality grading. *Computers and Electronics in Agriculture*, **31**, 17–29.

Jeliński, T., Du, C.-J., Sun, D.-W. and Fornal, J. (2007) Inspection of the distribution and amount of ingredients in pasteurized cheese by computer vision. *Journal of Food Engineering*, **83**, 3–9.

Jiang, J.-A., Chang, H.-Y., Wu, K.-H., Ouyang, C.-S., Yang, M.-M., Yang, E.-C., Chen, T.-W. and Lin, T.-T. (2008) An adaptive image segmentation algorithm for X-ray quarantine inspection of selected fruits. *Computers and Electronics in Agriculture*, **60**, 190–200.

Kröger, C., Bartle, C. M., West, J. G., Purchas, R. W. and Devine, C. E. (2006) Meat tenderness evaluation using dual energy X-ray absorptiometry (DEXA). *Computers and Electronics in Agriculture*, **54**, 93–100.

Lee, D.-J., Schoenberger, R., Archibald, J. and McCollum, S. (2008) Development of a machine vision system for automatic date grading using digital reflective near-infrared imaging. *Journal of Food Engineering*, **86**, 388–98.

Mahesh, S., Manickavasagan, A., Jayas, D. S., Paliwal, J. and White, N. D. G. (2008) Feasibility of near-infrared hyperspectral imaging to differentiate Canadian wheat classes. *Biosystems Engineering*, **101**, 50–7.

Moreda, G. P., Ortiz-Cañavate, J., García-Ramos, F. J. and Ruiz-Altisent, M. (2007) Effect of orientation on the fruit on-line size determination performed by an optical ring sensor. *Journal of Food Engineering*, **81**, 388–98.

Qin, J. and Lu, R. (2008) Measurement of the optical properties of fruits and vegetables using spatially resolved hyperspectral diffuse reflectance imaging technique. *Postharvest Biology and Technology*, **49**, 355–65.

Ridgway, C., Davies, E. R., Chambers, J., Mason, D. R. and Bateman, M. (2002) Rapid machine vision method for the detection of insects and other particulate bio-contaminants of bulk grain in transit. *Biosystems Engineering*, **83(1)**, 21–30.

Riquelme, M. T., Barreiro, P., Ruiz-Altisent, M. and Valero, C. (2008) Olive classification according to external damage using image analysis. *Journal of Food Engineering*, **87**, 371–9.

Stajnko, D., Lakota, M. and Hŏcevar, M. (2004) Estimation of number and diameter of apple fruits in an orchard during the growing season by thermal imaging. *Computers and Electronics in Agriculture*, **42**, 31–42.

Ureña, R., Rodríguez, F. and Berenguel, M. (2001) A machine vision system for seeds germination quality evaluation using fuzzy logic. *Computers and Electronics in Agriculture*, **32**, 1–20.

Webb, A. (2002) *Statistical Pattern Recognition*, 2nd edn. Chichester: Wiley.

Xing, J., Saeys, W. and De Baerdemaeker, J. (2007) Combination of chemometric tools and image processing for bruise detection on apples. *Computers and Electronics in Agriculture*, **56**, 1–13.

Zhang, G., Jayas, D. S. and White, N. D. G. (2005) Separation of touching grain kernels in an image by ellipse fitting algorithm. *Biosystems Engineering*, **92(2)**, 135–42.

7

Computer vision for foreign body detection and removal in the food industry

N. Toyofuku and R. P. Haff, Plant Mycotoxin Research Unit, USDA ARS WRRC, USA

Abstract: A broad range of solutions have been devised to remove foreign body contaminants, prompted by the many injuries and lawsuits that they cause each year. This chapter details the use of computer vision inspection systems to detect the wide variety of these contaminants and remove them from the product stream. Two major approaches are discussed: optical inspection systems, which use images generated using visible or near infrared wavelengths to inspect the exterior of the product, and X-ray inspection systems, which use X-ray imaging to non-destructively search the interior of packages and products.

Key words: computer vision, foreign body detection, optical inspection, X-ray inspection, contaminant removal.

7.1 Introduction

Recent outbreaks of bacterial contamination in food have led to an increased emphasis on food safety in the United States and elsewhere. Although bacterial contamination such as *E. coli* and *Salmonella* may present the greatest health threats and attract the most headlines, many other defects and contaminants found in food have been persistent problems both in terms of food safety and quality. Foreign body contaminants, such as pits and pit fragments, unwanted seeds, bones or bone fragments, sticks, rocks, metals, plastics, etc., continue to be a nuisance for producers and consumers alike, resulting in many injuries and lawsuits each year. Because of their variable nature, a wide range of solutions have been devised to detect and remove them.

In response to a number of food scares in Europe during the 1990s, the European Commission released the White Paper on Food Safety in July 2000 (European Commission, 2000). This report contained recommendations for ensuring food safety at all points in the food supply chain, known as the 'farm to table' approach, and a number of these recommendations had a direct impact on food processors. New regulations were introduced to cover all agriculture sectors across the EU, replacing a multitude of regulations in different countries. For food processors, the major impact of the new regulations is that they have the primary responsibility to ensure the safety of their product, and failure to use available technology leaves them liable for injuries or illnesses caused by their product.

Automating the inspection process would help ensure consistency and potentially increase processing speed. While some products still require human inspection, as imaging hardware evolves and decision algorithms grow more sophisticated, the number of items on the list that can feasibly be inspected automatically is growing. The challenges of this task have led to a huge expansion in the market for inspection systems and a wide range of inspection systems for food quality and safety. An excellent review of the many detection technologies, including metal detectors, magnets, optical systems, microwave reflectance, nuclear magnetic resonance (NMR), radar, electrical impedance, ultrasound and X-rays can be found in a book edited by Edwards *et al.* (2004). There are also a number of review articles available that summarize the various techniques specifically for the detection of foreign materials in food, usually combined with an automated processing and removal system (Abbott, 1999; Abbott *et al.*, 1997; Bee and Honeywood, 2004; Butz *et al.*, 2005; Graves *et al.*, 1998; Haff and Toyofuku, 2008; Jha and Matsuoka, 2000; Ruiz-Altisent *et al.*, 2010).

All of these automated systems involve computer vision. Computer vision can be a vague term because it encompasses a wide range of computer and hardware configurations. For the purposes of this chapter, we chose to define computer vision as any automated process that makes use of a computer (or the equivalent) to classify data extracted from a scene or object, whether in the form of formal images, hyperspectral images, or data values such as the absorbance of light. This is sometimes referred to as machine vision or optical sorting.

Computer vision inspection systems are often used for quality control, product grading, defect detection and other product evaluation issues. Details on these types of systems can be found in other chapters in this book. This chapter focuses on the use of computer vision inspection systems that detect foreign bodies and remove them from the product stream. Specifically, we will focus on the following two methods:

(i) *Optical inspection* relies on the generation of an image or other optical output (such as absorption of light) which is then used to sort the food stream for unwanted contaminants, such as unwanted varieties of grain in wheat processing streams, pits in cherries or unsalable pistachios. The image generation is typically accomplished using visible or near infrared (NIR) wavelengths.

(ii) *X-ray inspection* uses X-ray imaging to non-destructively search the interior of packages and products. Due to the way the X-rays pass through the product, certain foreign contaminants show up more readily under X-ray imaging, such as pits, bones and insects.

7.2   Optical inspection

Optical inspection is used extensively by the food processing industry to provide quality control, product feedback and to guide automated processes. While inspection can be, and often is, performed by human operators, some foreign contaminants are too difficult for humans to detect reliably or cost-effectively. Sablatnig (1997) has a detailed breakdown of the advantages and disadvantages to both human and computer inspection. Two of the most common issues are: the contaminant may be too small or subtle for the human eye to detect, or, the speed and volume at which the product needs to be inspected is more than a human inspector can handle. A computer vision optical inspection system can consistently and reliably inspect items as small as individual grains of wheat at real-time processing speeds. Advances in sensors, imaging capture devices and a range of computational devices have driven down the cost and enabled computer vision inspection to be used for an ever-increasing range of products.

7.2.1   Optical inspection components
While there are a wide range of specific devices used for an optical inspection system, they all generally have the same basic construction and parts. A generic system is illustrated in Fig. 7.1. Usually, each system consists of a type

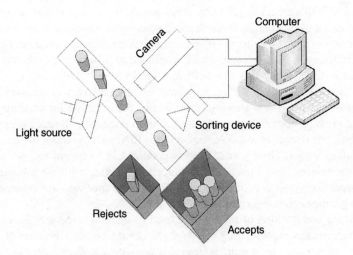

Fig. 7.1 Sample optical inspection system.

of image capture device, either a camera (CCD) or a sensor such as a complementary metal-oxide sensor (CMOS). The devices send data to a computer which processes that information, makes a decision based on pre-set conditions, and usually triggers a sorting device that removes any undesired product from the product stream. This chapter will focus on recent developments in this area and explore the computer vision aspect of this type of foreign body removal inspection in greater detail. For descriptions of earlier systems, please see Bee and Honeywood (2004).

7.2.2   Problems encountered with optical inspection

Several difficulties encountered in optical inspection have already been mentioned, such as irregularity of product size, shape or color. Some of these will be explored later in more detail in a commodity-by-commodity basis. However, there are some problems that occur at every stage of the system, independent of the type of product being inspected.

Accuracy
It is very important to understand what can be expected from the inspection system as a whole. Any sorting system, whether human, computer, or both, will require a compromise between accuracy and cost concerns (Bee and Honeywood, 2004). Changes in sensitivity can reduce misses but increase false alarms (good product identified as bad) to unacceptable levels. If the system is given more time to evaluate the product, it could be more accurate but most producers generally need a high rate of processing, so this type of delay would generally be unacceptable (Pearson *et al.*, 2008).

Image acquisition
Optical systems use cameras and sensors to capture images of the product. Because the product is often traveling at fairly rapid speeds, frame capture can be an issue. When passing pistachio nuts through a line-scan video camera system at a rate of 30–40 nuts per second, Pearson (1996) found that the cameras could only capture about six frames of data per nut, which combined to represent roughly 4% of the total nut surface. While this was sufficient for an 85–87% accuracy rate, being able to evaluate the whole surface may have led to an increase in accuracy.

Because the product is moving, there is also the potential for blurring, especially at the high speeds commonly used in food inspection. Reducing exposure time helps but decreases signal-to-noise ratios (SNRs), degrading the quality of the resulting image. Ding-ji *et al.* (1994) created a line-of-sight following technique that reduced blur, through use of a precisely rotating mirror. However, most food inspection systems concerned with motion blur appear to use cameras with motion-compensation algorithms built into them.

The type and position of the lighting used to illuminate the product can also complicate image capture. Regular daylight cannot be used because it is too variable, so artificial light must be used for consistency. Fluorescent lamps and low-voltage lamps can cause problems with speed 'beating' when the light is modulated at 50 or 60 Hz. This causes fluctuations in the image intensity due to difference between the camera shutter time and the light intensity. To solve this, systems need to incorporate high-frequency fluorescent lamps or low-voltage DC lamps (Domenech-Asensi, 2004). If the system designer is not aware of this potential problem, it could introduce sinusoidal fluctuations that could create artificial rejections and acceptances of product.

Wet produce can also be difficult to image, due to the reflective nature of the wet surface and complications from the potentially wet conveyor belt/background. Even algorithms have a hard time compensating for the specular highlights that can result. The solution is to use polarizing filters on the cameras, ideally two of them at 90° angles (Domenech-Asensi, 2004; Ruiz-Altisent et al., 2010).

7.2.3   Computer processing and algorithm development

The algorithm that evaluates the data sent from the imaging devices is the primary means of customizing each system to the product and contaminant being sorted. Advances in computer processing hardware and software have allowed systems to go beyond simple logic settings and automated sorting to more sophisticated sorting. Many can incorporate multiple criteria in their decision-making to reduce the chances of errors. Pearson et al. (2008) compared sorting performance based on a single feature (such as with the traditional color sorters) versus using multiple image features, and found that accuracy in classifying wheat kernels increased from an average of 85.75% using one feature to 97.75% with three features. The computer software can also help the system calibrate itself to account for irregularities, such as product color change over time, calibration drift errors, light source degradation, ambient light or environmental changes, dust accumulation and other real-world processing issues. This flexibility is vital in creating systems robust enough to handle the environments at packing plants and sorting facilities.

Because computer vision systems are often used to replace human inspectors, they need algorithms that can make decisions to mimic the results of human inspectors. Haff and Pearson (2007) compared algorithm performance to human sorting. The algorithm had 19.8% false negatives (missing an infested wheat kernel), 5.6% false positives (incorrectly identifying a kernel as infested) and an overall 14.4% error rate. The human data had a false negative error rate of 28.3% ± 5.7%, a false positive rate of 2.9% ± 2.34% and an overall combined error rate of 15.6% ± 2.3, showcasing the variability in performance that is common to human inspectors. Results like this illustrate how automated computer vision inspection can provide comparable or better results with lower variability than human inspection.

There are multiple steps in the software component of an inspection system; generally they consist of image enhancement, segmentation, feature extraction/ optimization and classification. Enhancement often involves standard techniques such as histogram stretching, normalization and pixel averaging. Segmentation can have multiple approaches and is vital as multiple products can appear in a single image or when the product is very similar to its background or surrounding environment. Visen *et al.* (2001) created an algorithm that separated overlapping grains of barley with 99% accuracy, allowing them to be inspected as a continuous stream, rather than needing to be separated into individual non-touching grains. An image can have a large number of possible features to use as a basis for classification, such as color, spectral band distribution, pixel intensity and distribution, edge detection and curvature, histogram shapes and many others. Once the features are extracted, they are examined for their predictive and sorting significance. There are several ways to do this. For example, Haff and Pearson (2006) developed a technique for optimizing NIR light band selection using a variation on a nearest neighbor classification scheme. They applied it to a dual-wavelength NIR-based inspection system that removes small in-shell and half-shell pistachios from the processing stream. They were able to achieve a 1.3x to 4.7x improvement in accuracy over the settings suggested by the manufacturer (error rates fell from 1.70% false negative for small in-shell, 2.40% false negative for half-shell and 0.70% false positives using the manufacturer's recommendations, to 1.20%, 1.80% and 0.15%, respectively). Additionally, optimal spectral bands for other defects were determined, to illustrate the flexibility of the technique.

Classification takes place once the optimal features and desirable criteria are determined. It usually involves some type of statistical classifier or multilayer neural network (Brosnan and Sun, 2004). Often, the system is trained with a set of data and then tested or validated on a second set of similar data. When building a system to remove closed-shell pistachios from regular pistachios, Pearson and Toyofuku (2000) trained the system and chose the optimal features using a set of 1188 nut images (one third closed shell, one third thin split and one third open shell), and then tested it with 3000 nuts (same ratios as training set) at a rate of 40 nuts/s. They were able to achieve an accuracy rate of 95% with this system.

7.2.4   Recent research in the optical inspection of food products

Fresh fruit

Optical inspection is used extensively in fruit processing; however, it is primarily focused on blemish and defect detection, rather than foreign body removal. Even items like leaves and stems can often be removed with simple color sorters (without computer algorithm enhancements) or mechanical methods (winnowing, water floatation, sieving, etc.). However, the removal and detection of pits and stones can still be complex enough to defeat these simpler methods. Qin and Lu (2005) used multiple band hyperspectral transmission imaging to detect pits in cherries and achieved error rates of 3.1–3.5%. Due to the small size of the pits and the fact that they are often occluded or entirely hidden by the flesh of the product, other methods such as X-ray have proven more effective for this type of foreign body removal, and will be discussed later in this chapter.

Almonds

Pearson and Young (2001) used laser transmittance imaging to detect embedded shell in almonds. This type of inspection was found to be superior over other types of inspection because the other systems were incapable of distinguishing the shell from the skin. Their algorithm used three image features and could process at a rate of 40 nuts/s. They attained accuracy levels of 88% and 82% for identifying regular and embedded shell almonds, respectively.

Pistachios

Most pistachios are sorted using a 'pricking' wheel sorter that separates in-shell pistachios with split shells from closed in-shell pistachios. This type of sorter causes small amounts of damage to the kernel of the pistachio and can result in consumer rejection due to the appearance of 'insect' damage. Pearson and Toyofuku (2000) developed a non-damaging system that used a color sorter. At a cost of $15 000 per channel, the color sorter reached an accuracy rate of 95% (comparable to the wheel sorter) while working at a rate of 40 nuts/s.

Using the sorter developed in 1996, Pearson *et al.* (2001) refined the algorithms to use discriminant functions that would use a set of image features to detect several pistachio defects in real time. They were able to successfully detect pistachios with oily stained hulls, adhering hulls and dark stained hulls with 96.3% accuracy at a throughput rate of 40 nuts/s. As these damaged kernels are highly associated with defects and contaminants such as navel orange worm, insect pupae, fungal decay and *aspergillus*, removing these defective nuts would significantly improve overall kernel quality in the final product stream.

Using the dual-band optimization technique reported by Haff and Pearson (2006), Haff and Jackson (2008) built a low cost sorter that cost less than $500 for parts and achieved an accuracy rate of 95% removing kernels from in-shell pistachio streams at real-time inspection rates.

Grains

Different wheat cultivars can occasionally become mixed during harvest. Separating them is important to breeders, seed foundations and seed companies which rely upon pure lots (Pearson, 2010). However, the differences between different cultivars are often very subtle, and the individual grains are very small. Additionally, wheat tends to be processed in large bulk quantities. All of these issues make hand-sorting or human inspection extremely expensive and difficult.

Majumdar and Jayas (2000a, b) wrote an algorithm that could classify various cereal grains (such as durum wheat, barley, oats and rye) with up to 99% accuracy. Wang *et al.* (2005) developed a neural network to detect grain characteristics with a 100% accuracy rate for non-vitreous kernels and 92.6% for mottled kernels. However, neither of these algorithms was combined with a complete sorting system.

In 2008, Pearson and his co-workers (Pearson *et al.*, 2008) created a system with a traditional camera and personal computer that achieved 95.5% and 94.7% accuracies (for 'easy' and 'difficult' samples, respectively), which was a 10%

improvement over commercial sorters (Pasikatan and Dowell, 2003). However, the throughput rate was 30 kernels per second, which would be too low for general commercial processing needs, but could be used in specialized applications, such as wheat breeding (Pearson *et al.*, 2008).

This underlines one of the major issues encountered when inspecting grain – the sheer volume of product that must be inspected to keep up with production rates overwhelms most computer vision inspection systems. To address the issue of throughput, Pearson (2009; 2010) developed a new system using a field-programmable gate array (FPGA) and CMOS combination that could have a throughput rate of 225 kernels/s, which would be fast enough to potentially be used for real-time grain sorting. Because the computer programs are hard-coded into the FPGA, it is extremely fast and robust to the strains of processing large amounts of data over a lengthy period of time. The CMOS's streamlined design lowers cost and makes it easier to integrate with a FPGA. Because the cost was even lower than that of the camera and personal computer system designed in 2008, it may be feasible for multiple systems to be set up in parallel, which would increase throughput even more. Pearson achieved an 88% accuracy for sorting red wheat from white, which is 10–20% higher than multiple passes through a commercial color sorter.

7.2.5   Future developments in optical inspection

The advances in FPGA and CMOS-based systems are very promising. With a resolution of 16 pixels/nm, even small products and foreign matter can be detected. Pearson (2010) took advantage of the parallel processing built into the FPGAs to inspect three streams at once, while using a single camera. The camera captured the image of all three streams but then the FPGA segmented the image into the three separate streams so that each could be processed separately in parallel. Work has already been done to expand this research to sort other products such as barley from durum wheat, and brown from yellow flax seeds (Pearson, 2010). The results prove that this system is flexible to different commodities and research is already beginning on adapting it to process almonds, pistachios and olives.

7.3   Fundamentals of X-ray inspection

In recent years, X-ray inspection has become increasingly common in certain segments of the food industry. This is particularly true for processed foods, including product that is packaged in cans, bottles, or jars, presumably due to the ever-increasing emphasis on food safety. It is now an accepted fact that X-ray inspection is superior to traditional metal detection technology for the detection of metallic contaminants, and adds the potential to eliminate other foreign non-metallic material such as bone, glass, wood, plastic and rocks. Technological advances in the areas of high-voltage power supplies, solid-state detectors and computation power and speed have made X-ray systems more affordable, reliable and easier to use while improving image quality and detection capabilities. A variety of improvements in sensor technology have improved resolution, including cesium iodide (CsI) crystals and improved CCD arrays.

As with optical inspection systems, the components used in X-ray inspection systems tend to be similar. Because some of the hardware involved is very specialized and a discussion of the problems associated with them require some background in the subject, we have included a more detailed discussion of the relevant hardware in this section.

7.3.1   X-ray system components

X-ray sources
X-rays are produced when high-energy electrons strike a target material, typically Tungsten. An X-ray tube is similar in design to a light bulb, except that the electrons shedding from the heated filament are subjected to a high voltage, causing them to accelerate and strike the target at high energies. As these high-energy electrons decelerate in the target material, electrons of target atoms are first excited to higher energy levels, and then decay to their ground states with the emission of X-ray photons. The size of the target area over which X-rays are generated is called the focal spot size, and has consequences for the characteristics of the imaging system as will be discussed in a later section. The X-rays themselves have two characteristics that are important in the operation of the X-ray machine: energy and current. The energy refers to the maximum energy that an X-ray photon can possess when exiting the tube (generally between 20 and 100 KeV for food inspection) and defines the penetrating power of the X-ray beam. The current, measured in mA, is associated with the number of X-ray photons being generated. The power supply has a maximum power (the product of the energy and the current) rating, and a balance is therefore required between the energy and current, which has consequences for the resulting image quality. The result of this power limitation is that most X-ray inspection systems are limited to less than 10 mA of current, many much less, and a discussion of the consequences of this will follow.

Detection and imaging
While the earlier X-ray detectors used film or even sheets of paper coated with barium platinocyanide, these are not suitable for real-time inspection. They are too slow in terms of exposure and developing time, so a digital solution was needed. While phosphors have been used since the time of Roentgen to display radiographic images, they could not compete effectively with film until the arrival of computer technology which could display digital images. Since then, phosphor-based X-ray detectors have become commonplace in line-scan type machines such as those seen in airports for luggage inspection. Phosphors are a class of luminescent material, which absorb electromagnetic radiation and re-emit it at a longer wavelength. When used as an X-ray detector, the phosphor will absorb X-ray photons and emit visible light photons which are subsequently detected by either

photodiodes or CCDs. Some line-scan detectors bypass the use of phosphors by using modern semiconductor materials that convert incident X-ray energy directly into an electric current. In a line-scan array, hundreds or thousands of detectors, either photodiodes overlaid with phosphor or semiconductor crystals, are placed in a row perpendicular to the direction of sample flow. While the sample moves over the array at a fixed rate, the output of the photodiodes are repeatedly read at a rate that is synchronized to the speed of the sample. The image is then constructed row by row. Since the creation of the image is dependent on the motion of the sample across the line of detectors, this arrangement is ideal for high-speed inspection. This is in contrast to camera capture, where the motion of the product at high speeds can cause blurring. Most high-speed applications employ a side view arrangement, as opposed to a top view system common in luggage inspection equipment. In the side view configuration, the conveyor belt is not included in the image, allowing for improved image quality.

Another class of line-scan X-ray imaging systems incorporates dual energy detection technology to differentiate between soft and hard materials. Two detector arrays are used, one on top of the other. The lower energy X-ray photons are absorbed by the first detector array while higher energy photons pass through to the second array. The difference between the two images represents the softer material. This technique is widely used in luggage inspection, allowing identification of softer organic material as well as harder metals.

A third class of X-ray imaging system, commonly known as X-ray fluoroscopy, involves the use of image intensifiers. These devices, which amplify light, are commonly used in low light situations such as night vision. A typical image intensifier uses a photocathode to convert incident light photons to electrons, which are accelerated across a large potential onto an appropriate phosphor material, which in turn converts the electrons back into visible light photons. The energy acquired by the electrons as they are accelerated results in photon gains of up to 30 000 in the most modern intensifiers. Unlike line-scan inspection systems, image intensifiers produce an image that can be observed directly or cast onto a screen. However, image capture applications require coupling to a CCD camera, which introduces some of the same complications noted in the optical inspection system discussion.

A variation on X-ray line-scan imaging that allows three-dimensional (3-D) images is computed axial tomography, or CT imaging. An X-ray source rotates around the sample with detectors positioned opposite the source. Multiple 'slices' are progressively imaged as the sample is gradually passed through the plane of the X-rays. These slices are combined using a mathematical procedure known as tomographic reconstruction to form a 3-D image. Helical or spiral CT machines incorporate faster computer systems and advanced software to process continuously changing cross sections. As the sample moves through the X-ray circle, 3-D images are generated that can be viewed from multiple perspectives in real time on computer monitors. CT scans produce images of superior quality to traditional X-ray systems, but the high cost and lengthy scanning and data processing times make it an impractical method for real-time food inspection at present. Nevertheless, there have been a number of research projects demonstrating the efficacy of using CT systems to non-destructively assess food quality. If technology and computer processing times improve, and system prices fall, CT scanning could replace traditional X-ray imaging as the predominant method for real-time inspection of food products.

While the three classes of X-ray systems described above represent the majority of systems currently in use, there are some other systems that have been investigated for their potential use in food inspection and are discussed in Haff and Toyofuku (2008) and on the website of the Safeline Corporation.

7.3.2   Problems with X-ray images

Generating digital images allows the X-ray system to be incorporated as part of a computer vision system. However, there are several image quality issues that are particular to X-ray images. Three important parameters associated with X-ray image quality are: resolution, SNR and contrast between the material of interest and its surroundings.

Resolution describes the differentiation of small, close objects, such as sets of bars on a resolution chart (Rosenfeld and Kak, 1982). There are several ways that resolution of an image can be expressed, the most common being line pairs per millimeter (lp/mm) as measured using an X-ray test pattern containing pairs of lead lines spaced with increasing frequency. Often the resolution of a system is simply expressed as the size in microns of half of a line pair.

There are several factors that affect the resolution of an X-ray imaging system. The size and spacing of the detectors determines the ideal resolution. For phosphor-based detectors, resolution is lost through light scattering in the phosphor material. The focal spot size of the X-ray tube can have an impact on the resolution of the output image in a number of ways, depending on the geometry of the imaging system. Changing the distance between the sample and the detectors has a magnification effect, affecting the resolution. However, this increase in resolution is offset to some degree, depending on the focal spot size, as shown in Fig. 7.2.

For X-ray tubes with relatively large focal spot sizes, the loss of resolution is minimized when the sample is placed directly against the detectors and the distance between the X-ray source and the detectors are kept as large as possible. X-ray tubes with very small focal spots are commercially available, which allow very high resolution via geometric magnification without the focal spot resolution effects. These 'fine focus' tubes, which are the basis for X-ray microscopes, are limited to low filament currents, as the electrons striking the anode are concentrated onto a small area and excessive heating can occur. The low tube current has the consequence of increased image noise, unless compensated with long exposure times, and X-ray microscopes are thus unsuited for high-speed inspection. On the other hand, they are suitable for sampling, as in a quality assurance operation.

Noise is broadly defined as an additive or multiplicative contamination of an image (Castleman, 1979). When the ratio of signal-to-noise is low, it can be

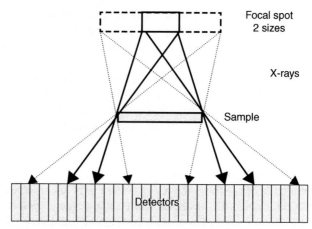

Fig. 7.2 Effect of focal spot size on resolution. Moving the sample toward the X-ray source increases the magnification but also increases the loss of resolution due to the focal spot size.

difficult to gain any meaningful data out of the image, and the subsequent classification becomes difficult. For X-ray imaging systems running at real-time speeds, the primary source of noise is generally quantum (shot) noise. The amount of quantum noise varies between imaging systems, and is predominantly governed by the length of time the detectors are exposed to the X-rays. For high-speed inspection, exposure times are necessarily low, and quantum noise is the predominant source of noise. Given an array of detectors in the vicinity of an X-ray source with no sample in between, or with a uniform sample, an equal number of X-ray photons would be expected to strike each detector, resulting in a uniform image. In reality, the number of photons striking each detector at any given time follows a Poisson distribution. Only with a large number of incident photons will the mean behavior result in the expected uniform distribution. Quantum noise decreases as the square root of the number of incident photons – that is, a fourfold increase in the photon count – corresponds to cutting the quantum noise in half. Alternatively, with incident intensity unchanged, a fourfold increase in the collection area of the detector will result in a twofold reduction in quantum noise. For this reason, there is a tradeoff between quantum noise and resolution in any X-ray detector.

The third critical image quality parameter of interest is contrast, defined as the difference in output for a given difference in input (Jain, 1989). Related is the dynamic range, which describes the ratio between the smallest and largest pixel values in an image. For an 8 bit image, maximum dynamic range (and contrast) occurs when the darkest image pixel has a value of zero and the brightest pixel has a value of 255. Matching the dynamic range of the scene (X-ray intensity) to that of the detector to achieve maximum contrast in the resulting image is an area of interest in all types of imaging, including X-ray and food inspection (Chen *et al.*, 2000; Kotwaliwale *et al.*, 2003). Increasing contrast within a particular area of interest in the X-ray image can improve detection of inclusions (Chen *et al.*,

2001a). It has also been shown that image contrast between soft materials is dependant on X-ray energy and spatial resolution of the imaging system, and for organic materials such as food best contrast is achieved at low energies and high resolution (Zwiggelaar *et al.*, 1997). This has been tested and verified by human recognition studies of insect larvae in X-ray images of infested grain (Schatzki and Keagy, 1991) as well as through computer simulation programs (Zwiggelaar *et al.*, 1996).

Image quality is dependent on the interaction of the different components of the X-ray system, including the X-ray source, conveyor mechanism, detector arrays, image capture, processing and display. It is therefore important to optimize the components to give the best image possible. Techniques for doing so have been investigated and reported (Edwards *et al.*, 2007; Graves *et al.*, 1994). Methods for correcting image deficiencies that are introduced by the X-ray system itself have also been described (Haff and Schatzki, 1997).

X-ray images of food products may contain deficiencies resulting from the curvature of the sample. For packaged goods, the curvature of the container may contribute to this effect as well. This effect is illustrated in Fig. 7.3 for the case of a spherical sample with no container. Since it is necessary to apply sufficient X-ray energy to penetrate the thick center of the sample, the edges can be washed out in the X-ray image due to saturation of either film or solid-state detectors at that location. The resulting image shows lower pixel intensity at the center of the sample compared to the edges. While this may not be a deterrent when inspecting for metal contaminants, which generally absorb all incident X-rays, less dense contaminants, such as wood, bone and even glass, may not be detected if they are situated along the edges. Figure 7.3 illustrates the additional effect of a container,

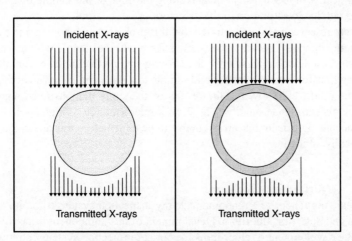

Fig. 7.3 X-rays of uniform intensity incident on a curved object (left) transmit X-rays of non-uniform intensity, resulting in an image with higher pixel intensity toward the edges. The object shown here is representative of a piece of spherical fruit. X-rays incident on an empty can (right) also transmit X-rays of non-uniform intensity, except with higher pixel intensity toward the center. In the case of a filled can, the two effects compete.

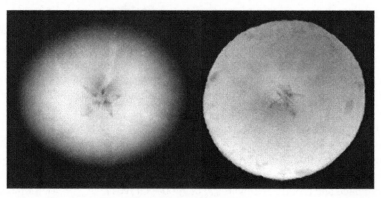

Fig. 7.4   Attenuated (right) and unattenuated (left) X-ray images of an apple (Haff, 2008).

which becomes thicker along the direction of X-ray attenuation toward the edges. For thin-walled containers with low X-ray density, such as aluminum cans, this effect is minimal. For containers with thicker walls or higher density, such as glass or steel, the effect becomes more dominant. Note that for the case of a filled container, the two effects are competing.

This variation in pixel intensity is commonly addressed with a software correction to normalize the image. This is useful when automatic recognition algorithms are used to drive a rejection mechanism, as the algorithms can be affected by the lack of uniformity of image brightness. However, software corrections cannot recover information lost in the imaging process, such as the presence of a small object with low density situated along the edge of the sample that has been washed out due to saturation of the detectors.

Recent research has demonstrated that this problem can be overcome by introducing an attenuator between the X-ray source and the sample with a shape and X-ray density that compensate for the varying sample thickness, essentially generating a transformation from a round sample to a flat sample as seen at the X-ray detectors (Haff, 2008). While the results of the study demonstrated significant improvement in image quality (Fig. 7.4), implementation at high speed is cumbersome and significant research remains to be done before this technique can be realistically applied.

7.3.3   Algorithm development

A critical component of an automatic X-ray inspection system is the image processing and detection algorithm. Development of this component is nearly identical to the development of algorithms for optical inspection systems, so it will not be discussed in detail here. The majority of detection algorithms for X-ray images of food products use a variation of discriminant analysis (Haff and Pearson, 2007), neural networks (Brosnan *et al.*, 1996; Casasent *et al.*, 1996; Talukder *et al.*, 1998), or simple/adaptive thresholding (Chen *et al.*, 2001b). Spectral filtering of the X-ray image has also been shown to be an effective tool (Krger *et al.*, 2004; Shahin *et al.*, 1999).

7.4   X-ray inspection of food products

The sections below discuss specific applications of X-ray technology, illustrating the versatile and varied uses to which it can be put.

7.4.1   Luggage inspection for contraband food products

While most X-ray applications of food products involve detecting defects or con-taminants within the product itself, there is also an important field of study involv-ing X-ray detection of food items concealed within a container, generally luggage. Efforts to protect agricultural crops from invasion by foreign pests make this area of study particularly crucial.

Currently, incoming passenger luggage is X-rayed upon arrival. If the inspec-tor finds anything in the image that looks suspicious, the luggage is diverted to a hand inspection station (Schatzki, 1991). Hand inspection of luggage is time-con-suming and expensive, requiring an inspector to remove the contents of a piece of luggage and often open every container found within to search for potential contraband items. Creating a computer vision inspection system that could either reduce the number of pieces of luggage that need to be inspected, or help guide the inspectors in identifying which type of contraband they should be looking for, would be very useful. However, designing computer algorithms that can handle the complex images generated by the X-ray machine is a formidable task. Luggage comes in a wide variety of sizes and can be packed with an even larger range of objects, most of which are not of interest to agricultural inspectors. Many items that are not contraband can look like agricultural items due to similar shape and density, causing high proportions of false positives. Additionally, X-ray images of agricultural products look very different to the way the products do under normal viewing conditions. An example with olives can be seen in Fig. 7.5. Generally these problems are mitigated with extensive X-ray inspection experience which is expensive and time-consuming (Toyofuku and Schatzki, 2007).

For a limited range of contraband items, research efforts have shown that auto-matic shape recognition software combined with dual energy imaging is an effec-tive tool (Schatzki, 1985, 1991; Schatzki and Young, 1988; Schatzki *et al.*, 1983, 1984). Due to software and hardware limitations at the time, the program focused on detecting round objects in the X-ray images of luggage. This feature was cho-sen because agricultural items tend to be round (fruits, canned foods and seeds). However, the software would still trigger false positives with items like compact discs, jars of cold cream and several other frequently seen luggage items that appear round on X-ray images.

More recently, Toyofuku and Schatzki (2007) examined the types of image fea-tures, other than just round objects, that would most reliably signal the presence

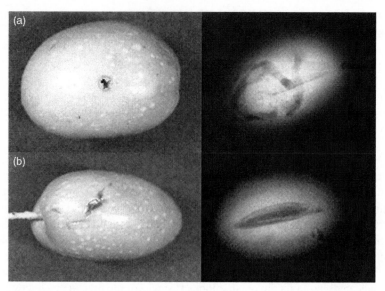

Fig. 7.5 Digital photographs and X-ray images of olives with fruit fly infestation (a) and without (b) (Jackson and Haff, 2006). Note that in the photographs both olives show what appears to be insect damage. For the first olive, X-ray imaging reveals that the internal damage is much greater than is apparent from visual inspection. For the second olive, external damage turns out to be uncorrelated with insects, as the X-ray image shows no interior infestation.

of agricultural products. They found that by expanding the feature set to include round, tubes and granular with subset modifiers of dense and moon cake (named because of the very distinct features of this product), they could produce very promising and consistent inspection results at real-time inspection rates. While this research was performed with human inspectors, the list and training was structured to allow for future adaptation in a computer recognition algorithm.

7.4.2   Packaged foods

It is in the packaged foods segment of the industry that X-ray inspection has traditionally been implemented to the greatest extent. Product packaged in bottles or cans is ideally suited for high-speed X-ray inspection, with line-scan units routinely processing more than 20 samples per second. The processing plant environment introduces the possibility of contamination of the product by metals, plastics, glass, bone fragments, etc. Line-scan X-ray units are rapidly replacing metal detectors because of their ability to detect all kinds of contaminants. The uniformity of packaged materials as compared to fresh produce makes it a better candidate for quality X-ray images. Finally, the types of foreign contaminants generally of interest in packaged foods (metal, plastic, glass, etc.) provide much higher contrast in X-ray images than do the typical defects or contaminants of interest that are found in fresh produce (insect infestation, physiological defects, etc.).

Areas of reported research include detection of common contaminants as described above through system development (Batchelor and Davies, 2004; Cambier and Pasiak, 1986; Penman *et al.*, 1992; Schatzki and Wong, 1989; Schatzki *et al.*, 1995) or rejection algorithm development (Dearden, 1996; Patel *et al.*, 1995). Simulation studies have indicated that CT scanning using CT numbers as opposed to X-ray absorption coefficient could give the ability to detect inclusions smaller than the physical resolution of the CT scanner due to the smooth transition of CT number between the inclusion and surrounding material (Ogawa *et al.*, 1998). These results were obtained through theoretical simulation and not verified using actual product. Furthermore, practical considerations such as cost and exposure time currently make CT an impractical detection method. Nonetheless, it does suggest that future technologies could increase the detection capabilities of X-ray systems in general.

7.4.3   Poultry inspection

Poultry inspection is another segment of the food industry that employs X-ray inspection on a routine basis. The inclusion of greatest interest is bone fragments, often left behind by the de-boning process. While standard line-scan systems are effective for detection of heavier contaminants such as metal or rock, detection of softer material is hampered by the irregular shape and non-uniform thickness of the product. From an algorithm approach, this problem has been addressed by applying adaptive thresholding to the image (Chen *et al.*, 2001b) or using local contrast enhancement (Chen *et al.*, 2001a). In addition, a new X-ray imaging system combining laser range images with X-ray image data is designed to determine thickness at any point in the image (Chen *et al.*, 2003; Jing *et al.*, 2003; Tao and Ibarra, 2000).

7.4.4   Grain inspection

Of all food commodities, the greatest amount of research found in the literature regarding X-ray inspection is devoted to grain. Most of this research effort has been devoted to the problem of insect infestation in wheat kernels. Scanned film images of wheat kernels infested by larvae of the granary weevil, *Sitophilus granarius* (L.), are shown in Fig. 7.6. In spite of considerable research effort spanning several decades, bulk grain is still not routinely inspected by X-ray. The reason for this is twofold: the size of grain kernels mandates inspection speeds beyond the capability of even the fastest computers, plus the inability of current high-speed X-ray systems to detect larvae at their earlier stages.

The challenges have not discouraged research efforts. For detection of insect infestations, X-ray has been shown to be superior to NIR detection, except in the case where species identification is required (Karunakaran *et al.*, 2005). CT scanning has been demonstrated to be an effective tool for rapid scanning of 100 g samples for insect infestation (Towes *et al.*, 2005, 2006), but is too slow for bulk inspection. X-ray system parameters for maximizing recognition of insects in wheat

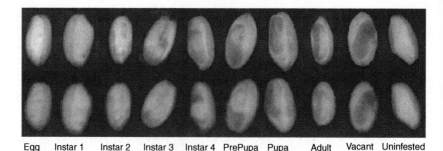

Egg   Instar 1   Instar 2   Instar 3   Instar 4   PrePupa   Pupa   Adult   Vacant   Uninfested

Fig. 7.6   Digital X-ray images (bottom) and scanned film images (top) for the various larval stages of the granary weevil in wheat kernels (Haff and Slaughter, 2004). The bottom images were acquired on a high-speed X-ray imaging system. Compared with the scanned film images, it is clear that with existing equipment the earliest larval stages cannot be detected at high speed due to lower resolution and higher noise levels.

kernels have been investigated (Keagy and Schatzki, 1991; Schatzki and Keagy, 1991). Human recognition studies based on scanned film images have shown that X-ray imaging is extremely reliable for quality control purposes (Haff and Slaughter, 2002, 2004; Haff and Toyofuku, 2000; Schatzki and Fine, 1989) but is obviously of no use for bulk inspection. While several effective automatic recognition algorithms have been reported (Brosnan et al., 1996; Haff, 2001; Haff and Pearson, 2007; Jayas et al., 2003; Karunakaran et al., 2003a, 2003b, 2005; Keagy and Schatzki, 1993), in most cases the images used for training and testing were obtained using high resolution X-ray systems, such as film, X-ray fluoroscopes, or X-ray microscopes that require exposure times, making bulk inspection unrealistic.

The fastest system reported is an X-ray image intensifier/CCD camera combination with an exposure time on the order of 143 ms, or 7 kernels per second (Haff, 2001; Haff and Slaughter, 2004). Although seemingly rapid, this is still far too slow for bulk inspection without introducing many channels at the same time, which is cost-prohibitive. The CCD camera in this study was the limiting factor in terms of speed, and digital cameras with much higher frame rates exist today. However, efforts to increase the frame rate lead to the problem of image deterioration due to quantum noise.

It would seem, then, that the limiting factor for high-speed automatic inspection of wheat for insect infestation is an X-ray tube with low energy, on the order of 20 KeV, coupled with very high current, on the order of hundreds of milliamps. Such a source installed in a high-speed line-scan system should make bulk inspection feasible. As equipment costs, especially in the area of power supplies, continue to decrease, such a system may become available.

Although the bulk of X-ray work devoted to grain inspection has been for insect infestation, a few studies have used X-ray technology to detect other defects or quality parameters in grain. X-ray imaging and machine vision have been combined to detect multiple defects in cereal grain (Jayas et al., 2004), and an X-ray fluoroscope system has been shown to be effective for mass determination in wheat kernels (Karunakaran et al., 2004).

7.4.5   Apples

Research involving X-ray imaging of apples for foreign contaminants has primarily concentrated on detection of codling moth damage. Detection of codling moth larvae in apples (Fig. 7.7) has been investigated using CT (Hansen *et al.*, 2005), film (Hansen *et al.*, 2005; Schatzki *et al.*, 1996, 1997), as well as line-scan (Lin *et al.*, 2005; Schatzki *et al.*, 1996, 1997) X-ray systems.

7.4.6   Cherries

Pits are difficult to detect with optical methods, but due to the difference in density, they are well suited for detection with X-ray systems. Jackson *et al.* (2009) built a system that non-destructively detected cherry pits with a total performance error rate of 3.5%. By using a system that combined the output of photodiode X-ray detectors in a line-scan configuration, they were able to reduce the costs of the system below that of most typical X-ray systems.

7.4.7   Tree nuts

X-ray images of pistachio nuts infested by the naval orange worm (NOW) have been used as training and validation sets for the development of algorithms for insect detection (Casasent *et al.*, 1998; Keagy *et al.*, 1996a, 1996b; Sim *et al.*, 1995) as well as separation of touching samples in X-ray images (Casasent *et al.*, 2001; Talukder *et al.*, 1998). Algorithm strategies include: neural networks that achieve 98% recognition with less than 1% false positives (good product classified as bad) on scanned film images, discriminant analysis routines achieving 89% accuracy in line-scan images and multiple feature extraction strategies. Although developed for insect detection in pistachio nuts, these algorithm strategies should be useful for other commodities as well, particularly the segmentation algorithms for separating touching samples. Algorithms have also been developed for the detection of NOW in almonds based on scanned film as well as line-scan images (Kim and Schatzki, 2001) and the burrowing activity of the pecan weevil has been studied using X-ray (Harrison *et al.*, 2007).

7.4.8   Miscellaneous food products

More limited research has been reported on the use of X-ray imaging for foreign contaminant detection in other food products. Thomas *et al.* (1995) looked for

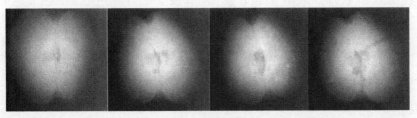

Fig. 7.7   X-ray images of codling moth infestation in an apple. The ages of infestation are, from left to right, 3, 12, 14, and 17 days after the apple was inoculated with eggs. This again illustrates the difficulty of detecting insects at the earlier life stages.

seed weevils in mango, Lin *et al.* (2005) investigated insect infestation in guava fruit and peaches and Jackson and Haff (2006) examined the olive fruit fly (see Fig. 7.5).

7.5   Conclusions

Recent advances in both optical and X-ray inspection have made computer vision inspection systems a viable option for foreign body detection and removal in a wide range of products in the food industry. While optical inspection has been used for many decades in the food industry, it is only in the last few decades that it has become a part of computer vision inspection systems that can automatically sort products using multiple criteria, while maintaining high levels of accuracy and working at real-time processing rates. The systems continue to evolve. Work with FPGAs and CMOS have overcome previous limitations of processing speed while maintaining a low implementation cost.

X-ray inspection of food products as part of real-time food product processing is relatively new, but due to its unique advantages of non-destructive internal inspection, it has motivated considerable research efforts. Based on past trends, it is likely that future technological improvements will see decreased sensor sizes, increased SNRs and increased speed of processing. These advances would make X-ray inspection available to inspect an even wider range of products.

Both approaches will reduce the variability, cost and inaccuracy problems encountered with other systems, such as human inspection or automated sorting using only simple logic circuits. Certain products and foreign contaminants that could not be processed automatically before due to their size or complexity may soon have that option. The flexibility demonstrated by the hardware and software developments thus far promises that a wide range of products could ultimately be processed with a computer vision system.

7.6   References

Abbott, J. A. (1999) Quality measurements of fruits and vegetables. *Postharvest Biology and Technology*, **15**, 207–25.

Abbott, J. A., Lu, R., Upchurch, B. L. and Stroshine, R. L. (1997) Technologies for non-destructive quality evaluation of fruits and vegetables. In Janick, J. (ed.), *Horticultural Reviews*. Chichester, UK: Wiley, 1–120.

Batchelor, B. G. and Davies, E. R. (2004) Using x-rays to detect foreign bodies. In M. Edwards (Ed.), *Detecting Foreign Bodies in Food*, 1st edn. Boca Raton, FL.: Woodhead, 226–64.

Bee, S. C. and Honeywood, M. J. (2004) Optical sorting systems. In Edwards, M. (ed.), *Detecting Foreign Bodies in Food*. Cambridge, England: Woodhead, 86–118.

Brosnan, T. and Sun, D.-W. (2004) Improving quality inspection of food products by computer vision – a review. *Journal of Food Engineering*, **61**, 3–16.

Brosnan, T. M., Daley, W. D. and Smith, M. J. T. (1996) Comparison of image analysis techniques for defect detection in food-processing applications. In *Proceedings for SPIE*

2907: Optics in Agriculture, Forestry, and Biological Processing II, November 19 1996, Boston, MA, USA, 186-194.

Butz, P., Hofmann, C. and Tauscher, B. (2005) Recent developments in noninvasive techniques for fresh fruit and vegetable internal quality analysis. *Journal of Food Science,* **70,** 131–41.

Cambier, J. L. and Pasiak, D. C. (1986) On-line X-ray inspection of packaged foods. In *Proceedings for Society of Manufacturing Engineers Conference on Vision,* Machine Vision Association of the Society of Manufacturing Engineers, Detroit, MI, USA, 6–21.

Casasent, D. P., Sipe, M. A., Schatzki, T. F., Keagy, P. M. and Le, L. C. (1996) Neural net classification of X-ray pistachio nut data. *Proceedings of SPIE, 2907,* 217–27.

Casasent, D. P., Sipe, M. A., Schatzki, T. F., Keagy, P. M. and Le, L. C. (1998) Neural net classification of X-ray pistachio nut data. *Lebensmittel-Wissenschaft und-Technologie,* **31,** 122–8.

Casasent, D. P., Talukder, A., Keagy, P. M. and Schatzki, T. F. (2001) Detection and segmentation of items in X-ray imagery. *Transactions of the ASAE,* **44,** 337–45.

Castleman, K. R. (1979) *Digital Image Processing.* Englewood Cliffs, NJ: Prentice Hall.

Chen, X., Jing, H., Tao, Y., Carr, L. and Wheaton, F. (2003) Real-time detection of physical hazards in de-bonded poultry using high resolution X-ray imaging. 2003 ASAE/CSAE Annual International Meeting Paper # 033084.

Chen, Z., Tao, Y., Chen, X. and Ying, Y. (2000) Contrast enhancement of X-ray imaging and postprocessing for food internal inspection. Proceedings of the ASAE Annual International Meeting, Milwaukee, WI, Paper # 003124.

Chen, Z., Tao, Y. and Xin, C. (2001a) Multiresolution local contrast enhancement of X-ray images for poultry meat inspection. *Applied Optics,* **40,** 1195–200.

Chen, Z., Tao, Y., Jing, H. and Walker, J. (2001b) Internal inspection of deboned poultry using X-ray imaging and adaptive thresholding. *Transactions of the ASABE,* **44,** 1005–9.

Commission of the European Communities (2000) White Paper on Food Safety. http://ec.europa.eu/dgs/health_consumer/library/pub/pub06_en.pdf [accessed 4/2/2007].

Dearden, R. (1996) Automatic X-ray inspection for the food industry. *Food Science and Technology Today,* **10,** 87–90.

Ding-ji, S., Shou-Yan, W., Han-Wen, H. and Bo-Xin, W. (1994) Reduce image blur due to motion by sight-line following. *Proceedings of the IEEE International Conference on Industrial Technology,* December 5–9, Guangzhou, China, 659–64.

Domenech-Asensi, G. (2004) Applying optical systems. In Edwards, M. (ed.), *Detecting Foreign Bodies in Food.* Woodhead, 119–31.

Edwards, M., Marsh, R. A., Angold, R. E., Gaze, R., Campbell, A., Hines, T., *et al.* (2004) *Detecting Foreign Bodies in Food.* Cambridge: Woodhead.

Edwards, M., Marsh, R. A., Angold, R. E., Gaze, R., Campbell, A., Hines, T., *et al.* (2007) *Detecting Foreign Bodies in Food.* Cambridge: Woodhead.

Graves, M., Smith, A. and Batchelor, B. G. (1998) Approaches to foreign body detection in foods. *Trends in Food Science and Technology,* **9,** 21–7.

Graves, M., Smith, A., Batchelor, B. G. and Palmer, S. C. (1994) Design and analysis of X-ray vision systems for high-speed detection of foreign body contamination in food. *SPIE Proceedings,* **2347,** 80–92.

Guide to X-ray Inspection by Safeline (2007) http://www.Safelinexray.com/about-x-ray.html [accessed 11/24/2007].

Haff, R. P. (2001) *X-ray Inspection of Wheat for Granary Weevils.* Davis: University of California.

Haff, R. P. (2008) Real-time correction of distortion in X-ray images of cylindrical or spherical objects and its application to agricultural commodities. *Transactions of the ASABE,* **51,** 341.

Haff, R. P. and Jackson, E. S. (2008) Low cost real-time sorting of in shell pistachio nuts from kernels. *Applied Engineering in Agriculture,* **24,** 487–90.

Haff, R. P. and Pearson, T. C. (2006) Spectral band selection for optical sorting of pistachio nut defects. *Transactions of the ASABE*, **49**, 1105–13.

Haff, R. P. and Pearson, T. C. (2007) An automatic algorithm for detection of infestations in X-ray images of agricultural products. *Journal of Sensing and Instrumentation for Food Quality and Safety*, **1**, 143–50.

Haff, R. P. and Schatzki, T. F. (1997) Image restoration of line-scanned X-ray images. *Optical Engineering*, **36**, 3288–96.

Haff, R. P. and Slaughter, D. C. (2002) X-ray inspection of wheat for granary weevils. Real-time digital imaging vs film. *2002 ASAE Annual International Meeting/CIGR XVth World Congress*. Chicago, Illinois, USA: ASAE, 531–7.

Haff, R. P. and Slaughter, D. C. (2004) Real-time X-ray inspection of wheat for infestation by the granary weevil, *Sitophilus Granarius (L.)*. *Transactions of the ASAE*, **47**, 531–7.

Haff, R. P. and Toyofuku, N. (2000) *An Economical Method for X-ray Inspection of Wheat* (Rep. No. Paper No. 003069).

Haff, R. P. and Toyofuku, N. (2008) X-ray detection of defects and contaminants in the food industry. *Sensing and Instrumentation for Food Quality and Safety*, **2**, 262–73.

Hansen, J. D., Haff, R. P., Schlaman, D. W. and Yee, W. L. (2005) Potential postharvest use of radiography to detect internal pests in deciduous tree fruits. *Journal of Entomological Science*, **40**, 255–62.

Harrison, R. D., Gardner, W. A., Tollner, E. W. and Kinard, D. J. (2007) X-ray computed tomography studies of the burrowing behavior of the fourth-instar pecan weevil(coleoptera:Curculionidae). *Journal of Economic Entomology*, **86**, 1714–19.

Jackson, E. S. and Haff, R. P. (2006) X-ray detection of and sorting of olives damaged by fruit fly. In Transactions of the ASABE, 2006 ASAE Annual Meeting, Portland, OR, USA (Paper #06-6062).

Jackson, E. S., Haff, R. P. and Gomez, J. (2009). Real-time methods for non-destructive detection of pits in fresh cherries. In *Proceedings of the ASABE*, American Society of Agricultural and Biological Engineers Annual International Meeting, June 21-24, 2009, Reno NV, USA. Paper # 096249

Jain, A. K. (1989) *Fundamentals of Digital Image Processing*. Englewood Cliffs, NJ: Prentice Hall.

Jayas, D. S., Karunakaran, C. and Paliwal, J. (2004) Grain quality monitoring using machine vision and soft x-rays for cereal grains. In *Proceedings of the 2004 International Quality Grains Conference*, July 19-21, Indianapolis, IN.

Jayas, D. S., Karunakaran, C. and White, N. D. G. (2003) Soft X-ray image analysis to detect wheat kernels damaged by *Plodia interpunctella* (Lepidoptera: Pyralidae). *Sciences des Aliments*, **23**, 623–31.

Jha, S. N. and Matsuoka, T. (2000) Non-destructive techniques for quality evaluation of intact fruits and vegetables. *Food Science and Technology Research*, **6**, 248–51.

Jing, H., Chen, X. and Tao, Y. (2003) Analysis of factors influencing the mapping accuracy of X-ray and laser range images in a bone fragment detection system. In *Proceedings from the 2003 ASAE Annual Meeting*, July 27-30, Las Vegas, NV, USA. (Paper #036191).

Karunakaran, C., Jayas, D. S. and White, N. D. G. (2003a) Soft X-ray inspection of wheat kernels infested by *Sitophilus oryzae*. *Transactions of the ASAE*, **46**, 739–45.

Karunakaran, C., Jayas, D. S. and White, N. D. G. (2003b) X-ray image analysis to detect infestations caused by insects in grain. *Cereal Chemistry*, **80**, 553–7.

Karunakaran, C., Jayas, D. S. and White, N. D. G. (2004) Mass determination of wheat kernels from X-ray images. *2004 ASAE/CSAE Annual International Meeting, Ontario, Canada*.

Karunakaran, C., Paliwal, J., Jayas, D. S. and White, N. D. G. (2005) Comparison of soft x-rays and NIR spectroscopy to detect insect infestations in grain. *2005 ASAE Annual International Meeting*. Tampa, Florida.

Keagy, P. M., Parvin, B. and Schatzki, T. F. (1996a) Machine recognition of naval orange worm damage in X-ray images of pistachio nuts. *Lebensmittel-Wissenschaft und-Technologie*, **29**, 140–5.

Keagy, P. M. and Schatzki, T. F. (1991) Effect of image resolution on insect detection in wheat radiographs. *Cereal Chemistry*, **68**, 339–43.

Keagy, P. M. and Schatzki, T. F. (1993) Machine recognition of weevil damage in wheat radiograms. *Cereal Chemistry*, **70**, 696–700.

Keagy, P. M., Schatzki, T. F., Le, L. C., Casasent, D. P. and Weber, D. (1996b) Expanded image database of pistachio X-ray images and classification by conventional methods. In *Proceedings for SPIE 2907: Optics in Agriculture, Forestry, and Biological Processing II*, November 19, 1996, Boston, MA, USA, 196-204.

Kim, S. and Schatzki, T. F. (2001) Detection of pinholes in almonds by X-ray imaging. *Transactions of the ASAE*, **44**, 997–1003.

Kotwaliwale, N., Subbiah, J., Weckler, P., Brusewitz, G. and Kranzler, G. A. (2003) Digital radiography for quality determination of small agricultural products: development and calibration of equipment. In *Proceedings from the 2003 ASAE Annual Meeting*, July 27-30, Las Vegas, NV, USA. (Paper #036231).

Krger, C., Bartle, C. M., West, J. G. and Tran, V.-H. (2004) Digital x-ray imaging and image processing for object detection in closed containers. In *Proceedings of Computer Graphics and Imaging*, August 17-19, Kauai, Hawaii, USA.

Lin, T.-T., Chang, H.-Y., Wu, K.-H., Jiang, J.-A., Ouyang, C.-S., Yang, M.-M., *et al.* (2005) An adaptive image segmentation algorithm for X-ray quarantine inspection of selected fruits. *ASABE Meeting Presentation Paper # 053123*.

Majumdar, S. and Jayas, D. S. (2000a) Classification of cereal grains using machine vision: II color models. *Transactions of the ASAE*, **43**, 1677–80.

Majumdar, S. and Jayas, D. S. (2000b) Classification of cereal grains using machine vision: IV combined morphology color and texture models. *Transactions of the ASAE*, **43**, 1689–94.

Ogawa, Y., Morita, K., Tanaka, S., Setoguchi, M. and Thai, C. N. (1998) Application of X-ray CT for detection of physical foreign materials in foods. *Transactions of the ASAE*, **41**, 157–62.

Pasikatan, M. C. and Dowell, F. E. (2003) Evaluation of high-speed color sorter for segregation of red and white wheat. *Applied Engineering in Agriculture*, **19**, 71–6.

Patel, D., Davies, E. R. and Hannah, I. (1995) Towards a breakthrough in the detection of contaminants in food products. *Sensor Review*, **15**, 27–8.

Pearson, T. C. (1996) Machine vision system for automated detection of stained pistachio nuts. *Lebensmittel-Wissenschaft und-Technologie*, **29**, 203–9.

Pearson, T. C. (2009) Hardware-based image processing for high-speed inspection of grains. *Computers and Electronics in Agriculture*, **69**, 12–18.

Pearson, T. C. (2010) High-speed sorting of grains by color and surface texture. *Applied Engineering in Agriculture*, **26**, 499–505.

Pearson, T. C. and Toyofuku, N. (2000) Automated sorting of pistachio nuts with closed shells. *ASAE Applied Engineering in Agriculture*, **16**, 91–4.

Pearson, T. C. and Young, R. (2001) Automated sorting of almonds with embedded shell by laser transmittance imaging. *Applied Engineering in Agriculture*, **18**, 637–41.

Pearson, T. C., Doster, M. A. and Michailides, T. J. (2001) Automated sorting of pistachio defects by machine vision. *Applied Engineering in Agriculture*, **17**, 729–32.

Pearson, T. C., Brabec, D. and Haley, S. (2008) Color image based sorter for separating red and white wheat. *Sensing and Instrumentation for Food Quality and Safety*, **2**, 280–8.

Penman, D. W., Olsson, O. J. and Beach, D. (1992) Automatic x-ray inspection of canned products for foreign material. In *Proceedings SPIE Machine Vision Applications, Architectures, and Systems Integration*, November 17, Boston, MA, USA, 342–7.

Qin, J. and Lu, R. (2005) Detection of pits in tart cherries by hyperspectral transmission imaging. *Transactions of the ASAE*, **48**, 1963–70.

Rosenfeld, A. and Kak, A. C. (1982) *Digital Image Processing*. Orlando, FL: Academic Press.

Ruiz-Altisent, M., Ruiz-Garcia, L., Moreda, G. P., Lu, R., Hernandez-Sanchez, N., Correa, E. C. *et al.* (2010) Sensors for product characterization and quality of specialty crops – a review. *Computers and Electronics in Agriculture*, **74**, 176–94.

Sablatnig, R. (1997) *A Highly Adaptable Concept for Visual Inspection.* Vienna University of Technology.

Schatzki, T. F. (1985) Detection of agricultural contraband in luggage. [4,539,648]. US. Ref Type: Patent.

Schatzki, T. F. (1991) Effectiveness of screening methods for contraband recovery. *Journal of Economic Entomology*, **84**, 489–95.

Schatzki, T. F. and Fine, T. A. (1989) Analysis of radiograms of wheat kernels for quality control. *Cereal Chemistry*, **65**, 233–9.

Schatzki, T. F. and Keagy, P. M. (1991) Effect of image size and contrast on the recognition of insects. *Proceedings of SPIE*, **1379**, 182–8.

Schatzki, T. F. and Wong, R. Y. (1989) Detection of submilligram inclusions of heavy metals in processed foods. *Food Technology*, **43**, 72–6.

Schatzki, T. F. and Young, R. (1988) X-Ray imaging of baggage for agricultural contraband. *Journal of Transportation Engineering*, **114**, 657–71.

Schatzki, T. F., Grossman, A. and Young, R. (1983) Recognition of agricultural objects by shape. *IEEE Transactions on Pattern Analysis and Machine Intelligence*, **5**, 645–53.

Schatzki, T. F., Young, R. and Duryea, R. (1984) Detection of agricultural items by image processing. In . St Joseph, MI: American Society of Agricultural Engineers, 631–9.

Schatzki, T. F., Young, R., Haff, R. P., Eye, J. G. and Wright, G. R. (1995) Visual detection of particulates in X-ray images of processed meat products. In *SPIE conference on Optics in Agriculture, Forestry, and Biological Processing*, November 1994, Boston, MA, USA, 348–53.

Schatzki, T. F., Haff, R. P., Young, R., Can, I., Le, L. C. and Toyofuku, N. (1996) Defect detection in apples by means of X-ray imaging. In *Proceedings for SPIE 2907: Optics in Agriculture, Forestry, and Biological Processing II*, November 19, Boston, MA, USA, 176-85.

Schatzki, T. F., Haff, R. P., Can, I., Le, L. C. and Toyofuku, N. (1997) Defect detection in apples by means of X-ray imaging. *Transactions of the ASAE*, **40**, 1407–15.

Shahin, M. A., Tollner, E. W. and Prussia, S. E. (1999) Filter design for optimal feature extraction from X-ray images. *Transactions of the ASABE*, **42**, 1879–87.

Sim, A., Parvin, B. and Keagy, P. M. (1995) Invariant representation and hierarchical network for inspection of nuts from X-ray images. In *IEEE International Conference on Neural Networks*, November 27 to December 1, Perth, WA, USA, 738–43.

Talukder, A., Casasent, D. P., Lee, H.-W., Keagy, P. M. and Schatzki, T. F. (1998) A new feature extraction method for classification of agricultural products from X-ray images. *Proceedings of SPIE*, **3543**, 53–64.

Tao, Y. and Ibarra, J. G. (2000) Thickness-compensated X-ray imaging detection of bone fragments in deboned poultry-model analysis. *Transactions of the ASABE*, **43**, 453–9.

Thomas, P., Kannan, A., Degwekar, V. H. and Ramamurthy, M. S. (1995) Non-destructive detection of seed weevil-infested mango fruits by X-ray imaging. *Postharvest Biology and Technology*, **5**, 161–5.

Towes, M. D., Pearson, T. C. and Campbell, J. F. (2005) Rapid detection of internal insect infestations in bulk grain. In *2005 Entomological Society of America Annual Meeting and Exhibition*, December 17, Fort Lauderdale, FL, USA. (Paper #1070).

Towes, M. D., Pearson, T. C. and Campbell, J. F. (2006) Imaging and automated detection of *Sitophilus oryzae* (Coleoptera: Curculionidae) pupae in hard red winter wheat. *Journal of Economic Entomology*, **99**, 583–92.

Toyofuku, N. and Schatzki, T. F. (2007) Image feature based detection of agricultural quarantine materials in X-ray images. *Journal of Air Transport Management*, **13**, 348–54.

Visen, N. S., Shashidhar, N. S., Paliwal, J. and Jayas, D. S. (2001) Identification and Segmentation of Occluding Groups of Grain Kernels in a Grain Sample Image. *Journal of Agricultural Engineering Research*, **79**, 159–66.

Wang, N., Zhang, N., Dowell, F. E. and Pearson, T. C. (2005) Determining vitreousness of durum wheat using transmitted and reflected images. *Transactions of the ASAE*, **48**, 219–22.

Zwiggelaar, R., Bull, C. R. and Mooney, M. J. (1996) X-ray simulations for imaging applications in the agricultural and food industries. *Journal of Agricultural Engineering Research*, **63**, 161–70.

Zwiggelaar, R., Bull, C. R., Mooney, M. J. and Czarnes, S. (1997) The detection of 'soft' materials by selective energy X-ray transmission imaging and computer tomography. *Journal of Agricultural Engineering Research*, **66**, 203–12.

8

Automated cutting in the food industry using computer vision

W. D. R. Daley, Georgia Institute of Technology, USA and O. Arif, King Abdullah University of Science and Technology, Saudi Arabia

Abstract: The processing of natural products has posed a significant problem to researchers and developers involved in the development of automation. The challenges have come from areas such as sensing, grasping and manipulation, as well as product-specific areas such as cutting and handling of meat products. Meat products are naturally variable and fixed automation is at its limit as far as its ability to accommodate these products. Intelligent automation systems (such as robots) are also challenged, mostly because of a lack of knowledge of the physical characteristic of the individual products. Machine vision has helped to address some of these shortcomings but underperforms in many situations. Developments in sensors, software and processing power are now offering capabilities that will help to make more of these problems tractable. In this chapter we will describe some of the developments that are underway in terms of computer vision for meat product applications, the problems they are addressing and potential future trends.

Key words: computer vision, machine vision, algorithms, machine learning, visible sensors, infrared (IR) sensors, X-ray.

8.1 Introduction

Natural products do not have 'design tolerances' (Graves and Batchelor, 2003).

Two areas that have hindered automation have been the lack of developed sensing technologies and the variability in the product being measured (Ring *et al.*, 2007).

These statements, the first by Graves and Batchelor and the latter by Ring and colleagues, sum up the challenges faced in sensing and handling natural products, of which the slaughter and handling of animals is an example. Almost all commercial slaughter operations have developed to a level where they are probably operating at their peak efficiency in terms of the ability of the equipment to accommodate the natural variability of the product. Meat, fish and poultry seemed to be natural candidates for automation (Khodabandehloo, 1993) but challenges still remain in terms of fielding systems today. As stated by Wilson 'although the packaging of finished food products is a relatively simple task, automating the processing (sic) of slaughtering and then butchering the animals is more complex' (Wilson, 2010). In a processing plant, initial processing (this encompasses killing to evisceration) is mostly all automated with decreasing levels of overall automation as the size of the animals increases – for example, from the avian to the bovine species. The problem becomes more acute in the second and further processing areas of the plant. In these areas, most of the value is added to the product and requires many specialized operations, depending on the customer requirements.

Today, the ability to conduct these specialized operations is dependent on whether or not the producer is able to find human labor that is able to provide the flexibility to carry out the required operations. These operations can be product-specific and in some cases require judgment. In order to meet the anticipated future needs of these industries, the equipment here and in other sections of the plant will have to be flexible to accommodate the incoming product mix. This will require that the devices used have the ability to sense and respond to the structure and variability of the product being processed. The ability to respond to this variability has been identified as a significant factor in our ability to automate these processes (Ring et al., 2007). Much work has been done in automating 'front end' processing for poultry, but secondary and further processing is where the profit margin is. The majority of producers now require a deboned product along with specialty cuts to meet the needs of their customers. Most of these operations are currently being done satisfactorily by people. This, however, creates bottlenecks and inconsistencies in the process and because of the relatively high-speed operations the line workers are placed at risk for repetitive motion injuries. Additionally, these workers typically also have inspection tasks that are sometimes not recognized until automation is inserted for that particular operation. These aspects all have to be borne in mind as we seek to streamline and improve these operations.

For the next generation of machinery to respond to the natural variability of products, the ability to properly sense and manipulate these products will be a necessity. Machine and computer vision combined with the appropriate data analysis will be essential elements of the next generation of solutions. In the end, the following functions have to be accommodated:

- flexibility needed to handle variability;
- non-uniform operations that people handle well and
- systems that behave like people (ability to handle variability and novel situations).

This might require human-like robots to provide the required functionality and adaptability; this results in key technology needs in the areas of sensing and manipulation. Sensing here is defined as detecting some measureable property and analyzing the measurements to make useful decisions. This chapter will not address the manipulation challenge but in most cases the proper handling of a wet, slippery product is still a challenging problem. Two technologies that will be beneficial in solving this problem are machine vision and machine learning. This is because so much of what we do is visual and our ability to learn and generalize plays a significant role in our ability to perform in these somewhat semi-unstructured situations. The outline for the rest of the chapter is as follows: a brief description of the relevant technologies will be presented, several applications and the approaches taken will follow and, finally, some thoughts for the future will be given.

8.2   Machine vision and computer vision

A differentiation is usually made between machine and computer vision in terms of the duties that they are intended to perform. Machine vision typically refers to the processing of sensor data that allows a machine to extract information of interest for achieving some specific function.

One of the early definitions of machine vision, coined by the Machine Vision Association (MVA) of the Society of Manufacturing Engineers was (Zuech, 1988): 'The use of devices for optical noncontact sensing to automatically receive and interpret an image of a real scene in order to obtain information and/or control machines or processes.' The MVA was one of the first professional groups formed around promoting machine vision and its applications. A more recent definition states (Davies, 2005): 'Machine vision is the study of methods and techniques whereby artificial vision systems can be constructed and usefully employed in practical applications. As such, it embraces both the science and engineering of vision.'

The second definition makes explicit what is implicit in the first definition: that the operations are usually part of production processes and have to operate in 'real time'. The data is not only optical but can come from a variety of sources such as ultrasonic, visible, X-rays or infrared. The other aspect of these solutions is the systems integration requirement that usually requires the skills of physicists, electrical and mechanical engineers and software developers. The overall goal is to build systems to address the myriad manufacturing sensing and process control problems. Food applications bring their own challenges because of the non-standard nature of the product. This manifests itself especially in the development of the software, as, in most cases, it is necessary to handle the natural variability in the products. This can be challenging to say the least.

We will describe the process using the human visual system as a model. In this context we will also briefly describe the elements and operating principles of the typical machine vision system. The word 'typical', however, is a misnomer as most successful applications tend to be domain-specific. We are still not at this time able to solve the general machine vision problem. So the question that is typically asked of machine vision systems is not 'What is this?' but rather 'Is this close to the way it should be?'

The human visual process is outlined in Fig. 8.1. Light leaving the scene being viewed goes through the lens in the eye and is imaged on the retina. This data is preprocessed to implement data compression and feature extraction algorithms. This reduced data set is then sent to the brain where further processing is done to highlight areas of interest and to extract information about these areas. Commands are then sent to the appropriate appendages to conduct whatever operation is deemed necessary.

A similar process is utilized by the artificial version of this system with the optics in the camera replacing the eye. Human systems work only in the visible unless augmented in some way. Also some compression is done on the data transferred to the brain. The sensor in the camera corresponds to the retina and the computer corresponds to the brain. The previous generation of systems required frame grabbers that converted the image sensor data into a digital representation in the memory of the computer. Today, cameras directly transfer digital data using common data transfer and interface protocols. The result has been to make the hardware portion of the systems significantly more cost-effective. To date, however, the capabilities of the brain have been difficult to match in the current generation of software and hardware and the generation of software for many applications continues to be a challenge. The end results from the processing of this image data are features that describe what is being seen and the main task is to use these features to determine if there is a close enough match between the

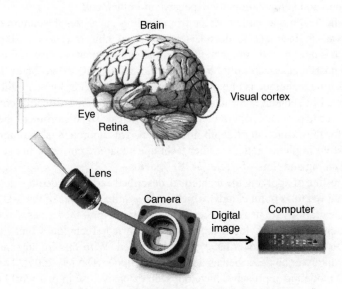

Fig. 8.1   Comparison of human and machine vision.

characteristics of interest and what we currently observe in the scene. We will now present an overview of some of the tools that are used to process these features. Most of the sensing problems that are faced in meat production require the use of three-dimensional (3-D) vision. This is not usually a problem for humans, especially for tasks that do not require high accuracy, but it can be a challenge for machines.

A variety of techniques are used to extract 3-D information for the use of automation and we will now briefly explore the different methods. Since humans have two eyes it has always been thought that binocular vision is an important technique for extracting depth information from a scene. It is obvious that other processes are at work, however, as people that have only one eye are still able to function in a 3-D world. It is thought, for example, that we discern shape from shading (Davies, 2005), where knowledge of the light source and its position is used. In addition, learning and a vast array of knowledge about the world play a role. The question that faces us, then, is: how are machines to be imbued with some of these same capabilities?

8.2.1   Machine vision techniques

One of the more common techniques for extracting 3-D information is stereo imaging. In this approach an image of the same scene is taken using two cameras in different positions. Knowing the geometric relationships between the cameras, as well as the image position of the same feature in both cameras, it is possible to determine depth using one of the cameras as a reference. The business of trying to find the same point in the images can be tricky, however, and is called the correspondence problem. Much work has gone into addressing this problem and it can be difficult and computationally expensive to implement.

Another technique, called 'shape from shading', analyzes the pattern of intensities in a single image to decode the shapes of objects (Davies, 2005). This approach is not typically implemented in practical systems because of the computational costs and susceptibility to lighting variations. One extension of the shape from shading technique is photometric stereo, where a set of known light sources are applied in sequence. Structured lighting also uses an artificially configured light source to project a pattern on the area of interest. The distortion of the pattern allows for the extraction of depth information. This approach is commonly used in industrial implementations. Other techniques that are currently in use include laser scanning and time-of-flight (TOF) systems.

The results of applying the techniques described earlier are depth maps (sometimes referred to as point clouds) which are x, y and z samples of the 3-D surface. Further operations are usually necessary to fully interpret the scenes. Approaches for manipulating point clouds are many and varied. Techniques that liken 3-D shapes to general cylinders have been developed. With this implementation, a sheep is likely to be described as a distorted cylinder (Davies, 2005). These general 3-D modeling approaches have not seen extensive use in real-world applications. One direction that has produced useful results is the use of partial models.

These utilize reference points and locate important points of interest relative to those reference points. This data provides features that tell us about the objects. While the focus here is on 3-D information, other visible surface cues such as texture and colors can generate features that aid in quality and safety determinations. We will now address techniques for handling the features generated.

8.3   Feature selection, extraction and analysis

The first step in utilizing features is to extract those that are distinctive and invariant from the sensor data (ultrasonic, visible, X-rays or infrared) that can discriminate between different classes of objects present in the data. From the standpoint of visible images, the features selected should be invariant to scale and the rotation of the objects in different classes and invariant to illumination changes from image to image. This will result in features that will remain consistent and highly distinctive under different image conditions. Below, we briefly explain some of the methods for feature extraction.

The objective usually is to build algorithms that use the empirical data to learn the underlying pattern in the data and make useful decisions. In this respect, the tasks can be broadly grouped into two categories (Huang *et al.*, 2006): (1) unsupervised learning and (2) supervised learning. In unsupervised learning, the data is not labeled. Problems that fit into this category are clustering, density estimation, etc. There are various approaches taken to solve these problems, such as K-means clustering and component analysis (PCA, KPCA). In supervised learning, the training data is labeled – that is, elements of the data are grouped into different classes. The task of the algorithm is to search a function that maps the individual elements of the data to their respective classes and then make predictions about future instances. For the problems at hand in this chapter, we are mostly concerned with supervised learning algorithms, where we want to recognize an object based on positive and negative samples obtained *a priori*.

Humans are able to learn fairly complex relationships from data through exposure and training. This is the reason why, in many food processing production applications, people are a key part of the production process as they are able to adjust to the vagaries of the process. They are also a drawback in most situations because while they use sight to easily handle the novel situations, they also easily get bored and distracted, which can lead to failures in the production process. Machine learning attempts to imbue machines with some of these desirable qualities while still handling in an acceptable manner the mundane and repetitive.

8.3.1   Scale invariant feature transform

Scale invariant feature transform (SIFT) (Lowe, 1999) features help in reliable matching between different views of the same object. The first step finds key points in the image. The key points are those locations in the image that are very distinct from the neighboring points, such as corners or edges. The algorithm does that by detecting the maxima and minima of a set of difference of Gaussian (DoG) filters applied to the image at different scales. An orientation is assigned to the key points based on the local image features and, finally, a local descriptor is computed based on image gradients. To find an object (whose SIFT features have already been calculated from some sample images) in a given image, SIFT features are extracted over the whole image region and matched to the SIFT features of the object extracted before. The features that match the closest define the located object.

8.3.2   Histogram of oriented gradients

Histogram of oriented gradients (HoG) (Dalal and Triggs, 2005) is a technique that utilizes the image gradients to aid in classification. The image window is divided into a number of small spatial regions (cells). For each cell a HoG is computed. For better invariance to illumination changes, the histogram is normalized using larger spatial regions (blocks). The histogram from all the cells is concatenated into one vector called the HoG descriptor of the window region. First, HoG features are computed for all the positive and negative samples in the training set. A supervised learning algorithm is then used to distinguish the positive and negative HoG features.

8.3.3   Haar-like features

Haar-like features (Viola and Jones, 2001) compute the oriented contrasts between regions in the image using simple rectangular features. The rectangular features consist of two sub-rectangles: one corresponds to a high interval and the other to low interval. The Haar feature is computed by subtracting the average intensity value of the high region from the low region. If the difference is higher than a threshold value, that feature is present in that region. Haar-like features are used in conjunction with Adaboost to select the most optimal image locations and the size and orientation of the Haar features.

8.4   Machine learning algorithms

'Artificial intelligence' was the name that was given to the area of computer science that sought to mimic in some way the ability of humans to tackle and solve novel problems. This activity is now included in an area that is called machine learning. In order to handle product variability the ability to 'learn' and draw inferences from the data will be a key element for any development to result in robust applications. We will now describe some of the techniques that are involved in generating machine learning solutions.

Once the features have been extracted, which reliably characterize different objects, learning algorithms are utilized to find a function (classifier) that captures the mapping from the input space (space of all feature descriptors) to the output space (different classes of objects). Usually, we are interested in a two-class problem only, where we want to determine if the object is present in the image or not. Below, we briefly explain some of the learning algorithms.

8.4.1   Support vector machine

Support vector machine (SVM) (Schölkopf *et al.*, 1999) is a supervised learning algorithm. Given a set of input/output sample pairs $\{(x_i, y_i) \mid x_i \in R^d, y_i \in R\}_{i=1}^n$, associated to a function, an SVM approximates a function $f: R^d \to R$, that assigns a value $y_i \in R$ to each input sample $x_i \in Rd$. It does that by constructing a hyperplane, or set of hyperplanes, that separate the input space into different classes. Given a new sample, the SVM projects the sample close to the appropriate hyperplane, which defines its classification. For most cases, the set of feature descriptors extracted from the object of interest (positive samples) and the feature descriptors extracted from negative samples, form the input space. The output space consists of [0, 1] – that is, representing negative and positive samples respectively.

8.4.2   K-nearest neighbor

K-nearest neighbor (k-NN) (Shakhnarovish *et al.*, 2005) is a method for classifying objects based on the distance to the k closest neighbors in the training sample. The object is classified as belonging to the class of its k-nearest neighbors. Many variants of k-NN algorithms exist. They differ mainly on the basis of the distance metric employed.

8.4.3   Adaboost

Adaboost (Viola and Jones, 2001) is another supervised learning algorithm, which combines multiple weak classifiers into a stronger one. It starts by training several weak classifiers successively, with each classifier trying to classify the training samples wrongly classified by the previous classifier. The final classifier is a combination of all the weak classifiers.

8.5   Application examples: sensing for automated cutting and handling

In meat processing, there is a sequence of operations necessary to get the product from the live state to what is marketed to the consumer. A significant number of the operations required involved cutting and handling. In the next sections, we describe some of the sensing necessary to guide these operations for poultry and beef products.

8.5.1   Poultry slaughter and production operations

The typical poultry slaughter facility supports the sequence of operations shown in Fig. 8.2. Similar procedures are implemented in other slaughter and processing

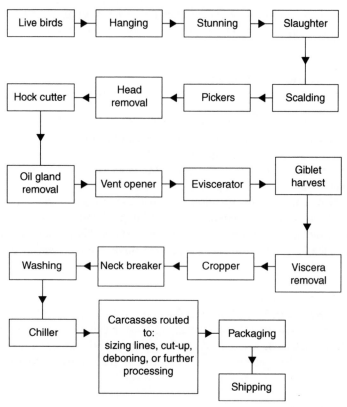

Fig. 8.2   Sequence of operations in a poultry slaughter facility (Northcutt and Russell, 2010).

facilities. For poultry applications it typically begins with the live handling of the birds to transfer them to a processing line where they are then stunned. With the introduction of gas stunning, the order of these events is being changed in many facilities where the animals are stunned using a mixture of gases before they are placed on the processing line. After stunning the birds are slaughtered and then prepared for picking which is handled by mechanical pickers. The birds are then transferred from what is called the 'kill line' to the 'evisceration line', where they are prepared for evisceration. After evisceration there is a process of visual inspection of the product that is usually carried out by the relevant authority. In the USA, this is done by the Food Safety Inspection Service (FSIS) of the United States Department of Agriculture (USDA). The next processes handle and harvest portions of the viscera after which the products called WOGs (i.e., without giblets) are chilled and aged. This process can take place from a few hours to about 24 h.

The next stage is what is called 'second and further processing'. This is where value is added to the product by conducting operations to meet the needs of different customers. This spans the spectrum of deboning, skin removal, specialty cuts, marination and specialty packaging. Currently, this area is where most of the challenges lie in terms of the implementation of automation. Up to this point equipment manufacturers have been able to mostly automate the processes (with the exception of live hanging at the start of the process) with relatively little human backup. In second and further processing, the operations are more varied in nature and thus require more accurate sensing, dexterity and flexibility in handling to meet both quality and product specifications. This chapter will describe some of the approaches being taken to address the sensing needs in this area.

8.5.2   Poultry cutting applications

The current poultry production system evolved over years but is designed to handle the animals with dispatch taking into consideration the welfare of the animals. The part of the system that is typically referred to as the 'front end', which describes the slaughter and immediate post-slaughter operations, is for the most part automated in that there are relatively few people involved in this part of the operation. The challenges present themselves in second and further processing. In this section, of most facilities, there is a significant increase in the number of people needed to conduct the required operations. This is because of the range of products and the operations necessary to achieve them cannot currently be done by existing automation. An operation that is symbolic of the difficulties involved in executing the processes required in most plants is that of deboning what is called a 'front half'. This is the half of the bird that includes the breast and wings. Currently this is done by both automated and manual systems, but the performance of both these modes leaves something to be desired based on yield, food safety and worker safety concerns. The worker safety concern is due to the repetitive motions; the food safety concern is due to the possibility of generating bone chips that could be a safety hazard. The major issues stem from the variability of the product, which can have a significant impact on yield in both manual and automated systems. Most approaches therefore focus on different techniques for tackling the problem.

An approach being explored by researchers at the Georgia Tech Research Institute involves the location of external reference points on the front half from which models for cutting trajectories can be derived. These models are then used to generate appropriate cutting paths to guide an automatic cutting device in real time (Daley et al., 2005). In the Georgia Tech approach, models are derived using CAT scan images to determine reference points and also the correlations between these points and the trajectories of interest. A sample CAT scan model is shown in Fig. 8.3 in which we identify three reference points: the right wing body point, the left wing body point and the keel point. These three points are used to define a coordinate system that is then used to guide a cutting device.

The correlations are derived by using the CAT scan data to go from the external features to the internal features, such as the joints and the various other structures of interest, that would enable us to optimize the cutting paths. Concurrently, with the external features we can also obtain internal features that are important for cutting. An example of the corresponding skeleton is shown in Fig. 8.4, which allows for

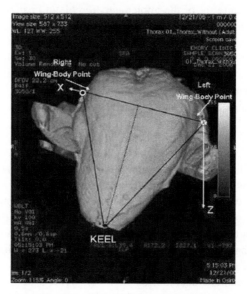

Fig. 8.3   Sample CAT scan model.

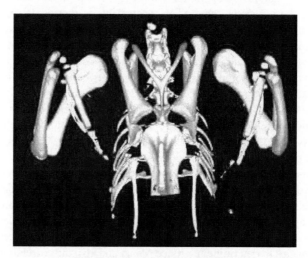

Fig. 8.4   Corresponding skeleton.

the identification of suitable trajectories offset from the reference points. In a similar vein it is also possible to locate the positions of the fan and clavicle bone. This data allows us to locate the relative positions of these structures so that we can implement process control by monitoring on the line for real-time feedback to the process.

As mentioned earlier we have developed correlations that require three reference points that would allow us to guide the cutting operations. We process images from stereo cameras to obtain the 3-D position of the points of interest. A sample image

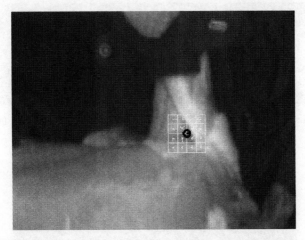

Fig. 8.5   Location of a SIFT feature at a fixed scale and orientation
for the left wing body point.

and the located point are shown in Fig. 8.5. The technique used for this implementation is as follows. A set of training images are obtained. The SIFT (Lowe, 1999) features are extracted from these training images at the three reference points. SIFT features are scale and rotation invariant. However, in the problem at hand, the bird scale and orientation do not change much from frame to frame. Therefore, the scale and the orientation of the features can be fixed, or, to make the process robust, few scales and orientations can be considered – for example, one scale and three orientations. Figure 8.5 shows an example of a SIFT feature at a particular scale and orientation. After the features have been obtained, they are grouped to three separate classes corresponding to the three reference points.

Given a new image, SIFT features are extracted from the image at the used scale and orientation. The search location is narrowed down using other image processing techniques. The extracted features are matched against the features generated from the training samples, using approximate nearest neighbor searching (ANN) (Arya *et al.*, 1998). The locations of the features that best match the three classes are taken as the reference points for the image. The geometry of the obtained reference points is matched against the desired geometry, to discard spurious feature points matches, and the next best match is taken. Figure 8.6 shows an example test image. The white rectangular regions indicate the narrowed search region for the left wing body point, keel point and the right wing body point, with the small circle indicating the found reference point. Once these points are located then it is possible to compute the required trajectories for executing the cuts that facilitate removal of the 'butterfly', which gets its name from the shape but consists of the breast meat.

8.5.3   Breast meat harvesting

Another important aspect of this sort of operation is the ability to monitor the process in order to provide feedback control. Vision sensing also provides this

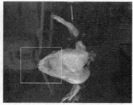

Fig. 8.6   Boxes indicate the search regions and the dots indicate the located wing and body points.

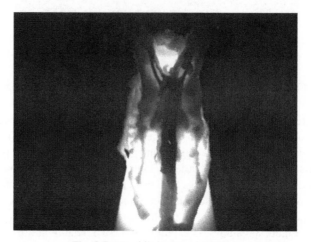

Fig. 8.7   Backlit image of a frame.

opportunity. The approach being evaluated is to examine the skeleton of the carcass immediately after the meat removal operations. This is a backwards approach where, instead of examining the product for the presence of bone, we want to ensure that the bones remain on the skeleton. In order to conduct this operation, we use a backlit cone to illuminate the frame (skeleton); we then examine the frame for the presence or the absence of bones to determine whether or not we have a potential problem. We illustrate below how this operation would function. Figure 8.7 shows the backlit image of a skeleton on the debone line. Somewhat like an X-ray image, the bones that are of concern can be observed. If these bones are still on the skeleton then it can be assumed that they are not in the finished product. Using correlations similar to the ones developed to locate the reference points for cutting we can identify areas of interest for the bones of concern as in Fig. 8.8 for the clavicle region and Fig. 8.9 for the fan bone region. These areas can then be further processed to detect the presence or absence of these bones.

To detect the presence or absence of the clavicle and fan bone (shown in Figs 8.8 and 8.9) we use a machine learning approach to recognize the shape of the bone. We employ HoG features (Dalal and Triggs, 2005) for this task. HoG features are well suited for objects that have well-defined boundaries or silhouettes.

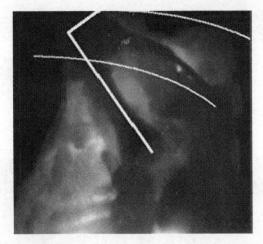

Fig. 8.8   Location of clavicle region.

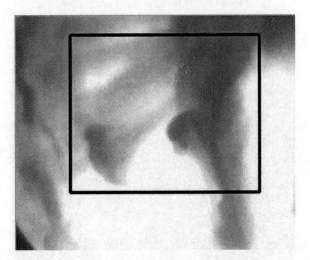

Fig. 8.9   Location of fan bone region.

The process consists of two phases. First, image regions containing the bone are extracted from a set of training samples. These extracted regions form the positive samples (Fig. 8.10 shows a few positive samples for the fan bone region). Negative samples regions are generated randomly from the images not containing the bone. HoG features are computed over the positive and negative samples, and, finally, a binary SVM is used to distinguish the positive samples from the negative. Samples are also presented in Fig. 8.11.

Second, during the detection phase, a search window is moved across the image. At each location the HoG features are computed for the window region. The features are passed onto the SVM, which determines whether the bone is present in the window or not. The area over which the bone is to be searched

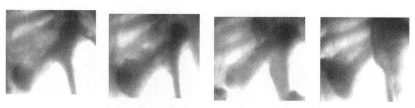

Fig. 8.10   Positive fan bone sample region.

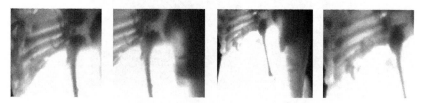

Fig. 8.11   Sample non-fan bone regions.

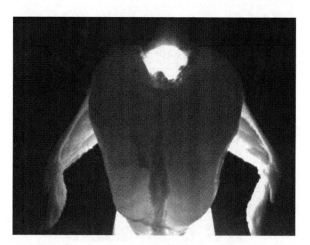

Fig. 8.12   Backlit image for yield determinations.

is narrowed down, based on the setup of the experiment. Figure 8.10 shows an example of the found fan bone regions. Preliminary results show accuracies of about 85% in correctly detecting the presence or absence of these bones using this approach. An additional benefit of these approaches is the monitoring of yield during the operations. This would be done by monitoring also the amount of meat that was left on the frame (skeleton), as shown in Fig. 8.12, using the correlations presented in Figs 8.13 and 8.14.

The techniques developed also support the handling of products, which requires the design of grippers. This is an area that will also require further development and research.

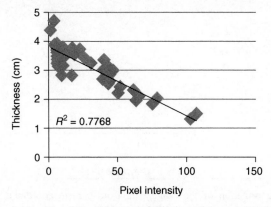

Fig. 8.13 Correlation between thickness and intensity (thick).

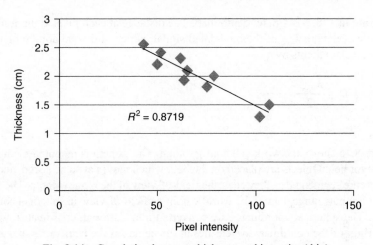

Fig. 8.14 Correlation between thickness and intensity (thin).

8.5.4   Time-of-flight (TOF) sensor for automatic chicken tracking

One of the steps required in automating second processing in poultry production is the ability to automatically pick up the product from the belts. This is a ubiquitous operation in second and further processing. Proper grasping of the product by a robot or other smart automation requires knowledge of its orientation. This section describes a prototype system that can estimate the orientation of WOGs placed on a moving conveyor belt. A WOG is a carcass that has been chilled without giblets.

In TOF systems, the time taken for light to travel from an active illumination source to the objects in the field-of-view and back to the sensor is measured. From the speed of light c, the distance can be determined directly from this round-trip time. The TOF sensor modulates its illumination LEDs with a modulation frequency of f. A CMOS/CCD (complementary metal oxide semiconductor/charge-coupled device) sensor chip with the associated electronics is positioned to receive and measure the phase of the returned modulated signal at each pixel,

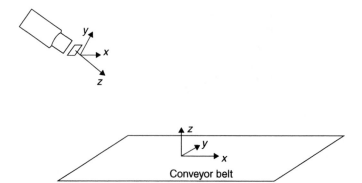

Fig. 8.15   Transformation of the range data from camera coordinate system to belt coordinate system.

resulting in 176×144 pixel depth map. The distance at each pixel is the fraction of the wavelength λ_{mod} of the modulated signal, which limits the non-ambiguous range measurements by

$$D = \frac{\lambda_{mod}}{2} = \frac{c}{2f}$$

Setup
Figure 8.15 shows the work cell configuration. The camera is mounted facing the conveyor belt. Objects are placed on the belt, which moves at some speed, and the objects are segmented as they enter the field-of-view of the range camera. The TOF sensor outputs range data of the world within its field-of-view in spherical coordinates. The spherical coordinates are converted into Cartesian coordinates, where the origin of the coordinate system lies at the center of the camera, as explained in the next section. The 3-D point cloud in the camera coordinate system is transformed to the belt coordinate system with the origin at a pre-determined reference point. The coordinate transformation is required for the following two reasons:

- The robot and the range camera need to agree on a coordinate system so that the measurements from the range camera can guide the robot.
- In the belt coordinate system, the z-axis is normal to the plane of the belt. Aligning the coordinate axis in such a manner simplifies extraction of object point clouds on the belt by referencing points above the belt plane and within the confines of the belt boundaries.

Estimating the belt frame
The following procedure is used to estimate the belt coordinate system, and is carried out only once at the start. Let $u \in R^2$ be the 3-D position of a point in the camera coordinate frame. The belt is manually segmented using the intensity image as shown in Fig. 8.16a. The corresponding range data for the belt is

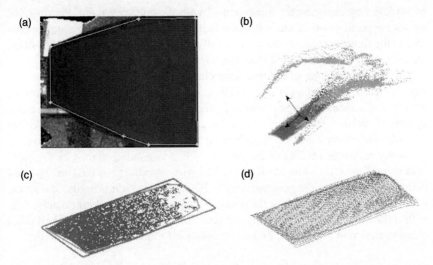

(a)   (b)

(c)   (d)

Fig. 8.16   Estimating the belt frame: (a) belt selection using intensity image; (b) point cloud in camera coordinate system; (c) convex hull and smallest bounding rectangle; and (d) occupancy grid.

obtained. Figure 8.16b shows the point cloud in the camera coordinate frame, with the points corresponding to the belt shown as grey. Let S be the set of 3-D positions of the pixels belonging to the belt. The normal e of the plane fitting the set of points is obtained by the following equation:

$$e = \arg_e \min \sum_{u \in S} e^T u + d$$

where d is the distance from the origin to the plane, which is assumed to be 0. The above equation is a non-linear optimization problem whose solution is the eigenvector associated to the smallest eigenvalue of the covariance matrix C, which is defined to be

$$C = \frac{1}{n} \sum_{u \in S} (u - \bar{u})(u - \bar{u})^T,$$

where $\bar{u}$ is the mean of the points in S. Let E be the eigenvectors of the covariance matrix. The eigenvector corresponding to the smallest eigenvalue provides the normal e to the plane of the belt. A point u is transformed to the conveyor belt coordinate system $v \in R^3$ by the following equation:

$$v = E^T \cdot (u - \bar{u}).$$

All points lying on the same plane as the belt plane have $z \cong 0$ in the belt coordinate system. All the subsequent computations are performed in the belt coordinate system. The coordinate system is then translated to a known reference point. This can be achieved by changing the mean point $\bar{u}$ with the reference point. Figure 8.16c shows the point cloud of the belt in the belt coordinate system. The interior contour in the figure represents the convex hull of the belt region as obtained from the initialization polygon selected from the intensity image.

After transformation of the belt point cloud to the belt coordinate system, it is represented by a rectangular occupancy grid. Occupancy grids are based on the principle of allocating range points to a prepared world that is divided into a grid system of variable or fixed voxels. In this case, a 2-D occupancy grid is used for the belt. The convex hull of the belt point cloud is computed and the minimum bounding rectangle (MBR) of the convex hull is computed, shown as the outer contour in Fig. 8.16c. The MBR is meshed with a grid size of 0.01 m. The belt points are mapped to the occupancy grid. Each grid location is assigned a value, which is the average z value of the points falling within the block. If no point falls within the block, the block is assigned zero. This representation of the belt point cloud is shown in Fig. 8.16d. This representation has the following properties:

- Range data for objects close to the camera is denser than for objects that are further from the camera. The variable density as a function of distance is apparent in Fig. 8.16c. The occupancy grid converts the range points to a surface, given by a height function defined over the occupancy grid as shown in Fig. 8.16d.
- The data representation of the scene is reduced and hence the memory and computational time.
- The effect of noise is also reduced because values of all the points falling with a block are averaged.

The resulting rectangular occupancy grid can be visualized as an image, leading to the use of image processing techniques to process the objects on the belt.

The range measurements are based on measuring the phase of the reflected signal. The signal may get reflected by multiple surfaces before returning to the sensor. In this situation, the light travels by the direct as well as indirect path and the distance is then the weighted average of the path distances. When this happens, the belt point cloud may not lie on the estimated plane. To account for that, we learn the surface of the belt using a polynomial approximation and evaluate the surface at each grid location.

Extracting without giblets (WOG) point cloud
The steps explained in the previous section are carried out once using the first frame. In subsequent frames, the following steps are carried out to extract the point cloud of the objects on the belt. For processing, only the 3-D range data is used; the intensity information is ignored.

- Range data obtained from the sensor is transformed to the belt coordinate system. All the points falling outside the convex hull of the belt are discarded.
- Points within the convex hull are mapped to height values defined over the occupancy grid, which is the average z value of all range points falling within

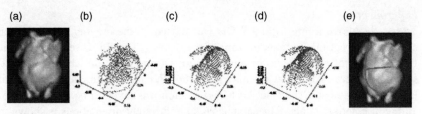

Fig. 8.17   (a) Intensity image; (b) extracted point cloud; (c) polynomial approximation; (d) principal components and (e) principal components mapped to the image domain.

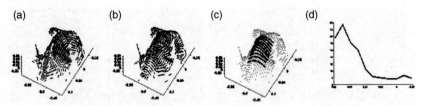

Fig. 8.18   (a) Maximum principal curvature κ_1; (b) minimum principal curvature κ_2; (c) the mean curvature $\kappa_1 + \kappa_2$ computed along the second principal component and (d) the average of mean curvature along each transverse line.

a block. If no point falls within the block, the block is assigned the value zero. The value assigned to each grid location is the distance of the point cloud to the belt.

• The distance value at each grid location is subtracted from the polynomial approximation of the belt evaluated at that grid location. If the result is bigger than a predefined threshold, the point belongs to the object on the belt; otherwise it belongs to the belt.

Orientation of WOG

After extracting the point cloud of the WOG using the procedure explained above, a polynomial surface is fitted to the point cloud of the chicken. Figure 8.17b and 8.17c show the extracted point cloud and the polynomial fitted surface. To find the orientation of the WOG, principal component analysis (PCA) is performed on the extracted point cloud. Modeling the back of the chicken as an ellipsoid, PCA computes the direction of the major and minor axes. Figure 8.17d shows the principal components found using PCA. These points, when mapped back to the image plane, are shown in Fig. 8.17e.

The next step is to identify the wings and the legs. Surface curvatures are employed for this task. Figure 8.18a and 8.18b shows the maximum and minimum principal curvatures, κ_1 and κ_2. The mean curvature $\kappa_1 + \kappa_2$ is computed along the second principal component as shown in Fig. 8.18c. The average value of mean curvature along the transverse lines is computed and shown in Fig. 8.18d. This figure shows that the value of the curvature is higher at legs than at wings. This information is used to distinguish legs from wings along the first principal component.

Experiment

To test the above framework, a WOG was placed on the moving conveyor belt at eight different orientations in front of the camera. The WOG remained in the field-of-view of the camera for about 0.5 m. During that time about 180 measurements were taken. The orientation of the WOG was estimated for each of these measurements. The final WOG orientation was taken to be the median of all the orientation values. The sample WOG at different orientations is shown in Fig. 8.19, together with the estimated orientations.

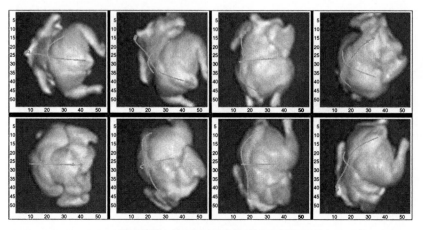

Fig. 8.19   Tracked WOG orientations.

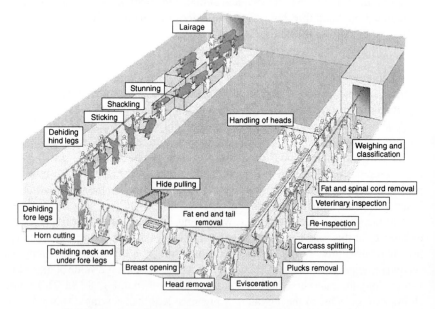

Fig. 8.20   Steps in beef slaughter process (Madsen *et al.*, 2007).

8.5.5   Larger animal applications

Larger animals such as cows, sheep and pigs face similar challenges in terms of the application of automation. The processing rate for cattle is about ten animals per minute. While somewhat slower than poultry (approximately 230 birds per minute), the operations required are similar except on a much larger scale. The organization of a typical abattoir is shown in Fig. 8.20. There are similar steps involved, such as live handling, and the same concerns for animal welfare.

Researchers and other commercial entities have explored a variety of sensors for guiding processing operations in beef, sheep and pork processing. Applications typically require 3-D information, but 2-D sensing combined with appropriate material handling can also achieve some of the desired outcomes. Color sensing is necessary in many cases for quality control operations. The main challenge, as noted before, is the ability to handle product variability. It would be nice if systems were able to handle lots of one – that is, accommodate each individual animal. Using beef as an example, carcass lengths can range from 1200 to 3000 mm and variation in weight can be hundreds of kilograms (Ring *et al.*, 2007), further underscoring the need for the automation to accommodate the product.

A range of sensors have been tested for use in slaughter and processing operations for sensing external features including: RGB and black and white cameras, thermal cameras, stereo vision and laser scanners. For internal features, X-ray, ultrasound and subsurface radar have been evaluated (Ring *et al.*, 2007). RGB cameras have been used for detecting and highlighting surface differences, but can also pose some problems due to the natural variation of coloring of the animals. Thermal cameras, while more expensive, have also proven useful in similar operations where the difference in body temperature of the animal compared to the environment can be exploited (Ring *et al.*, 2007). Stereo vision has also been applied to locate the 3-D position of surface features but the correspondence issue can be a problem in production operations. Surface mapping using laser scanners has also produced useful results. A system that has been demonstrated for the cutting of sheep briskets (Condie *et al.*, 2007) utilizes a combination of a color camera and a Sick LMS 400 laser scanner to obtain 3-D information to control an industrial robot to make the cut of the brisket.

In some cases, external features are not able to provide all the necessary information for automation; in those cases ultrasound and X-ray have been utilized. For a splitting operation, ultrasound was found to be effective for the detection of the feather bone. Subcutaneous air can prove to be a problem in these applications as it presents barriers to the passage of the sound waves and the data no longer is representative of the internal structure of the animal. X-ray imaging has been demonstrated for location of joints and other cut points. These implementations, while possible, tend to be somewhat expensive and there are also some safety concerns with their use. Subsurface radar has also been explored but it was determined that further development is necessary, especially in the area of antennas for automated slaughtering tasks (Ring *et al.*, 2007).

Similar systems are being implemented and tested in pork processing facilities that utilize robots guided by vision and 3-D sensing systems to automate a variety of operations needed for processing. As an example, a system for conducting the debunging (removal of the rectum) operation for pigs has been implemented. It utilizes a 3-D laser scanner; the point cloud data is processed to locate the position of the rectum, which is then removed using an industrial robot (Wilson, 2010).

8.6   Future trends

It is predicted that the world will have to produce 70% more food by 2050 to feed an extra 2.3 billion people (Kaku, 2011). This is going to require us to be as efficient as possible to meet this anticipated need using all the traditional measures (energy, labor, environmental effects). One aspect of this will be the development of automated processes to support the overall production system. A key part of these developments will be sensing, particularly visual sensing. Many of the advances in vision systems technology in the past leveraged developments that occurred to meet the needs of the consumer market. The early machine vision industry, for example, relied heavily on television technology. We expect a similar trend to continue in the future. Current systems are building on advances made for handling media on the web and interfaces to game systems. One of the significant requirements for the continued improvement in automation will be the ability to rapidly obtain 3-D information to guide devices. Developments such as TOF cameras that provide both visual and 3-D information in one integrated device are illustrative of the components that will be necessary. Demonstration algorithms for locating WOGs using a combination of a TOF camera fused with a visible camera are shown in Fig. 8.21. A device such as the Kinect™ that is currently marketed as a game controller also has the capability for providing visible and 3-D information at a significantly reduced cost. Similar processing done with data from the Kinect™ device is shown in Fig. 8.22. This example illustrates the cross-pollination of ideas that will fuel needed developments.

While the ability to obtain external information on product and correlation between internal and external features can be utilized for some applications, there will be a need in some cases for direct information on the internal structure. Systems using ultrasonic holography, for example, might be a cost-effective way to obtain this kind of information (Garlick and Garlick, 2007). Example images showing the structure of joints in poultry products using this technology are shown in Figs 8.23 and 8.24.

Fig. 8.21   TOF camera with visible imaging to locate position and orientation of WOGs.

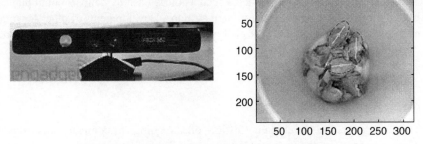

Fig. 8.22   Kinect™ device and processing to locate WOGs.

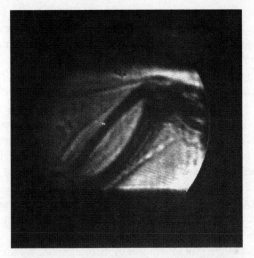

Fig. 8.23   Ultrasonic holographic image of wing joint.

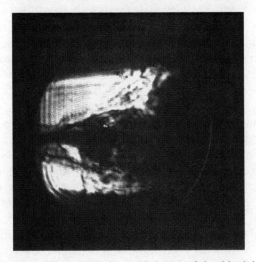

Fig. 8.24   Ultrasonic holographic image of shoulder joint.

Similarly, compact X-ray systems such as those used by dentists could play a role in sensing for the next generation of systems. Developments of other sensing modes and advances in software will make many of the current problems more tractable.

8.7   Conclusions

The innovation and development officer of an equipment company was asked: what is needed for the next generation of machines? His answer was succinct: smarter sensing and lighter, more robust machines (robust to the operational environment). This sounds as if we need more humans on the lines. Practically speaking this is the goal in terms of the performance required of these systems. They will have to handle variability and novelty while at the same time being low cost. What humans bring to the game, as stated by Michio Kaku, is the ability to undertake pattern recognition along with exercising common sense (Kaku, 2011). The current generation of technology is allowing us to move in this direction. Sensing systems are being developed for consumer uses that can be leveraged, and computational power is being delivered in smaller and smaller packages (good examples are smart phones), allowing processing to be embedded in more devices. Vision continues to be an important part of these systems. While the developments are slower than hoped for in the early years, progress is being made. Costs for equipment, for example, have fallen significantly due to mass production to meet the consumer market. The software to control these systems continues to be a challenge, but developments in image understanding and machine learning will continue to provide useful tools.

Another aspect of the problem is that of manipulation. We are yet to develop a manipulator with the flexibility of the human hand. It could be argued, however, that we could reduce the requirements for these devices if we were able to do a better job of sensing and scene understanding. What we have in most meat production operations is a repetitive task with variation. Improved sensing along with the development of appropriate models to accommodate the variation will result in more practical automation. Expect, in the near future, to see some humanoid devices working on processing lines with skill levels comparable to their human counterparts.

As we look to the future, automation in the food production sector will continue to be a need, especially as we meet the challenge of feeding the world. The current sensing technologies are not meeting all needs, but advances are being made in the development of sensors and the software to support them. We can begin to anticipate more integrated systems capable of operating in real time with sensors and manipulators working in concert to address both automation and food safety concerns. We cannot yet, as hoped by Khodabandeloo, attach the label 'food untouched by human hands' (Khodabandehloo, 1993), but we are getting closer. There is discussion in the manufacturing arena now about how to cost-effectively manufacture 'lots of one' (as opposed to the more typical process of manufacturing identical products). The question for us in the food industry is somewhat the inverse – it is: how do we cost-effectively accommodate each one? This is the grand challenge we face. We will have to move from the evolutionary process that drove us in the past to more of a transformational paradigm that seeks solutions through a more integrated and multidisciplinary approach. We believe industry, government and academia have begun the journey.

8.8   Acknowledgments

The authors would like to acknowledge the support of the State of Georgia that sponsored this work through the Agricultural Technology Research Program at Georgia Tech, in cooperation with the Georgia Poultry Federation and its member companies, for the support of the poultry-related work described herein. The authors would also like to acknowledge the work done by Sergio Grullon and Colin Usher in the development and testing of the prototype devices described. Also, thanks to the many contributions of our colleagues in the Food Processing Technology Division of the Aerospace, Transportation and Advanced Systems Lab in the Georgia Tech Research Institute. Many thanks also to Stuart Shaw of Machinery Automation and Robotics in Australia for his comments and insight into the developments to support large animal processing.

8.9   References

Arya, S., Mount, D. M., Netanyahu, N. S., Silverman, R. and Wu, A. (1998) An optimal algorithm for approximate nearest neighbor searching. *Journal of the ACM*, **45**, 891–923.

Condie, P., MacRae, P. R. and Boyce, P. (2007) *Automated Sheep Brisket Cutting.* Available from http://www.araa.asn.au/acra/acra2007/papers/paper188final.pdf (accessed 15 May 2011).

Dalal, N. and Triggs, B. (2005) Histograms of oriented gradients for human detection. *Computer Vision and Pattern Recognition*, **1**, 886–93.

Daley, W., Britton, D., Usher, C., Diao, M. and Ruffin, K. (2005) 3D sensing for machine guidance in meat cutting applications. *SPIE Optics East*, Boston, **5996**, 112–19.

Davies, E. R. (2005) *Machine Vision.* San Francisco, CA: Elsevier.

Garlick, T. F., and Garlick, G. F. (2007) Volumetric imaging using acoustical holography. *Acoustical Imaging*, **28**, 317–29.

Graves, M. and Batchelor, B. (2003) *Machine Vision for the Inspection of Natural Products.* London: Springer-Verlag.

Huang, T.-M., Kopriva, I. and Kecman, V. (2006) *Kernel Based Algorithms for Mining Huge Data Sets, Supervised, Semi-supervised, and Unsupervised Learning.* Berlin, Heidelberg: Springer-Verlag.

Kaku, M. (2011) *Physics of the Future.* New York: Random House.

Khodabandehloo, K. (1993) *Robotics in Meat, Fish and Poultry Processing.* Glasgow: Blackie Academic and Professional.

Lowe, D. G. (1999) Object recognition from local scale-invariant features. *International Conference on Computer Vision*, **2**, 1150–7.

Madsen, N., Nielsen, J. and Monsted, K. (2007) Automation in the meat production. In iMAC (ed.), *International Meat Automation Conference, Australasia*, November 28–29, Sydney.

Northcutt, J. K. and Russell, S. M. (2010) General Guidelines for Implementation of HACCP in a Poultry Processing Plant, The University of Georgia Cooperative Extension, Athens, GA, Bulletin 1155, January 2010.

Ring, P., MacRae, K. and Condie, P. (2007) Sensors adapted for automating tasks in beef and sheep slaughter. In Dunbabin, M. and Srinivasan, M. (eds), *Proceedings of the 2007 Australasian Conference on Robotics and Automation*, Brisbane, Australia: Australian Robotics and Automation Association.

Schölkopf, B., Burges, C. and Smola, A. (1999) *Advances in Kernel Methods: Support Vector Learning.* Cambridge, MA: MIT Press.

Shakhnarovish, G., Darrell, T. and Indyk, P. (2005) *Nearest-Neighbor Methods in Learning and Vision.* Cambridge, MA: MIT Press.

Viola, P. and Jones, M. (2001) Rapid object detection using a boosted cascade of simple features. *Computer Vision and Patteren Recognition,* , I-511–I-518.

Wilson, A. (2010) Pork process: Robots and vision team to automate pork production. *Vision Systems Design,* 3.

Zuech, N. (1988) *Applying Machine Vision.* New York: Wiley-Interscience.

9

Image analysis of food microstructure

J. C. Russ, North Carolina State University, USA

Abstract: Quality control measurement of foods, and determination of structural parameters for correlation to processing variables and product properties, relies on measurements. These are often performed using digital imaging and computer-based image analysis. This chapter reviews stereological procedures based on geometric principles that are capable of efficiently determining meaningful structural values. These include metric properties such as volumes, surface areas, lengths and spatial distributions, as well as topological properties such as number density. Characterization of surface roughness is also described.

Key words: image analysis, stereology, structure measurement, volume, surface area, metric properties, topological properties.

9.1   Introduction

Many properties of food depend on microstructure. These include appearance and texture, as well nutritive properties. For some foods, the structure is a direct result of genetics and agricultural practices, but for many foods subsequent processing is the principal effect that controls structure. The perceived structure of the food may be further modified in human consumption, in the mouth and by digestion.

Measuring the structure of food, to obtain data that can be correlated with product quality, processing history and consumer acceptance, generally requires microscopic examination. The light microscope is the principal tool, but confocal microscopy, electron microscopy, atomic force microscopy, magnetic resonance or computer tomographic imaging, and other methods, all have their uses. The images that these techniques generate are usually digitized and displayed using computer technology. Measurement of the images to obtain microstructural information can be efficiently carried out using the techniques described in this chapter.

9.2   Quality control applications of digital imaging

The general subject of computer vision, as implemented using digitized images that are processed and measured in the computer, is well covered in the preceding chapters. Typically, it is applied to one of two tasks:

- on-line process or quality control in which deviations from an acceptable norm are monitored; or
- research in which correlations are sought between structure and food processing, acceptance or behavior.

These two applications place different emphases on some aspects of the task, such as the need for fast and simple algorithms and high-speed cameras and processors in the first case, or the use of appropriate sample acquisition and preparation procedures in the second. Many quality control tasks are more concerned with the stability and reproducibility (i.e., precision) of the measurement than with the absolute accuracy or a firm geometric meaning for the result. However, both are similarly based on measurements performed on the digitized images, and these measurements are made much simpler by control of the sample presentation and lighting.

For example, a camera viewing a production line at perpendicular orientation with backlighting that shows objects or particulates as silhouettes makes measurement much easier than viewing from an angle, or the use of top or reflected lighting. If particles can be presented as a single layer with minimal pileups, overlaps, or touching, the measurement of size and shape is greatly facilitated. Of course, this is not always practical and so computer software may be required to extract meaningful data from more difficult images. As an example of a straightforward measurement task, Fig. 9.1 shows rice grains. They have been dispersed on a desktop scanner by holding a mechanical vibrator against the scanner to cause the grains to separate. By leaving the scanner lid open, the image shows the grains against a dark background.

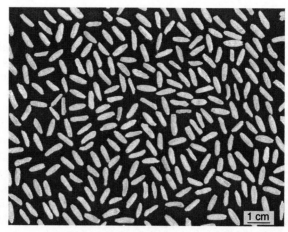

Fig. 9.1   Image of dispersed rice grains whose lengths are to be measured.

The spatial calibration is defined by the scanner setting, so the relationship between pixel size and spacing and the real-world dimensions of the grains is known.

In addition, the analysis task is simplified by having an established criterion for the measurement and its interpretation. Long-grain rice must not contain more than 10% of short or broken kernels. The length of a rice grain is the maximum distance between any two points on the periphery, and is conveniently measured in the digitized image by calculating the projected length in a various directions.

As shown in Fig. 9.2, the coordinates in a rotated set of axes can be computed for each pixel along a feature's boundary, and the maximum difference stored as a projected dimension in that direction. If angle steps of, for example, 10° are used, a table of sines and cosines can be predetermined. The maximum error due to the finite angle steps is given by the cosine of 5° (the worst case misalignment between the measurement direction and the actual maximum length). The cosine of 5° is 0.996, meaning that the worst case error is an underestimate of 0.4%. For an object whose actual length is 250 pixels, this amounts to one pixel, or no more than the precision with which the image can be captured to begin with.

Applying this procedure to the image in Fig. 9.1, and not measuring any grains that intersect the edge of the image, produces the result shown in Fig. 9.3. From this size distribution, it is evident that there are only two grains shorter than 6 mm and that the sample is acceptable. Of course, in a real application there would also be a requirement for the minimum number of grains to be measured, in order to achieve the desired statistical precision. Combining data from multiple images is straightforward.

There is one more issue to deal with – that is, the problem of objects that intersect the edge of the image and cannot be measured. Simply ignoring these objects will result in biased results, because it is the larger objects that are more likely to touch the image borders and be unmeasurable. This can be dealt with statistically, by calculating an adjusted count for each feature that is measured. The adjusted count compensates for the rejection of other similar objects that intersect the boundaries (Russ, 2005).

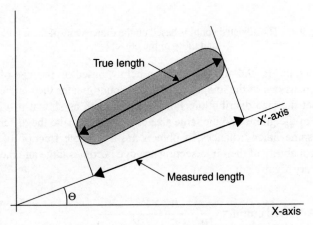

Fig. 9.2 Measuring length as a projected distance, as described in the text.

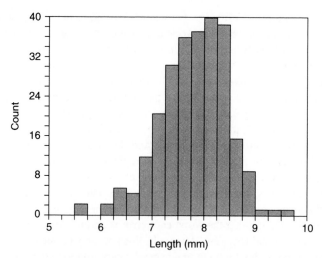

Fig. 9.3   Distribution of the length values for the rice grains in Fig. 9.1.

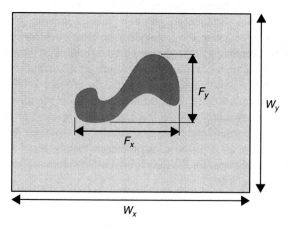

Fig. 9.4   The adjusted count is based on the dimensions of the image (W) and those of the object (F).

As shown in Fig. 9.4 and the equation below, based on the size of the object and the dimensions of the image, an adjusted count greater than one can be used to construct the size distribution. For an image such as shown in Fig. 9.1, the small size of the grains and the large size of the image make the difference fairly small. In some other instances, if objects are a sizeable fraction of the image size, the probability of their intersecting an edge becomes large and the correction becomes very significant.

$$\text{Adjusted count} = \frac{W_x \times W_y}{(W_x - F_x) \times (W_y - F_y)} \quad\quad [9.1]$$

Fig. 9.5   Bubbles rising in beer. The multiple reflections, refraction, positioning of some bubbles entirely or partially in front of others and problematic lighting preclude the automatic measurement of the size distribution.

The size distribution of objects ranging from fruits on a conveyor belt (monitored with a video camera) to the ice crystals in ice cream (imaged using a light microscope) can all be measured in essentially the same way. In some cases, the measurement parameter may be area, or an equivalent circular diameter, or some other defined dimension. In all of these cases, the dimension is an external one, corresponding to the exterior size. And, in addition, the measurements are of discrete objects, even if some image processing must be applied to deal with touching or slightly overlapping boundaries for the multiple objects present. In more extreme cases, such as shown in Fig. 9.5, automatic measurement cannot be achieved.

9.3   Characterizing the internal structure

Viewing the external surfaces or silhouettes of objects provides important, easily understood, but very limited information. Indentations in the surface are not seen or measured. This also biases measurements of shape, as objects appear smoother in their projected outline. However, the greatest limitation is the inability to assess internal structure. The heart of structure-processing and structure–property relationships in food science depends upon robust and meaningful characterization of the structures, often at the level of light or even electron microscopy.

In some applications to quality control, simply cutting or breaking open a good product and visually comparing the appearance to a set of standards is sufficient. However, it is only with the availability of measurement data that science can begin to find correlations and quantitative relationships. The problem is in deciding what to measure. Human instincts often fail in these situations, because we are conditioned to interpret external images rather than internal ones.

As an example, consider the bubbles that form in breads. These are variously called pores or cells, and arise from the evolution of gases produced by yeast (or in some cases due to chemical processes). The size of each pore that is seen on the cut surface of a bread slice is not the three-dimensional (3-D) size of the pore. The chance of the slice cutting a pore through its maximum diameter is very low, and most of the visible sections will be smaller than the true pore size. In addition, large pores are more likely to be intersected so that any measurements of size would be biased and not representative.

This effect was understood nearly 100 years ago, and various techniques have been proposed to deal with it. The classical approach, called 'unfolding', proposes a shape for the pores (most straightforwardly, a sphere) and calculates the probability of obtaining circles of various sizes from a unit sphere (Weibel, 1979). This probability function can then be applied to a measured distribution of circle sizes to calculate the size distribution of spheres that must have been present to produce that observed distribution.

There are several difficulties with this approach, which is now generally not favored. The first is that the calculation is mathematically ill-conditioned. This means that the answer is sensitive to small variations in the input data, so that a very large number of circles must be measured to achieve adequate statistical precision to generate a stable result. That is worrisome, but not as serious as the major objection: the entire process depends on knowing exactly what the shape of the 3-D objects really is, and that all of them have the same shape.

If the pores are actually ellipsoidal, or have flat faces (as in a foam), there are extensive published tables (Wasen and Warren, 1990–96) that can be used to perform the unfolding. However, if the shapes vary, and particularly if they vary as a function of size (which is often the case), then the method fails.

More recent techniques, described below, are preferred for measuring microstructural parameters. These have been developed with rigorous mathematical underpinnings based on geometrical probability, and are described as stereological methods. The name comes from the Greek words, *stereos* meaning 'solid' and *logos* often interpreted as 'study of', but based originally on the idea of words, principles or reason (as in 'logic'). An excellent summary of the geometric basis for modern stereological procedures, and a guide to their use and interpretation, can be found in Baddeley and Vedel Jensen (2005).

9.4   Volume, surface and length

Microstructural analysis is usually concerned with the internal, 3-D arrangement of surfaces, voids, or phases, although sometimes the geometry of external surfaces is included. The latter is typically described as 'two-and-a-half'-D, because of the use of local elevation values relative to the nominal surface. The measurement of surface geometry is discussed below. Food microstructure is present at many different scales, from the macroscopic (e.g., the distribution of pepperoni slices on a pizza) to microscopic, which may require light or electron microscopes,

(a)

(b)

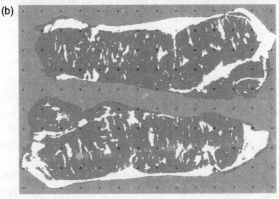

Fig. 9.6   Beef steaks, showing the meat and fat: (a) original image and (b) thresholded binary with superimposed grid of points (shown as small circles for visibility).

atomic force microscopes and other technologies. Because they are based on geometric principles, the same stereological measurement procedures can be applied to images at any of these scales (indeed, they have been used in the study of the shape and distribution of galaxies).

Internal microstructure is intimately connected with processing history, composition and consumer acceptance of food products. For example, Fig. 9.6 shows the surface of beef steaks in which the fat is visible (these were imaged with a desktop scanner). The area fraction of the fat in the steaks can be measured by counting pixels, which results in a value of 19.9%. The stereological relationship between this area fraction and the volume fraction of the fat is shown in the following equation using the usual stereological notation in which the subscripts identify the encompassing structure, in this case volume (V) and area (A) of the steaks:

$$V_v = A_A \qquad\qquad [9.2]$$

Determining the volume fraction this way is convenient in an imaging system, but does not provide any estimate of the measurement precision. A preferred method uses a grid of points superimposed on the image as shown in Fig. 9.6b. The points must be far enough separated that rarely do two of them fall on the same region of fat. Counts of the number of points (P) that fall on the steaks (84) and the number that fall on the fat (16) allow calculating the volume fraction according to the equation below, resulting in a value of 19%.

$$V_V = P_P$$

[9.3]

At first glance, it might seem that this result is a poor substitute for one based on all of the points, used to determine the area fraction. However, the advantage is that its precision can be determined. Counting of independent events (the number of 'hits' made by points in the grid) produces a standard deviation equal to the square root of the number of counts. The square root of 16 is 4, so the precision is 25%.

Since it is always desirable to measure multiple samples or images to characterize a specimen or population, this assists in experimental design. Based on a target precision of 5% relative, for example (i.e., 20±1% fat in the steaks), then if this image is reasonably typical, examining a total of 25 similar images (a total of 50 steaks) would be expected to produce about 400 hits. The square root of 400 is 20, or 5%.

The ease of counting and its ability to predict the precision of the result makes the use of properly designed grids and counting preferred over other types of measurements whenever possible. Many stereological measurements have been performed with manual counting, often using an appropriate grid mounted directly in the microscope eyepiece. However, the trend is to use computers to capture and process images, generate the grids and perform the counting automatically.

Surfaces in a 3-D structure may be of several different kinds. In the beef example shown above, there is a surface between fat and meat. However, in a plant, the cell walls are surfaces between identical types of material. And in a foam, the surfaces of the pores are boundaries that separate the material from air. In many complex structures, containing several different types of material or objects, there may be many different kinds of interfaces. Measuring the area of these surfaces is important, because the surfaces are the locations where much of the chemistry, rheology and other mechanical properties are controlled.

In a planar section through a 3-D structure, internal surfaces appear as lines where the section plane intersects the surfaces. One approach to measuring the amount of surface area begins by measuring the length of those lines, as shown in Fig. 9.7. The stereological relationship between the length of the boundary lines (B) per unit area of image (A) and the amount of surface area (S) per unit volume (V) of structure is given in equation [9.4]. The assumption in the calculation is that the surfaces and the section plane are randomly oriented. The numerical factor $4/\pi$ arises from the fact that the surfaces can intersect the section plane at many angles.

$$S_V = \frac{4}{\pi} \cdot B_A$$

[9.4]

Unlike the measurement of volume fraction based on either the area fraction or the point fraction, the surface area per unit volume has dimension (e.g., $\mu m^2/\mu m^3$). This means that the image scale must be calibrated. In the example of Fig. 9.7, the total area of the image is 10.69 in^2 and the total length of boundary line is 166.07 in. The calculated surface area per unit volume is 19.78 in^2/in^3.

As for the volume measurement, it is desirable to substitute a counting experiment for a measurement task. Placing a grid of lines of known length (L) on the image and counting the number of intersections (N) between the boundary lines and the grid is also shown in Fig. 9.7. The relationship between the surface area per unit volume and the number of intersections and length of grid line is shown in equation [9.5]. The numerical factor two arises from the angles at which the grid lines may intersect the surfaces. The grid shown has total length (L) of 23.13

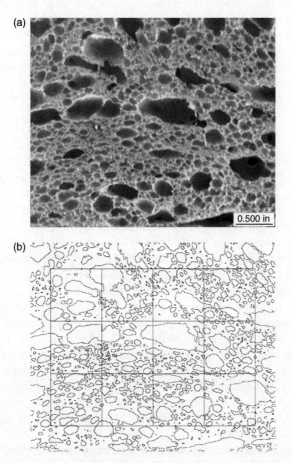

(a)

0.500 in

(b)

Fig. 9.7   Surface of a bread slice: (a) original image; (b) line grid superimposed on the outlines of the pores.

inches, and produces 233 'hits' (N), for a calculated surface area per unit volume of 20.15 in²/in³.

$$S_V = 2 \cdot N_L \tag{9.5}$$

In a more complex structure with many different types of surfaces, the area of each may be measured separately. Some types of interface may be absent (for example, in biological tissue there are no surfaces separating one cell nucleus from another). If the surfaces are not isotropic (for example, muscle tissue in animals or cells in the stems of plants) then it is either necessary to perform randomized sampling, or to use specialized sampling methods and grids to eliminate bias. A sampling technique known as 'vertical sectioning' (Baddeley et al., 1986), with grids consisting of cycloidal arcs, is often used in such cases.

Some structures in 3-D space can be approximated as lines, if the lateral dimensions are much smaller than the overall length. Examples are blood vessels or nerves in animals, branching patterns in tree limbs and even pore structures through which liquids migrate. Measuring the length of these structures in a volume can be accomplished by counting the number of intersections (N) that they make with a randomly oriented section plane, and the area (A) of the plane. The relationship is given in equation [9.6], where the numerical factor two arises from the angles at which the structure may intersect the plane.

$$L_V = 2 \cdot N_A \tag{9.6}$$

In some cases, it is possible to count the number of intersections on a plane as it is viewed normally. Another technique often used in light microscopy is to examine a transmission image through a section of known thickness, and to draw a grid of lines on that image. The lines represent planes extending down through the cut section. Counting the number of intersections that the structure of interest makes with the lines gives the number of times the structure intersects that plane. The calculation then proceeds as in equation [9.6], except that the area (A) is the product of the length of the grid line and the section thickness.

In the example of Fig. 9.8, the fibers are thinned to their skeletons by image processing, to facilitate automatic counting of the number of intersections. There are 94 'hits', with a total length of grid line of 990 μm and a section thickness of 5 μm. The result is a total length of fibers of 0.038 μm/μm³. Lest this seem like a small value, it is equivalent to 38 000 mm/mm³. It is possible to pack a very large amount of line length or surface area into quite small volumes. Once again, the underlying assumption is that the structure is isotropic and the planes are randomly oriented. If this is not the case, either randomly orienting many samples or a structured sectioning procedure with cycloidal grid lines can be used to obtain an unbiased average result.

In some applications, it is the degree of anisotropy that is of interest. Sampling by cutting section planes in selected orientations, and measurement using a grid of parallel lines that can be rotated to different orientations, allows counting the

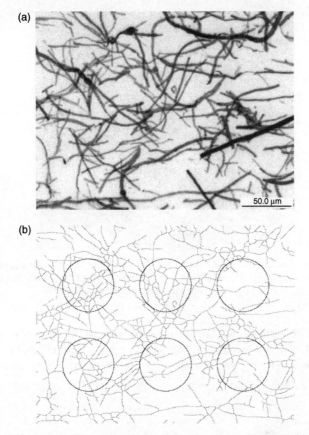

Fig. 9.8   Fibers viewed in a transmission light microscope: (a) original image and (b) line grid superimposed on the skeletons of the fibers.

number of intersections as a function of orientation. This is sometimes plotted as shown in Fig. 9.9 to indicate the average degree of elongation. The plot shows the mean intercept length (the length of the grid lines divided by the number of intersections) for grid lines drawn in steps of 10°. The overall elongation of the cells in this growing corn plant is 1.97.

9.5   Number and spatial distribution

The measurements of volume, surface area and length calculated in the previous section are all metric properties. There is no information about whether the volumes, surfaces or lengths come from one single object that connects the various regions seen on the planar section, or from many separate ones. Number is a topological property, not so readily determined, but often of great interest. Estimating number from images of section planes through a structure requires independent

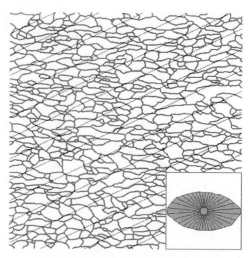

Fig. 9.9   Thresholded and skeletonized cell walls in plant tissue, with a grid of parallel lines drawn at an angle of 25°. The inset shows the plot of mean intercept length as a function of grid orientation, determined in steps of 10°.

information or assumptions. The relationship between the number of objects per unit area (A) that are revealed on the section planes and the number of objects per unit volume (V) is given below, which requires knowing the mean diameter (D_{mean}) of the objects:

$$N_V = \frac{N_A}{D_{mean}} \qquad [9.7]$$

Figure 9.10 shows a case in which this relationship can be applied. The structure shows bubbles in a foamed food product, and because of the constant pressure used and the design of the nozzle, it is assumed that all of the bubbles are spheres of the same size. That being the case, since many sections through the bubbles are visible, it is likely that the largest circle seen is close to the equatorial diameter of the spheres. This value is 31 μm.

Counting the number of features per unit area (N_A) requires including in the count those features that intersect two edges of the image (e.g., the bottom and right) and not counting those that intersect the other two. This produces an unbiased count since other adjacent fields would report only those features that are not counted in this one. For the image shown, the number of bubbles is 156, the total image area is 6.82×10^4 μm², and the number of bubbles per unit volume is 7.38×10^4/mm³.

If the objects present are not spherical, the mean diameter is the caliper or projected dimension averaged over all orientations, which is not easily determined. Furthermore, the assumption is that all of the objects are identical in size and shape. In the general case, the method shown in equation [9.7] is not readily

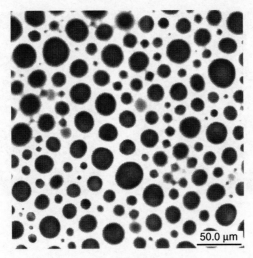

Fig. 9.10   Plane section through a foamed food product, showing circular intersections through spherical bubbles.

applied (in fact, it may be used in the reverse direction to calculate a mean diameter once the number of features per unit volume is determined).

To determine topological properties such as the number of objects per unit volume, regardless of their size or shape, it is necessary to examine a volume. Of course, this can be done with a complete 3-D volumetric image, such as those provided by tomography or serial sectioning, but they are expensive and tedious procedures. A simpler method known as the disector (Sterio, 1984) requires comparing the images of two parallel section planes at a known distance apart. The spacing between the planes must be small enough so that no object of interest can hide between them without being intersected by at least one.

The method is simply to count the number of objects whose bottom lies between the planes. Since each object has one and only one bottom, it gives a direct value for the number in the volume represented by the area (A) of each image and the spacing between them. For convex objects, counting the bottoms is done by comparing features in the two images and by counting only those that appear in the top section image. Any feature that appears in both, even if the size has changed, does not have its bottom in the test volume and is not counted.

$$N_V = \frac{(N_{A1} + N_{A2})}{2 \cdot \text{Spacing}}$$ [9.8]

In practice, counting tops as well (those features that appear in the bottom section image and not in the top one) and dividing by two gives an improvement in statistics (equation [9.8]). That is the procedure shown in Fig. 9.11. A confocal microscope was used to acquire optical section images, which are shown. The images were processed to delineate the objects visible in each, and automatically

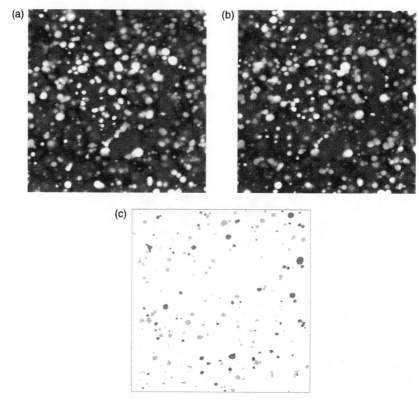

Fig. 9.11   Two optical sections (a and b) and the features present in each that
are not matched by a feature in the other (c).

compared to detect those that are not present in both, as shown. The number (N) of
unmatched objects in the images (111 and 114) are combined with the image area
($2.13 \times 10^6 \ \mu m^2$) and the spacing between the planes (5 μm) to obtain the number
per unit volume ($1.056 \times 10^4/mm^3$).

The number of objects present does not reveal how they are spatially distributed, which may be characterized as random, clustered or self-avoiding, depending on chemical and physical effects. A convenient measurement is the mean
nearest neighbor distance. For a random spatial distribution, the nearest neighbor
distances between objects in an image produce a histogram with a Poisson shape
(Schwarz and Exner, 1983). The mean value depends only on the number per unit
area, as given below:

$$\text{Mean NND} = \frac{0.5}{\sqrt{N_A}} \qquad [9.9]$$

If the measured mean nearest neighbor distance is greater than the mean value
for a random distribution, it indicates self-avoidance in which the objects try to

maintain separation from each other (e.g., the spacing of cacti in the desert or starch granules in potatoes). Conversely, a measured mean value less than that for a random distribution indicates clustering. Figure 9.12 shows two custards in which the lipid droplets have been stained. The mean nearest neighbor distances for the custard in Fig. 9.12a have an approximately Poisson-shaped distribution with a mean value of 28.89 μm. The predicted mean based on the number per unit area is 28.14 μm and so the spatial distribution is close to random. The custard in Fig. 9.12b, however, has a distribution of nearest neighbor values that is much narrower and has a mean value of 15.43 μm (the predicted mean based on the number per unit area is 26.50). The distribution is clustered, and the ratio of 15.43/26.50 = 0.58 provides a measure of the degree of clustering. It is similarly possible to measure the distance of objects from a boundary or surface, to determine whether chemical or physical effects influence their spatial distribution.

Many other stereological procedures are available, relying on geometrical probability to extract meaningful relationships from relatively straightforward measurements. For example, Fig. 9.13a shows the thresholded cross-sections through layers of ingredients produced by stirring. Determining the thickness of the layers is complicated by the fact that a section plane will not necessarily

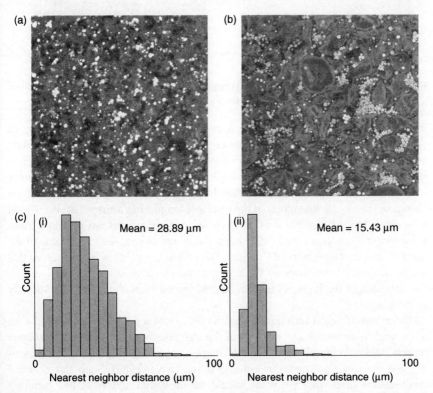

Fig. 9.12   Lipid distribution in custard: (a, b) images of two custards and (c) histograms of the nearest neighbor distances, with the mean values.

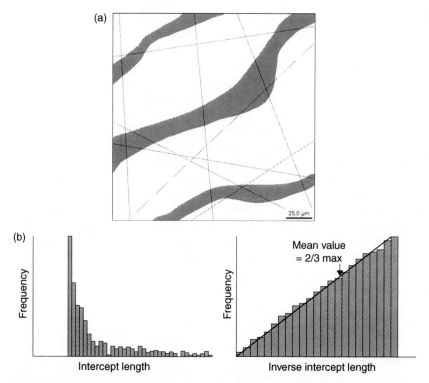

Fig. 9.13   Measuring layer thickness: (a) random lines on a random section and (b) histograms of intercept lengths and inverse intercept lengths.

intersect the layers at right angles, so that the dimension revealed in the image is not the true 3-D thickness. This is an example of a measurement that could be made, but would not provide any useful structural information. However, if random section planes are used to sample the layers, and a grid of random lines is drawn on the images, the intersection lengths do produce the correct result. As shown in Fig. 9.13b, histogram of the intercept lengths has a shape that is difficult to analyse, but the histogram of the inverse intercept lengths has a simple triangular shape (Gundersen *et al.*, 1978). For the image shown, the mean value of the inverse intercept lengths is 0.0708/µm. The maximum value is just 1.5 times this mean, because of the shape of the distribution. The inverse of this (9.41 µm) is the thickness of the layer, which can be determined even if it is not seen directly in the images.

   Other stereological techniques, such as the use of a point grid combined with a line grid to determine the variance of the size distribution of irregularly shaped objects, make it possible to obtain information about object sizes without assuming that the shape is known or uniform (Cruz-Orive, 1989). Such stereological relationships often indicate what should be measured and what has meaning rather than what can be measured but may not (e.g., the number or size of features seen on a section).

Efficient experimental design is also important (Gundersen and Osterby, 1981). Rather than a purely random sectioning strategy, the use of an ordered sectioning procedure with appropriate grid design to avoid bias is generally a preferred approach. Also, rather than randomly selecting sections or fields of view for analysis, a structured but randomized approach is significantly more efficient. The derivation of basic stereological relationships is non-trivial, based on geometric probability and integral mathematics. However, the application of the results, as shown in the earlier equations, is straightforward. Most of the measurements can be performed using grids applied to processed and thresholded images, followed by automatic counting. When this is not practical, manual counting may be performed.

9.6   Surfaces and fractal dimensions

Relationships between processing, properties and structural characteristics present within 3-D volumes are not the only ones of importance. Surfaces play a large role in the properties of food products, including their consumer acceptance. Some surface geometries are smooth, controlled by surface tension, but others are not. In many instances, the roughness of natural surfaces has a fractal geometry that can be measured to determine a fractal dimension.

For a surface, this will be a numerical value between 2.0 (the topological dimension of a surface, corresponding to one that is ideally smooth) and 2.999..., approaching the topological dimension of a volume. Most typical surfaces have dimensions less than 2.5, often in the range of 2.05–2.25. The examples in Fig. 9.14 have the same total amplitude of roughness, but the relationship between nearby locations is different, so that a higher fractal dimension corresponds to a greater perceived 'roughness' of the surface.

There are several ways to measure the dimension (Russ, 1994), but generally the simplest uses the Fourier transform of the elevation data. This is shown in Fig. 9.15 (for the data corresponding to Fig. 9.14a). A plot of the log of the amplitude versus the log of the frequency is linear over much of the range. Deviations from linearity at low frequencies generally correspond to the overall form of the surface, while those at high frequencies arise from noise in the measurement data. These data may come from contacting scanners (profilometers, atomic force microscopes, etc.), from optical devices such as interference or confocal microscopes, or from other types of instruments.

$$\text{Fractal Dim.} = \frac{6 - |\text{Slope}|}{2} \qquad [9.10]$$

The slope of the linear portion of the plot in Fig. 9.15b is related to the fractal dimension according to equation [9.10]. For the surfaces shown in Fig. 9.14, the fractal dimensions are 2.05 and 2.25, respectively.

Other methods for measuring the surface fractal dimension include intersecting the surface with a plane parallel to its nominal orientation and adding one to

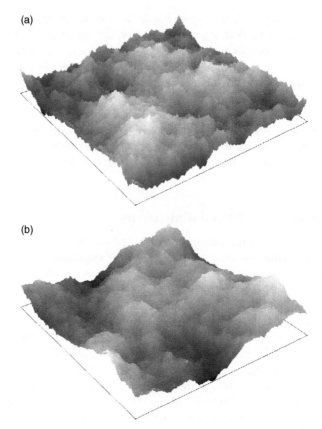

Fig. 9.14   Two fractal surfaces: (a) dimension = 2.25 and (b) dimension = 2.05.

the measured dimension of the outlines of islands. This is the same technique used to measure the roughness of geographic coastlines, which correlate with erosion (e.g., Norwegian fjords versus Florida's beaches). Fractal dimensions are also applied to the characterization of branching patterns in networks, agglomeration of particulates and other natural structures.

9.7   Conclusions

Structure-property and structure-processing relationships are the keys to understanding many foods, both natural and manufactured. Design and control depend on the ability to measure the key structural parameters. Stereology is a mature science based on geometric probability that is most appropriate for internal 3-D structural measurements. Other computer-based measurement procedures are well developed for measuring dispersed particles and surface geometry.

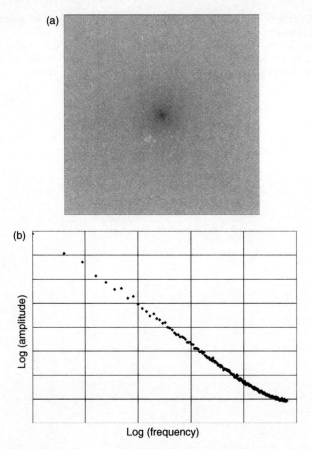

Fig. 9.15   Measuring the fractal surface: (a) Fourier transform power spectrum of the elevation data from Fig. 9.14a. (b) Plot of log(amplitude) versus log(frequency).

9.8   References

Baddeley, A. and Vedel Jensen, E. B. (2005) *Stereology for Statisticians.* Boca Raton, FL: Chapman & Hall.
Baddeley, A., Gundersen, H. and Cruz-Orive, L. M. (1986) Estimation of surface area from vertical sections. *Journal of Microscopy*, **142**, 259–76.
Cruz-Orive, L. M. (1989) Second-order stereology: estimation of second moment volume measures. *Journal of Microscopy*, **153**, 315–33.
Gundersen, H., Jensen, E. B. and Osterby, R. (1978) Distribution of membrane thickness determined by lineal analysis. *Journal of Microscopy*, **113**, 27.
Gundersen, H. and Osterby, R. (1981) Optimizing sampling efficiency or stereological studies in biology: or 'Do more less well!'. *Journal of Microscopy*, **121**, 65–74.
Russ, J. C. (1994) *Fractal Surfaces.* New York: Plenum.
Russ, J. C. (2005) *Image Analysis of Food Microstructure.* Boca Raton, FL: CRC Press.
Schwarz, H. and Exner, H. E. (1983) The characterization of the arrangement of feature centroids in planes and volumes. *Journal of Microscopy*, **129**, 155.

Sterio, D. C. (1984) The unbiased estimation of number and sizes of arbitrary particles using the disector. *Journal of Microscopy*, **134**, 127–36.

Wasen, J. and Warren, R. (1990–96) *Catalogue of Stereological Characteristics of Selected Solid Bodies*. volumes 1–4, Göteborg, Sweden: Chalmers University.

Weibel, E. R. (1979) *Stereological Methods*. London: Academic Press.

Part III

Current and future applications of computer vision for quality control and processing of particular products

10

Computer vision in the fresh and processed meat industries

P. Jackman and D.-W. Sun, University College Dublin, Ireland

Abstract: Computer vision has proven itself to be a viable alternative to expert grading of meat by human inspectors in recent years. Grading by human experts does have a number of critical shortcomings that can be effectively mitigated with computer vision technology. Computer vision systems offer a basic, affordable and ergonomic option that, while needing some expertise is not very demanding technologically. Computer vision technology has demonstrated some absolutely vital attributes of flexibility and ease of compatibility, allowing a variety of meat quality assessment problems to be tackled. The simplest approaches have often proven to the best, with basic visible light imaging and traditional statistical modelling proving successful on the vast majority of occasions. For some of the more difficult tasks, non-visible wavelength imaging and implicit data processing are required. Truly automatic image segmentation still remains an awkward problem. Advances in hardware and software are allowing more computationally demanding texture characterisation algorithms to be applied. Finally, properly calibrating and validating a computer vision system will require a comprehensive array of image and independent meat quality data.

Key words: palatability, grading, classification, computer vision, tenderness, image processing, image analysis.

10.1 Introduction

Of all foods, meat products have particularly high added value and, hence, precise and robust quality control is essential to ensure correct meat product classification. Thus, the meat industry is an ideal candidate for the application of an objective classification system. By replacing the human expert grader substantial cost savings both in cash terms and in terms of merchant goodwill can be made by reducing misclassification due to subjective and inconsistent assessment. However, the essential advantages of human grading – that is, speed and non-destruction – must be preserved.

A possible alternative to human expert grading would be to extract a sample and prepare it for a taste panel. This would give a direct measurement of public opinion in the case of a consumer panel or an indirect measurement in the case of a trained panel. However, these are impractical options due to time and cost as well as causing product damage. The attribute of tenderness could be estimated with mechanical equipment such as Instron (Instron, 2008), flavour could be estimated objectively (Scanlan, 1977) and juiciness could also be estimated objectively (Pearson and Dutson, 1995). However, these tests are also time consuming, costly and cause product damage. Thus, a rapid, cheap and non-destructive alternative is required. Computer vision systems are the most viable option as they can deliver all the required attributes for effective classification.

Meat products are normally classified based on features that are amenable to computer vision systems, such as colour, fat distribution, texture and morphology (USDA Agricultural Marketing Service, 2012). In many instances, the features to be judged can be observed without the need for non-visible or penetrative imaging, which prevents excessive hardware costs. The prime example would be the beef grading system (USDA, 1997) that uses features including colour, marbling, texture and ossification, which can all be measured with a basic colour camera and simple lighting. The first three of these features have been extensively investigated in recent years (Li et al., 1999, 2001; Tian et al., 2005; Jackman et al., 2008, 2009a, 2009b, 2009c, 2009d, 2010a, 2010b). There have been many other successful applications of computer vision systems in many areas of meat classification. Some of these are discussed in depth by Aguilera and Briones (2005), Brosnan and Sun (2004), Du and Sun (2004, 2006a), Sun (2008) and Tan (2004). With hardware advancements in recent years the cost of non-visible and penetrative imaging equipment is falling; thus, there are emerging opportunities to both widen and deepen the application of computer vision systems in meat classification.

10.2   Meat image features

Choice of equipment will be dictated by what is being measured, which will in turn be dictated by what expert grading system is being recreated. Thus, computer vision systems tend to be tailor-made rather than prefabricated as grading systems can be different both in terms of what is being measured and the point in the production process where measurement takes place.

Fresh meat products are derived from carcasses consisting of flesh, fat and bone. Thus, various attributes are important: colour, as it reflects the muscle pH; surface texture, as it reflects the connective tissue distribution; and the geometry of muscle fibres and morphological features, as they reflect fat distributions and carcass development. It is features like these that will be part of an expert grading systems, such as those published by the United States Department of Agriculture (USDA Agricultural Marketing Service, 2012). These features are particularly ergonomic for large animal carcasses as large carcasses normally need to be cut into pieces for further processing; an example is shown in Fig. 10.1. Measurement

Fig. 10.1   A cut beef carcass (USDA Meat Animal Research Centre,
Clay Centre, NE, USA).

of features like these are well suited to basic computer vision systems consisting of simple cameras recording visible light under fixed illumination; thus, costly non-visible imaging equipment is not normally required.

For processed meats all bone has been removed and some or all of the fat may have been removed or blended in. Thus, texture features become even more important. The colour features remain important, but for a different reason as it can indicate the presence of pale exudative meat which is of poor quality (Jackman *et al.*, 2010c). As with fresh meat products, non-visible imaging equipment is not normally required as palatability indicators are well suited to basic systems.

Computer vision can and does generate huge amounts of data. For example, a basic 500 × 500 pixel colour image contains 750 000 pieces of information. This will include a huge amount of noise and redundancy. Noise removal is relatively straightforward as there are numerous algorithms for effective noise removal that have been rigorously tested by the computer science community. However, eliminating redundancy is far more challenging. The first and most important step in removing redundancy is accurate and robust segmentation to correctly identify the region of interest (ROI) within the object under observation. This is a particularly difficult challenge for images of meat (Tan, 2004). Colour can be highly variable, making robust simple thresholding very difficult; the structure of meat leaves many local gradients that are not muscle boundaries, making gradient-based segmentation also very difficult; furthermore, local variations can mean that too many clusters are identified by a cluster-based algorithm. Hence, elaborate segmentation methods are often required (Gao, 1995; Li *et al.*, 1999; Jackman *et al.*, 2009b). Alternatively, automatic segmentation can be avoided by manual intervention or by restricting the field of view to a part of the ROI; however, these approaches risk skewing the subsequent data. This point is covered in detail by Subbiah *et al.* (2004).

Once the ROI has been identified it must be succinctly described with parameters that are meaningful. For colour this is simple and straightforward, as histograms of red, green and blue will describe the colour and colour distribution (Jackman *et al.*, 2008). However, the food science community are used to expressing colour in the device independent L*a*b* colourspace, developed by the Commission Internationale d'Eclairage (CIE) in 1976 (HunterLab, 2008). This colour system avoids the differences in colour sensitivity of various brands of digital camera and any variations in lighting. Hence, for direct comparison of image data with colorimeter readings, the image acquisition equipment must be specially calibrated to a standard red, green and blue colourspace (sRGB). A convenient format for the sRGB colourspace was developed by Hewlett-Packard© and Microsoft© (Stokes *et al.*, 1996). A wide variety of algorithms for transforming RGB data into L*a*b* data were analysed by Leon *et al.* (2006). For any colour conversion process to be reliable it must not introduce errors of greater than 2.2 colour units, otherwise a noticeable error will be introduced (Brainard, 2003). A robust model for colour conversion was developed by Valous *et al.* (2009a) to do this successfully.

Describing texture (the localised spatial variation) succinctly is straightforward but far from simple as highly complex and computationally demanding algorithms are required to condense the texture information efficiently; furthermore, magnification may be required to properly observe the texture patterns (Jackman *et al.*, 2009d). With a wide choice of colourspace transformations readily available, there are many possible greyscales that could be used for texture analysis. The saturation channel was put forward by Li *et al.* (2001) to best express beef surface texture; the Cb channel from the digital video YCbCr colourspace was proposed by Jackman *et al.* (2010a) as an alternative to best express beef surface texture. The chroma channels of the L*a*b* colourspace were used by Jackman *et al.* (2010c) to best express ham surface texture.

Unlike with colour, there is no prospect of describing texture with a few simple summary properties as a large number of variables are required to describe the subtle patterns that are present. One of the oldest and still one of the best options for texture characterisation is the classic pixel co-occurrence algorithm of Haralick *et al.* (1973), as well as pixel run lengths (Li *et al.*, 1999). These are discussed in more detail by Zheng *et al.* (2006a). A simpler classic algorithm is the pixel difference histograms (PDHs), which track the difference between neighbouring pixels and build up a histogram (Chandraratne *et al.*, 2006a). The above classical algorithms have some key shortcomings that have led to the consideration of alternative texture algorithms, such as the Fourier transform, wavelet transform and fractal dimensions. Each of these seeks to condense texture information more efficiently. These are explained in detail by Stein and Shakarchi (2003), Kaiser (1994) and Soille and Rivest (1996), respectively. Texture analysis offers an important reminder of the vital importance of correct camera exposure, focus and focal length.

Morphological features pertinent to fresh meats are the size and shape of intermuscular fat, intramuscular fat, a particular muscle or the entire carcass. For processed meats the thickness of edges, eccentricity, inclusions or porosity would be important. To properly measure shape and size, first the object outline must be found by image segmentation. Then shape and size features can be extracted that are indicative of quality. This is an area where visible wavelength imaging may not be sufficient due to a need to highlight an object that is obscured or to identify an object beneath the meat surface. The importance of image segmentation becomes clear when shape and size features are being measured. In the absence of a reliable outline, the extracted features could be heavily skewed. This becomes obvious when considering the shape and size features listed by Du and Sun (2004). Morphology analysis also offers a timely reminder of the need for correct camera focus and focal length.

The above image features can be supplemented with additional data that can be expected to provide information on the state of the meat product. A leading option is hyperspectral imaging, which can help identify the composition of material within a pixel boundary based on the reflectance spectra (this has led to the term 'chemical imaging') and can also provide alternative greyscales for texture analysis. A comprehensive discussion on the utility of hyperspectral imaging for meat quality assessment is given by ElMasry and Sun (2010).

It is also possible to perform some internal structural analysis with hyperspectral imaging by introducing light from a point source and recording the backscatter pattern as the light reflects off the connective tissue and sarcomeres inside the meat structure (Ranasinghesagara et al., 2010). The pattern of the spectral contours recorded is indicative of the amount of connective tissue and sarcomere length. Figure 10.2 shows such spectral contours. Other options for internal structural analysis are computed tomography (CT) and magnetic resonance imaging (MRI); while these will generate genuine three-dimensional (3-D) images, they are much slower and much more costly.

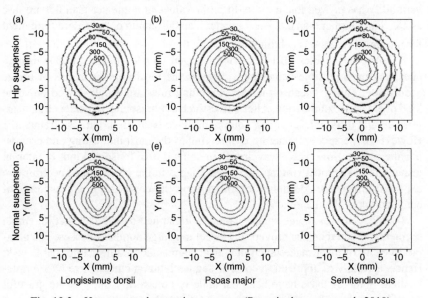

Fig. 10.2   Hyperspectral scattering contours (Ranasinghesagara et al., 2010).

With the ROI described, a relationship must be established between the image features and meat attributes expected by the consumer. For meat there are three essential components of palatability: flavour, juiciness and tenderness (Warriss, 2000). These three attributes can be summarised by overall acceptability or likeability. The relationship between meat image features and palatability attributes can be complex and non-linear, thus complex statistical or artificial intelligence modelling might be required.

Simple statistics are preferable as a tractable model is derived. This approach will be used if sufficient classification accuracy is possible by it. Neural networks are the most common artificial intelligence option used to boost correct classification rates and reduce prediction error. It is highly likely that some type of dimensionality reduction or feature selection will be needed to make the model usable, given that image processing can and does generate a huge number of data (Du and Sun, 2006a). Fortunately, the data processing can be assisted by many user-friendly software packages available at reasonable cost, such as Unscrambler (Woodbridge, NJ, USA), SAS (Cary, NC, USA), Statistica (Tulsa, OK, USA) and Matlab (Natick, MA, USA).

10.3   Application and implementation

Recent applications of computer vision in the meat industry are shown in Tables 10.1–10.4. It can be easily seen that computer vision has been most commonly applied to beef, lamb, pork and hams. Most of the correct classification rates are good with some cases of perfect classification. Classification rates of 90% or higher are found regularly. This demonstrates that computer vision is capable of giving good predictions of meat quality attributes. Modelling meat variability has proved much more difficult. Values of r^2 greater than 0.9 are rare as are values higher than 0.8. An r^2 greater than 0.8 should be considered excellent for multiple regressions (Shiranita *et al.*, 2000).

As shown in Tables 10.1–10.4, the charged coupled device (CCD) camera operating in the visible wavelength range is the most popular means of image acquisition. This is because the expert grading rules are usually based on some kind of external examination. Alternatives such as ultrasound, hyperspectral, near-infrared and ultraviolet imaging equipment have been used, but less frequently. However, hyperspectral imaging is emerging as the most interesting and promising alternative. Non-visible wavelength imaging is used for some kinds of internal inspection of the meat structure or when images in the non-visible spectrum are expected to provide additional information on the composition and patterns of composition of the meat.

The difficulty of image segmentation is illustrated by the fact that no single segmentation method has proved successful in a large number of cases. The natural diversity of meat makes such universal segmentation a near impossible task. Hence, a variety of application-specific segmentation methods have been implemented. Some of these have been particularly noteworthy, such as the graceful degradation algorithm of Borggaard *et al.* (1996), the active contour algorithm of

Table 10.1 Recent applications of computer vision for assessing beef quality

Author	Acquisition	Segmentation	Learning	Application	Best results
Li et al. (2001)	CCD	Window	NN	Predicting tenderness from wavelet surface texture	83.3% CC
Li et al. (1999)	CCD	Gao's method	SS and NN	Predicting tenderness	$r^2 = 0.7$
Shiranita et al. (1998)	Photo	Window	SS	Grading meat from marbling and texture	100% CC ±1 grade
Shiranita et al. (2000)	Photo	NN	SS	Grading beef	$r^2 = 0.93$
Demos et al. (1996)	CCD	n/a	SS	Assessing surface metmyoglobin	$r^2 = 0.93$
Basset et al. (2000)	CCD and UV	Window	SS	Classifying beef from surface texture	96.3% CC
Whittaker et al. (1992)	Ultrasound	Window	SS	Determining intramuscular fat	$r^2 = 0.66$
McCauley et al. (1994)	Ultrasound	Window	NN	Identifying intramuscular fat	78.8% CC
Hwang et al. (1997)	BW Camera	NN	NN	Segmenting beef robustly	RMSEP = 0.044
Borggaard et al. (1996)	Video	Graceful degradation	NN	Grading beef carcasses	$r^2 = 0.93$
McDonald and Chen (1991)	Video	Thresholding	SS	Predicting taste from marbling score	$r^2 = 0.13$
Gerrard et al. (1996)	CCD	Component labelling	SS	Predicting beef colour and marbling score	$r^2 = 0.86$
Huang et al. (1997)	Ultrasound	Window	SS	Predicting beef shear tenderness from wavelets	$r^2 = 0.7–0.9$
Vote et al. (2003)	CCD	n/a	SS	Predicting beef shear tenderness	80% CC
Subbiah et al. (2004)	Video	FCM and convex hull	Expert analysis	Grading beef	<2% class error
Park et al. (2001)	NV	Manual	SS	Predicting beef shear tenderness	$r^2 = 0.69$
Lee et al. (2001)	CCD	Contour generation	NN	Segmenting lean beef from image	n/a
Tian et al. (2005)	CCD	Modified Gao method	NN	Predicting sensory tenderness of grazing beef	$r^2 = 0.62$
Steiner et al. (2003)	Video	n/a	SS	Online carcass grading	$r^2 = 0.63$

(Continued)

Table 10.1 Continued

Author	Acquisition	Segmentation	Learning	Application	Best results
Lu and Tan (2004)	CCD	Gao-methods	NN	Predicting beef lean yield	MSE = 4.54 testing
Zheng et al. (2006b)	CCD	Threshold	SS	Predicting beef joint shrinkage during cooling	$r^2 = 0.99$
Hatem et al. (2003)	CCD	Manual	NN	Determining beef skeletal maturity	75% CC
Jackman et al. (2008, 2009a, 2009b, 2009c, 2009d, 2010a, 2010b)	CCD	Manual or clustering and contrast enhancement	GA and SS	Predicting beef palatability	r^2 up to 0.95
Du and Sun (2009)	CLSM	Thresholding	n/a	Building 3-D images	Shading eliminated
Naganathan et al. (2008a)	CCD and NV	Window	SS	Classifying shear tenderness	96.4% CC
Ranasinghesagara et al. (2010)	CCD and NV	n/a	SS	Predicting shear tenderness	$r^2 = 0.5$
Cluff et al. (2008)	CCD and NV	n/a	SS	Predicting shear tenderness	$r = 0.67$
Naganathan et al. (2008b)	CCD and NV	Window	SS	Predicting shear tenderness	77% CC
Zheng et al. (2006c)	CCD	Thresholding	NN	Predicting moisture from colour	$r^2 = 0.77$
Xia et al. (2007)	CCD and NV	n/a	SS	Predicting shear tenderness	$r^2 = 0.59$
Xia et al. (2008)	CCD and NV	n/a	n/a	Monitor structural changes during heating	Monitoring possible
Alvarez et al. (2010)	CCD and NV	n/a	SS	Predicting emulsion fat/lean ratio	$r^2 = 0.76$
Alvarez et al. (2009)	CCD and NV	n/a	SS	Monitoring emulsion stability	Significant correlations found
Goni et al. (2008)	MRI	B-Splines	n/a	Segmenting muscle	Successful segmentation

Notes: CCD = charged coupled device, MRI = magnetic resonance imaging, UV = ultraviolet, BW = black and white, NV = non-visible, CLSM = confocal laser scanning microscope, FCM = fuzzy-C-means, NN = neural network, SS = standard statistics, GA = genetic algorithms, CC = correct classifications, RMSEP = root mean square error of prediction, MSE = mean square error.

Table 10.2   Recent applications of computer vision for assessing lamb quality

Author	Acquisition	Segmentation	Learning	Application	Best results
Chandraratne et al. (2006b)	CCD	Thresholding	SS	Classifying carcass grade	85% overall CC
Chandraratne et al. (2007)	CCD	n/a	NN	Classifying carcass grade	97% CC
Chandraratne et al. (2006a)	CCD	Thresholding	NN	Predicting tenderness from surface texture	$r^2 = 0.75$
Kongsro et al. (2009)	CT and VR	n/a	SS	Classifying carcass grade	r up to 0.98
Goni et al. (2008)	MRI	B-Splines	n/a	Segmenting the carcass	Successful segmentation

Notes: CCD = charged coupled device, VR = visible reflectance, CT = computed tomography, MRI = magnetic resonance imaging, NN = neural network, SS = standard statistics, CC = correct classifications.

Antequera et al. (2007) and the B-splines algorithm of Goni et al. (2008). Image segmentation can in fact be so difficult that manual intervention was required to find the ROI or where the segmentation issue was bypassed by keeping just the ROI within the camera's view. Generally, the segmentation strategies employed have depended on established techniques with only a few segmentation algorithms standing out as being radically different.

Most of the feature characterisation has changed very little recently; this is not surprising as the expert grading rules on which the computer vision system is based have also changed little. There is little that can be added in terms of colour characterisation and morphology characterisation that has not already been discovered. Thus, the only features where radically new methods are being used and investigated relate to the characterisation of texture.

There is a strong preference for statistical learning algorithms and classic multi-linear statistics such as discriminant analysis, clustering and multiple regressions. These remain attractive due to their simplicity and tractability. Regression models have an additional edge in that the relative contribution of image features to the overall prediction can be seen in the resulting regression equation. Neural networks occur most frequently as the chosen artificial intelligence learning algorithm.

10.3.1   Beef
Beef is the highest value meat (USDA Office of the Chief Economist, 2012). Thus, it is not surprising that accurate and robust beef classification has attracted

Table 10.3 Recent applications of computer vision for assessing pork quality

Author	Acquisition	Segmentation	Learning	Application	Best results
Antequera et al. (2007)	MRI	Active contours	SS	Correlating pixel area with joint weight	$r^2 = 0.99$
Lu et al. (2000)	CCD	Not stated	NN	Grading of surface colour	$r^2 = 0.75$
Tan et al. (2000)	CCD	Pixel comparison	NN	Predicting colour score	86% CC
Fortin et al. (2003)	Ultrasound	Manual	SS	Grading pork carcasses	$r^2 = 0.82$
O'Sullivan et al. (2003)	CCD	n/a	SS	Predicting pork visual sensory quality	Significant correlations
Liu et al. (2010)	CCD and NV	Window	SS	Predicting pork class	84% overall CC
Qiao et al. (2007a)	CCD and NV	Window	NN	Classifying pork quality	75–80% CC
Qiao et al. (2007b)	CCD and NV	Window	NN	Predicting colour, pH and drip loss	r up to 0.86
Magowan and McCann (2006)	Ultrasound	n/a	SS	Correlating ultrasonic and conventional backfat probes	r up to 0.94
Jia et al. (2010)	CCD	n/a	SS	Prediction of composition	$r = 0.74$–0.81
Cernadas et al. (2005)	MRI	Thresholding	SS	Prediction of sensorial traits	70%+ sensitivity
Sanchez et al. (2008)	CCD	Contrast enhancement &and k-NN	SS	Identifying fatty, lean and connective tissue and salt gain	$r = 0.8$ for salt concentration
Fulladosa et al. (2010)	CT	n/a	SS	Monitoring Salting	0.3% error in salt content
Vestergaard et al. (2005)	CT	Thresholding	SS	Predicting salt content	$r^2 = 0.94$
Wakamatsu et al. (2006)	Purple LED	n/a	n/a	Measuring Zn protoporphyrin	Successfully observed
Fantazzini et al. (2009)	MRI	Molecular relaxation time	SS	Monitoring ham maturation	$r^2 = 0.90$
Goni et al. (2008)	MRI	B-Splines	n/a	Segmenting the carcass	Successful segmentation

Notes: CCD = charged coupled device, CT = computed tomography, MRI = magnetic resonance imaging, NV = non-visible, NN = neural network, SS = standard statistics, k-NN = k nearest neighbours, CC = correct classifications.

Table 10.4 Recent applications of computer vision for assessing processed meat quality

Author	Acquisition	Segmentation	Learning	Application	Best results
Valous et al. (2009b)	CCD	Thresholding	n/a	Characterising ham texture	Significant differences found
Du and Sun (2006b)	CCD	Thresholding and watershed algorithm	SS	Characterising pores	r = −0.96
Jackman et al. (2010c)	CCD	Window	GA and SS	Predicting ham class	100% CC
Du and Sun (2006c)	CCD	Gradient and splines	SS	Predicting morphology features	<5% error
Mendoza et al. (2009a)	CCD	Thresholding and edge detection	SS	Classifying by directional fractal dimensions	82.2% CC
Iqbal et al. (2010)	CCD	Thresholding	SS	Classifying by colour and classic texture features	91–99% CC
Valous et al. (2010a)	CCD	Window	NN	Classifying by quaternionic singular values	86.1% CC
Valous et al. (2010b)	CCD	Window	n/a	Expressing texture with lacunarity	Effective texture expression
Valous et al. (2010c)	CCD	Window	n/a	Expressing texture with detrended fluctuation analysis	Effective texture expression
Mendoza et al. (2009b)	CCD	Window, thresholding and morphological operations	n/a	Characterising fat-connective tissue with multi-fractal dimensions	Effective expression

Notes: CCD = charged coupled device, NN = neural network, SS = standard statistics, GA = genetic algorithms, CC = correct classifications.

considerable interest in recent years. A number of studies have considered transposing the USDA expert grading system for beef carcasses (USDA, 1997) into a computer vision system (Li *et al.*, 1999; Tian *et al.*, 2005; Jackman *et al.*, 2008, 2009a, 2009b, 2009c, 2009d, 2010a, 2010b). These studies have had varying degrees of success with some accurate and robust modelling of beef palatability proven possible.

Prediction of beef palatability
Application of computer vision for predicting tenderness has been a topic of many investigations in recent years. Table 10.1 lists a large number of these. Obtaining a very strong prediction or a very high classification rate has proven to be very difficult. An r^2 value in the region of 0.6–0.7 has been typical of the better attempts to predict beef tenderness. A value of 0.69 was achieved by Park *et al.* (2001) in predicting shear force tenderness; research by Li *et al.* (1999) achieved a value of 0.7 for predicting sensory panel tenderness and a similar study by Tian *et al.* (2005) was slightly less successful. A variety of predictive models of tenderness by Jackman *et al.* (2008, 2009a, 2009c, 2009d, 2010a, 2010b) achieved values in the range of 0.7–0.75. More successful efforts were made in the same research to predict acceptability ($r^2 = 0.95$) and juiciness ($r^2 = 0.88$). A very low r^2 of 0.13 was found by McDonald and Chen (1991) in attempting to predict shear force tenderness but this was from just marbling scores. The difficulty of accurate predictions of beef tenderness is discussed in depth by Thompson (2002). Segregating beef carcasses into a small number of discrete tenderness classes has been proven to be a more achievable goal with 96% accuracy found by Naganathan *et al.* (2008a) for trinary classification. The learning algorithms employed in these studies have nearly always been neural networks or statistical.

Most data were acquired in the visible wavelength range with rare exceptions such as the use of near-infrared by Park *et al.* (2001), hyperspectral imaging (Naganathan *et al.*, 2008a, 2008b) and hyperspectral scattering imaging by Ranasinghesagara *et al.* (2010) and Cluff *et al.* (2008). A similar approach of optical scattering in both visible and near-infrared ranges was used by Xia *et al.* (2007, 2008), but a point receiver was used rather than a camera recording the backscatter data.

The prediction of beef palatability is a prime example of the need for accurate segmentation. This is because evaluation of muscle colour and marbling fat could be substantially skewed if part of the ribeye muscle was missed during segmentation. Errors can also be introduced into the data if intramuscular fat or other muscle was mistakenly included as part of the ribeye muscle. Even greater errors can be introduced if marbling fat and intermuscular fat are mistaken for each other (Subbiah *et al.*, 2004). Complex segmentation methods are often needed for images of beef. For beef ribeyes, automatic segmentation algorithms such as the Gao method (Gao, 1995) have been successful (Li *et al.*, 1999; Tian *et al.*, 2005). An alternative was developed by Jackman *et al.* (2009b). This could produce colour features with a very strong correlation to those that were produced with manual segmentation of the same images (r up to 1) and marbling features with a strong correlation with those manually produced (r up to 0.96). Figure 10.3

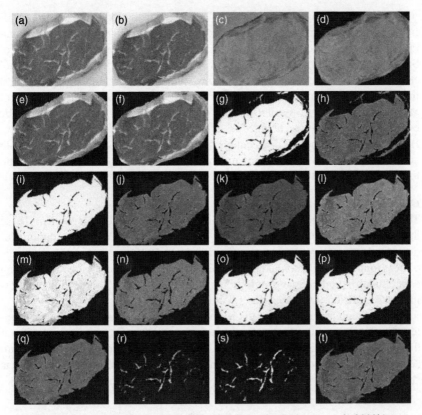

Fig. 10.3 Automatic segmentation of beef muscle (Jackman *et al.*, 2009b).

shows an example of this segmentation algorithm. Some avoided the problems of segmentation by keeping the field of view inside the ROI or imposing a window that lies within the ribeye (Naganathan *et al.*, 2008a, 2008b).

Prediction of beef carcass grade
The other main area of interest is predicting beef carcass grade. Predicting carcass grade has generally proved an easier challenge than predicting tenderness. Borggaard *et al.* (1996) and Shiranita *et al.* (2000) had the most successful attempts at carcass classification, with both finding an r^2 value of 0.93 for carcass conformation and grade respectively. Shiranita *et al.* (1998) did manage good classification plus or minus a single grade-based muscle texture and marbling. Basset *et al.* (2000) was able to achieve very good results with correct classifications as high as 96.3%. Subbiah *et al.* (2004) were also able to achieve good results with a low pixel misclassification rate of less than 2%. Statistical learning algorithms were the norm as seen in Table 10.1. As with predictions of beef palatability, the data acquisition was normally at visible wavelengths as can be seen in Table 10.1. Where visible light was insufficient, Basset *et al.* (2000) used an ultraviolet light and McCauley *et al.* (1994) used ultrasound.

As with predicting beef palatability, image segmentation again was very difficult. A host of segmentation algorithms were used. Of interest was the neural network method of Shiranita *et al.* (2000) and, in particular, the algorithm of Subbiah *et al.* (2004) that merged the fuzzy-c-means algorithm and convex hull erosion, allowing the ribeye to be found by successive erosion and dilation until the smallest polygon (convex hull) is found. Another segmentation algorithm of particular interest was developed by Borggaard *et al.* (1996). This worked by evaluating many features, thus reducing its vulnerability to exceptional carcasses. Hence, the term 'graceful degradation' applies when the algorithm is faced with an exceptional carcass. The segmentation problem has, on occasion, been found to be so difficult that manual intervention was required (Hatem *et al.*, 2003).

Other beef applications
There have been some other notable applications in beef quality assessment. These include the prediction of surface metmyglobin (Demos *et al.*, 1996), the prediction of colour and marbling scores (Gerrard *et al.*, 1996), modelling the shrinkage of beef joints with ellipsoidal curves while they are being air blast cooled (Zheng *et al.*, 2006b), predicting moisture from colour (Zheng *et al.*, 2006c) and monitoring beef emulsions (Alvarez *et al.*, 2009, 2010).

At very high magnifications the issue of shadows in the image becomes problematic. Thus, a means of retrospectively mitigating the effect of shadows without the loss of information is required. Such a solution was found by Du and Sun (2009). This removed shadows from confocal microscope images. An example is shown in Fig. 10.4.

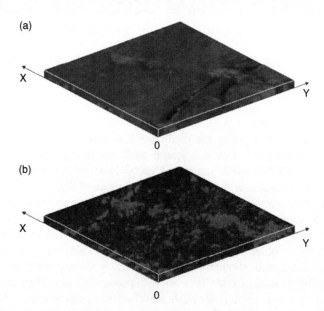

Fig. 10.4   Beef muscle images corrected for shading: (a) before thresholding and (b) after thresholding. (Du and Sun, 2009).

Data acquisition in the visible wavelength range is the norm (McDonald and Chen, 1991; Demos *et al.*, 1996; Gerrard *et al.*, 1996; Hwang *et al.*, 1997; Lee *et al.*, 2001; Hatem *et al.*, 2003; Lu and Tan, 2004; Zheng *et al.*, 2006b). Ultrasound data were used by Whittaker *et al.* (1992) and McCauley *et al.* (1994). Spectral scattering was used by Alvarez *et al.* (2009, 2010).

Image segmentation generally proved very difficult. A wide variety of procedures were used, as seen in Table 10.1. The segmentation algorithms that stand out are the neural network method of Hwang *et al.* (1997) and also the contour generation method of Lee *et al.* (2001). Other algorithms of particular interest were the morphological labelling algorithm (Gerrard *et al.*, 1996) and the B-spline curves method of Goni *et al.* (2008) for MRI of full beef carcasses, allowing carcass muscle assessment in a fully non-destructive way. Only in two cases was simple thresholding found to be sufficient (McDonald and Chen, 1991; Zheng *et al.*, 2006b).

10.4   Application and implementation for lamb, pork and other processed meats

The sections below discuss the different issues and demands raised by the application of this technology to types of meat other than beef.

10.4.1   Lamb

Some very similar work has been done with lamb as has been done with beef, mainly to predict tenderness from surface texture and geometric properties (Chandraratne *et al.*, 2006a) and to predict carcass grade (Chandraratne *et al.*, 2006b, 2007; Kongsro *et al.*, 2009) as seen in Table 10.2. The attempt at tenderness predictions was largely good but not complete ($r^2 = 0.75$). The attempts at predicting the carcass grade were very successful with the rate of correct classifications as high as 97% (Chandraratne *et al.*, 2007) and strong correlations with measured 'EUROP' carcass criteria (Kongsro *et al.*, 2009). In these studies, the lamb surface texture was characterised with classic texture algorithms of grey-level co-occurrence, grey-level run lengths and grey-level difference histograms. Visible imaging was the normal means of image acquisition with visible reflectance (VR) and with CT (Kongsro *et al.*, 2009) as an alternative. To segment lamb images thresholding proved adequate. Statistical learning or neural networks were preferred for data analysis.

10.4.2   Pork

Computer vision has been used for the analysis of pork colour (Lu *et al.*, 2000; Tan *et al.*, 2000; Qiao *et al.*, 2007b), marbling (Qiao *et al.*, 2007b), backfat thickness (Magowan and McCann, 2006), pH and drip loss (Qiao *et al.*, 2007a), observation of Zn protoporphyrin (Wakamatsu *et al.*, 2006), the grading of carcasses (Fortin *et al.*, 2003), prediction of sensorial characteristics (Cernadas *et al.*, 2005), monitoring salt content (Vestergaard *et al.*, 2005; Sanchez *et al.*, 2008; Fantazzini *et al.*, 2009; Fulladosa *et al.*, 2010) and the prediction of carcass composition (Jia *et al.*, 2010) as shown in Table 10.3.

The results for the prediction of carcass grade were very good ($r^2 = 0.82$). The prediction of salt content by Fulladosa *et al.* (2010) proved highly successful, with a very low prediction error of 0.3%, as did predictions by Vestergaard *et al.* (2005) and Fantazzini *et al.* (2009), with an r^2 of 0.94 and 0.9, respectively. However, efforts by Sanchez *et al.* (2008) were less successful, with an r value of 0.8. Some of the results found for predicting the surface colour grade ($r^2 = 0.75$, Lu *et al.*, 2000) and for predicting the colour score (86% correct classifications, Tan *et al.*, 2000) were quite good. The prediction of composition was reasonable at $r = 0.74$–0.81 (Jia *et al.*, 2010), as were sensitivities for pork sensorial traits (Cernadas *et al.*, 2005). The ripening process in pork carcass hams can be tracked over time (Antequera *et al.*, 2007), where the observed muscle shrinkage correlated strongly with the experimental weight loss ($r^2 = 0.99$). Statistical learning or neural networks were preferred for model building.

Image data were normally acquired in the visible wavelength range for pork (Lu *et al.*, 2000; Tan *et al.*, 2000, Jia *et al.*, 2010). Ultrasound imaging has been used as an alternative (Fortin *et al.*, 2003), as has MRI (Cernadas *et al.*, 2005; Antequera *et al.*, 2007; Fantazzini *et al.*, 2009) and CT (Vestergaard *et al.*, 2005; Fulladosa *et al.*, 2010). Hyperspectral imaging was used by Liu *et al.* (2010) to create greyscale images for Gabor filter-based surface texture characterisation. Hyperspectral imaging was also used by Qiao *et al.* (2007a, 2007b). A novel imaging approach of purple LED was used by Wakamatsu *et al.* (2006). No single segmentation approach has been adequate for pork. Very complex or elaborate segmentation was rare; however, a notable exception was the B-spline curves method of Goni *et al.* (2008) and the active contours method of Antequera *et al.* (2007).

10.4.3   Processed meats

The processed meats that are best suited to computer vision are hams, as seen in Table 10.4. While colour and texture are important indicators of ham yield and were successfully analysed by Jackman *et al.* (2010c) for their ability to do so, porosity is an absolutely vital indicator. Thus, pore density and area need to be evaluated too. Research on characterising ham porosity was performed by Du and Sun (2006b). This developed correlations between porosity and a variety of parameters (processing time, moisture content and texture profile analysis). Ham morphology features of surface area and volume were computed by Du and Sun (2006c) by using a gradient and spline-based segmentation algorithm, resulting in a low prediction error.

The problem of telling different types of hams apart was investigated extensively by Jackman *et al.* (2010c), Mendoza *et al.* (2009a, 2009b), Valous *et al.* (2009b, 2010a, 2010b, 2010c) and Iqbal *et al.* (2010). Hams have very similar visual appearance, making classification by human observation extremely difficult; thus, attempts were made to find subtle image features that could differentiate hams of a variety of different classes apart. This was centred on finding efficient expression of surface texture that allowed strong discrimination, along with considering other important ham quality indicators such as porosity, heterogeneity, connective tissue distributions and colour. In this series of research, the ham features were expressed via a range of advanced algorithms, such as wavelets, fractal dimensions, multifractal dimensions quaternions, lacunarity and detrended fluctuation analysis.

10.5   Future trends

The application of computer vision in meats will continue to be based around visible wavelength imaging as this is what can simulate the expert grading processes that were originally designed for a human observer. Thus, the established simple and low-cost vision systems will continue to be viable. Adding magnification equipment to the image acquisition device, allowing for a variety of modalities to be considered, is a simple option for improving basic vision systems.

The emergence of hyperspectral imaging in recent years offers a means of supplementing and improving the visible wavelength systems by providing additional information on the chemical composition at the meat surface or on the internal structure of the meat. Hyperspectral imagers are costly, but this cost will fall over time and become more reasonable; furthermore, the additional technological burden is manageable. Powerful penetrative imagers such as MRI and CT will remain at the fringes due to their high cost and technological burden, which is tolerable in fields such as medicine that has enormous financial backing and where image processing time is not a major concern.

Image segmentation has drawn mainly on established techniques. Segmentation algorithms standing out as being particularly radical or different were rare and will continue to be rare as fresh meat products will not change substantially in the near future; there is also little scope for processed meats to change substantially. Future work on describing the ROI will be centred on developing more efficient texture features as there is very little to discuss about colour or morphology that has not been discussed before.

Similarly, image data processing is not likely to change significantly in coming years as the same problems needed solving – that is, palatability prediction, quality classification or defect detection. Thus, the tried and trusted learning algorithms of multivariate statistics and artificial intelligence will continue to be applied. There are likely to be novel and inventive learning algorithms developed in the future, particularly by the mathematical and computer science communities; however, these will be less tractable and slower, thus making them less valuable even though they might slightly improve classification or prediction.

10.6   Conclusions

Based on the above discussion, it is clear that computer vision technology will continue to be a very useful tool in tackling a wide variety of challenges in meat classification and quality prediction. Good predictions and classification results have been found on a large number of occasions using simple, rapid and affordable technology and the new challenges encountered will follow similar lines. This is because the meats being examined will not change radically and neither will the expert grading rules.

The most promising opportunities for taking computer vision to a higher level of performance lie with hyperspectral imaging. Detailed internal inspections with technology such as MRI or CT are not yet practical. Alongside this, new algorithms developed in other scientific disciplines aiming at more efficient image texture characterisation should be evaluated for potential application to meat images. Advances in artificial intelligence are likely to be of limited value.

10.7   References

Aguilera, J. M. and Briones, V. (2005) Computer vision and food quality. *Food Australia*, **57(3)**, 79–87.

Alvarez, D., Castillo, M., Payne, F. A. and Xiong, Y. L. (2009) A novel fiber optic sensor monitor beef meat emulsion stability using visible light scattering. *Meat Science*, **81(3)**, 456–66.

Alvarez, D., Castillo, M., Xiong, Y. L. and Payne, F. A. (2010) Prediction of beef meat emulsion quality with apparent light backscatter extinction. *Food research International*, **43(6)**, 1260–6.

Antequera, T., Caro, A., Rodriguez, P. G. and Perez, T. (2007) Monitoring the ripening process of Iberian ham by computer vision on magnetic resonance imaging. *Meat Science*, **76(3)**, 561–7.

Basset, O., Buquet, B., Aboulekaram, S., Delachartre, P. and Culioli, J. (2000) Application of texture image analysis for the classification of bovine meat. *Food Chemistry*, **69(4)**, 437–45.

Brainard, D. (2003) Color appearance and color difference specification. InShevell, S. K. (ed.), *The Science of Color*, 2nd edn, Amsterdam: Elsevier Science.

Borggaard, C., Madsen, N. T. and Thodberg, H. H. (1996) In-line image analysis in the slaughter industry, illustrated by beef carcass classification. *Meat Science*, **43(S1)**, 151–63.

Brosnan, T. and Sun, D.-W. (2004) Improving quality inspection of food products by computer vision – a review. *Journal of Food Engineering*, **61(1)**, 3–16.

Cernadas, E., Carrion, P., Rodriguez, P. G., Muriel, E. and Antequera, T. (2005) Analyzing magnetic resonance images of Iberian pork loin to predict its sensorial characteristics. *Computer Vision and Image Understanding*, **98(3)**, 345–61.

Chandraratne, M. R., Samarasinghe, S., Kulasiri, D. and Bickerstaffe, R. (2006a) Prediction of lamb tenderness using image surface texture features. *Journal of Food Engineering*, **77(3)**, 492–9.

Chandraratne, M. R., Kulasiri, D., Frampton, C., Samarasinghe, S. and Bickerstaffe, R. (2006b) Prediction of lamb carcass grades using features extracted from lamb chop images. *Journal of Food Engineering*, **74(1)**, 116–24.

Chandraratne, M. R., Kulasiri, D. and Samarasinghe, S. (2007) Classification of lamb carcass using machine vision: Comparison of statistical and neural network analyses. *Journal of Food Engineering*, **82(1)**, 26–34.

Cluff, K., Naganathan, G. K., Subbiah, J., Lu, R., Calkins, C. R. and Samal, A. (2008) Optical scattering in beef steak to predict tenderness using hyperspectral imaging in the VIS-NIR region. *Sensors and Instrumentation in Food Quality*, **2(1)**, 189–96.

Demos, B. P., Gerrard, D. E., Gao, X., Tan, J. and Mandigo, R. W. (1996) Utilization of image processing to quantitate surface metmyglobin on fresh beef. *Meat Science*, **43(3)**, 265–74.

Du, C.-J. and Sun, D.-W. (2004) Recent developments in the applications of image processing techniques for food quality evaluation. *Trends in Food Science & Technology*, **15(5)**, 230–49.

Du, C.-J. and Sun, D.-W. (2006a) Learning techniques used in computer vision for food quality evaluation: a review. *Journal of Food Engineering*, **72(1)**, 39–55.

Du, C.-J. and Sun, D.-W. (2006b) Automatic measurement of pores and porosity in pork ham and their correlations with processing time, water content and texture. *Meat Science*, **72(2)**, 294–302.

Du, C.-J. and Sun, D.-W. (2006c) Estimating the surface area and volume of ellipsoidal ham using computer vision. *Journal of Food Engineering*, **73(2)**, 260–8.

Du, C.-J. and Sun, D.-W. (2009) Retrospective shading correction of confocal laser scanning microscopy beef images for three-dimensional visualization. *Food and Bioprocess Technology*, **2(2)**, 167–76.

ElMasry, G. and Sun, D.-W. (2010) Meat quality assessment using a hyperspectral imaging system. In D.-W. Sun (ed.), *Hyperspectral Imaging for Food Quality Analysis and Control*. Amsterdam, Netherlands: Academic Press (Elsevier).

Fantazzini, P., Gombia, M., Schembri, P., Simoncini, N. and Virgili, R. (2009) Use of magnetic resonance imaging for monitoring parma dry-cured ham processing. *Meat Science*, **82(2)**, 219–27.

Fortin, A., Tong, A. K. W., Robertson, W. M., Zawadaski, S. M., Landry, S. J., Robinson, D. J., Liu, T. and Mockford, R. J. (2003) A novel approach to grading pork carcasses: computer vision and ultrasound. *Meat Science*, **63(2)**, 451–62.

Fulladosa, E., Santos-Garces, E., Picouet, P. and Gou, P. (2010) Prediction of salt and water content in dry-cured hams by computed tomography. *Journal of Food Engineering*, **96(1)**, 80–5.

Gao, X. O. (1995) Vision-based methodology development for food quality evaluation. Ph.D. thesis, University of Missouri, Colombia, MO.

Gerrard, D. E., Gao, X. and Tan, J. (1996) Beef marbling and colour score determination by image processing. *Journal of Food Science*, **61(1)**, 145–8.

Goni, S. M., Purlis, E. and Salvadori, V. O. (2008) Geometry modelling of food materials from magnetic resonance imaging. *Journal of Food Engineering*, **88(4)**, 561–7.

Haralick, R. M., Shanmugam, K., and Dinstein, I. (1973). Textural features for image classification. *IEEE Transactions on Systems, Man & Cybernetics*, **6(3)**, 610–21.

Hatem, I., Tan, J. and Gerrard, D. E. (2003) Determination of animal skeletal maturity by image processing. *Meat Science*, **65(4)**, 999–1004.

HunterLab (2008) *Insight on Color: CIE L*a*b* Color Scale*. Reston, VA, USA.

Huang, Y., Lacey, R. E., Moore, L. L., Miller, R. K., Whittaker, A. D. and Ophir, J. (1997) Wavelet textural features from ultrasonic elastograms for meat quality prediction. *Transactions of the ASAE*, **40(6)**, 1741–8.

Hwang, H., Park, B., Nguyen, M. and Chen, Y.-R. (1997) Hybrid image processing for robust extraction of lean tissue on beef cut surfaces. *Computers and Electronics in Agriculture*, **17(1)**, 281–94.

Instron (2008) *5500 Series Advanced Materials Testing Systems*. Instron Inc., Norwood, MA, USA.

Iqbal, A., Valous, N. A., Mendoza, F., Sun, D.-W. and Allen, P. (2010) Classification of pre sliced pork and turkey ham qualities based on image colour and textural features and their relationships with consumer responses. *Meat Science*, **84(2)**, 455–65.

Jackman, P., Sun, D.-W., Du, C.-J., Allen, P. and Downey, G. (2008) Prediction of beef eating quality from colour, marbling and wavelet texture features. *Meat Science*, **80(4)**, 1273–81.

Jackman, P., Sun, D.-W., Du, C.-J., and Allen, P. (2009a) Prediction of beef eating qualities from colour, marbling and wavelet surface texture features using homogenous carcass treatment. *Pattern Recognition*, **42(5)** 751–63.

Jackman, P., Sun, D.-W. and Allen, P. (2009b) Automatic segmentation of beef *longissimus dorsi* muscle and marbling by an adaptable algorithm. *Meat Science*, **83(2)**, 187–94.

Jackman, P., Sun, D.-W. and Allen, P. (2009c) Comparison of various wavelet texture features to predict beef palatability. *Meat Science*, **83(1)**, 82–7.

Jackman, P., Sun, D.-W. and Allen, P. (2009d) Comparison of the predictive power of beef surface wavelet texture features at high and low magnification. *Meat Science*, **82(3)**, 353–6.

Jackman, P., Sun, D.-W. and Allen, P. (2010a) Prediction of beef palatability from colour, marbling and surface texture features of longissimus dorsi. *Journal of Food Engineering*, **96(1)**, 151–65.

Jackman, P., Sun, D.-W., Allen, P., Brandon, K. and White, A. M. (2010b) Correlation of consumer assessment of longissimus dorsi beef palatability with image colour, marbling and surface texture features. *Meat Science*, **84(3)**, 564–8.

Jackman, P., Sun, D.-W., Allen, P., Valous, N. A., Mendoza, F. and Ward, P. (2010c) Identification of important image features for pork and turkey ham classification using colour and wavelet texture features and genetic selection. *Meat Science*, **84(4)**, 711–17.

Jia, J., Schinckel, A. P., Forrest, J. C., Chen, W. and Wagner, J. R. (2010) Prediction of lean and fat composition in swine carcasses from ham area measurements with image analysis. *Meat Science*, **85(2)**, 240–4.

Kaiser, G. (1994) *A Friendly Guide to Wavelets*. Boston, MA: Birkhauser .

Kongsro, J., Roe, M., Kvaal, K., Aastveit, A. H. and Egelandsdal, B. (2009) Prediction of fat, muscle and value in Norwegian lamb carcasses using EUROP classification, carcass shape and length measurements, visible light reflectance and computer tomography (CT). *Meat Science*, **81(1)**, 102–7.

Lee, C. H., Lee, S. H. and Hwang, H. (2001) Automatic lean tissue generation of carcass beef via color computer vision. ASAE Meeting presentation, Paper 01-6122.

Leon, K., Mery, D., Pedreschi, F. and Leon, J. (2006) Color measurement in $L^*a^*b^*$ units from RGB digital images. *Food Research International*, **39(10)**, 1084–91.

Li, J., Tan, J., Martz, F. A. and Heymann, H. (1999) Image texture features as indicators of beef tenderness. *Meat Science*, **53(1)**, 17–22.

Li, J., Tan, J. and Shatdal, P. (2001) Classification of tough and tender beef by image texture analysis. *Meat Science*, **57(4)**, 341–6.

Liu, L., Ngadi, M. O., Prasher, S. O. and Gariepy, C. (2010) Categorization of pork quality using Gabor filter-based hyperspectral imaging technology. *Journal of Food Engineering*, **99(2)**, 284–93.

Lu, W. and Tan, J. (2004) Analysis of image-based measurements and USDA characteristics as predictors of beef lean yield. *Meat Science*, **66(2)**, 483–91.

Lu, J., Tan, J., Shatadal, P. and Gerrard, D. E. (2000) Evaluation of pork color by using computer vision. *Meat Science*, **56(1)**, 57–60.

Magowan, E. and McCann, M. E. E. (2006) A comparison of pig backfat measurements using ultrasonic and optical instruments. *Livestock Science*, **103(1)**, 116–23.

McCauley, J. D., Thane, B. R. and Whittaker, A. D. (1994) Fat estimation in beef ultrasound images using texture and adaptive logic networks. *Transactions of the ASAE*, **37(3)**, 997–1002.

McDonald, T. P. and Chen, Y. R. (1991) Visual characterization of marbling in beef ribeyes and its relationship to taste parameters. *Transactions of the ASAE*, **34(6)**, 2499–504.

Mendoza, F., Valous, N. A., Allen, P., Kenny, T. A., Ward, P. and Sun, D.-W. (2009a) Analysis and classification of commercial ham slice images using directional fractal dimension features. *Meat Science*, **81(2)**, 313–20.

Mendoza, F., Valous, N. A., Sun, D.-W. and Allen, P. (2009b) Characterization of fat-connective tissue size distribution in pre-sliced pork hams using multifractal analysis. *Meat Science*, **83(4)**, 713–22.

Naganathan, G. K., Grimes, L. M., Subbiah, J., Calkins, C. R., Samal, A. and Meyer, G. E. (2008a) Visible/near-infrared hyperspectral imaging for beef tenderness prediction. *Computers and Electronics in Agriculture*, **64(2)**, 225–33.

Naganathan, G. K., Grimes, L. M., Subbiah, J., Calkins, C. R., Samal, A. and Meyer, G. E. (2008b) Partial least squares analysis of near-infrared hyperspectral images for beef tenderness prediction. *Sensors and Instrumentation in Food Quality*, **2(1)**, 178–88.

O'Sullivan, M. G., Byrne, D. V., Martens, H., Gidskehaug, L. H., Andersen, H. J. and Martens, M. (2003) Evaluation of pork colour: prediction of visual sensory quality of meat from instrumental and computer vision methods of colour analysis. *Meat Science*, **65(7)**, 909–18.

Park, B., Chen, Y. R., Hruschka, W. R., Shackleford, S. D. and Koohmaraie, M. (2001) Principle component regression of near-infrared reflectance spectra for beef tenderness prediction. *Transactions of the ASAE*, **44(3)**, 609–15.

Pearson, A. M. and Dutson, T. R. (1995) *Quality Attributes and Their Measurement in Meat, Poultry and Fish Products*. New York: Springer.

Qiao, J., Wang, N., Ngadi, M. O., Gunenc, A., Monroy, M., Gariepy, C. and Prasher, S. O. (2007a) Prediction of drip-loss, pH, and color for pork using a hyperspectral imaging technique. *Meat Science*, **76(1)**, 1–8.

Qiao, J., Ngadi, M. O., Wang, N., Gariepy, C. and Prasher, S. O. (2007b). Pork quality and marbling level assessment using a hyperspectral imaging system. *Journal of Food Engineering*, **83(1)**, 10–16.

Ranasinghesagara, J., Nath, T. M., Wells, S. J., Weaver A. D., Gerrard, D. E. and Yao, G. (2010) Imaging optical diffuse reflectance in beef muscles for tenderness prediction. *Meat Science*, **84(3)**, 413–21.

Sanchez, A. J., Albarracin, W., Grau, R., Ricolfe, C. and Barat, J. M. (2008) Control of ham salting by using image segmentation. *Food Control*, **19(1)**, 135–42.

Scanlan, R. A. (ed) (1977) *Flavor Quality: Objective Measurement*. Washington, DC: American Chemical Society .

Shiranita, K., Hayashi, K., Otsubo, A., Miyajima, T. and Takiyama, R. (2000) Grading meat quality by image processing. *Pattern Recognition*, **33(1)**, 97–104.

Shiranita, K., Miyajami, T. and Takiyama, R. (1998) Determination of meat quality by texture analysis. *Pattern Recognition Letters*, **19(6)**, 1319–24.

Soille, P. and Rivest, J. F. (1996) On the validity of fractal dimension measurements in image analysis. *Journal of Visual Communication and Image Representation*, **7(3)**, 217–29.

Stein, E. M. and Shakarchi, R. (2003) *Fourier Analysis: An Introduction*. Princeton, NJ: Princeton University Press .

Steiner, R., Wyle, A. M., Vote, D. J., Belk, K. E., Scanga, J. A., Wise, J. W., Tatum, J. D. and Smith, G. C. (2003) Real-time augmentation of USDA yield grade application to beef carcasses using video image analysis. *Journal of Animal Science*, **81(6)**, 2239–46.

Stokes, M., Anderson, M., Chandrasekar, S. and Motta, R. (1996) *A Standard Default Color Space for the Internet: sRGB*. Reston, VA: International Color Consortium.

Subbiah, J., Ray, N., Kranzler, G. A. and Acton, S. T. (2004) Computer vision segmentation of the *longissimus dorsi* for beef quality grading. *Transactions of the ASAE*, **47(4)**, 1261–8.

Sun, D.-W. (ed.) (2008) *Computer Vision Technology for Food Quality Evaluation*. San Diego, CA: Academic Press/Elsevier.

Tan, J. (2004) Meat quality evaluation by computer vision. *Journal of Food Engineering*, **61(1)**, 27–35.

Tan, F. J., Morgan, M. T., Ludas, L. I., Forrest, J. C. and Gerrard, D. E. (2000) Assessment of fresh pork color with color machine vision. *Journal of Animal Science*, **78(2)**, 3078–85.

Thompson, J. (2002) Managing meat tenderness. *Meat Science*, **62(2)**, 295–308.

Tian, Y. Q., McCall, D. G., Dripps, W., Yu, Q. and Gong, P. (2005) Using computer vision technology to evaluate the meat tenderness of grazing beef. *Food Australia*, **57(8)**, 322–6.

USDA Agricultural Marketing Service (2012) *Grading, Certification and Verification: USDA Quality Standards.* Washington, DC: USDA.

USDA Office of the Chief Economist (2012) *World Supply and Demand Estimates.* Washington, DC: USDA.

USDA (1997) *United States Standards for Grades of Carcass Beef.* Washington, DC: USDA.

Valous, N. A., Mendoza, F., Sun, D.-W. and Allen, P. (2009a) Colour calibration of a laboratory computer vision system for quality evaluation of pre-sliced hams. *Meat Science*, **81(1)**, 132–41.

Valous, N. A., Mendoza, F., Sun, D.-W. and Allen, P. (2009b) Texture appearance characterization of pre-sliced pork ham images using fractal metrics: Fourier analysis dimension and lacunarity. *Food Research International*, **42(3)**, 353–62.

Valous, N. A., Mendoza, F., Sun, D.-W. and Allen, P. (2010a) Supervised neural network classification of pre-sliced cooked pork ham images using quaternionic singular values. *Meat Science*, **84(3)**, 422–30.

Valous, N. A., Sun, D.-W., Allen, P. and Mendoza, F. (2010b) The use of lacunarity for visual texture characterization of pre-sliced cooked pork ham surface intensities. *Food Research International*, **43(2)**, 387–95.

Valous, N. A., Drakakis, K. and Sun, D.-W. (2010c) Detecting fractal power-law long-range dependence in pre-sliced cooked pork ham surface intensity patterns using detrended fluctuation analysis. *Meat Science*, **86(2)**, 289–97.

Vestergaard, C., Erbou, S. G., Thauland, T., Adler-Nissen, J. and Berg, P. (2005) Salt distribution in dry-cured ham measured by computed tomography and image analysis. *Meat Science*, **69(1)**, 9–15.

Vote, D. J., Belk, K. E., Tatum, J. D., Scanga, J. A. and Smith, G. C. (2003) Online prediction of beef tenderness using a computer vision system equipped with a Beefcam module. *Journal of Animal Science*, **81(2)**, 457–65.

Wakamatsu, J., Odagiri, H., Nishimura, T. and Hattori, A. (2006) Observation of the distribution of Zn protoporphyrin IX (ZPP) in Parma ham by using purple LED and image analysis. *Meat Science*, **74(4)**, 594–9.

Warriss, P. D. (2000). *Meat Science: An Introductory Text.* Wallingford, UK: CABI.

Whittaker, A. D., Park, B., Thane, B. R., Miller, R. K. and Savell, J. W. (1992) Principles of ultrasound and measurement of intramuscular fat. *Journal of Animal Science*, **70(4)**, 942–52.

Xia, J., Berg, E. P., Lee, J. W. and Yao, G. (2007) Characterizing beef muscles with optical scattering and absorption coefficients in VIS-NIR region. *Meat Science*, **75(1)**, 78–83.

Xia, J., Weaver, A., Gerrard, D. E. and Yao, G. (2008) Heating induced optical property changes in beef muscle. *Journal of Food Engineering*, **84(1)**, 75–81.

Zheng, C., Sun, D-W. and Zheng, L. (2006a) Recent applications of image texture for evaluation of food qualities – a review. *Trends in Food Science and Technology*, **17(3)**, 113–28.

Zheng, C., Sun, D.-W. and Du, C.-J. (2006b) Estimating shrinkage of large cooked beef joints during air-blast cooling by computer vision. *Journal of Food Engineering*, **72(1)**, 56–62.

Zheng, C., Sun, D-W. and Zheng, L. (2006c) Correlating colour to moisture content of large cooked beef joints by computer vision. *Journal of Food Engineering*, **77(4)**, 858–63.

11

Real-time ultrasound (RTU) imaging methods for quality control of meats

S. R. Silva, CECAV, University of Trás-os-Montes e Alto Douro, Portugal and V. P. Cadavez, CIMO, ESA, Instituto Politécnico de Bragança, Portugal

Abstract: In this chapter the use of real-time ultrasonography to predict *in vivo* carcass composition and meat traits will be reviewed. The chapter begins by discussing background and principles of ultrasound. Then aspects affecting the suitability of real-time ultrasonography and image analysis for predicting carcass composition and meat traits of meat producing species and fish will be presented. This chapter also provides an overview of the present and future trends in the application of real-time ultrasonography in the meat industry.

Key words: meat quality, ultrasound, carcass, image analysis.

11.1 Introduction

Carcass composition and meat traits are important aspects of animal science relating to food production. This knowledge is fundamental to the study of genetics, nutrition, physiology, to marketing based on carcass value, as well as for monitoring body fat reserves. Dissection and chemical analysis have traditionally been used as the standard methods for determining carcass composition. However, these procedures are expensive, laborious and destructive (i.e., an animal or carcass can be used only once). Non-destructive techniques are often required to test valuable animals or when sequential study of the animals is necessary or desirable (Fuller *et al.*, 1990). The search for non-destructive methods of estimating carcass composition or meat traits has led to the evaluation of numerous techniques such as real-time ultrasound (RTU), computer tomography (CT), magnetic resonance imaging (MRI), dual-energy X-ray absorptiometry (DXA), whole-body ^{40}K count-ing, total body electrical conductivity (TOBEC), dilution techniques, bioelectrical impedance and neutron activation analysis. These techniques have been reviewed by several researchers, including Allen (1990), Fuller *et al.* (1994), Stanford *et al.* (1998), Szabo *et al.* (1999), De Campeneere *et al.* (2000), Mitchell and Scholz (2005) and Teixeira (2009). Of the techniques mentioned above, only those based on RTU will be outlined in this chapter. Techniques based on ultrasound have had great success in the fields of medical and animal science, as they are non-invasive, non-destructive and do not cause pain to the animal. For over 50 years, ultrasound techniques have been used to predict carcass composition and meat traits *in vivo*. Since its initial use, and especially in the last two decades, RTU has been demonstrated to be a valuable tool for the estimation of carcass composition and meat traits in living animals. The recent interest in the technique is almost cer-tainly a result of the application of technology originally developed for comput-ers, whereby a digital image formation process provides good quality black and white images. Furthermore, modern equipment is robust, easy to use and portable, and offers accurate imaging with great repeatability at relatively low cost, while also being well accepted by the public (Allen, 1990; Stanford *et al.*, 1998). This chapter presents an overview of the use of RTU in predicting carcass composition and meat traits in meat-producing species and fish.

11.2   Historical background on ultrasound use for carcass composition and meat traits evaluation

The roots of the use of ultrasound techniques for animal science purposes can be identified in several discoveries throughout history and are closely connected to the same developments in the medical field. The discovery of the piezoelectric properties of certain crystals in 1880 by the Curie brothers is one major milestone in the development of ultrasound (Woo, 2006) (Table 11.1). Since then, the appli-cations of ultrasound have expanded rapidly in the fields of navigation, medicine (Thwaites, 1984; Szabo, 2004) and non-destructive testing in industry (Bray and McBride, 1992; Chen, 2007).

Mankind has always had a fascination with the idea that it might be possible to look inside objects and, with sonar and radar as models, it was established that pulse-echo techniques had the potential to be used for medical purposes (Szabo, 2004). After five years of work on ultrasound principles and equipment develop-ment, the first diagnostic ultrasound study was published (Dussik, 1942). Some years later, Wild (1950) presented the first use of ultrasonic pulses for the mea-surement of biological tissues and the detection of tissue density with ultrasound equipment. This equipment contained a transducer that sent a sequence of repeti-tive ultrasonic pulses into a material or a body (Wells, 1991; Whittaker *et al.*, 1992). Echoes from different target objects and boundaries were received and amplified so that they could be displayed on an oscilloscope as an amplitude-versus-time record (Thwaites, 1984; Wells, 1991; Whittaker *et al.*, 1992). This

Table 11.1 Milestones for ultrasound development and application in animal science

Discovery or application	Year	Author	Description	Reference
Eco-localization	1790	Lazzaro Spallanzani	In 1790 Lazzaro Spallanzani experimented with bats and found that they manoeuvred through the air using their hearing rather than sight	Kane *et al.* (2004)
Piezoelectric proprieties	1881	Pierre Curie	In 1881 Pierre Curie found a connection between electrical voltage and pressure on crystalline material. This was the breakthrough that was needed to create the modern ultrasound transducer	Turner *et al.* (1994)
Using ultrasound in submarine warfare	1916		The first recorded detection and subsequent sinking of a German U-boat (UC-3) using a hydrophone	Kane *et al.* (2004)
First report of ultrasound use for medical purposes	1942	Karl Dussik	Karl and Friederich Dussik used the first medical application of ultrasound when they localized brain tumours by measuring the sound transmission through the skull and brain	Weinstein *et al.* (2006)
Ultrasound propagation speed in body tissues	1950	Ludwig	Ludwig (1950) made a number of time-of-flight measurements of sound speed through arm, leg, and thigh muscles. He found the average to be $1540\ ms^{-1}$, which is the standard value still used today	Ludwig (1950)
First article showing the utility of ultrasounds for soft tissues	1950	Wild	The first scientific proof of sonic energy reflection from within soft tissue histological elements, using 'A' mode readout	Wild (1950)
First B-mode	1952	Wild and Reid	The B-mode scanner became one of the first to differentiate between abnormal tissue	Wild and Reid (1952)
First animal evaluation publication using ultrasounds	1956	Temple *et al.*	First ultrasound animal evaluation publication in the United States	Temple *et al.* (1956)

(*Continued*)

Table 11.1 Continued

Discovery or application	Year	Author	Description	Reference
First B-mode study with animals	1959	Stouffer	Cross-sectional image of beef rib eye produced by an early mechanical B-scan system improved by mounting the transducer on a carriage that moved along a fixed, shaped, curved guide	Stouffer *et al.* (1959)
A reference study	1961	Stouffer	Study with hogs, cattle and sheep showing the superior performance of mechanical B-scan over the A-mode	Stouffer *et al.* (1961)
First real time	1965		Appearance of Vidoson from Siemens, the first real-time mechanical commercial scanner	Szabo (2004)
Scanogram	1969	Stouffer	Scanogram, commercial mechanical B-scanner, second generation, produced by Ithaco Inc, 1969	Stouffer (2004)
First RTU application for carcass traits evaluation	1976	Hans Busk	Use of the RTU Danscanner in breeding programmes for pigs, cattle and sheep	Busk (1984)
First 3-D	1987		First 3-D ultrasound	Szabo (2004)

type of display became known as the A-mode, with A standing for amplitude (Szabo, 2004). The distances between successive peaks represent the thickness of the different tissues. In animal science, the horizontal axis of the oscilloscope is calibrated in millimetres, allowing for a direct reading.

After the first medical applications, this ultrasound technique was immediately recognized as a potential tool for animal science, as shown by the work published in the late 1950s on animal carcass evaluation (Stouffer, 2004). These publications reported research results showing the feasibility of using ultrasound to evaluate carcass composition in live cattle (Temple et al., 1956), swine (Claus, 1957; Dumont, 1957; Hazel and Kline, 1959) and sheep (Campbell et al., 1959).

Despite the encouraging findings, the accuracy of the early A-mode, a single-transducer device, was often quite variable (Stouffer, 2004). Moreover, the A-mode display has limited use, because it lacks anatomical information, meaning that it is difficult to identify the anatomical sources of the echoes (Ophir and Maklad, 1979), and that it is impossible to trace area measurements from images of organs or tissues (Thwaites, 1984). To overcome these limitations, the B-mode presentation ('B' meaning brightness) was introduced. B-mode is an image display created by integrating multiple A-mode signals (Amin, 1995). In B-mode, the brightness of the dots is proportional to the amplitude of the echoes. The display consists of time traces running vertically (top to bottom) to indicate depth.

By the early 1960s a pioneering technique in the use of ultrasound for animal science purposes was introduced: a continuous mechanical scanning procedure (Stouffer et al., 1961). An electric motor was mounted on a thick rubber belt that was placed on the animal's back. The motor moved a transducer horizontally as it was held vertically by an operator, and was synchronized to keep the lens open for the duration of the 10 s scan in order to capture the image. The image on Polaroid film was developed in about 1 min, and the image data was then evaluated and measured. In the same period, the first commercialized contact B-mode mechanical scanners became available for medical purposes (Szabo, 2004). At the end of the 1960s, a commercial unit of the primary system was introduced for live animal evaluation, using similar technology (Stouffer, 2004). This equipment – the Scanogram – produced in 1969 by Ithaco Inc. (Ithaca, NY), was in use until the mid-1980s for the majority of in vivo carcass evaluation studies using ultrasound (Miles et al., 1972; Shelton et al., 1977; Kempster et al., 1982; Andersen et al., 1983; Simm et al., 1983). However, one of the major limitations of B-mode mechanical scanners for animal applications was the movement of the animal, which, being random, was the cause of inaccuracy of images and low repeatability of measurements (Hedrick et al., 1962; Gooden et al., 1980; Stouffer, 2004).

The launch of RTU systems with good image quality marked the end of the mechanical B-scanners, which had completely disappeared by the late 1980s (Klein, 1981; Szabo, 2004). RTU systems are based on the B-mode technique, and use multiple-crystal transducers to display an image on the screen that is constantly updated. The entire image frame must be displayed in 33 ms or less in order to be able to update the information at real-time frame rates (Insana, 2006).

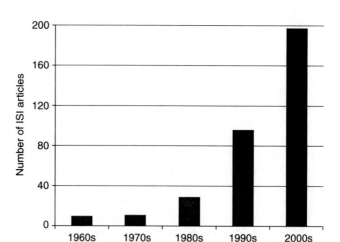

Fig. 11.1   Number of articles in main animal science journals (ISI indexed) using ultrasound to evaluate carcass traits.

RTU reduces the time required to produce and record an image, which greatly enhances the use of this technique in live animal evaluation.

Since the first attempts to image carcass traits using the RTU technique in the late 1970s (Kempster *et al.*, 1979) and early 1980s (Kempster *et al.*, 1982), significant advances in hardware and software have allowed the initial obstacles – animal movement and long acquisition times – to be overcome. In recent years, RTU has become a crucial tool in many routine carcass evaluations for animal production, and offers the advantage of providing data not only on carcass traits but also on a multitude of meat and fat deposits, which are similar to or even superior to those provided by more expensive imaging tools. The features of the ultrasound equipment, combined with the possibility of differentiating tissues and organs in the image, form the basis of the huge success that this technique has achieved in medicine and in animal science.

There has been a radical increase in the use of RTU imaging in animal science in the last 20 years, due to significant technological improvements and the availability of more accurate and less costly equipment. During this period, RTU has been widely used in the prediction of carcass traits, as shown by the numerous works found in the major animal science journals (Fig. 11.1).

11.3   Basic ultrasound imaging principles

Ultrasound is sound waves that have a frequency beyond the range of human hearing (above 20 kHz). These acoustic waves propagate through body tissues via compression and expansion of the tissues, and during propagation small particles of the material move back and forth in order to generate the compressions and expansions of the acoustic wave (Mannion, 2006). In soft tissues (biological tissue), these particles move back and forth in the same direction that the acoustic wave is travelling (Prince and Links, 2006). The particles themselves merely oscillate or are displaced locally; it is the wave that travels from source to detector, not the particles (Leighton, 2007). The ultrasonic waves have different propagation properties and can be characterized by the following formula

$$v = f \lambda \qquad [11.1]$$

where λ is the wavelength, f is frequency and v is velocity. The acoustic impedance Z is a fundamental property of the tissue and is related to the density ρ and the velocity

$$Z = \rho v \qquad [11.2]$$

The fraction of energy reflected, R, at the normal interface of two different tissue types is

$$R = \left(\frac{Z_2 - Z_1}{Z_2 + Z_1} \right)^2 \qquad [11.3]$$

where Z_1 and Z_2 are the impedance of the tissues in the interface (Seidband, 1998). The acoustic signals decrease as a result of attenuation by the medium, and the signal intensity I is given by

$$I = \frac{I_o e^{-\alpha r}}{r^2} \qquad [11.4]$$

where α is the coefficient of attenuation, I_0 is the incident signal intensity and r is the distance (Seidband, 1998). Most biological tissues have high coefficients of attenuation, which increases as frequency increases (Seidband, 1998). Thus, it is important to establish the thickness of the tissue at which the attenuation of the medium decreases the signal by half (half-value layer, HVL) (Seidband, 1998). Ultrasound of higher frequencies provides higher resolution, yet the increased HVL reduces the depth of penetration. The acoustic properties of some tissues at 1.0 MHz are presented in Table 11.2.

In soft tissues, the ultrasound waves propagate at a velocity of about 1500 m/s. Each change of tissue type causes a reflection and the greater the difference in acoustic impedance between the tissues the greater the proportion of the ultrasound wave to be reflected. For example, more energy is reflected in the passage from muscle to bone than from muscle to fat. The time taken for the echoes to reach the transducer is directly proportional to the thickness of the medium and inversely proportional to the velocity of ultrasound in that particular tissue. Thus, the time delay between the transmitted pulse and its echo is a measure of the depth

Table 11.2   Acoustic proprieties of some tissues at 1.0 MHz

Material	v (m/s)	Z	HVL (cm)	Interface	R
Water	1496	1.49	4100	Air/water	0.999
Fat	1476	1.37	3.8	Water/fat	0.042
Muscle	1568	1.66	2.5	Water/muscle	0.054
Brain	1521	1.58	2.5	Water/brain	0.029
Bone	3360	6.20	0.23	Water/bone	0.614
Air	331	4.13	1.1	Tissue/air	0.999

Source: adapted from Seidband (1998).

of the tissue interface (Seidband, 1998). The tissue thickness can be estimated on the basis of the time difference between the generation of the ultrasonic wave and the reception of the echoes (Thwaites, 1984).

11.3.1   Ultrasound transducers

Ultrasound transducers make use of the piezoelectric properties of ceramics such as barium titanate: such ceramics have the ability to generate an electric potential when mechanically strained, and conversely an electric potential can cause physical deformation of the ceramic (Peura and Webster, 1998). This ability to transform electrical energy into mechanical energy and vice versa is called the piezoelectric effect (Szabo, 2004).

The piezoelectric effect enables an ultrasound transducer to act simultaneously as a transmitter and a receiver of ultrasound energy. The transducer converts electrical signals to acoustic signals (ultrasound), which are sent to the animal's body. The tissue boundaries then produce echoes by reflecting and scattering the ultrasound waves, which turn back and are detected by the transducer, which in turn converts this acoustic signal to an electric signal (Prince and Links, 2006). Thus, with appropriate electronic circuits, the ceramic can be pulsed to transmit a short burst of ultrasonic energy as a miniature loudspeaker and then switched to act as a microphone to receive signals reflected from the interfaces of various tissue types (Seidband, 1998).

11.3.2   Ultrasound imaging of tissues

When ultrasonic waves are generated by a transducer and applied to the skin of an animal, the transducer receives the reflected waves and converts them into electrical power, which is then displayed on a screen in several ways, as outlined previously. Thus, ultrasound imaging systems capture the reflected energy (echoes), which is used as an indication of the position of the interface between two tissues (Goddard, 1995). The reflected sound can be used to obtain a spatial distribution of the tissues through which the ultrasonic wave has passed, and of the interfaces at which part of the ultrasonic wave was reflected. The ultrasound imaging systems process the echoes and present an image of the tissue anatomy on a display, in which each point in the image corresponds to the anatomical location of an echo-generating structure, with its brightness corresponding to the echo strength (Prince and Links, 2006).

The spatial resolution of an image produced by ultrasound is limited by the wavelength of the ultrasound. The wavelength decreases with the increase in frequency: for example, at 2, 5 and 7.5 MHz the wavelength is approximately 0.77, 0.31 and 0.21 mm, respectively (Mannion, 2006). The best resolution is obtained with higher frequencies, since these are associated with a higher attenuation by biological tissues (Goddard, 1995). Thus, the choice of ultrasonic frequency should be based on two factors: (1) the desired resolution – the minimum number of elements to be differentiated – as the resolution power varies along with the ultrasound frequency; and (2) energy absorption by the medium, which increases very rapidly with the increase in ultrasound frequency, and high-frequency waves are less penetrative (Goddard, 1995; Mannion, 2006). The choice of ultrasound frequency must take into account the compromise between the type and thickness of the tissue to be analysed. For animal science, the ultrasound frequency range used is between 1 and 10 MHz (Stouffer, 2004; Silva *et al.*, 2006a).

11.4   Applications of real-time ultrasound (RTU) to predict carcass composition and meat traits in large animals

The ability of RTU to measure carcass composition and meat traits in cattle, swine, sheep and goats has been the subject of a number of studies. This section presents a comprehensive review of the methods used, degree of precision achieved and the factors affecting the use of RTU for predicting the carcass composition and meat traits of those species *in vivo*.

11.4.1   Use of RTU to predict carcass composition and meat traits in cattle and swine

Since the first reports on the use of ultrasound to predict carcass composition and meat traits in cattle (Temple *et al.*, 1956) and swine (Claus, 1957; Dumont, 1957), it was understood that these two species would be the main target for this technology. Although cattle and swine are very different species, it has been shown that RTU can be used to assess carcass composition and meat traits in both, as will be discussed in this section.

The RTU imaging system allows the collection of anatomical measurements in live animals; these measurements, when combined with other sources of information, represent a good basis from which to estimate carcass composition (Paisley *et al.*, 2007). During the last decade, RTU has increased in popularity and, today, has a great impact on the beef cattle and swine industry through two principal applications: as a selection tool for genetic programmes to improve the quality of the carcass and meat traits (e.g., Wilson, 1992) and as a management tool to optimize time of slaughter (Hassen *et al.*, 1998, 1999a; DuPonte and Fergerstrom, 2006).

The use of ultrasound as a selection tool involves the collection of data relating to the carcass and meat traits of cattle (yearling bulls and heifers) and swine. This information is then used to select the best breeding animals in genetic improvement programmes (Wilson, 1992). Traits such as *Longissimus thoracis et lumborum* muscle area (LMA) and subcutaneous fat depth (SFD), usually over the 12th–13th rib, and intramuscular fat have been used as selection criteria. However, in this section, attention will be focused on the use of RTU as a management tool to improve carcass and meat quality for consumers, who are the ultimate evaluators of meat quality.

The use of RTU to measure SFD and LMA in live animals has been thoroughly documented (Perkins *et al.*, 1992a; Greiner *et al.*, 2003, and these two carcass traits are good estimators of lean meat yield in beef cattle (Hamlin *et al.*, 1995; Griffin *et al.*, 1999; Hassen *et al.*, 1999a; May *et al.*, 2000; Suguisawa *et al.*, 2003) and in swine (McLaren *et al.*, 1989; Moeller, 1990; Gresham *et al.*, 1994; Morlein *et al.*, 2005; Olsen *et al.*, 2007). Similarly, ultrasound has been used to estimate intramuscular fat (marbling), which is the principal criterion determining meat quality in beef cattle and swine, and will be discussed later in this chapter. The ability to model and predict the composition of the carcass is the basis for a decision support system that allows the producers to adjust animal feeding and handling strategies according to their specific needs.

Several studies have been carried out to analyse the efficacy of ultrasound as a predictor of carcass composition prior to slaughtering in beef cattle (Perkins *et al.*, 1992b; Smith *et al.*, 1992; Delehant *et al.*, 1996; Ragland *et al.*, 1997; Griffin *et al.*, 1999; Wall *et al.*, 2004) and in swine (McLaren *et al.*, 1989; Terry *et al.*, 1989; Gresham *et al.*, 1992, 1994; Ragland *et al.*, 1997; Newcom *et al.*, 2002). Optimum composition means the highest lean meat proportion, and optimum organoleptic properties. When a carcass meets these requirements, it should be sold for the highest price; if, on the other hand, the composition and organoleptic properties are not optimal, its price will be lowered. Predicting the composition of a carcass is therefore key in determining its value at the slaughter line. A low-cost and expeditious method for predicting carcass composition can be used for carcass classification at the slaughter line (Smith *et al.*, 2008a), and for determining the price along the commercialization chain.

The methodology used to predict carcass composition should be accurate, fast and automated, and the first step in developing these prediction models is to achieve accurate measurement of the SFD and LMA, since these are the most frequently used predictors of carcass composition and quality (Perkins *et al.*, 1992b; Delehant *et al.*, 1996; Griffin *et al.*, 1999; Hassen *et al.*, 1999a; Suguisawa *et al.*, 2003; Wall *et al.*, 2004), and are the main price drivers for the value-based marketing system used by the meat industry. The use of RTU to measure both SFD and LMA in live animals has been well documented for both swine and beef cattle (e.g., Houghton and Turlington, 1992), and several studies have shown that it is an accurate method if the images are taken and interpreted by a trained technician (McLaren *et al.*, 1991; Perkins *et al.*, 1992b; Herring *et al.*, 1994; Hassen *et al.*, 1998). SFD is principally used to predict the lean meat content of carcasses of similar weights (Faulkner *et al.*, 1989), while LMA is also used to predict the carcass composition (Perkins *et al.*, 1992b; Hassen *et al.*, 1998). The evaluation of the accuracy of ultrasound measurements of SFD and LMA is the first step in assessing the applicability of the technology (Robinson *et al.*, 1992) for this purpose. Several articles have focused on the accuracy of the measurement of these carcass traits in beef cattle and in swine. Table 11.3 summarizes the values of correlation coefficients (r) and confidence interval (CI) between SFD and LMA measured by ultrasound and the homologous measurements taken on the carcass for beef cattle and swine.

In cattle, the correlation coefficients between SFD and LMA measured by ultrasound and the homologous measurements taken on the carcass range from 0.70

Table 11.3   Correlation coefficients (r) and confidence interval (CI) between SFD and LMA measured by ultrasound and the homologous measurements taken on the carcass attained by several authors for cattle and swine

Species	Reference	SFD		LMA	
		r	CI (95%)	r	CI (95%)
Cattle	Hedrick et al. (1962)	0.71	0.56–0.81	0.88	0.81–0.92
	Davis et al. (1964)	0.90	0.84–0.93	0.87	0.79–0.82
	Henderson-Perry et al. (1989)	0.86	0.82–0.89	0.76	0.69–0.81
	Brethour (1992)	0.90	0.88–0.91	0.58	0.48–0.66
	Perkins et al. (1992a)	0.75	0.71–0.78	0.60	0.54–0.65
	Robinson et al. (1992)	0.91	0.74–0.93	0.88	0.76–0.94
	Smith et al. (1992)	0.82	0.75–0.87	0.63	0.52–0.72
	Hassen et al. (1998)	0.70	0.61–0.77	0.48	0.35–0.59
	Griffin et al. (1999)	—	—	0.52	0.10–0.78
	May et al. (2000)	0.81	0.76–0.85	0.61	0.52–0.69
	Silva et al. (2004)	0.86	0.76–0.92	—	—
	Fixed effect model	0.84	0.78–0.88	0.66	0.63–0.68
	Random effects model	0.84	0.78–0.88	0.72	0.62–0.80
Swine	Busk (1986)	0.90	0.86–0.93	—	—
	McLaren et al. (1989)	0.55	0.40–0.67	0.61	0.48–0.72
	McLaren et al. (1991)	0.86	0.74–0.93	0.80	0.63–0.89
	Gresham et al. (1992)	0.49	0.34–0.62	—	—
	Ragland et al. (1997)	0.84	0.81–0.87	0.74	0.69–0.78
	Moeller and Christian (1998)	0.87	0.85–0.88	0.74	0.71–0.77
	Fixed effect model	0.85	0.84–0.86	0.72	0.70–0.74
	Random effects model	0.82	0.74–0.88	0.70	0.64–0.75

to 0.91 and from 0.48 to 0.88, respectively. For swine, the correlation between ultrasound and the homologous carcass measurements are very similar (0.49–0.90 and 0.61–0.80 for SFD and LMA, respectively).

Although these correlations are generally significant, the data presented in Table 11.3 shows some variation in the correlation coefficients between different studies and species. This variation is influenced by several factors, namely the ultrasound equipment used, differences between animal and carcass position after slaughter, methods for RTU image analysis and operator training (Robinson *et al.*, 1992; Herring *et al.*, 1994; Stouffer, 2004). All these factors contribute to reducing the accuracy of the RTU technique (Houghton and Turlington, 1992). However, in recent years, important advances have been made in ultrasound technology, which allow for increasing accuracy and reliability in measuring the SFD and LMA (Lusk *et al.*, 2003). Moreover, efforts have been made to improve image acquisition protocols by certified independent technicians (Stouffer, 2004). Additionally, the captured RTU images can be sent to a central laboratory and analysed by trained staff (Greiner *et al.*, 2003). Similar procedures could be adopted to optimize the use of the RTU technique in swine (Moeller, 2002; Schwab *et al.*, 2010).

11.4.2   Use of RTU to predict carcass composition and meat traits in sheep and goats

One of the first ultrasound scanning examinations to predict the composition of sheep carcasses was reported in the late 1950s (Campbell *et al.*, 1959). Since then, numerous studies have been carried out with the aim of predicting carcass composition and meat traits in small ruminants (Table 11.4). The application of RTU technology to small ruminants has been centred on the development of genetic improvement programmes for fat reduction and on the prediction of carcass composition (e.g., Simm, 1987; McEwan *et al.*, 1989), which also proved useful in marketing decisions (Alliston, 1980; Leeds *et al.*, 2007). Particularly in sheep, an excess of fat in the carcass is a major problem that reduces the commercial value of the animal (Sañudo *et al.*, 2000). To overcome this problem, the use of RTU technology has been shown to be very effective in the evaluation of carcass fat levels (Simm *et al.*, 2002; MacFarlane and Simm, 2008). In addition, for sheep and goats, RTU shows great potential for *in vivo* evaluation of carcass composition and meat traits. In recent years, several works have been published that clearly show that the RTU technique allows good estimates of the composition of the carcass to be obtained. These studies aimed to predict carcass composition in adult animals (Hopkins *et al.*, 2007; Teixeira *et al.*, 2008), market lambs (Teixeira *et al.*, 2006; Leeds *et al.*, 2008; Orman *et al.*, 2008; Thériault *et al.*, 2009; Emenheiser *et al.*, 2010; Orman *et al.*, 2010) or light carcasses (Ripoll *et al.*, 2009). Although the results obtained were generally good, attention must be paid to the factors that lead to inaccuracy in the RTU technique when used for small ruminants. These factors include wool or hair, identification of measurement points, fat level and image interpretation and analysis.

Table 11.4 Summary of studies conducted with sheep and goat for predicting *in vivo* carcass traits using the RTU technique

Species	Reference	Equipment	Objective
Sheep	Kempster *et al.* (1982)	Danscanner	Predicting carcass composition
	McEwan *et al.* (1989)	Aloka; 3 MHz and Toshiba SAL 22; 5 MHz	Accuracy of RTU measurements and predicting carcass composition
	Ramsey *et al.* (1991)	Toshiba SAL 32B; 5 MHz	Predicting carcass composition
	Young *et al.* (1992)	Aloka SSD-210 DXII; 5 MHz	Predicting carcass composition
	Delfa *et al.* (1995a)	Toshiba SAL 32B; 5 MHz	Accuracy of RTU measurements and predicting carcass composition
	Stanford *et al.* (1995a)	Aloka SSD 500V; 2 MHz	Accuracy of RTU measurements and predicting carcass composition
	Glasbey *et al.* (1996)	Vetscan MKI; 5 MHz	Accuracy of RTU measurements
	Hopkins *et al.* (1996)	Aloka 500V; 3.5 MHz	Accuracy of RTU measurements and predicting carcass composition
	Fernández *et al.* (1997)	Toshiba SAL 32B; 5 MHz	Accuracy of RTU measurements
	Silva *et al.* (2005)	Aloka 500V; 7.5 MHz	Predicting body and carcass chemical composition
	Silva *et al.* (2006a)	Aloka 500V; 5 and 7.5 MHz	Predicting carcass composition
	Teixeira *et al.* (2006)	Aloka 500V; 5 and 7.5 MHz	Predicting carcass composition
	Hopkins *et al.* (2007)	Honda HS-1201; 5 MHz	Accuracy of RTU measurements and predicting carcass composition
	Silva *et al.* (2007a)	Aloka 500V; 7.5 MHz	Predicting carcass composition
	Leeds *et al.* (2008)	Aloka 500V; 3.5 MHz	Accuracy of RTU measurements and predicting yields
	Orman *et al.* (2008)	Dynamic imaging; 7.5 MHz	Accuracy of RTU measurements
	Ripoll *et al.* (2009)	Aloka SSD 900; 7.5 MHz	Accuracy of RTU measurements and predicting carcass composition
	Thériault *et al.* (2009)	Ultrascan 50; 3.5 MHz	Accuracy of RTU measurements
	Emenheiser *et al.* (2010)	Aloka 500V; 3.5 MHz	Accuracy of RTU measurements
	Orman *et al.* (2010)	Dynamic Imaging; 7.5 MHz	Accuracy of RTU measurements
	Ripoll *et al.* (2010)	Aloka SSD 900; 7.5 MHz	Accuracy of RTU measurements and predicting carcass composition

(Continued)

Table 11.4 Continued

Species	Reference	Equipment	Objective
Goat	Delfa *et al.* (1995b)	Toshiba SAL 32B/5 MHz	Accuracy of RTU measurements and predicting carcass composition
	Stanford *et al.* (1995b)	Keikei CS-3000/3.5 MHz	Accuracy of RTU measurements
	Delfa *et al.* (1996)	Toshiba SAL 32B/5 MHz	Accuracy of RTU measurements and predicting carcass composition
	Mesta *et al.* (2004)	Aloka 500, 5 MHz	Accuracy of RTU measurements
	Teixeira *et al.* (2008)	Toshiba SAL 32B/5 MHz	Accuracy of RTU measurements and predicting carcass composition
	Monteiro (2010)	Aloka 500, 5 MHz	Accuracy of RTU measurements and predicting carcass composition

The value of clipping or shearing the wool or hair at the measuring points of sheep and goats is a controversial issue. One perspective is that the procedure is useful, as it helps to avoid aberrant echoes caused by air bubbles trapped between the conductive medium and wool or hair (Kempster *et al.*, 1982; Stouffer, 1991, Silva *et al.*, 2006a; Leeds *et al.*, 2008). Air bubbles are the cause of low quality images because the ultrasonic beams can dissipate quickly in the air. The necessity of shearing animals was verified by McLaren *et al.* (1991): after studying data from seven sheep that had not been shorn, an increase from 0.15 ($P > 0.05$) to 0.59 ($P < 0.01$) was observed for the correlation between fat thickness measured with ultrasound and fat thickness measured in carcass. The main argument against shearing is the need for the whole ultrasound examination process to be carried out quickly (Hopkins *et al.*, 1996; Teixeira *et al.*, 2008). This issue can be of great economic importance when RTU examinations are performed in a large number of animals, as is the case in commercial herds. Poor acoustic contact between the probe and skin may cause RTU measurements to be underestimated, and it is therefore necessary to increase the pressure on the probe, which causes a deformation of superficial tissues (McEwan *et al.*, 1989; McLaren *et al.*, 1991). To overcome this problem, Ramsey *et al.* (1991) and Young and Deaker (1994) pointed out the possibility of following the tissue deformation through the image on the monitor, allowing it to be immediately corrected. Using a greater amount of conductive medium and the use of a standoff pad between probe and skin are other procedures that can reduce the tissue deformation problem.

In most studies involving sheep and goats, ultrasonic measurements are carried out along the midline of the thoracic and lumbar regions, usually between the 12th thoracic vertebra and 4th lumbar vertebra, using the *Longissimus thoracis et lumborum* (LTL) muscle and thoracic and lumbar vertebrae for orientation. Subcutaneous fat depth above the LTL muscle and muscle depth and area are the parameters usually measured in RTU examinations. A large number of studies (McEwan *et al.*, 1989; Silva *et al.*, 2006a; Teixeira *et al.*, 2006; Hopkins *et al.*, 2007; Silva *et al.*, 2007a; Leeds *et al.*, 2008; Teixeira *et al.*, 2008; Thériault *et al.*, 2009; Emenheiser *et al.*, 2010; Orman *et al.*, 2010) have found a significant correlation ($r > 0.6$; $P < 0.01$) between both fat and muscle RTU measurements and the corresponding carcass measurements in those regions. The placement of the probe at reference points must be correct, since SFD and LTL muscles vary significantly over short distances either cranio-caudally (Delfa *et al.*, 1991; Silva *et al.*, 2007a) or medium-laterally (Simm, 1983; Korn *et al.*, 2005). In addition, anatomical distortions arising from the position of the animals and skin flexibility may also contribute to discrepancies between RTU and carcass measurements (Thwaites, 1984; Leeds *et al.*, 2008). This problem is more evident in young animals, in which the skin is more flexible and the thickness of tissues is lower (Thwaites, 1984; Silva *et al.*, 2006a; Ripoll *et al.*, 2009).

Ultrasound measurements of other regions are also carried out, including over the sternum (Delfa *et al.*, 1996, 2000; Silva *et al.*, 2005; Teixeira *et al.*, 2008) and grade rule (GR) measurement, which is taken between the 11th and 12th ribs at a lateral distance of 11 cm from the spine (Hopkins *et al.*, 1993; Thériault *et al.*,

2009). The sternum region is particularly appropriate for the assessment of fat thickness in goats (Delfa *et al.*, 1996, 2000). In fact, in this species the subcutaneous fat layer in the thoracic and lumbar regions is usually very thin, and is therefore more difficult to measure (Teixeira, 2009).

It has been observed that the accuracy of subcutaneous fat thickness measurements is higher in fatter animals within the same species and between different species (McLaren *et al.*, 1991; Stouffer, 2004; Silva *et al.*, 2006a). In general, RTU measurements obtained in lambs (Young and Deaker, 1994; Silva *et al.*, 2006a; Ripoll *et al.*, 2010) or in kids (Stanford *et al.*, 1995b; Monteiro, 2010) lead to difficulties in SF measurement due to the lower thickness of this tissue. One way to overcome the problem of thin subcutaneous fat is the use of higher-frequency probes, as demonstrated by Silva *et al.* (2006a) and Teixeira *et al.* (2006), who showed that a frequency of 7.5 MHz outperformed 5 MHz in measuring SFD. The 7.5 MHz probe showed higher resolution and lower penetration (Silva *et al.*, 2006a). Generally, the first 6 mm are the focus of attention, comprising skin and subcutaneous fat (Gooden *et al.*, 1980). Therefore, the interface between skin and subcutaneous fat should be located with precision, because the fat is only a few millimetres thick (Gooden *et al.*, 1980; McEwan *et al.*, 1989). The difficulty of determining the interface between skin and subcutaneous fat led some authors to include the skin in the measurement of SFD (Kempster *et al.*, 1982; Silva *et al.*, 2005; Thériault *et al.*, 2009).

Accurate depth measurements require clear identification of the tissues and their interfaces. RTU equipment usually contains an internal measurement system that typically has a resolution of 1 mm (McEwan *et al.*, 1989; Fernández *et al.*, 1997). This resolution, as discussed previously, undermines the accuracy of SF measurements when lean animals are examined (Fernández *et al.*, 1997; Silva *et al.*, 2006a) or when it is necessary to monitor variations of tissue thickness in growing animals (Hamby *et al.*, 1986; Silva *et al.*, 2005). This resolution issue was reported by Young *et al.* (1992) who took fat and muscle thickness measurements in a group of sheep using an RTU associated with a video system for image recording and an image analysis programme. Young *et al.* (1992) observed superior measurement repeatability with their approach, including image recording and analysis, compared with measurements performed directly on the equipment monitor. The difference in repeatability of the two approaches was connected to the resolution, which was 0.1 mm on the image analysis system and 1 mm on the monitor. Similar results were also observed by Silva *et al.* (2005) in a study with growing lambs. They report that better results were obtained when a high-frequency probe (7.5 MHz) was used, allowing an image resolution of 0.2 mm, which was capable of detecting differences in SFD between animals. This is undoubtedly a strong justification for recording ultrasonic images for later analysis. Other factors that justify the recording of images and their subsequent analysis are the shorter time required to evaluate the animals (Glasbey *et al.*, 1996; Silva *et al.*, 2006a); the possibility of obtaining several measurements from the same image including irregular areas (Silva *et al.*, 2007a); and improvements in repeatability, since the interpretation of the images is more important than its acquisition (McLaren *et al.*, 1991).

Problems in identifying tissue interfaces may also arise in fat animals. Indeed, in fat lambs, two or even three layers of subcutaneous fat can be formed, which can be problematic for the interpretation of images taken at the interface between skin and subcutaneous fat (Miles *et al.*, 1972; Silva *et al.*, 2005; Thériault *et al.*, 2009) and can lead to an underestimation of SFD. This underestimation can have serious implications when the RTU measurements aim to select animals with lean carcasses (Gibson and Alliston, 1983; Brethour, 1992).

Over the years, several reports have shown that RTU is a suitable technique for predicting carcass composition both in sheep (Kempster *et al.*, 1982; McEwan *et al.*, 1989; Ramsey *et al.*, 1991; Young and Deaker, 1994; Silva *et al.*, 2005, 2006a; Teixeira *et al.*, 2006; Hopkins *et al.*, 2007; Ripoll *et al.*, 2009) and goats (Delfa *et al.*, 1995a, 1996; Teixeira *et al.*, 2008). In general, these studies develop models which are able to explain the variation in carcass composition in terms of muscle, fat and bone content. Very often, the best models include RTU measurements and body weight. For example, Silva *et al.* (2006a) and Ripoll *et al.* (2009) used models for lambs that included the body weight and one or two RTU measurements, which demonstrate 59–99% and 51–98% of the variation in muscle and fat content, respectively. For goats, Teixeira *et al.* (2008) also observed the value of body weight in combination with RTU measurements to predict carcass muscle content ($r^2 = 0.90$; $P < 0.01$), carcass fat ($r^2 = 0.92$; $P < 0.01$) and total body fat ($r^2 = 0.92$; $P < 0.01$).

11.5   Applications of RTU to predict carcass composition and meat traits in small animals and fish

This section presents an overview of the research conducted with RTU to measure carcass composition and meat traits in poultry, rabbits and fish.

11.5.1   Use of RTU to predict carcass composition and meat traits in poultry

Intensive research into the quality of poultry meat began after World War II, mainly in industrialized countries (Grashorn, 2010). Since then, the successful application of science (health, management, nutrition and genetics) to business in a challenging industry has led to astounding changes in the final product (Boyle, 2006). In the 1950s, to grow a 1.8 kg broiler, it took 90 days (20 g body weight gain day^{-1}), with a consumption of 3.6 kg of feed kg^{-1} of body weight gain, producing about a breast meat yield of approximately 12%. Today, to grow a broiler with the same body weight takes less than 39 days (46 g body weight gain day^{-1}), with less than half the feed used previously (1.7 kg of feed kg^{-1} of body weight gain) and with a breast meat yield of 19% (Arthur and Albers, 2003; Boyle, 2006). These changes have mostly been implemented through selection methods that efficiently improved the yield of the carcass, particularly the breast and leg muscles (Berri *et al.*, 2005; Duclos *et al.*, 2006). As stated in several papers, the breast is the most valuable part of a poultry carcass (Silva *et al.*, 2006b; Larivière *et al.*, 2009) and the breast muscle thickness is a good indicator of poultry carcass composition (Michalik *et al.*, 1999, Rymkiewicz and Bochno, 1999). Poultry carcass traits have therefore been measured with both invasive (slaughtering and dissection of progenies/sibs) or *in vivo* non-invasive methods (Zerehdaran *et al.*, 2005; Larivière *et al.*, 2009). The latter are particularly relevant when serial determinations of carcass traits in the same animal are required. For poultry, as for other species, numerous studies have described non-invasive methods for the evaluation of carcass composition and meat and fat traits. For example, methods, such as TOBEC (Latshaw and Bishop, 2001), DXA (Mitchell *et al.*, 1997; Swennen *et al.*, 2004), MRI (Mitchell *et al.*, 1991; Kallweit *et al.*, 1994; Kövér *et al.*, 1998a; Scollan *et al.*, 1998; Davenel *et al.*, 2000) or CT (Bentsen and Sehested, 1989; Svihus and Katle, 1993; Andrassy-Baka *et al.*, 2003) have been shown to be useful for poultry research. Among these techniques, MRI and CT have been identified as being particularly accurate. However, the high cost of the equipment required for these techniques, combined with the fact that the equipment is not portable, severely limits their routine application in poultry research. Real-time ultrasonography has been used for several years in poultry science studies to predict carcass composition and meat traits (Bochno *et al.*, 2000; Melo *et al.*, 2003; Silva *et al.*, 2006b; Larivière *et al.*, 2009). Ultrasound studies on broiler chickens to predict carcass traits were focused mainly on abdominal fat (e.g., Melo *et al.*, 2003; Arceo *et al.*, 2009) and breast measurements (e.g., Silva *et al.*, 2006b; Kleczek *et al.*, 2009). The results obtained using RTU to predict poultry carcass traits are given in Table 11.5.

In general, the results show that RTU measurements are useful for developing models of broiler breast and leg cuts and lean tissue. It has also been observed in several studies that the best models combine body weight (BW) with RTU measurements (Konig *et al.*, 1998; Melo *et al.*, 2003; Silva *et al.*, 2006b; Oviedo-Rondón *et al.*, 2007). The use of BW in combination with RTU measurements in models for predicting carcass composition is a common practice since BW is closely related to key carcass traits and there are minimal costs associated with its measurement. Using BW combined with RTU measurements for chicken breast, it was possible to account for between 85% and 97% of the observed variation in breast yield and between 63% and 99% of the observed variation in total lean meat content. Based on these results, it is understandable that some reports recommend the use of this non-invasive technique as a valuable tool for selection schemes in broiler breeding (Zerehdaran *et al.*, 2005; Oviedo-Rondón *et al.*, 2007). Ultrasound is also recognized as being sufficiently accurate to monitor the changes in breast yield that occur over the course of a bird's growth; it will therefore prove to be a powerful tool for making the necessary adjustments in feeding regimes to enhance productivity (Dixson and Teeter, 2001; Oviedo-Rondón *et al.*, 2007) and in deciding the optimal weight for slaughter (Oviedo-Rondón *et al.*, 2007). Nevertheless, ongoing study of the use of RTU in poultry research is necessary in order to optimize methods and equipment. Several RTU variables such as breast thickness (e.g., Dixson and Teeter, 2001), breast area obtained both

Table 11.5 Summary of trials with broilers for prediction of carcass traits from breast measurements obtained by RTU alone or associated with body weight (BW)

Reference	Equipment	Probe	n	Dependent variable	Independent variables	r^2
Konig et al. (1998)	Aloka 500V	5 MHz, linear	150 male	Breast yield, %	2 Breast area measurements	0.54
			108 female	Breast yield, %	2 Breast area measurements	0.50
Michalik et al. (1999)			77 male	Total lean, g	BW + Breast thickness	0.64
			76 female	Total lean, g	BW + Breast thickness	0.59
Rémignon et al. (2000)	Toshiba SAL38B	5 MHz, linear	48	Breast, g	BW+ Breast cross-sectional area	0.89
				Breast yield, %	BW+ Breast cross-sectional area	0.63
			104	Breast, g	BW+ Breast cross-sectional area	0.80
				Breast yield, %	BW+ Breast cross-sectional area	0.60
Dixson and Teeter (2001)				Breast, g	Breast thickness	0.90
				Total lean, g	Breast thickness	0.90
Melo et al. (2003)	Ekhoson 500V	7.5 MHz, linear	96	Breast, g	BW + Breast thickness	0.85
Silva et al. (2006b)	Aloka 500V	7.5 MHz, linear	103	Breast yield, %	BW + Breast volume	0.52
				Breast, g	BW + Breast volume	0.92
Oviedo-Rondón et al. (2007)	Aloka 500V	3.5 MHz, linear		Breast, g	BW + Breast area	0.97
				Legs, g	BW + Breast area	0.98
				Total meat, g	BW + Breast area	0.99

(*Continued*)

Table 11.5 Continued

Reference	Equipment	Probe	n	Dependent variable	Independent variables	r^2
Kleczek et al. (2009)	Dramiński Animal Scanner	7 MHz, sector	40 male	Breast, g	Breast thickness	0.22
			40 female	Breast, g	Breast thickness	0.45
Larivière et al. (2009)	Pie Medical 100	5 MHz, linear	24	Breast, g	Breast thickness	0.62

by perpendicular (e.g., Silva *et al.*, 2006b) or longitudinal scanning (e.g., Oviedo-Rondón *et al.*, 2007), or breast volume (e.g., Silva *et al.*, 2006b) have been used as independent variables in linear regression equations for *in vivo* estimation of the total breast muscle content. The fact that these variables are of different origins underlines the need for the selection of one site and a specific scanning procedure. There are also concerns related to the choice of equipment: some studies use a linear probe whereas others use a sector probe. The images resulting from breast scanning with a sector probe (Fig. 11.2a) are very different from those obtained with a linear probe (Fig. 11.2b). This difference causes additional difficulties in the interpretation of tissue interfaces. The frequency and length of the probe are also sources of inaccuracy. Using a 3.5 MHz with 17.5 cm length probe, as is usually employed in large animals, allows a wider ultrasonic window to be examined, which in turn allows better identification of the anatomical site and hence a consistent measurement of the muscle area (Oviedo-Rondón *et al.*, 2007). If small probes are used, the anatomical site must be correctly identified before ultrasound image acquisition can be carried out (Konig *et al.*, 1997; Silva *et al.*, 2006b). On the other hand, the high probe frequency reported by Silva *et al.* (2006b) is potentially more useful in monitoring small changes in breast muscle thickness, particularly in smaller birds, because, as a result of a direct relationship between frequency and attenuation, a lower-frequency probe is more appropriate for deep tissue examinations, whereas a high-frequency probe is better suited to the examination of superficial structures (Goddard, 1995; Silva *et al.*, 2006b).

From a practical point of view, the time needed to acquire a RTU image is very important (Silva *et al.*, 2006b; Oviedo-Rondón *et al.*, 2007). The correct placement of the probe at the anatomical site, along with proper acoustic contact between the probe and bird are crucial for image quality (Konig *et al.*, 1997, 1998; Silva *et al.*, 2006b; Oviedo-Rondón *et al.*, 2007). In the studies just listed, 50–76 birds were examined per hour, and the RTU images were captured with minimum stress, as only manual restraint was necessary, with no detached feathers. Additional time is necessary for image analysis. Two different procedures can be followed: the first takes advantage of the equipment callipers and software, with

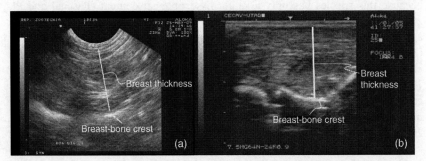

Fig. 11.2   RTU images obtained from cross-sectional view of broiler chicken breast muscle: (a) with a 7.5 MHz sector probe and (b) with a 7.5 MHz linear probe. Breast bone crest and breast thickness were represented.

measurements taken during the image capturing session (Kleczek *et al.*, 2009); in the second, images are stored for subsequent image analysis (Konig *et al.*, 1997; Rémignon *et al.*, 2000; Silva *et al.*, 2006b; Oviedo-Rondón *et al.*, 2007; Larivière *et al.*, 2009). In the latter process, a higher-resolution power is expected, with the result that the measurements obtained are more accurate (Young *et al.*, 1992).

11.5.2   Use of RTU to predict carcass composition and meat traits in rabbit

Rabbit meat is an important product in the Mediterranean areas of Europe, and is also popular in other parts of the world (FAOSTAT, 2010). In recent years, several studies have focused on rabbit meat and carcass traits (Hernández *et al.*, 2004; Larzul *et al.*, 2005). A relationship has been established between growth rate on the one hand, and carcass characteristics and meat quality on the other (Gondret *et al.*, 2005; Pascual and Pla, 2007). Moreover, rabbit is a good experimental model for meat carcass traits because experiments on rabbits can be performed more quickly and at a lower cost than those on other species (Hernández *et al.*, 2006). To further current understanding of rabbit carcass and meat traits, several studies have called for techniques that can evaluate these features *in vivo* (Szabo *et al.*, 1999). Although several non-invasive techniques have been successfully used to evaluate carcass composition in rabbits, such as CT (Szendrö *et al.*, 1992, 2008; Romvári *et al.*, 1996), MRI (Kövér *et al.*, 1998b) and TOBEC (Fortun-Lamothe *et al.*, 2002), only a few studies have been conducted using RTU. Some of these studies are related to the prediction of fat deposits (Pascual *et al.*, 2000, 2002, 2004; Dal Bosco *et al.*, 2003; Castellini *et al.*, 2006; Quevedo *et al.*, 2006) while others are related with carcass and meat traits (Silva *et al.*, 2007b, 2008a, 2008b, 2009). In the pioneering study by Pascual *et al.* (2000), it was shown that RTU was suitable for fat deposit evaluation. A method based on the measurement of perirenal fat thickness at a fixed anatomical location (8th–9th thoracic vertebrae) was developed and accurate results for predicting carcass perirenal fat weight ($r^2 = 0.95$; $n = 42$) and total fat weight ($r^2 = 0.93$; $n = 42$) were obtained.

In rabbits, the ability to assess the fat content of the carcass is of little value because this species has only a small dissectible fat content (Pascual and Pla, 2007). For rabbit carcasses, the evaluation was mainly focused on meat percentage and muscularity, defined as the ratio between meat and bone (Lukefahr and Ozimba, 1991). Several reports have shown that muscularity or cutability attributes may be improved through selection programmes (Lukefahr *et al.*, 1982, 1983).Thus, the development of *in vivo* measurements of muscularity and carcass composition in rabbits using RTU has potentially useful applications in genetic improvement programmes or simply in economic carcass evaluation (Silva *et al.*, 2009). In rabbits, good results were achieved in studies using live body measurements to predict the muscle percentage and muscularity in the carcass (Lukefahr *et al.*, 1982; Lukefahr and Ozimba, 1991; Michalik *et al.*, 2006). Lukefahr and Ozimba (1991); it was found that the lean cut weight (measure of cutability) of the loin was accurately predicted using body weight and loin width ($r^2 = 0.797$).

Several recent studies have assessed the suitability of *in vivo* RTU measurements for assessing rabbit carcass composition and muscularity of loin and leg (Silva *et al.*, 2007b, 2008a, 2008b, 2009). The results reported by these studies clearly showed that measurements obtained from RTU images could account for a large amount of the variation observed in carcass composition and muscularity traits. By employing a 7.5 MHz probe and carrying out image analysis using Image J software for RTU images, 51–94% of the variation in carcass chemical composition (Silva *et al.*, 2007b) and 49–77% of the variation in carcass meat and bone weight (Silva *et al.*, 2009) could be explained with LTL muscle measurements (Fig. 11.3). These results are close to those obtained with CT (Szendrö *et al.*, 1992). Silva *et al.* (2007b, 2009) pointed out that with this system it was possible to estimate the amount of loin muscle ($r = 0.80$; $P < 0.01$).

Moreover, after RTU image analysis, it was possible to predict the LTL muscle volume of the carcass from the *in vivo* LTL volume ($r^2 = 0.81$), which can be calculated from area measurements obtained with multiple scanning images and by using Cavalieri's principle (Silva *et al.*, 2008a). As stated previously, muscularity is an important trait in rabbit carcasses and *in vivo* RTU is able to accurately estimate loin muscularity (r between 0.76 and 0.81; $P < 0.01$) (Silva *et al.*, 2009). However, for leg muscularity, lower coefficients of correlation (r from 0.15; $P > 0.05$–0.46; $P < 0.01$) were found (Silva *et al.*, 2008b). These results highlighted the need to improve the procedures related to RTU and carcass measurements so that increased accuracy can be achieved when using RTU to determine hind leg

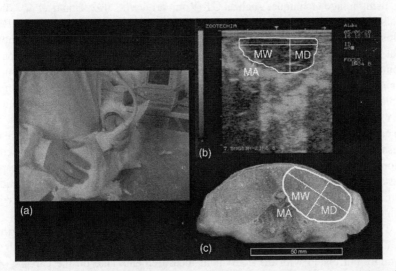

Fig. 11.3 (a) RTU image acquisition procedure with a linear probe placed over loin region between the 6th and 7th lumbar vertebrae. Note the hair clipped close to the skin. (b) RTU image taken *in vivo* showing the representations of *longissimus thoracis et lumborum* muscle area (MA), width (MW) and depth (MD) measurements. (c) Carcass cut section at homologous anatomical position showing the representations of equivalent *longissimus thoracis et lumborum* muscle measurements.

muscularity. Nevertheless, the results obtained from these studies are encouraging as far as the use of *in vivo* RTU to predict rabbit carcass traits is concerned; further research is necessary to improve the practicability of RTU and image analysis for extensive use in the evaluation of rabbit carcasses, since other attributes such as animal restraint, equipment mobility, ease of use and non-invasive nature have already been well established for this technique. The need for the removal of hair from the ultrasound measurement site is also inconvenient, and this problem will need to be addressed in order to improve the practicability of the technique in rabbits.

11.5.3   Use of RTU to predict carcass composition and meat traits in fish

An understanding of the carcass composition, and particularly the fat content, of live fish is important for feeding, breeding and genetics, and for increasing the meat yield of the carcass. It is also an important factor in consumer acceptance (Probert and Shannon, 2000; Romvári *et al.*, 2002; Veliyulin *et al.*, 2005). Traditionally, carcass composition in fish was determined by comparative slaughtering followed by chemical analysis (Oberle *et al.*, 1997). Other methods such as ultrasound velocity (Suvanich *et al.*, 1998; Sigfusson *et al.*, 2000) or near-infrared techniques (Wold and Isaksson, 1997) are based on fillet samples or dead fish. However, fish production is heavily dependent on quick, accurate and, above all, non-invasive methods to predict carcass composition in live fish (Probert and Shannon, 2000; Veliyulin *et al.*, 2005; Silva *et al.*, 2010a). Comprehensive studies using image techniques such as CT (Romvári *et al.*, 2002; Hancz *et al.*, 2003; Kolstad *et al.*, 2004), MRI (Collewet *et al.*, 2001; Veliyulin *et al.*, 2005) and RTU (Bosworth *et al.*, 2001; Rodrigues *et al.*, 2010; Silva *et al.*, 2010a) have shown that these techniques are able to predict carcass traits in fish. From a practical point of view, CT has several characteristics that make it the preferred technique for *in vivo* evaluation of carcass composition in fish. In fact, RTU is a simple, rapid and reasonably priced technique (Stouffer, 2004). Additionally, as water is an excellent coupling medium between transducer and fish, the RTU images can be captured when the fish are in the water (Crepaldi *et al.*, 2006). Over the years, this technique has been shown to be sufficiently precise and accurate to be used as a tool for carcass composition studies (e.g., De Campeneere *et al.*, 2001). Despite these attributes, little information is available about the use of RTU to predict carcass composition in fish. Examples of studies that use RTU images to predict fish carcass composition traits are summarized in Table 11.6.

The results of these studies are reliable. In farm-raised catfish, Bosworth *et al.* (2001) reported that ultrasound measurements of muscle area in live fish are strongly correlated with the equivalent measurements take on the carcass ($r = 0.84$–0.94; $P < 0.001$), but in meat yield measurements there was only moderate correlation between the two. A single transverse ultrasound scan accounted for 40–50% of the variation in meat yield traits in female catfish, and 16–23% in male catfish (Bosworth *et al.*, 2001). In the same study, using multiple regressions, Bosworth *et al.* (2001) found a three-variable model using ultrasound and

Table 11.6 Examples of studies for predicting fish carcass traits using RTU technique

Reference	Objective	Results	Equipment	Notes
Probert and Shannon (2000)	Low intensity ultrasound to determine fish composition, particularly the fat content	Encouraging results	2.0 MHz convex	Freshly killed fish
Bosworth et al. (2001)	Determine the relationships between meat yield traits with body shape traits and transverse ultrasound images of muscle area measured in live catfish	A single ultrasound measurement explained 40–50% and 16–23% of the variation in meat yield traits of females and males, respectively. The best three variable models using ultrasound and body shape traits explained 48–56% and 31–38% of the variation in meat yield traits in females and males, respectively	Toshiba Echocee; 7.5 MHz convex	Fish were tranquilized
Bosworth et al. (2001)	Study with 30 market weight channel catfish to compare muscle area measured from transverse ultrasound images with muscle area measured in fish	Correlations between 0.84 and 0.94 for ultrasound muscle area with equivalent carcass measurement	Aloka 1700; 5 MHz linear	

(Continued)

Table 11.6 Continued

Reference	Objective	Results	Equipment	Notes
Silva *et al.* (2010a)	Develop a rapid non-destructive and non invasive method to predict fillet volume of *Solea senegalensis* individuals from volume measurements obtained *in vivo* after RTU image analysis	The best model explains 98% of the fillet volume variation and was obtained by stepwise procedure with S3, S2 and S4 cross-sectional slices volumes	Aloka 500V; 7.5 MHz linear	Fishes under anaesthesia
Rodrigues *et al.* (2010)	Relationship between the traditional solvent-extraction fat determination method with RTU measurements	Preliminary results with RTU images clearly support the preferential accumulation of fat in subcutaneous tissues of Senegalese sole	Aloka 500V; 7.5 MHz linear	

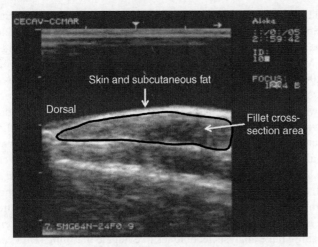

Fig. 11.4   Example of RTU image showing a fillet cross-section area obtained after image analysis.

body shape variables that accounted for 48–56% and 31–38% of the variation in meat yield traits in females and males, respectively. On the other hand, Silva *et al.* (2010a) scanned flat fish (*Solea senegalensis*) with an RTU ultrasound with a 7.5 MHz probe to capture ten cross-sectional slices (Fig. 11.4), from which fillet volume measurements were obtained after image analysis. The best model accounted for 98% of the fillet volume variation and was obtained through volume measurements of three cross-sectional slices.

In fish species processing errors combined with potential errors in RTU image capture and image analysis may limit the accuracy of models for predicting tissues and yield in live fish. For example, fish movement during ultrasound scanning is one drawback that limits the use of RTU in fish farms (Probert and Shannon, 2000). In general, the use of an anaesthetic (Bosworth *et al.*, 2001; Silva *et al.*, 2010a) or, more radically, the use of freshly killed fish (Probert and Shannon, 2000) were reported to reduce fish movement. The need to use these procedures, especially the latter, has restricted the use of this technique in fish farms. Fish size is another limiting factor: the RTU technique is less effective with small fish (Bosworth *et al.*, 2001). However, the use of high-frequency probes (7.5–10 MHz) overcomes this problem because clearer images can be obtained from proximal anatomical structures.

11.6   Using real-time ultrasonography to predict intramuscular fat (IMF) *in vivo*

It is recognized that fat plays an important role in the eating quality of meat (Wood *et al.*, 2008; Kouba and Sellier, 2011). Intramuscular fat (IMF) content, particularly in cattle and swine, affects meat quality, especially the sensory properties of juiciness and flavour (Huff-Lonergan *et al.*, 2002; Thompson, 2004; Skiba, 2010). Some studies have shown that IMF or marbling is essential for meat acceptability by consumers (Shi-Zheng and Su-Mei, 2009) and for meat industry grading (Smith *et al.*, 2008a). Before reviewing the use of ultrasound technology to examine these features, the differences between IMF and marbling should be outlined. IMF refers to the chemically extractable fat in a muscle (Shi-Zheng and Su-Mei, 2009) and is an objective measurement, whereas marbling, assessed visually, refers to the appearance of evenly distributed white flecks or streaks of fatty tissue between bundles of muscle fibres (Tume, 2004) and can be subjectively assessed with grading scores or objectively assessed when image analysis is used (e.g., Faucitano *et al.*, 2005; Jackman *et al.*, 2008). Both are relevant for meat quality evaluation and are closely related to each other (correlation coefficients of up to 0.8 in Savell *et al.*, 1986; Devitt and Wilton, 2001; Kemp *et al.*, 2002). In general, the percentage of IMF is taken as the reference trait (Brethour, 1994), while marbling proves useful in understanding the size and distribution of IMF deposit in meat (Ferguson, 2004). These attributes are relevant and reinforce the value of using image analysis techniques to evaluate marbling (Du *et al.*, 2008).

The IMF trait has been extensively studied in swine and cattle (Pethick *et al.*, 2006). It is now generally agreed that the IMF content accounts for a significant amount of the genetic variation in the eating quality of meat of these species (Shi-Zheng and Su-Mei, 2009; Schwab *et al.*, 2009, 2010). In addition, IMF is one of the meat quality traits that has the potential to be measured in live animals (Newcom *et al.*, 2002; Aass *et al.*, 2009). Thus, a cost-effective and accurate method for quantifying IMF *in vivo* is needed, because repeated measurements are necessary on one animal, if it is intended for breeding (Williams, 2002; Parnell, 2004). Furthermore, it is possible to establish the optimal point at which the animal should be sold with the greatest economic benefit (Houghton and Turlington, 1992; Rimal *et al.*, 2006). To achieve this goal, experimental work was conducted to predict IMF *in vivo* through RTU and image analysis (e.g., Brethour, 1990; Amin *et al.*, 1997). This technique was found to be particularly promising since it is relatively cheap, easy to use and animal-friendly (Stouffer, 2004). Even though the majority of the studies using ultrasonography to estimate the percentage of IMF date from the 1990s (e.g., Brethour, 1990; Sather *et al.*, 1996), the technique has been used since the 1960s to estimate fat thickness in the back and the rib eye area, and by then it was already perceived as a technique with the potential for use in IMF prediction (Hedrick *et al.*, 1962; Davis *et al.*, 1964).

The use of ultrasound image analysis to predict marbling in beef cattle was first attempted by Haumschild and Carlson (1983). This study had only marginal success and was considered too inefficient to have any practical significance. Later on, a number of authors (e.g., Brethour, 1994; Hassen *et al.*, 1999b; Newcom *et al.*, 2002) reported results which undoubtedly suggest that IMF was accurately predicted with RTU and image analysis.

In the early 1990s, Wilson (1992) stated that considerable research and development was needed before ultrasound could be effectively employed in cattle production and breeding. Since then, ultrasound technology has become a well-established and widely accepted method for predicting IMF in live cattle and swine (i.e., Brethour, 1994; Herring *et al.*, 1998; Hassen *et al.*, 1999b, 2001; Chambaz *et al.*,

2002; Newcom *et al.*, 2002; Bahelka *et al.*, 2009; Schwab *et al.*, 2010). Recently, the RTU technology for predicting IMF was chosen as one of the *100 Innovations from Academic Research to Real-World Application* (AUTM, 2007). This report recognized the work developed by Professor John Brethour from Kansas State University, which changed the beef industry by allowing producers to employ a cost-efficient method for measuring intramuscular fat in livestock. Nonetheless, ultrasound technology can still be further optimized for IMF prediction, by improving RTU image analysis and image acquisition (Shi-Zheng and Su-Mei, 2009).

11.6.1   Using RTU image analysis for IMF prediction

The IMF is primarily determined by the distribution pattern of fat flecks in a cross-section of the LTL muscle, usually between the 12th and the 13th thoracic vertebrae (Fig. 11.5a). Although IMF is present in other muscles, the assessment generally is performed on a LTL muscle section. The IMF consists of deposits that occur within the muscle, which are irregular either in form or in their dispersal. These deposits represent a cluster of IMF cells. Individual cells can be very small (40–60 μm) and are not visible to the human eye (Anon., 2004). The rough surface and small size of IMF deposits cause sound waves to scatter (Brethour, 1990; Whittaker *et al.*, 1992), producing spots on RTU images that are referred to as speckles (Fig. 11.5b). This is why ultrasound techniques have the potential to predict IMF *in vivo* after RTU image analysis (Brethour, 1990; Whittaker *et al.*, 1992).

The RTU image analysis for predicting IMF or marbling has been carried out in a number of ways over the years. Early studies were conducted to predict marbling scores from a subjective analysis of the RTU image features (coherent speckle, attenuating and reverberation) from which a speckle score was obtained (Harada and Kumazaki, 1979; Brethour, 1990). Speckle scores were estimated visually and corresponded subjectively to a point classification scheme. This procedure had the benefit of allowing an immediate estimation of the marbling score and, thanks to the portability of the ultrasound equipment portability, could be used for farm animals (Brethour, 1990). However, it is subjective, and dependent on beam geometry and machine calibration. Furthermore, an understanding of the classification scheme and calculation of the score can be difficult for a technician to acquire (Brethour, 1990). These negative aspects led Brethour (1990) to observe that ultrasound speckle was a 'quick and dirty' way to estimate the marbling score of a carcass and that, consequently, further improvements were necessary to reduce the subjectivity of RTU images. Although a skilled ultrasound technician can visually interpret an RTU image and estimate marbling in a live animal with fair accuracy (Brethour, 1990, 1994), it was recognized that research using mathematical models for RTU image analysis was imperative (Amin *et al.*, 1993; Kim *et al.*, 1998).

11.6.2   Mathematical modelling approaches from RTU image analysis

Since the early studies (Harada and Kumazaki, 1979; Brethour, 1990), several papers have dealt with the assessment of IMF content and marbling by RTU

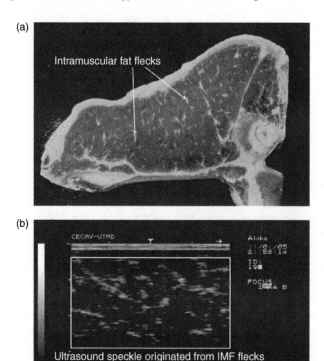

Fig. 11.5   (a) Image from a cattle lumbar cut section showing LTL muscle and intramuscular fat flecks and (b) RTU image of the LTL muscle showing speckle originated from IMF.

computer image analysis (Brethour, 1994; Hassen *et al.*, 2001; Harron and Dony, 2009). On the whole, results obtained with mathematical procedures were superior to the subjective RTU image evaluation, even when this task was conducted by an experienced individual (Raeth *et al.*, 1985; Couto *et al.*, 2011). Since the first attempt was made to predict IMF using RTU image analysis, significant advances have been possible as a result of developments in equipment and software. Table 11.7 summarizes some studies that have used RTU images to predict IMF and marbling score.

The algorithms used to predict the IMF percentage of live animals were based on regression analysis (Whittaker *et al.*, 1992; Amin *et al.*, 1993; Newcom *et al.*, 2002; Li *et al.*, 2009); neural network (Brethour, 1994; Amin *et al.*, 1992; Harron and Dony, 2009; Li *et al.*, 2009) or support vector machine (Harron and Dony, 2009), among others. These algorithms were developed from textural RTU image features such as a histogram of pixel grey levels, Fourier-based

Table 11.7 Summary of trials to predict intramuscular far percentage or marbling score from RTU image analysis in cattle and swine

Reference	RTU equipment	Probe	n	Species	Anatomical position	Y	X	r^2	RSD	Statistical analysis
Brethour (1990)	Aloka 210	3 MHz, 107 mm	40	Cattle	12th rib, parallel and perpendicular	Marbling score	Speckle patterns	0.45	0.36%	Regression
Amin et al. (1993)	Aloka 500	3.5 MHz, 172 mm	126	Cattle	Across 12th and 13th ribs	IMF%	Image texture features		1.39 and 1.42%	Regression
Brethour (1994)	Aloka 210	3.5 MHz, 125 mm	53 and 108	Cattle	12th rib, parallel and perpendicular	Marbling score	Image texture features	0.53		Neural network
Kim et al. (1998)	Aloka 500	3.5 MHz, 125 mm	207	Cattle	Across the 11th and 13th rib	IMF%	Image texture features		1.4%	
Hassen et al. (1999b)	Aloka 500	3.5 MHz, 172 mm	144	Cattle	Across 11th –13th ribs	IMF%	Image texture features			
Hassen et al. (2001)	Aloka 500	3.5 MHz, 172 mm	500	Cattle	Across the 11th and 13th rib	IMF%	Image texture features	0.72	0.84%	Regression
Hassen et al. (2001)	Pie 200	3.5 MHz, 180 mm	500	Cattle	Across the 11th and 13th rib	IMF%	Image texture features	0.70	0.85%	Regression
Chambaz et al. (2002)	Pie 200	3.5 MHz, 180 mm		Cattle	Across 12th and 13th ribs	IMF%	RTU, hide thickness and liveweight		0.96%	Regression
Aass et al. (2006)	Pie 200	3.5 MHz, 180 mm	145	Cattle	12th thoracic and 1st lumbar vertebrae	IMF%	Image texture features	0.48	0.46%	Regression
Aass et al. (2009)	Pie 200	3.5 MHz, 180 mm	172	Cattle	12th thoracic and 1st lumbar vertebrae	IMF%	Image texture features	0.80	0.66%	Regression

(Continued)

Table 11.7 Continued

Reference	RTU equipment	Probe	n	Species	Anatomical position	Y	X	r^2	RSD	Statistical analysis
Harron and Dony (2009)	Aloka 500	3.5 MHz, 172 mm	75	Cattle	Across 12th and 13th	IMF%	Image texture features		1.37%	Recursive least squares filter
Harron and Dony (2009)	Aloka 500	3.5 MHz, 172 mm	75	Cattle	Across 12th and 13th	IMF%	Image texture features		2.67%	Support vector machine
Harron and Dony (2009)	Aloka 500	3.5 MHz, 172 mm	75	Cattle	Across 12th and 13th	IMF%	Image texture features		1.36%	Linear neural network
Harron and Dony (2009)	Aloka 500	3.5 MHz, 172 mm	75	Cattle	Across 12th and 13th	IMF%	Image texture features		1.36%	Multilayer perceptron network
Sather et al. (1996)	LS-1000	3.5 MHz	149	Swine	3rd and 4th lumbar	IMF%	Percent object area of muscle	0.012		Regression
Sather et al. (1996)	CS-3000	3.5 MHz	240	Swine	3rd and 4th lumbar	IMF%	Percent object area of muscle	0.04		Regression
Newcom et al. (2002)	Aloka 500	3.5 MHz, 125 mm	207	Swine	Across 10th to the 13th ribs	IMF%	Image texture features	0.32	1.02%	
Bahelka et al. (2009)	Aloka 500	3.5 MHz, 172 mm	144	Swine	Last rib	IMF%		0.38	0.52%	Regression
Schwab et al. (2010)	Aloka 500	3.5 MHz, 125 mm		Swine	Across the 10th to 13th ribs	IMF%	Image texture features	0.81		Regression

parameters, gradient-based parameters and co-occurrence parameters (Amin *et al.*, 1997; Hassen *et al.*, 1999b; Newcom *et al.*, 2002; Harron and Dony, 2009). Generally, image features can be calculated after selecting a region of interest (ROI) over the RTU image; the image analysis software then provides an ROI parameter file with the image parameters (Amin *et al.*, 1997; Silva *et al.*, 2010b).

In parallel to research on RTU image analysis for IMF evaluation, several studies aimed at establishing practical and usable software image analysis for livestock production (Amin *et al.*, 1997; Aass *et al.*, 2006). For example, the software proposed by Amin *et al.* (1997) was frequently used in both swine (Newcom *et al.*, 2002; Schwab *et al.*, 2010) and cattle (Hassen *et al.*, 1999b). An RTU image acquisition protocol must be followed in order to use this software. For swine, a minimum of four longitudinal images were collected at 7 cm off-midline across the 10th–13th ribs (Newcom *et al.*, 2002; Schawab *et al.*, 2009, 2010). With cattle, four to six images were taken longitudinally without a wave guide (standoff block) across the 11th–13th ribs of the animal at a position three-quarters of the distance from the medial end of the rib eye area to the lateral end (Hassen *et al.*, 1999b, 2001). Currently, most swine and cattle scanning for IMF prediction is carried out using an Aloka 500 V with a 17 cm linear array 3.5 MHz transducer (e.g., Hassen *et al.*, 2001; Newcom *et al.*, 2002; Schawab *et al.*, 2010) or with a Pie 200 SLC with a 18 cm linear array 3.5 MHz transducer (e.g., Aass *et al.*, 2009). In either case, ultrasound images of the highest quality must be collected. In fact, it is well established that image quality has an impact on measurement accuracy (Houghton and Turlington, 1992; Spangler and Moser, 2009). For both swine and cattle, a typical image with acceptable quality includes the following features: clearly visible hide and subcutaneous fat layer(s) without any sign of uneven couplant or poor transducer contact; LTL muscle area taken from across the 10th–13th ribs with clearly visible rib-shadows; even speckle or texture pattern in the muscle area; and ROI box area completely free of deficiencies (Amin *et al.*, 1997; Hassen *et al.*, 1999b; Newcom *et al.*, 2002). It is also important to correctly distinguish the various tissue types – subcutaneous fat, muscle, blood capillaries, intramuscular fat and bones – for ROI box selection and subsequent use of computer image analysis (Amin *et al.*, 1997). During the image analysis process, attention must be paid to all these aspects since they affect the nature of the ultrasonic backscattered signal and consequently the quality of the image (Amin *et al.*, 1992). For each image, parameters were generated using texture analysis from a 100 × 100 pixels ROI box (Amin *et al.*, 1997; Hassen *et al.*, 2001; Newcom *et al.*, 2002).

Two observations may be made on the basis of the reports on RTU image analysis and IMF presented in Table 11.7. First, the developments in ultrasound technology offer an opportunity to better predict carcass and meat quality with regard to IMF. Second, the developing technology of using ultrasound image analysis for IMF prediction has been successfully transferred from research to the beef and swine industry, which has allowed improvements in carcass quality.

11.7 Optimization of production system and market carcass characteristics

The most important attribute of meat quality is its overall eating satisfaction, which is a function of the combined effects of tenderness, juiciness and flavour (Ferguson, 2004). However, today's meat is usually criticized for its lack of succulence, due to the low levels of intramuscular fat (marbling), which has been the outcome of years of genetic selection that has aimed to reduce the fat content of the carcass. Moreover, the slaughtering of animals either before or after the optimum point is responsible for significant economic losses (Brethour, 2000), and the prediction of the optimum slaughter date is key in maximizing the quality of the meat and the income of the producer (Williams, 2002). RTU can be used to develop a decision support system to sort animals into management groups prior to feedlot feeding, and to predict the optimum slaughter point using computer-based models to assist the management decisions.

11.7.1 Predicting optimum slaughter date

Meat tenderness and juiciness are positively correlated with the proportion of fat in the carcass (Wood, 1990; Bruns et al., 2004). Marbling fat has no direct effect on meat tenderness (Renand et al., 2003; Thompson, 2004); however, it plays an important role in meat juiciness and overall eating satisfaction. In fact, marbling leads to greater palatability in panel scores (McPeake, 2001) and lower shear force values (Dolezal et al., 1982). Carcasses with higher marbling content also have a higher subcutaneous and intermuscular fat content, thus insulating the muscles during chilling and preventing the phenomenon of cold shortening.

The production of carcasses with excessive weight, excessive subcutaneous fat and only a small degree of marbling, as well as a lack of uniformity, is a common problem in meat production systems. The production of carcasses with the correct weight and an optimum amount of subcutaneous fat therefore ensures that the meat is protected during the cooling process and also maximizes the organoleptic properties.

Fatter carcasses undergo a faster drop in pH, which is associated with more tender meat; and slower cooling of fatter carcasses contributes to an increase in the activity of ageing enzymes, leading to greater tenderness (Wood, 1995). Even under normal chilling conditions, carcasses with less than 13 mm of SFD over LTL display reduced tenderness due to the cold shortening effect (Wood, 1995). Ageing a carcass affected by cold shortening will not alleviate the detrimental effects on tenderness. Thus, the SFD is a very important attribute, because it protects the meat from thermal shock during refrigeration, which prevents cold shortening, oxidation of muscles, browning and microbial contamination of meat during skinning.

As stated previously, the ability to measure the SFD, LMA and marbling using ultrasound images taken from live animals provides an opportunity to study the relationship between animal growth and the development of various tissue types.

Thus, for a known feeding strategy, it is possible to monitor the growth of SFD and LMA and the deposition of marbling as the animals grow and during the finishing phase. These data can then be used to project the slaughter date, for a pre-defined subcutaneous fat level (Brethour, 2000). For example, Delehant *et al.* (1996) showed that ultrasound measures, taken on cattle prior to feedlot feeding, combined with performance data collected during the finishing phase, could effectively predict LMA, SFD and IMF percentage at any point during the finishing phase. The ability to predict the optimum slaughter date of a particular animal is an attractive use of ultrasound technology (Lusk *et al.*, 2003). The RTU can be used to develop models to predict the number of days necessary to reach a target carcass composition under a defined feeding regime (Hassen *et al.*, 1999a), or to develop a feeding regime that maximizes the production of carcasses with a higher-yield or higher-quality grade (Basarab *et al.*, 1999).

So far, the use of RTU to optimize the slaughter date has been focused on beef production; however, RTU can also be successfully used for meat species such as swine. RTU is also useful in predicting market-weight slaughter characteristics and in predicting the percentage of lean cuts in market-weight swine. The ability to predict market-weight slaughter characteristics was investigated by Robinson *et al.* (1992). Similarly, McLaren *et al.* (1989) studied 110 barrows and gilts, which were scanned every two weeks from 42 days old up to the point of slaughter to measure SFD at the first rib, last rib and last lumbar vertebrae, and to measure LMA at the 10th rib. They showed that ultrasound measurements were able to estimate lean gain a day early (up to 53 kg BW) immediately prior to slaughter. These authors (McLaren *et al.*, 1989) concluded that ultrasound data were useful in early selection decisions and for selections made at market weight for carcass merit in swine. Olsen *et al.* (2007) also used ultrasound for online classification of swine carcasses and showed that live animal ultrasound measurements could predict retail product yield after slaughter.

The implementation of RTU in meat production systems can help to reduce the production of carcasses with either too little or too much subcutaneous fat. This is beneficial for producers – first, as it can lead to reduced feeding costs, and also because it improves the quality of the product presented to consumers.

11.7.2   Sorting animals prior to feedlot feeding

The study by Green *et al.* (2000) found a total of 280 inefficiencies in the US beef industry, and proved that the majority of losses occurred due to excessive fat production, leading to poor consistency in taste. Green *et al.* (2000) concluded that the beef cattle industry needed to improve carcass quality by improving feeding and management practices, as well as by genetic improvement.

Ultrasound technology provides information on the optimum sorting of animals into feedlot groups, based on their body composition predicted by RTU measurements of SFD at the 12th rib level (Houghton and Turlington, 1992; Hassen *et al.*, 1999a). This approach for beef cattle was shown to be more accurate than simple visual appraisal (Delehant *et al.*, 1996). Sorting meat animals into uniform groups based on frame size, SFD and LMA can help to obtain carcasses with uniform slaughter weight and consistent composition, which can then be sold at the optimal time (Houghton and Turlington, 1992; Wall *et al.*, 2004; Rimal *et al.*, 2006). When cattle have an average initial SFD of more than 3 mm, ultrasound is useful in projecting the number of days required to reach a target SFD level, which allows animals to be clustered into groups for more effective marketing (Brethour, 2000).

Ultrasound provides information about the carcass of each animal individually, and if the data are collected when animals start on feed, it can also provide information that can be used to sort animals into adequate feeding regimes (Lusk *et al.*, 2003). Therefore, RTU contributes to reducing the problem of overfeeding, improves the efficiency of the production system and increases the income of the meat producers (Lusk *et al.*, 2003; Pyatt *et al.*, 2005). For example, Basarab *et al.* (1999) used ultrasound to sort beef cattle three to four months before slaughter into more uniform groups, and this strategy displayed positive effects on growth rate, feed efficiency, carcass yield and quality grade, as well as increasing the net return by \$15–\$27 per head slaughtered. Similarly, with swine, Gresham *et al.* (1992) showed that RTU was able to separate either live animals or carcasses using a single SFD measurement along with live or carcass weight. Gresham *et al.* (1992) concluded that RTU can be used in a commercial environment to achieve accurate measurements of carcass value or of compositional differences between the carcasses.

Variation among cattle within a pen diminishes opportunities for precision feeding. If cattle within a pen are more uniform in their characteristics, they can be fed more precisely according to requirements; this is preferable to using an average measurement to determine feeding, as this can overfeed or underfeed a portion of the cattle (Trenkle and Williams, 1997). The costs and the labour required to operate the system remain the main barriers to the adoption of this technology (Basarab *et al.*, 1999). However, recent developments in ultrasound equipment, along with remote sensing and infrared technologies, may make the system of sorting cattle for feeding purposes completely non-invasive and also less labour intensive Li, 2010.

11.7.3   Optimizing marketing strategies

A genuine value-based marketing system will necessarily result in some premiums as well as discounts (Trenkle and Williams, 1997); management optimization can contribute to increased economic returns and carcass desirability in the marketplace. The optimization of management can include strategies such as energy concentration in the diets used during growth and in the finishing phase, and the length of feeding, among many others. However, the prediction of carcass composition pre-slaughter allows the identification of animals with higher carcass cutability (Paisley *et al.*, 2007), and beef producers are able to provide carcasses according to consumer preferences (Williams and Trenkle, 1997). Thus, RTU will enhance the profitability of meat producers, as it will allow them to raise meat animals that directly correspond to the desired attributes of consumers

(Rimal *et al.*, 2006). The ultrasound data collected from live animals can be used to predict carcass yield and quality grades (Lusk *et al.*, 2003), and to enhance the meat marketing decisions by optimally targeting carcasses to specific market needs. Lusk *et al.* (2003) studied the potential use of ultrasound measurements taken in the feedlot in guiding pricing decisions for cattle. They found that actual carcass merits were reasonably accurately predicted; and that when cattle were sorting for live, dressed, or grid-based pricing, an increase of returns of $25 per head was achieved compared with marketing all cattle on a live-weight basis.

11.8   The future of RTU imaging in the meat industry

There have been remarkable achievements in the development of ultrasound as a tool for the prediction of carcass composition and meat traits in animals since its first application in the late 1950s. The advent of RTU and image analysis have made ultrasound a valuable and reliable tool in animal research and production, with major applications in genetics, nutrition, carcass value-based marketing and monitoring for body fat reserves (Moeller, 2002; Williams, 2002; Parnell, 2004; Schröder and Staufenbiel, 2006). For all these applications, ultrasound technology will continue to expand as a tool for management practices that affect the productivity and profitability of the meat industry (Moeller, 2002; Li, 2010). One good example of this is the use of ultrasound technology, coupled with current selection methods and molecular tools, to speed up genetic progress in meat traits (MacNeil *et al.*, 2010; Nalaila *et al.*, 2011). Despite the impressive advances in ultrasound systems, mainly in the last decade for all meat species, some technological aspects have the potential for further improvement in the near future. As in the past, current developments in ultrasound technology originate in the field of medicine, particularly from the very dynamic and expanding field of image diagnostics (Stouffer, 2004; Thompson, 2010). Although the needs of the medical sector are quite different from those of the animal science sector, the same medical ultrasound equipment can still be easily used animal science protocols (Stouffer, 2004). Therefore, some of the developments in the medical field may be potentially useful for animal science too. Increasing ultrasound processing capability (King, 2006; Wells, 2006), improvements in image quality (Szabo, 2004; Smith *et al.*, 2008b; Whitsett, 2009), better portability (Ault and Rosen, 2010; Thompson, 2010; Bret *et al.*, 2011), capacity for online analysis and image storage (Whitsett, 2009; Li, 2010) and reduction of equipment and operational costs (Szabo, 2004) are the features that will have the biggest impact on the evaluation of carcass composition and meat traits. In fact, these improvements will allow an increase in the speed of the RTU data collection process either on the farm or at the slaughterhouse. Additionally, faster image analysis and accurate results lead to more information being available along the entire production chain (from stable to table), which helps the industry to better understand and accurately describe the meat products, hence driving improvements in meat quality and productivity (Bindon, 2002; Li, 2010).

11.8.1   Prospects for the development of novel ultrasound scanning techniques

Advances in ultrasound such as synthetic aperture focusing, also known as zone sonography (Wells, 2000; Lyons, 2004), and elastrography (O'Brien and Holmes, 2007; Whitsett, 2009) are likely to be employed in animal science in the future. The zone sonography technology allows faster image acquisition and high image quality, which will prove particularly useful in situations in which multiple images of a subject are required during a scanning session. Elastrography is a technology with the potential to improve the accuracy with which marbling can be predicted. The use of three-dimensional (3-D) ultrasonography is another imaging technique with promising applications in the evaluation of carcass composition and meat traits. As stated by several authors (Mitchell *et al.*, 2001; Kvame and Vangen, 2006; Monziols *et al.*, 2006; Alston *et al.*, 2009), the use of volume measurements, along with image techniques such as MRI and CT, is an attractive approach for predicting carcass composition and meat traits. Although 3-D ultrasound is more costly than conventional ultrasound, it is not prohibitively expensive when incorporated into large breeding programmes.

Despite the impressive advances in ultrasound systems, the software and algorithms still need to be constantly reviewed, and comparisons between systems will still be necessary to find the most suitable means of predicting carcass composition and meat traits for all producing species (Williams, 2002). Future developments in molecular genetics, together with more efficient data collection and dissemination using web-based databases, will increase the value of RTU technology for acquiring information on carcass composition and meat traits (MacFarlane and Simm, 2008; Bertrand, 2009).

11.9   Conclusion

Real-time ultrasonography imaging is a versatile and dynamic technology with many current and potential applications in animal science research and animal production. The attributes of the RTU technique have led to its current widespread use in animal science for the *in vivo* prediction of carcass composition and meat traits in several species. The results obtained with RTU are likely to play a major role in the meat industry by providing accurate and objective carcass and meat traits information in live animals. In the future, it is probable that modern ultrasound techniques will continue to be used in animal science, bringing about further advances in value-based marketing and in precision meat production systems. Research will be focused on developments in ultrasound practicability, portability, cost and public acceptability, and the rapidly advancing field of molecular genetics and the dissemination of web-based databases will further expand the capabilities of RTU as a tool for evaluating carcass composition and meat traits.

11.10 References

Aass, L., Gresham, J. D. and Klemetsdal, G. (2006) Prediction of intramuscular fat by ultrasound in lean cattle. *Livestock Science*, **101**, 228–41.

Aass, L., Fristedt, C. G. and Gresham, J. D. (2009) Ultrasound prediction of intramuscular fat content in lean cattle. *Livestock Science*, **125**, 177–86.

Allen, P. (1990) Measuring body composition in live meat animals. In Wood, J. D. and Fisher, A. V. (eds), *Reducing Fat in Meat Animals*. London: Elsevier, 201–54.

Alliston, J. C. (1980) Evaluation of carcass quality in live animal. In Haresign, W. (ed.), *Sheep Production*. London: Butterworth, 75–96.

Alston, C. L., Mengersen, K. L. and Gardner, G. E. (2009) A new method for calculating the volume of primary tissue types in live sheep using computed tomography scanning. *Animal Production Science*, **49**, 1035–42.

Amin, V. (1995) *An Introduction to Principles of Ultrasound*. Study guide, Iowa State University.

Amin, V. R., Doerr, V. J., Ani, P. R. and Carlson, D. L. (1992) Application of neural network to ultrasound tissue characterization using backscattered signal parameters. *Proceedings of the IEEE Medical Imaging Conference*, Orlando, Florida, 1357–9.

Amin, V., Roberts, V. R., Patel, A., Wilson, D. and Rouse, G. (1993) Tissue characterization for beef grading using texture analysis of ultrasonic images. *Proceedings of the IEEE Ultrasonics Symposium*, **2**, 969–72.

Amin, V., Wilson, D. E. and Rouse, G. H. (1997) USOFT: An ultrasound image analysis software for beef quality research beef research report. *Beef Research Report AS Leaflet R1437*, Iowa State University.

Andersen, B. B., Busk, H., Chadwick, J. P., Cuthbertson, A., Fursey, G. A. J., Jones, D. W., Lewin, P., Miles, C. A. and Owen, M. G. (1983) Comparison of ultrasonic equipment for describing beef carcass characteristics in live cattle (report on a joint ultrasonic trial carried out in the UK and Denmark). *Livestock Production Science*, **10**, 133–47.

Andrassy-Baka, G., Romvári, R., Milisits, G., Suto, Z., Szabo, A., Locsmandi, L. and Horn, P. (2003) Non-invasive body composition measurement of broiler chicken between 4 and 18 weeks of age by computer tomography. *Arch Tierernahr*, **46**, 585–95.

Anon. (2004) Visual assessment of marbling and meat colour, *Meat Technology Update* Newsletter, **2**.

Arceo, M., Fassa, V., Conte, A., Iglesias, G., Demarco, A., Romano, E., Huguet, M., Lamouroux, F., Canet, Z., Binda, V. F., Trasorras, V., Caldevilla, M. and Melo, J. (2009) Prediction of weight and proportion of abdominal fat from live animal measurements in Campero-Inta broilers. *Rev Argent Prod Anim*, **29**, 69–73.

Arthur, J. A. and Albers, G. A. A. (2003) Poultry genetics, breeding and biotechnology. In Muir, W. M. and Aggrey, S. E. (eds), *Industrial Perspective on Problems and Issues Associated with Poultry Breeding*. Wallingford: CAB International, 1–12.

Ault, M. J. and Rosen, B. T. (2010) Portable ultrasound: the next generation arrives. *Critical Ultrasound Journal*, **2**, 39–42.

AUTM (2007) *Ultrasound Technology Helps Maximize Beef Production Technology Transfer Works: 100 Innovations from Academic Research to Real-World Application*, Technical Report. Association of University Technology Managers, Better World project, 10.

Bahelka, I., Oravcova, M., Peskovicová, D., Tomka, J., Hanusová, E., Lahucky, R. and Demo, P. (2009) Demo comparison of accuracy of intramuscular fat prediction in live pigs using five different ultrasound intensity levels. *Animal*, **3**, 1205–11.

Basarab, J. A., Brethour, J. R., ZoBell, D. R. and Graham, B. (1999) Sorting feeder cattle with a system that integrates ultrasound backfat and marbling estimates with a model that maximizes feedlot profitability in value-based marketing. *Canadian Journal of Animal Science*, **79**, 327–34.

Bentsen, H. B. and Sehested, E. (1989) Computerized tomography of chickens. *British Poultry Science*, **30**, 575–85.

Berri, C., Debut, M., Lebihan-Duval, E., Sante-Lhoutellier, V., Hattab, N., Jehl, N. and Duclos, M. J. (2005) Technological quality of broiler breast meat in relation to muscle hypertrophy. *Arch Tierz*, **48**, 131.

Bertrand, J. K. (2009) Using actual and ultrasound carcass information in beef genetic evaluation programs. *Revista Brasileira de Zootecnia*, **38**, 58–63.

Bindon, B. M. (2002) Measuring and managing productivity-what do we know; what do we need?. *Proceedings of the World Brahman Congress*, Rockhampton, Australia, 1–14.

Bochno, R., Rymkiewicz, J. and Szeremeta, J. (2000) Regression equations for in vivo estimation of the meat content of duck carcasses. *British Poultry Science*, **41**, 313–17.

Bosworth, B. G., Holland, M. and Brazil, B. L. (2001) Evaluation of ultrasound imagery and body shape to predict carcass and fillet yield in farm-raised catfish. *Journal of Animal Science*, **79**, 1483–90.

Boyle, M. (2006) The modern poultry yield (r)evolution. *International Poultry Production*, **13**, 7–11.

Bray, D. E. and McBride, D. (1992) *Nondestructive Testing Techniques*. New York: John Wiley and Sons.

Bret, P. N., Melnick, E. R. and Li, J. (2011) Portable ultrasound for remote environments, Part I: feasibility of field deployment. *Journal of Emergency Medicine*, **40**, 190–7.

Brethour, J. R. (1990) Relationship of ultrasound speckle to marbling score in cattle. *Journal of Animal Science*, **68**, 2603–13.

Brethour, J. R. (1992) The repeatability and accuracy of ultrasound in measuring backfat of cattle. *Journal of Animal Science*, **70**, 1039–44.

Brethour, J. R. (1994) Estimating marbling score in live cattle from ultrasound images using pattern recognition and neural network procedures. *Journal of Animal Science*, **72**, 1425–32.

Brethour, J. R. (2000) Using serial ultrasound measures to generate models of marbling and backfat thickness changes in feedlot cattle. *Journal of Animal Science*, **78**, 2055–61.

Bruns, K. W., Pritchard, R. H. and Boggs, D. L. (2004) The relationships among body weight, body composition, and intramuscular fat content in steers. *Journal of Animal Science*, **82**, 1315–22.

Busk, H. (1984) Improved Danscanner for cattle, pigs and sheep. In Lister, D. (ed.), *In vivo Measurement of Body Composition in Meat Animals*. London: Elsevier, 158–62.

Busk, H. (1986) Measure carcass quality on live pigs. *World Review of Animal Production*, **22**, 35–8.

Campbell, D., Stonaker, H. H. and Esplin, A. L. (1959) The use of ultrasonics to estimate the size of longissimus dorsi muscle in sheep. *Journal of Animal Science*, **24**, 364–7.

Castellini, C., Dal Bosco, A. and Cardinali, R. (2006) Long term effect of post-weaning rhythm on the body fat and performance of rabbit doe. *Reproduction Nutrition Development*, **46**, 195–204.

Chambaz, A., Dufey, P. A., Kreuzer, M. and Gresham, J. (2002) Sources of variation influencing the use of realtime ultrasound to predict intramuscular fat in live beef cattle. *Canadian Journal of Animal Science*, **82**, 133–9.

Chen, C. H. (2007) Ultrasonic and advanced methods for non-destructive testing and material characterization. PhD Thesis, University of Massachusetts, Dartmouth.

Claus, A. (1957) The measurement of natural interfaces in the pig's body with ultrasound. *Fleischwirtschaft*, **9**, 552–7.

Collewet, G., Toussaint, C., Davenel, A., Akoka, S., Médale, F., Fauconneau, B. and Haffray, P. (2001) Magnetic resonance imaging as a tool to quantify the adiposity distribution in fish. In Webb, G. A., Belton, P. S., Gil, V. and Delgadillo, I. (eds), *Magnetic Resonance in Food Science*. Cambridge: Royal Society of Chemistry, 252–7.

Couto, P., Silva, S. R., Barrenechea, E., Santos, V. and Melo-Pinto, A. (2011) Fuzzy based subcutaneous fat assessment in real-time ultrasound images. *Proceedings of the European Society for Fuzzy Logic and Technology*, Aix-les-Bains, France, 350–7.

Crepaldi, D. V., Teixeira, A. E., Faria, P. M. C., Ribeiro, P. L., Melo, S. H., Daniela, M., Sousa, A. B. and Carvalho, D. C. (2006) Ultrasonography in fish culture (in Portuguese). *Rev Bras Reprod Anim*, **30**, 174–81.

Dal Bosco, A., Castellini, C. and Mugnai, C. (2003) Evaluation of body condition in pregnant rabbit does by ultrasound scanner. *Proceedings of the ASPA*, Parma, Italy, 480–2.

Davenel, A., Seigneurin, F., Collewet, G. and Reamignon, H. (2000) Estimation of poultry breast meat yield: magnetic resonance imaging as a tool to improve the positioning of ultrasonic scanners. *Meat Science*, **56**, 153–8.

Davis, J. K., Long, R. A., Saffe, R. L., Warren, E. P. and Carmon, J. L. (1964) Use of ultrasonics and visual appraisal to estimate total muscling in beef cattle. *Journal of Animal Science*, **23**, 638–44.

De Campeneere, S., Fiems, L. O. and Boucqué, C. V. (2000) In vivo estimation of body composition in cattle. *Nutrition Abstracts & Reviews*, **70**, 495–508.

Delehant, T. M., Dahlke, G. R., Hoffman, M. P., Iiams, J. C., Rouse, G. H. and Wilson, D. E. (1996) Using real-time ultrasound during the feeding period to predict cattle composition. *Beef Research Report AS Leaflet R1433*, Iowa State University.

Delfa, R., Teixeira, A., Blasco, I. and Colomer-Rocher, F. (1991) Ultrasonics estimates of fat thickness, c measurement and longissimus dorsi depth in rasa Aragonesa ewes with same body condition score. *Options Mèditerranéennes*, **13**, 25–30.

Delfa, R., Teixeira, A., Gonzalez, C. and Blasco, A. (1995a) Ultrasonic estimates of fat thickness and longissimus dorsi muscle depth for predicting carcass composition of live aragon lambs. *Small Ruminant Research*, **16**, 159–64.

Delfa, R., Teixeira, A. and González, C. (1995b) Ultrasonic measurements of fat thickness and longissimus dorsi depth for predicting carcass composition and body fat depots of live goats. *Proceedings of the 46th Annual Meeting of the EAAP*, Prague, Czech Republic, 276.

Delfa, R., Gonzalez, A. T. C. and Vijil, E. (1996) Ultrasonic measurements in live goats prediction of weight of carcass joints. *Proceedings of the 47th Annual Meeting of the EAAP*, Lillehammer, Norway, 272.

Delfa, R., Teixeira, A., Cadavez, V., Gonzalez, C. and Sierra, I. (2000) Relationships between ultrasonic measurements in live goats and the same measurements taken on carcass. *Proceedings of the 7th International Conference Goats*, Tours, France, 833–4.

Devitt, C. and Wilton, J. (2001) Genetic correlation estimates between ultrasound measurements on yearling bulls and carcass measurements on finished steers. *Journal of Animal Science*, **79**, 2790–7.

Dixson, S. J. and Teeter, R. G. (2001) Fast heat production and body composition of broiler breeder females ranging from 5 to 50 weeks of age. *Animal Science Research Report P986*, Oklahoma Agricultural Experiment Station, Oklahoma.

Dolezal, H. G., Smith, G. C., Savell, J. W. and Carpenter, Z. L. (1982) Comparison of subcutaneous fat thickness, marbling and quality grade for predicting palatability of beef. *Journal of Food Science*, **47**, 397–401.

Du, C. J., Sun, D. W., Jackman, P. and Allen, P. (2008) Development of a hybrid image processing algorithm for automatic evaluation of intramuscular fat content in beef M. longissimus dorsi. *Meat Science*, **80**, 1231–7.

Duclos, M. J., Molette, C., Guernec, A. and Berri, H. R. C. (2006) Cellular aspects of breast muscle development in chicken with high or low growth rate. *Arch Tierernahr*, **49**, 147–51.

Dumont, B. L. (1957) Nouvelles methods pour l'estimation de la qualite des carcasses sur les porcs vivants. Joint Food and Agriculture Organization of the United Nations/EAAP, Meeting on Pig Progeny Testing, Copenhagen, Denmark, 23.

DuPonte, M. W. and Fergerstrom, M. L. (2006) Application of ultrasound technology in beef cattle carcass research and management: Frequently asked questions. *Livestock Management*, **13**, 1–3.

Dussik, K. T. (1942) Ube die moglichkeit hochfrecluente mechanische Schwingungen als diagnostiches Hilfsmittel zu verwenden. *Z Gesamte Neurol Psychiatr*, **174**, 153–68.

Emenheiser, J. C., Greiner, S. P., Lewis, R. M. and Notter, D. R. (2010) Longitudinal changes in ultrasonic measurements of body composition during growth in Suffolk ram lambs and evaluation of alternative adjustment strategies for ultrasonic scan data. *Journal of Animal Science*, **88**, 1341–8.

FAOSTAT (2010) Available from: http://faostatfaoorg/site/291/defaultaspx (accessed 6 May 2011).

Faucitano, L., Huff, P., Teuscher, F., Gariepy, C. and Wegner, J. (2005) Application of computer image analysis to measure pork marbling characteristics. *Meat Science*, **69**, 537–43.

Faulkner, D. B., McKeith, F. K., Berger, L. L., Kesler, D. J. and Parrett, D. F. (1989) Effect of testosterone propionate on performance and carcass characteristics of heifers and cows. *Journal of Animal Science*, **67**, 1907.

Ferguson, D. M. (2004) Objective on-line assessment of marbling: a brief review. *Australian Journal of Experimental Agriculture*, **44**, 681–5.

Fernández, C., Gallego, L. and Quintanilla, A. (1997) Lamb fat thickness and longissimus muscle area measured by a computerized ultrasonic system. *Small Ruminant Research*, **26**, 277–82.

Fortun-Lamothe, L., Lamboley-Gaüzère, B. and Bannelier, C. (2002) Prediction of body composition in rabbit females using total body electrical conductivity (TOBEC). *Livestock Production Science*, **78**, 133–42.

Fuller, M. F., Fowler, P. A., McNeill, G. and Foster, M. A. (1990) Body composition: the precision and accuracy of new methods and their suitability for longitudinal studies. *Proceedings of the Nutrition Society*, **49**, 423–36.

Fuller, M. F., Fowler, P. A., McNeill, G. and Foster, M. A. (1994) Imaging techniques for the assessment of body composition. *Journal of Nutrition*, **124**, 1546S–1550S.

Gibson, J. P. and Alliston, A. C. (1983) Some sources of errors and possible bias in Danscan ultrasonic measurements of cattle. *Journal of Animal Production*, **37**, 61–71.

Glasbey, C. A., Abdalla, I. and Simm, G. (1996) Towards automatic interpretation of sheep ultrasound scans. *Animal Science*, **62**, 309–15.

Goddard, P. R. (1995) General principles. In Goddard, P. J. (ed.), *Veterinary Ultrasonography*. Wallingford: CAB International, 1–19.

Gondret, F., Larzul, C., Combes, S. and Rochambeau, H. (2005) Carcass composition, bone mechanical properties, and meat quality traits in relation to growth rate in rabbits. *Journal of Animal Science*, **83**, 1526–35.

Gooden, J. M., Beach, A. D. and Purchas, R. W. (1980) Measurements of subcutaneous backfat depth in live lambs with an ultrasonic probe. *New Zealand Journal of Agricultural Research*, **23**, 161–5.

Grashorn, M. A. (2010) Research into poultry meat quality. *British Poultry Science*, **51**, 60–7.

Green, R. D., Field, T. G., Hammett, N. S., Ripley, B. M. and Doyle, S. P. (2000) Can cow adaptability and carcass acceptability both be achieved?. *Journal of Animal Science*, **77**, 1–20.

Greiner, S. P., Rouse, G. H., Wilson, D. E., Cundiff, L. V. and Wheeler, T. L. (2003) The relationship between ultrasound measurements and carcass fat thickness and longissimus muscle area in beef cattle. *Journal of Animal Science*, **81**, 676–82.

Gresham, J. D., McPeake, S. R., Bernard, J. K. and Henderson, H. H. (1992) Commercial adaptation of ultrasonography to predict pork carcass composition from live animal and carcass measurements. *Journal of Animal Science*, **70**, 631–9.

Gresham, J. D., McPeake, S. R., Bernard, J. K., Riemann, M. J., Wyatt, R. W. and Henderson, H. H. (1994) Prediction of live and carcass characteristics of market hogs by use of a single longitudinal ultrasonic scan. *Journal of Animal Science*, 72, 1409–16.

Griffin, D. B., Savell, J. W., Recio, H. A., Garrett, R. P. and Cross, H. R. (1999) Predicting carcass composition of beef cattle using ultrasound technology. *Journal of Animal Science*, 77, 889–92.

Hamby, P. L., Stouffer, J. R. and Smith, S. B. (1986) Muscle metabolism and real-time ultrasound measurement of muscle and subcutaneous adipose tissue growth in lambs fed diets containing a beta-agonist. *Journal of Animal Science*, 63, 1410–17.

Hamlin, K. E., Green, R. D., Cundiff, L. V., Wheeler, T. L. and Dikeman, M. E. (1995) Real-time ultrasonic measurement of fat thickness and longissimus muscle area: II Relationship between real-time ultrasound measures and carcass retail yield. *Journal of Animal Science*, 73, 1725–34.

Hancz, C., Romvári, R., Szabo, A., Molnár, T., Magyary, I. and Horn, P. (2003) Measurement of total body composition changes of common carp by computer tomography. *Aquaculture Research*, 34, 1–7.

Harada, H. and Kumazaki, K. (1979) Estimating fat thickness, cross sectional area of M. Log. Thoracis and marbling score by use of ultrasonic scanning scope on live beef cattle. *Japan Journal of Zoological Science*, 50, 305–9.

Harron, W. and Dony, R. (2009) Predicting quality measures in beef cattle using ultrasound imaging more options. *IEEE Symposium on Computational Intelligence for Image Processing*, 96–103.

Hassen, A., Wilson, D. E., Willham, R. L., Rouse, G. H. and Trenkle, A. H. (1998) Evaluation of ultrasound measurements of fat thickness and longissimus muscle area in feedlot cattle: Assessment of accuracy and repeatability. *Canadian Journal of Animal Science*, 78, 277–86.

Hassen, A., Wilson, D. E. and Rouse, G. H. (1999a) Evaluation of carcass, live, and real-time ultrasound measures in feedlot cattle: II Effects of different age end points on the accuracy of predicting the percentage of retail product, retail product weight, and hot carcass weight. *Journal of Animal Science*, 77, 283–90.

Hassen, A., Wilson, D., Amin, V. and Rouse, G. (1999b) Repeatability of ultrasound predicted percentage of intramuscular fat in feedlot cattle. *Journal of Animal Science*, 77, 1335–40.

Hassen, A., Wilson, D. E., Amin, V. R., Rouse, G. H. and Hays, C. L. (2001) Predicting percentage of intramuscular fat using two types of real-time ultrasound equipment. *Journal of Animal Science*, 73, 11–18.

Haumschild, D. J. and Carlson, D. L. (1983) An ultrasonic Bragg scattering technique for the quantitative characterisation of marbling in beef. *Ultrasonics*, 21, 226–33.

Hazel, L. N. and Kline, E. A. (1959) Ultrasonic measurements of fatness in swine. *Journal of Animal Science*, 18, 815–19.

Hedrick, H. B., Meyer, W. E., Alexander, M. A., Zobrisky, S. E. and Naumann, H. D. (1962) Estimation of rib-eye area and fat thickness of beef cattle with ultrasonics. *Journal of Animal Science*, 21, 362–5.

Henderson-Perry, S. C., Corah, L. R. and Perry, R. C. (1989) The use of ultrasound in cattle to estimate subcutaneous fat thickness and ribeye area. *Journal of Animal Science*, 67(Suppl 1), 433.

Hernández, P., Aliaga, S., Pla, M. and Blasco, A. (2004) The effect of selection for growth rate and slaughter age on carcass composition and meat quality traits in rabbits. *Journal of Animal Science*, 82, 3138–43.

Hernández, P., Ariño, B., Grimal, A. and Blasco, A. (2006) Comparison of carcass and meat characteristics of three rabbit lines selected for litter size or growth rate. *Meat Science*, 73, 645–50.

Herring, W. O., Miller, D. C., Bertrand, J. K. and Benyshek, L. L. (1994) Evaluation of machine, technician, and interpreter effects on ultrasonic measures of backfat and longissimus muscle area in beef cattle. *Journal of Animal Science*, 72, 2216–26.

Herring, W., Kriese, L., Bertrand, K. and Crouch, J. (1998) Comparison of four real-time ultrasound systems that predict intramuscular fat in beef cattle. *Journal of Animal Science*, **76**, 364–70.

Hopkins, D. L., Pirlot, K. L., Roberts, A. H. K. and Beattie, A. S. (1993) Changes in fat depths and muscle dimensions in growing lambs as measured by real-time ultrasound. *Australian Journal of Experimental Agriculture*, **33**, 707–12.

Hopkins, D. L., Hall, D. G. and Luff, A. F. (1996) Lamb carcass. 3. Describing changes in carcasses of growing lambs using real-time ultrasound and the use of these measurements for estimating the yield of saleable meat. *Australian Journal of Experimental Agriculture*, **36**, 37–43.

Hopkins, D. L., Stanley, D. F. and Ponnampalam, E. N. (2007) Relationship between real-time ultrasound and carcass measures and composition in heavy sheep. *Australian Journal of Experimental Agriculture*, **47**, 1304–8.

Houghton, P. L. and Turlington, L. M. (1992) Application of ultrasound for feeding and finishing animals: A review. *Journal of Animal Science*, **70**, 930–41.

Huff-Lonergan, E. T., Baas, T. J., Malek, M., Dekkers, J. C. M., Prusa, K. and Rothschild, M. F. (2002) Correlations among selected pork quality traits. *Journal of Animal Science*, **80**, 617–27.

Insana, M. F. (2006) Ultrasonic imaging. In Akay, M. and Hoboken, N. J. (eds), *Encyclopedia of Biomedical Engineering*. New York: John Wiley and Sons, 3640–8.

Jackman, P., Sun, D.-W., Du, C.-J., Allen, P. and Downey, G. (2008) Prediction of beef eating quality from colour, marbling and wavelet texture features. *Meat Science*, **80**, 1273–81.

Kallweit, E., Wesemeier, H. H., Smidt, D. and Baulain, U. (1994) Application of magnetic-resonance-measurements in animal research. *Arch Tierernahr*, **31**, 105–20.

Kane, D., Grassi, W., Sturrock, R. and Balint, P. (2004) A brief history musculoskeletal ultrasound: From bats and ships to babies and hips. *Rheumatology*, **43**, 931–3.

Kemp, J. D., Herring, W. O. and Kaiser, C. J. (2002) Genetic and environmental parameters for steer ultrasound and carcass traits. *Journal of Animal Science*, **80**, 1489–96.

Kempster, A. J., Cuthbertson, A. and Owen, M. G. (1979) A comparison of four ultrasonic machines (Sonatest, Scanogram, Ilis observer, and Danscanner) for predicting the body composition of live pigs. *Journal of Animal Production*, **29**, 175–81.

Kempster, A. J., Arnall, D., Alliston, J. C. and Barker, J. D. (1982) An evaluation of two ultrasonic machines (Scanogram and Danscanner) for predicting the body composition of live sheep. *Journal of Animal Production*, **34**, 249–55.

Kim, N., Amin, V., Wilson, D., Rouse, G. and Upda, S. (1998) Ultrasound image texture analysis for characterizing intramuscular fat content of live beef cattle. *Ultrasonic Imaging*, **20**, 191–205.

King, A. M. (2006) Development, advances and applications of diagnostic ultrasound in animals. *Veterinary Journal*, **171**, 408–20.

Kleczek, K., Wawro, K., Wilkiewicz-Wawro, E., Makowski, W. and Konstan-Tynowicz, D. (2009) Relationships between breast muscle thickness measured by ultrasonography and meatiness and fatness in broiler chickens. *Arch Tierernahr*, **52**, 538–45.

Klein, H. G. (1981) Are B-scanners' days numbered in abdominal diagnosis?. *Diagnostic Imaging*, **3**, 10–11.

Kolstad, K., Vegusdal, A., Baeverfjord, G. and Einen, O. (2004) Quantification of fat deposits and fat distribution in Atlantic halibut (hippoglossus hippoglossus l) using computerised x-ray tomography (CT). *Aquaculture*, **229**, 255–64.

Konig, T., Grashorn, M. A. and Bessei, W. (1997) Estimation of breast meat yield in living broilers using b-scan sonography first report: Defining sites of measurement. *Arch Gelugelkd*, **61**, 227–31.

Konig, T., Grashorn, M. A. and Bessei, W. (1998) Estimation of breast meat yield in living broilers using b-scan sonography second report: Accuracy of the method. *Arch Gelugelkd*, **62**, 121–5.

Korn, S. V., Baulain, U., Arnold, M. and Brade, W. (2005) Nutzung von magnet-resonanz-tomographie und ultraschall-technik zur bestimmung des schlachko-rperwertes beim schaf. *Zuchtungskunde*, **77**, 382–93.

Kouba, M. and Sellier, P. (2011) A review of the factors influencing the development of intermuscular adipose tissue in the growing pig. *Meat Science*, **88**, 213–20.

Kövér, G., Romvári, R., Horn, P., Jensen, E. B. J. F. and Sorensen, P. (1998a) In vivo assessment of breast muscle, abdominal fat and total fat volume in meat type chickens by magnetic resonance imaging. *Acta Vet Hung*, **46**, 135–44.

Kövér, G., Szendrö, Z., Romvári, R., Jensen, J. F. and Milisits, G. (1998b) In vivo measurement of body parts and fat deposition in rabbits by MRI. *World Rabbit Science*, **6**, 231–5.

Kvame, T. and Vangen, O. (2006) In vivo composition of carcass regions in lambs of two genetic lines, and selection of CT positions for estimation of each region. *Small Ruminant Research*, **66**, 201–208.

Larivière, J. M., Michaux, C., Verleyen, V., Hanzen, C. and Leroy, P. (2009) Non-invasive methods to predict breast muscle weight in slow-growing chickens. *International Journal of Poultry Science*, **8**, 689–91.

Larzul, C., Gondret, S., Combes, S. and Rochambeau, H. (2005) Divergent selection on 63-old body weight in the rabbit: Response on growth, carcass and muscle traits. *Genetics Selection Evolution*, **37**, 105–22.

Latshaw, J. D. and Bishop, B. L. (2001) Estimating body weight and body composition of chickens by using noninvasive measurements. *Poultry Science*, **80**, 868–73.

Leeds, T. D., Mousel, M. R., Notter, D. R. and Lewis, G. S. (2007) Ultrasound estimates of loin muscle measures and backfat thickness augment live animal prediction of weights of subprimal cuts in sheep. *Proceedings, Western Section, American Society of Animal Science*, **58**, 97–100.

Leeds, T. D., Mousel, M. R., Notter, D. R., Zerby, H. N., Mollet, C. A. and Lewis, G. S. (2008) B-mode, real-time ultrasound for estimating carcass measures in live sheep: Accuracy of ultrasound measures and their relationships with carcass yield and value. *Journal of Animal Science*, **86**, 3203–14.

Leighton, T. G. (2007) What is ultrasound?. *Progress in Biophysics and Molecular Biology*, **93**, 3–83.

Li, C. (2010) A web service model for conducting research in image processing. *Journal of Computing in Small Colleges*, **25**, 294–9.

Li, C. C., Zheng, Y. F. and Kwabena, A. (2009) Prediction of IMF percentage of live cattle by using ultrasound technologies with high accuracies. In Qui, R. and Zhao, H. (eds), *WASE International Conference on Information Engineering (ICIE 2009)*, II, 474–8.

Ludwig, G. D. (1950) The velocity of sound through tissues and the acoustic impedance of tissues. *Journal of the Acoustical Society of America*, **22**, 862–6.

Lukefahr, S. D. and Ozimba, C. E. (1991) Prediction of carcass merit from live body measurements in rabbits of four breed-types. *Livestock Production Science*, **29**, 323–34.

Lukefahr, S. D., Hohenboken, W. D., Cheeke, P. R., Patton, N. M. and Kennick, W. H. (1982) Carcass and meat characteristics of flemish giant and New Zealand white purebred and terminal-cross rabbits. *Journal of Animal Science*, **54**, 1169–74.

Lukefahr, S. D., Hohenboken, W. D., Cheeke, P. R. and Patton, N. M. (1983) Appraisal of nine genetic groups of rabbits for carcass and lean yield traits. *Journal of Animal Science*, **57**, 899–907.

Lusk, J. L., Little, R., Williams, A., Anderson, J. and McKinley, B. (2003) Utilizing ultrasound technology to improve livestock marketing decisions. *Review of Agricultural Economics*, **25**, 203–17.

Lyons, E. (2004) Zone sonography: The next major advance in medical ultrasound. Winnipeg, University of Manitoba. Available from: http://www.radicansa.com/radicansa1/images/pdf1/Zone_Sonography_Technology_Clinical_White_Paper.pdf (accessed 14 May 2011).

Macfarlane, J. M. and Simm, G. (2008) Genetic improvement programme for meat type sheep: An experience from the United Kingdom. *Tecnologia e Ciência Agropecuária*, **2**, 15–22.

MacNeil, M. D., Nkrumah, J. D., Woodward, B. W. and Northcutt, S. L. (2010) Genetic evaluation of angus cattle for carcass marbling using ultrasound and genomic indicators. *Journal of Animal Science*, **88**, 517–22.

Mannion, P. (2006) Principles of diagnostic ultrasound. In Mannion, P. (ed.), *Diagnostic Ultrasound in Small Animal Practice*. Oxford: Blackwell, 1–19.

May, S. G., Miles, W. L., Edwards, J. W., Harris, J. J., Morgan, J. B., Garrett, R. P., Williams, F. L., Wise, J. W., Cross, H. R. and Savell, J. W. (2000) Using live estimates and ultrasound measurements to predict beef carcass cutability. *Journal of Animal Science*, **78**, 1255–61.

McEwan, J. C., Clarke, J. N., Knowler, M. A. and Wheeler, M. (1989) Ultrasonic fat depths in romney lambs and hoggets from lines selected for different production traits. *Proceedings of the New Zealand Society of Animal Production*, **49**, 113–19.

McLaren, D. G., McKeith, F. M. and Novakofski, J. (1989) Prediction of carcass characteristics at market weight from serial real-time ultrasound measures of backfat and loin eye area in the growing pig. *Journal of Animal Science*, **67**, 1657–67.

McLaren, D. G., Novakofski, J., Parrett, D. F., Lo, L. L., Singh, S. D., Neumann, K. R. and McKeith, F. K. (1991) A study of operator effects on ultrasonic measures of fat depth and longissimus muscle area in cattle, sheep and pigs. *Journal of Animal Science*, **69**, 54–66.

McPeake, C. A. (2001) USDA marbling and carcass physiological maturity related differences for beef tenderness and palatability characteristics. PhD Thesis, Oklahoma State University.

Melo, J. E., Motter, M. M., Morão, L. R., Huguet, M. J., Canet, Z. and Miquel, C. M. (2003) Use of in vivo measurements to estimate breast and abdominal fat content of a free-range broiler strain. *Journal of Animal Science*, **77**, 23–31.

Mesta, C. G., Will, P. A. and Gonzalez, J. M. (2004) The measurement of carcass characteristics of goats using the ultrasound method. *Texas Journal of Agriculture and Natural Resources*, **17**, 46–52.

Michalik, D., Bochno, R., Janiszewska, M. and Brzozowski, W. (1999) In vivo assessment of meatiness and fatness in broiler chickens using ultrasonography. *Pr Mater Zootech*, **54**, 77–83.

Michalik, D., Lewczuk, A., Wilkiewicz-Wawro, E. and Brzozowski, W. (2006) Prediction of the meat content of the carcass and valuable carcass parts in French lop rabbits using some traits measured in vivo and post mortem. *Czech Journal of Animal Science*, **51**, 406–15.

Miles, C. M., Pomeroy, R. W. and Harries, J. M. (1972) Some factories affecting reproductibility in ultrasonic scanning of animals. *Journal of Animal Production*, **15**, 239–49.

Mitchell, A. D. and Scholz, A. M. (2005) Body composition: Indirect measurements. In Pond, W. and Bell, A. (eds), *Encyclopedia of Animal Science*. New York: Marcel Dekker, 166–9.

Mitchell, A. D., Wang, P. C., Rosebrough, R. W., Elsasser, T. H. and Schmidt, W. F. (1991) Assessment of body composition of poultry by nuclear magnetic resonance imaging and spectroscopy. *Poultry Science*, **70**, 2494–500.

Mitchell, A. D., Rosebrough, R. W. and Conway, J. M. (1997) Body composition analysis of chickens by dual energy x-ray absorptiometry. *Poultry Science*, **76**, 1746–52.

Mitchell, A. D., Scholz, A. M. D., Wange, P. C. and Song, H. (2001) Body composition analysis of the pig by magnetic resonance imaging. *Journal of Animal Science*, **79**, 1800–13.

Moeller, S. J. (1990) Serial real-time ultrasonic evaluation of fat and muscle deposition in market hogs. PhD Thesis, Iowa State University.

Moeller, S. J. (2002) Evolution and use of ultrasonic technology in the swine industry. *Journal of Animal Science*, **80**(E. Suppl 2), E19–E27.

Moeller, S. J. and Christian, L. L. (1998) Evaluation of the accuracy of real-time ultrasonic measurements of backfat and loin muscle area in swine using multiple statistical analysis procedures. *Journal of Animal Science*, **76**, 2503–14.

Monteiro, A. (2010) Methods to predict carcass composition of Serrana kids. PhD Thesis, University of Trás-os-Montes and Alto Douro, Vila Real, Portugal.

Monziols, M., Collewet, G., Bonneau, M., Mariette, F., Davenel, A. and Kouba, M. (2006) Quantification of muscle, subcutaneous fat and intermuscular fat in pig carcasses and cuts by magnetic resonance imaging. *Meat Science*, **72**, 146–54.

Morlein, D., Rosner, F., Brand, S., Jenderka, K. V. and Wicke, M. (2005) Non-destructive estimation of the intramuscular fat content of the longissimus muscle of pigs by means of spectral analysis of ultrasound echo signals. *Meat Science*, **69**, 187–99.

Nalaila, S. M., Stothard, P., Moore, S. S., Wang, Z. and Li, C. (2011) Whole genome fine mapping of quantitative trait loci for ultrasound and carcass merit traits in beef cattle. *Canadian Journal of Animal Science*, **91**, 61–73.

Newcom, D. W., Baas, T. J. and Lampe, J. F. (2002) Prediction of intramuscular fat percentage in live swine using real-time ultrasound. *Journal of Animal Science*, **80**, 3046–52.

O'Brien, R. T. and Holmes, S. P. (2007) Recent advances in ultrasound technology. *Clinical Techniques in Small Animal Practice*, **22**, 93–103.

Oberle, M., Schwarz, F. J. and Kirchgessner, M. (1997) Growth and carcass quality of carp (cyprinus carpio l) fed different cereals, lupin seed or zooplankton. *Arch Tierernahr*, **50**, 75–86.

Olsen, E. V., Candek-Potokar, M., Oksama, M., Kien, S., Lisiak, D. and Busk, H. (2007) On-line measurements in pig carcass classification: Repeatability and variation caused by the operator and the copy of instrument. *Meat Science*, **75**, 29–38.

Ophir, J. and Maklad, N. (1979) Digital scan converters in diagnostic ultrasound imaging. *IEEE Proceedings*, **67**, 654–64.

Orman, A., Calfskan, G. U., Dikmen, S., Ustuner, H., Ogan, M. and Caliskan, C. (2008) The assessment of carcass composition of Awassi male lambs by real-time ultrasound at two different live weights. *Meat Science*, **80**, 1031–6.

Orman, A., Caliskan, G. U. and Dikmen, S. (2010) The assessment of carcass traits of Awassi lambs by real-time ultrasound at different body weights and sexes. *Journal of Animal Science*, **88**, 3428–38.

Oviedo-Rondón, E. O., Parker, J. and Clemente-Hernández, S. (2007) Application of real-time ultrasound technology to estimate in vivo breast muscle weight of broiler chickens. *British Poultry Science*, **48**, 151–61.

Paisley, S., Loehr, C. and Niemela, F. (2007) Ultrasound-based selection: Pitfalls and rewards. *Proceedings of the Range of Beef Cow Symposium XX Fort Collins*, Colorado.

Parnell, P. F. (2004) Industry application of marbling genetics: A brief review. *Australian Journal of Experimental Agriculture*, **44**, 697–703.

Pascual, J. J., Castella, F., Cervera, C., Blas, E. and Fernández-Carmona, J. (2000) The use of ultrasound measurement of perirenal fat thickness to estimate changes in body condition of young female rabbits. *Animal Science*, **70**, 435–42.

Pascual, J. J., Motta, W., Cervera, C., Quevedo, F., Blas, E. and Fernández-Carmona, J. (2002) Effect of dietary energy source on the performance and perirenal fat thickness evolution of primiparous rabbit does. *Animal Science*, **75**, 267–73.

Pascual, J. J., Blanco, J., Piquer, O., Quevedo, F. and Cervera, C. (2004) Ultrasound measurements of perirenal fat thickness to estimate the body condition of reproducing rabbit does in different physiological status. *World Rabbit Science*, **12**, 7–22.

Pascual, M. and Pla, M. (2007) Changes in carcass composition and meat quality when selecting rabbits for growth rate. *Meat Science*, **77**, 474–81.

Perkins, T. L., Green, R. D. and Hamlin, K. E. (1992a) Evaluation of ultrasonic estimates of carcass fat thickness and longissimus muscle area in beef cattle. *Journal of Animal Science*, **70**, 1002–10.

Perkins, T. L., Green, R. D., Hamlin, K. E., Shepard, H. H. and Miller, M. F. (1992b) Ultrasonic prediction of carcass merit in beef cattle: evaluation of technician effects on ultrasonic estimates of carcass fat thickness and longissimus muscle area. *Journal of Animal Science*, **70**, 2758–65.

Pethick, D. W., Harper, G. S., Hocquette, J. F. and Wang, Y. (2006) Marbling biology – what do we know about getting fat into muscle?. Proceedings of Australian beef – the leader! The Impact of Science on the Beef Industry, University of New England, Armidale, NSW, 103–10.

Peura, A. R. and Webster, J. G. (1998) Basic sensors and principles. In Webster, J. G. (ed.), *Medical Instrumentation: Application and Design*. New York: John Wiley and Sons, 45–90.

Prince, J. L. and Links, J. M. (2006) *Medical Imaging Signals and Systems*. New Jersey: Pearson Prentice Hall.

Probert, P. and Shannon, R. (2000) Wideband ultrasound to determine lipid concentration in fish. In Halliwell, M. and Wells, P. N. T. (eds), *Acoustical Imaging*. Oxford: Springer, 381–8.

Pyatt, N. A., Berger, L. L., Faulkner, D. B., Walker, P. M. and Rodriguez-Zas, S. L. (2005) Factors affecting carcass value and profitability in early-weaned Simmental steers: II. Days on feed endpoints and sorting strategies. *Journal of Animal Science*, **83**, 2926–37.

Quevedo, F., Cervera, C., Blas, E., Baselga, M. and Pascual, J. J. (2006) Long-term effect of selection for litter size and feeding programme on the performance of reproductive rabbit does. 1. Pregnancy of multiparous does. *Journal of Animal Science*, **82**, 739–50.

Raeth, U., Schlaps, D., Limberg, B., Zum, I., Lorenz, A., van Kaick, G., Lorenz, W. T. and Kommerell, B. (1985) Diagnostic accuracy of computerized b-scan texture analysis and conventional ultrasonography in diffuse parenchymal and malignant liver disease. *Journal of Clinical Ultrasound*, **13**, 87–9.

Ragland, K. D., Christian, L. L. and Baas, T. J. (1997) Evaluation of real-time ultrasound and carcass characteristics for assessing carcass composition in swine. *Beef Research Report AS Leaflet R1424*, Iowa State University.

Ramsey, C. B., Kirton, A. H., Hogg, B. and Dobbie, J. L. (1991) Ultrasonics, needle and carcass measurements for predicting chemical composition of lamb carcasses. *Journal of Animal Science*, **69**, 3655–64.

Rémignon, H., Seigneurin, F. and Moati, F. (2000) In vivo assessment of the quality of breast muscle by sonography in broilers. *Meat Science*, **56**, 133–8.

Renand, G., Larzul, C., Le Bihan-Duval, E. and Le Roy, P. (2003) Genetic improvement of meat quality in the different livestock species: Present situation and prospects. *Productions Animales*, **16**, 159–73.

Rimal, A., Perkins, T. and Paschal, J. C. (2006) Ultrasound technology for better beef price. *Acta Agriculturae Scandinavica C*, **3**, 99–104.

Ripoll, G., Joy, M., Alvarez-Rodriguez, J., Sanz, A. and Teixeira, A. (2009) Estimation of light lamb carcass composition by in vivo real-time ultrasonography at four anatomical locations. *Journal of Animal Science*, **87**, 1455–63.

Ripoll, G., Joy, M. and Sanz, A. (2010) Estimation of carcass composition by ultrasound measurements in 4 anatomical locations of 3 commercial categories of lamb. *Journal of Animal Science*, **88**, 3409–18.

Robinson, D. L., McDonald, C. A., Hammond, K. and Turner, J. W. (1992) Live animal measurement of carcass traits by ultrasound: Assessment and accuracy of sonographers. *Journal of Animal Science*, **70**, 1667–76.

Rodrigues, V., Dias, J., Rema, P. and Silva, S. R. (2010) Mapping body fat distribution in farmed senegalese sole (solea senegalensis). *Proceedings of EAS Aquaculture Europe*, World Aquaculture Society, Porto, Portugal, 1120–1.

Romvári, R., Hancz, C. S., Petrási, Z. S., Molnár, T. and Horn, P. (2002) Non-invasive measurement of fillet composition of four freshwater fish species by computer tomography. *Aquaculture International*, **10**, 231–40.

Romvári, R., Milisits, G., Szendrö, Z. and Sorensen, P. (1996) Non invasive method to study the body composition of rabbits by x-ray computerised tomography. *World Rabbit Science*, **4**, 219–24.

Rymkiewicz, J. and Bochno, B. R. (1999) Estimation of breast muscle weight in chickens on the basis of live measurements. *Arch Gefluelkd*, **63**, 229–33.

Sañudo, C., Enser, E., Nute, M. C. R., Maria, G., Sierra, I. and Wood, J. D. (2000) Fatty acid composition and sensory characteristics of lamb carcasses from Britain and Spain. *Meat Science*, **54**, 339–46.

Sather, A. P., Bailey, D. R. C. and Jones, S. D. M. (1996) Real-time ultrasound image analysis for estimation of carcass yield and pork quality. *Canadian Journal of Animal Science*, **76**, 55–62.

Savell, J. W., Cross, H. R., and Smith, G. C. (1986) Percentage ether extractable fat and moisture content of beef longissimus muscle as related to USDA marbling score. *Journal of Food Science*, **51**, 838–9.

Schröder, U. J. and Staufenbiel, R. (2006) Invited review: Methods to determine body fat reserves in the dairy cow with special regard to ultrasonographic measurement of back-fat thickness. *Journal of Dairy Science*, **89**, 1–14.

Schwab, C. R., Baas, T. J., Stalder, K. J. and Nettleton, D. (2009) Results from six generations of selection for intramuscular fat in duroc swine using real-time ultrasound. I. Direct and correlated phenotypic responses to selection. *Journal of Animal Science*, **87**, 2774–80.

Schwab, C. R., Baas, T. J. and Stalder, K. J. (2010) Results from six generations of selection for intramuscular fat in duroc swine using real-time ultrasound. II. Genetic parameters and trends. *Journal of Animal Science*, **88**, 69–79.

Scollan, N. D., Caston, L. J., Liu, Z., Zubair, A. K., Leeson, S. and McBrid, B. W. (1998) Nuclear magnetic resonance imaging as a tool to estimate the mass of the pectoralis muscle of chickens in vivo. *British Poultry Science*, **39**, 221–4.

Seidband, M. P. (1998) Medical imaging systems. In Webster, J. G. (ed.), *Medical Instrumentation: Application and Design*. New York: John Wiley and Sons, 518–76.

Shelton, M., Smith, G. C. and Orts, F. (1977) Predicting carcass cutability of Rambouillet rams using live animal traits. *Journal of Animal Science*, **44**, 333–7.

Shi-Zheng, G. and Su-Mei, Z. (2009) Physiology, affecting factors and strategies for control of pig meat intramuscular fat. *Recent Patents on Food Nutrition & Agriculture*, **1**, 59–74.

Sigfusson, H., Decker, E. A. and McClements, D. J. (2000) Rapid prediction of Atlantic mackerel (scomber scombrus) composition using a hand-held ultrasonic device. *Journal of Aquatic Food Production and Technology*, **9**, 27–38.

Silva, S. L., Leme, P. R., Putrino, S. M., Martello, L. S., Lima, C. G. and Lanna, D. P. D. (2004) Prediction of backfat at slaughter, by ultrasound, in Nellore and Brangus young bulls. *Revista Brasileira de Zootecnia*, **33**, 511–17.

Silva, S. R., Gomes, M. J., Dias-da-Silva, A. and Azevedo, J. M. T. (2005) Estimation in vivo of the body and the carcass chemical composition of growing lambs by real-time ultrasonography. *Journal of Animal Science*, **83**, 350–7.

Silva, S. R., Afonso, J. J., Santos, V. A., Monteiro, A., Guedes, C. M., Azevedo, J. M. T. and Dias-da-Silva, A. (2006a) In vivo estimation of sheep carcass composition using real-time ultrasound with two probes of 5 and 75 MHz and image analysis. *Journal of Animal Science*, **84**, 3433–9.

Silva, S. R., Pinheiro, V. M., Guedes, C. M. and Mourão, J. L. (2006b) Prediction of carcass and breast weights and yields in broiler chickens using breast volume determined in vivo by real-time ultrasonic measurement. *British Poultry Science*, **46**, 1–6.

326 Computer vision technology in the food and beverage industries

Silva, S. R., Guedes, C. M., Santos, V. A., Lourenço, A. L., Azevedo, J. M. T. and Dias-da-Silva, A. (2007a) Sheep carcass composition estimated from longissimus thoracis et lumborum muscle volume measured by in vivo real-time ultrasonography. *Meat Science*, **76**, 708–14.

Silva, S. R., Guedes, C. M., Mourão, J. and Pinheiro, V. (2007b) Rabbit carcass chemical composition predicted by real-time ultrasonography. *Proceedings of the XXXII Symposium de ASESCU*, Vila Real, Portugal, 9–12.

Silva, S. R., Mourão, J. L., Guedes, C. M., Pio, A. and Pinheiro, V. (2008a) In vivo rabbit carcass composition and longissimus dorsi muscle volume prediction by real-time ultrasonography. *Proceedings of the 9th World Rabbit Congress*, Verona, Italy, 1449–53.

Silva, S. R., Guedes, C. M., Mourão, J. L., Venâncio, C. and Pinheiro, V. (2008b) Estimation of rabbit hind leg muscle weight and muscularity by real-time ultrasonography. *Proceedings of the 9th World Rabbit Congress*. Verona, Italy, 1443–7.

Silva, S. R., Mourão, J. L., Pio, A. and Pinheiro, V. M. (2009) The value of in vivo real-time ultrasonography in assessing loin muscularity and carcass composition of rabbits. *Meat Science*, **81**, 357–63.

Silva, S. R., Guedes, C. M., Loureiro, N., Mena, E., Dias, J. and Rema, P. (2010a) Prediction in vivo of the fillet volume in senegalese sole (solea senegalensis) by multiple consecutive transverse real-time ultrasonography images, *Proceedings of the 6th International Conference on Simulation and Modelling in the Food and Bio-industry*, Bragança, Portugal, 219–23.

Silva, S. R., Patrício, M., Guedes, C. M., Mena, E., Silva, A. and Jorge, V. S. A. (2010b) Assessment of muscle longissimus thoracis et lumborum intramuscular fat by ultrasonography and image analysis. *Proceedings of the 6th International Conference on Simulation and Modelling in the Food and Bio-industry*, Bragança, Portugal, 211–15.

Simm, G. (1983) The use of ultrasound to predict the carcass composition of live cattle – a review. *Animal Breeding Abstracts*, **51**, 853–75.

Simm, G. (1987) Carcass evaluation in sheep breeding programmes. In Marai, I. F. and Owen, J. F. (eds), *New Techniques in Sheep Production*. London: Butterworth, 125–44.

Simm, G., Alliston, J. C. and Sutherland, R. A. (1983) Comparison of live animal measurements for selecting lean beef sires. *Journal of Animal Production*, **37**, 211–19.

Simm, G., Lewis, R. M., Grundy, B. and Dingwall, W. S. (2002) Responses to selection for lean growth in sheep. *Journal of Animal Science*, **74**, 39–50.

Skiba, G. (2010) Effects of energy or protein restriction followed by re-alimentation on the composition of gain and meat quality characteristics of musculus longissimus dorsi in pigs. *Archives of Animal Nutrition*, **64**, 36–46.

Smith, G. C., Tatum, J. D. and Belk, K. E. (2008a) International perspective: characterisation of united states department of agriculture and meat standards Australia systems will be assessing beef quality. *Australian Journal of Experimental Agriculture*, **48**, 1465–80.

Smith, L., Perron, A., Persico, A., Stravinskas, E. and Cournoyea, D. (2008b) Enhancing image quality using advanced signal processing techniques. *Journal of Diagnostic Medical Sonography*, **24**, 72–81.

Smith, M. T., Oltjen, J. W., Dolezal, H. G., Gill, D. R. and Behrens, B. D. (1992) Evaluation of ultrasound for prediction of carcass fat thickness and longissimus muscle area in feedlot steers. *Journal of Animal Science*, **70**, 29–37.

Spangler, M. L. and Moser, D. W. (2009) Real-time ultrasound: What does image quality mean to genetic evaluations?. *Proceedings of the Beef Improvement Federation 41st Annual Research Symposium*, Sacramento, California, 145–50.

Stanford, K., Clark, I. and Jones, S. D. M. (1995a) Use of ultrasound in prediction of carcass characteristics in lambs. *Canadian Journal of Animal Science*, **75**, 185–9.

Stanford, K., McAllister, T. A., McDougall, M. and Bailey, D. R. C. (1995b) Use of ultrasound for prediction of carcass characteristics in Alpine goats. *Small Ruminant Research*, **15**, 195–201.

Stanford, K., Jones, S. D. M. and Price, M. A. (1998) Methods of predicting lamb carcass composition: A review. *Small Ruminant Research*, **29**, 241–54.

Stouffer, J. R. (1959) Status of the application of ultrasonics in meat animal evaluation. *Proceedings of the Reciprocal Meat Conference*, **12**, 161–9.

Stouffer, J. R. (1991) Using ultrasound to objectively evaluate composition and quality of livestock. 21st Century Concepts Important to Meat-Animal Evaluation, University of Wisconsin, Madison, 49–54.

Stouffer, J. R. (2004) History of ultrasound in animal science. *Journal of Ultrasound in Medicine*, **23**, 577–84.

Stouffer, J. R., Wallentine, M. V., Wellington, G. H. and Diekmann, A. (1961) Development and application of ultrasonic methods for measuring fat thickness and rib-eye area in cattle and hogs. *Journal of Animal Science*, **20**, 759–67.

Suguisawa, L., Mattos, W. R. S., Oliveira, H. N., Silveira, A. C., Arrigoni, M. B., Haddad, C. M., Chardulo, L. A. L. and Martins, C. L. (2003) Ultrasonography as a predicting tool for carcass traits of young bulls. *Scientia Agricola*, **60**, 779–84.

Suvanich, V., Ghaedian, R., Chanamai, R., Decker, E. A. and McClements, D. J. (1998) Prediction of proximate fish composition from ultrasonic properties: Catfish, cod, flounder, mackerel and salmon. *Journal of Food Science*, **63**, 966–8.

Svihus, B. and Katle, J. (1993) Computerised tomography as a tool to predict composition traits in broilers comparisons of results across samples and years. *Acta Agriculturae Scandinavica A*, **43**, 214–18.

Swennen, Q., Janssens, G. P. J., Geers, R., Decuypere, E. and Buyse, J. (2004) Validation of dual-energy x-ray absorptiometry for determining in vivo body composition of chickens. *Poultry Science*, **83**, 1348–57.

Szabo, C., Babinszky, L., Verstegen, M. W. A., Vangen, O., Jansman, A. J. M. and Kanis, E. (1999) The application of digital imaging techniques in the in vivo estimation of body composition of pigs: A review. *Livestock Production Science*, **60**, 1–11.

Szabo ,T .L. (2004) *Diagnostic Ultrasound Imaging: Inside Out*. Connecticut: Academic Press Series in Biomedical Engineering.

Szendrö, Z., Horn, P., Kövér, G., Berenyl, E., Radnai, I. and Biróné-Németh, E. (1992) In vivo measurement of the carcass traits of meat type rabbits by x-ray computerised tomography. *Journal of Applied Rabbit Research*, **15**, 799–809.

Szendrö, Z., Metzger, S., Romvári, R., Szabó, A., Locsmándi, L., Petrási, Z., Nagy, I., Nagy, Z., Biró-Németh, E., Radnai, I., Matics, Z. and Horn, P. (2008) Effect of divergent selection based on CT measured hind leg muscle volume on productive and carcass traits of rabbits. *Proceedings of the 9th World Rabbit Congress*, Verona, Italy, 249–53.

Teixeira, A. (2009) Basic composition: Rapid methodologies. In Nollet, L. M. L. and Toldra, F. (eds), *Handbook of Muscle Foods Analysis*. Boca Raton,FL: CRC, Taylor and Francis Group, 291–314.

Teixeira, A., Matos, S., Rodrigues, S., Delfa, R. and Cadavez, V. (2006) In vivo estimation of lamb carcass composition by real-time ultrasonography. *Meat Science*, **74**, 289–95.

Teixeira, A., Joy, M. and Delfa, R. (2008) In vivo estimation of goat carcass composition and body fat partition by real-time ultrasonography. *Journal of Animal Science*, **86**, 2369–76.

Temple, R. S., Stonaker, H. H., Howry, D., Posakony, G. and Hazaleus, M. H. (1956) Ultrasonic and conductivity methods for estimating fat thickness in live cattle. *Proceedings, Western Section, American Society of Animal Science*, **7**, 477–81.

Terry, C. A., Savell, J. W., Recio, H. A. and Cross, H. R. (1989) Using ultrasound technology to predict pork carcass composition. *Journal of Animal Science*, **67**, 1279–2884.

Thériault, M., Pomar, C. and Castonguay, F. (2009) Accuracy of real-time ultrasound measurements of total tissue, fat, and muscle depths at different measuring sites in lamb. *Journal of Animal Science*, **87**, 1801–13.

Thompson, J. M. (2004) The effects of marbling on flavour and juiciness scores of cooked beef, after adjusting to a constant tenderness. *Australian Journal of Experimental Agriculture*, **44**, 645–52.

Thompson, M. R. (2010) The future of portable ultrasound: Business strategies for survival. MSc Thesis of Science in engineering and management, Massachusetts Institute of Technology.

Thwaites, C. J. (1984) Ultrasonic estimation of carcass composition – a review. *Australian Meat Research Committee*, **47**, 1–32.

Trenkle, A. H. and Williams, J. C. (1997) Potential value of ultrasound to sort feeder cattle into more uniform groups for finishing and marketing, *Beef Research Report AS Leaflet R1432*, Iowa State University.

Tume, R. K. (2004) The effects of environmental factors on fatty acid composition and the assessment of marbling in beef cattle: A review. *Australian Journal of Experimental Agriculture*, **44**, 663–8.

Turner, R. C., Fuierer, P. A., Newnham, R. E. and Shrout, T. R. (1994) Materials for high temperature acoustic and vibration sensors: A review. *Applied Acoustic*, **41**, 299–324.

Veliyulin, E., Zwaag, C., Burk, W. and Erikson, U. (2005) In vivo determination of fat content in Atlantic salmon (salmo salar) with a mobile NMR spectrometer. *Journal of the Science of Food and Agriculture*, **85**, 1299–304.

Wall, P. B., Rouse, G. H., Wilson, D. E., Tait Jr, R. G. and Busby, W. D. (2004) Use of ultrasound to predict body composition changes in steers at 100 and 65 days before slaughter. *Journal of Animal Science*, **82**, 1621–9.

Weinstein, S. P., Conant, E. F. and Chadra, S. (2006) Technical advances in breast ultrasound imaging seminars in ultrasound imaging. *Seminars in Ultrasound CT and MRI*, **27**, 273–83.

Wells, P. N. T. (1991) The description of animal form and function. *Livestock Production Science*, **27**, 19–33.

Wells, P. N. T. (2000) Current status and future technical advances of ultrasonic imaging. *IEEE Engineering in Medicine and Biology*, **19**, 14–20.

Wells, P. N. T. (2006) Ultrasound imaging. *Physics in Medicine and Biology*, **51**, R83–R98.

Whitsett, C. M. (2009) Ultrasound imaging and advances in system features. *Ultrasound Clinics*, **4**, 391–401.

Whittaker, A. D., Park, B., Thane, B. R., Miller, R. K. and Savell, J. W. (1992) Principles of ultrasound and measurement of intramuscular fat. *Journal of Animal Science*, **70**, 942–52.

Wild, J. J. (1950) The use of ultrasonic pulses for the measurement of biologic tissues and the detection of tissue density changes. *Surgery*, **27**, 183–8.

Wild, J. J. and Reid, J. M. (1952) Application of echo-ranging techniques to the determination of structure of biological tissues. *Science*, **115**, 226–30.

Williams, A. R. (2002) Ultrasound applications in beef cattle carcass research and management. *Journal of Animal Science*, **80** (E-Suppl 2), E183–E188.

Williams, R. E. and Trenkle, A. (1997) Sorting feedlot steers using ultrasound estimates of backfat at the 12th and 13th rib prior to the finishing phase. *Journal of Animal Science*, **75**(Suppl 1), 55.

Wilson, D. E. (1992) Application of ultrasound for genetic improvement. *Journal of Animal Science*, **70**, 973–83.

Wold, J. P. and Isaksson, T. (1997) Nondestructive determination of fat and moisture in whole Atlantic salmon by near-infrared diffuse spectroscopy. *Journal of Food Science*, **62**, 734–6.

Woo, J. D. (2006) A short history of the development of ultrasound in obstetrics and gynecology. Available from: http://www.ob-ultrasound.net/history1.html (accessed 6 March 2011).

Wood, J. D. (1990) Consequences for meat quality of reducing carcass fatness. In Wood, J. D. and Fisher, A. V. (eds), *Reducing Fat in Meat Animals*. London: Elsevier, 344–97.

Wood, J. D. (1995) The influence of carcass composition on meat quality. In Jones, S. M. (ed.), *Quality and Grading of Carcasses of Meat Animals*, Boca Raton, FL: CRC Press, 131–55.

Wood, J. V., Enser, M., Fisher, A. V., Nute, G. R., Sheard, P. R., Richardson, R. I., Hughes, S. I. and Whittington, F. M. (2008) Fat deposition, fatty acid composition and meat quality: A review. *Meat Science*, **78**, 343–58.

Young, M. J. and Deaker, J. M. (1994) Ultrasound measurements predict estimated adipose and muscle weights better than carcass measurements. *Proceedings of the New Zealand Society of Animal Production*, **54**, 215–17.

Young, M. J., Deaker, J. M. and Logan, C. M. (1992) Factors affecting repeatability of tissue depth determination by real-time ultrasound in sheep. *Proceedings of the New Zealand Society of Animal Production*, **52**, 37–9.

Zerehdaran, S., Vereijken, A. L. J., van Arendonk, J. A. M., Bovenhuis, H. and van der Waaij, E. H. (2005) Broiler breeding strategies using indirect carcass measurements. *Poultry Science*, **84**, 1214–21.

12

Computer vision in the poultry industry

K. Chao, Henry A. Wallace Beltsville Agricultural Research Center, USA, B. Park, Richard B. Russell Research Center, USA and M. S. Kim, Henry A. Wallace Beltsville Agricultural Research Center, USA

Abstract: Computer vision is becoming increasingly important in the poultry industry due to increasing use and speed of automation in processing operations. Research to develop the technology has included multiple approaches including spectroscopy and spectral imaging. Case studies are presented that focus on recent research targeting line-scan imaging for two food safety applications: automated wholesomeness inspection and fecal contamination detection for chicken carcasses.

Key words: food safety inspection, chicken processing, hyperspectral imaging, multispectral imaging.

12.1 Introduction

Computer vision has become a critical part of automated high-speed processing in the food industry. To satisfy increasing consumer demand for poultry products, the chicken industry depends on automation to effectively process high product volumes for a variety of markets. More recently, research has sought the development of automated computer vision technologies that can effectively address food safety concerns to facilitate or complement the work performed by human inspectors, utilizing and combining the benefits of spectroscopic analysis and image processing to develop computer vision systems for food safety inspection. Inspection of birds for wholesomeness and fecal contamination are two important areas currently included in the growing list of responsibilities performed at processing plants by food safety inspectors. To help address these concerns, research scientists of the US Department of Agriculture have been working over the past 15 years to develop spectroscopy and spectral imaging technologies suitable for high-speed chicken processing lines. The work has led to rapid line-scan imaging methods of hyperspectral/multispectral inspection that is now a feasible technology for automated online safety inspection of chickens. This chapter presents a brief discussion of automation currently used in poultry processing operations, early research to develop spectroscopic inspection methods, and the development of imaging inspection methods from whole-target filter-based and common-aperture systems to rapid spectral line-scan imaging systems. Two case studies are presented regarding online wholesomeness inspection and fecal contamination detection for freshly slaughter chickens. Future trends for spectral imaging and automation for food safety inspection are also discussed.

12.2 Poultry processing applications

Over the past 20 years, automation has become an increasingly important part of poultry processing operations. Automated kill lines and evisceration lines have advanced to very high speeds, with some plants operating as quickly as 200 birds per minute (bpm) in Europe and South America. In addition to the automation of basic functions, such as head and foot removal, defeathering and scalding, secondary operations have also benefited greatly from mechanical automation, such as portioning of whole birds into parts and deboning to recover what would otherwise be wasted meat for processed products.

In particular, the poultry industry is looking to incorporate further automation in the form of computer vision to improve many processing operations. Computer vision technologies currently available to industry focus on specific quality attributes. While a single grade evaluation for a whole carcass has typically been used to make a decision for the use of an individual bird, computer vision systems can now further specify grades for individual parts – for example, from a single bird, one leg given an A grade can be cut and packed for a tray of leg parts, while the other leg given a B grade can be left intact with thigh and drumstick together for deboning at a point further down the processing line. Similarly, individual wings can be evaluated for damaged wingtips – by handling damaged and intact wings differently, processors can maximize the added value of their final products by customizing the cut parts to best use. Computer visions systems can also estimate product weight based on images of surface area for applications.

Further advances in computer vision technologies are still sought to improve to the quality assessment and cutting/deboning operations. In addition to the consideration of the quality of the product output at the end of the line, there is also potential for bird data from online computer vision systems to be used for real-time operation monitoring or feedback to operation parameters. For example, a vision system developed by Georgia Tech Research Institute (Usher *et al.*, 2005) has been under long-term testing for identifying safety and quality defects. The system records bird defects such as cadavers, overscalding, bruises, skin tears and broken wings. Some of this defect information can be used as system feedback to adjust processing line operations, such as unusual line speeds, excessively high scalding tank temperatures and dull or improperly positioned neck cuts that produce inadequate bleeding. Bird information can also be used as feedback to guide adjustments in production practices for the growers that supply flocks to the processor.

Of significant note is the fact that there are not yet any automated computer visions systems currently in use that effectively address food safety inspection applications. In the USA, the 1957 Poultry Product Inspection Act (PPIA) mandates postmortem inspection of every bird carcass processed by a commercial facility for human consumption. The US Department of Agriculture employs inspectors from its Food Safety and Inspection Service (FSIS) agency to conduct on-site organoleptic inspection of all chickens processed in US poultry plants for indications of disease or defect conditions. The human inspectors visually and manually inspect the body and viscera of birds on the processing lines, to identify any unwholesome carcasses – these may include conditions such as septicemia/toxemia (septox), airsacculitis, ascites, cadaver and inflammatory process (IP), and defects such as bruises, tumors, sores and scabs. By law, the human inspectors may work at a maximum speed of 35 bpm, which results in multiple inspection stations along a single line. In this way, for example, one inspector examines every fourth chicken on a 140 bpm line and the line is equipped with four inspection stations. Increasing poultry consumption places pressure on the industry to increase output, but human inspection capability is becoming the limiting factor to the maximum achievable production throughputs.

With its 1996 final rule on Pathogen Reduction and Hazard Analysis and Critical Control Point (HACCP) systems (USDA, 1996), the FSIS implemented the HACCP and Pathogen Reduction programs in meat and poultry processing plants throughout the country to prevent food safety hazards. Instead of focusing only on product safety inspection, HACCP systems focus on prevention and monitoring of potential hazards at critical control points (CCPs) throughout a food production process. FSIS has also tested the HACCP-Based Inspection Models Project (HIMP) in a few volunteer plants (USDA, 1997). In this project, food safety performance standards are set by FSIS while the processing plants hold primary responsibility for conducting inspections and processing so that their products satisfy FSIS standards; FSIS inspectors perform carcass verification along and at the end of the processing line, before the birds enter the final chill step. FSIS inspectors do not perform bird-by-bird inspection in these HIMP plants, which number 20 plants out of over 400 federally inspected plants nationwide. FSIS food safety standards for HIMP include two major zero-tolerance requirements, one for birds exhibiting septicemia/toxemia and one for the presence of fecal contamination. Computer vision systems that can address these standards are one key area for research for the development of science-based computer vision inspection technologies for the industry.

12.3   Development of spectral imaging for poultry inspection

Significant advances began in the early 1990s for the development of automated poultry inspection systems based on spectroscopy and spectral imaging. Focusing on high-speed methods suitable for use on commercial processing lines, early research by the USDA Agricultural Research Service sought to address issues of chicken wholesomeness inspection and fecal contamination detection by utilizing visible/near-infrared reflectance spectroscopy with two approaches: one was the selection of specific wavelength data to implement in the development of filter-based multispectral imaging methods, and the other was for direct inspection of poultry based on broad-spectrum measurements. The filter-based imaging approach led to the investigation of using multiple-camera systems and then common-aperture camera systems, with all these systems using interference filters to narrow the spectral wavebands used for analysis, but the approach was not successfully demonstrated beyond the laboratory environment due to speed and image quality problems. The approach of directly using broad-spectrum measurements on bird carcasses eventually yielded a visible/near-infrared spectroscopy-based prototype system that successfully differentiated wholesome and unwholesome birds during in-plant trials at a commercial poultry plant through the use of a neural-network classification model.

In the late 1990s, hyperspectral imaging analysis gained a foothold as an analysis method for wavelength selection for spectral analysis of food products – due to its capacity to discern spatial features and spectral features of target samples, significant advantages could be gained compared with spectroscopy-only methods. Hyperspectral imaging was adapted from other research fields, such as remote geographical sensing where the sensing module (camera in an aircraft) traveled in straight-line directions to pass over the target area being imaged. Adapted to food products such as chickens, hyperspectral imaging of food samples was initially implemented by use of multiple filters or a tunable filter system to allow a fixed-position camera to acquire images at multiple wavebands of a fixed-position sample. USDA Agricultural Research Service scientists pursued 'line-scan' hyperspectral imaging methods, which used a spectrograph rather than a filter-wheel or tunable filter system, allowing for a fixed-position camera to acquire line-images of a sample as the sample was moved across the linear field of view (FOV). The spectrograph produced a full-spectrum measurement for every pixel along a single line-image, and thus a series of such images could be compiled to produce a three-dimensional (3-D) 'hypercube' image, which could be considered as a two-dimensional (2-D) spatial image for which every pixel contains an additional data dimension consisting of broad-spectrum data points.

Implemented in laboratory bench-top systems, hyperspectral line-scan imaging was effective as an alternative method to spectroscopy-based analysis for analyzing food samples in which both spatial features and constituent composition were of interest. Initially only used for laboratory acquisition of data to be analyzed for selection of wavelengths that could be implemented in real-time multispectral imaging applications such as those performed using multiple-camera or common-aperture camera systems, the available equipment for line-scan hyperspectral imaging soon evolved to become capable of more rapid data acquisition and analysis. In addition, the line-scan format readily lent itself to imaging of products normally handled on linear processing lines such as conveyor belts and shackle lines. Researchers began developing multispectral line-scan algorithms for real-time imaging and analysis, using hyperspectral imaging equipment that allowed for flexible and adjustable selection of wavebands to customize the spectral data and minimize unhelpful spectral redundancy. For poultry inspection, real-time hyperspectral/multispectral imaging inspection methods have now been successfully demonstrated during in-plant testing for chicken wholesomeness inspection and for fecal contamination detection. The following is a brief review of the research that has led to these current developments in computer vision for poultry inspection.

12.3.1   Spectroscopy

Visible/near-infrared (Vis/NIR) light reflected from chicken carcass carries information about the color, surface texture and chemical constituents of chicken skin and muscle tissue. Unwholesome chicken carcasses often exhibit changes in skin and tissue that can be detected with Vis/NIR reflectance techniques. Chen and Massie (1993) developed a laboratory photodiode-array spectrophotometer system to measure visible/near-infrared spectra of stationary chicken samples in the 471–964 nm spectral region. A bifurcated fiber-optic assembly provided sample illumination from quartz-tungsten halogen lamps and collected reflectance measurements with the probe end positioned at a 2 cm distance from the surface of the chicken sample. Spectra were measured for wholesome carcasses and septicemic/cadaver carcasses, with an acquisition time of 2 s for each measurement. Analysis based on principal component analysis (PCA) achieved classification accuracies of 93.3% and 96.2% for the chicken samples in the wholesome and the septicemic/cadaver classes, respectively.

Based on the laboratory work, Chen et al. (1995) developed a transportable pilot-scale Vis/NIR system and later acquired Vis/NIR reflectance spectra for chickens on a 70 bpm commercial evisceration line. Acquired over 0.32 s, each spectral measurement targeted an area approximately 10 cm^2 across the breast area of one bird as it passed the measurement end of the fiber-optic probe and consisted of 1024 raw data points. Reflectance measurements acquired for wholesome and unwholesome carcasses were identified by a veterinary medical officer who observed the birds on the line as they approached the system. Offline spectral analysis included running-mean smoothing and second-difference calculations and spectrum reduction to produce 190-point processed spectra spanning 486–941 nm that were then subjected to PCA. The PCA scores of the first 50 principal components were input to a feed-forward-back-propagation neural network of 50 input nodes, seven hidden nodes and two output nodes. The output node determined the classification of birds as either wholesome or unwholesome, and the offline classification achieved an average accuracy of 95% for the 1174 wholesome and 576 unwholesome bird spectra that were collected (Chen et al., 2000).

Windham *et al.* (2003) investigated spectroscopy-based detection of fecal contaminates on chicken carcasses. Using a commercially available monochrometer, Vis/NIR spectra were measured for uncontaminated and fecal-contaminated chicken skin samples and then four dominant wavelengths at 434, 517, 565 and 628 nm were determined through multivariate analysis (partial least squares analysis) of the spectra. The spectral ratio using the 565 and 517 nm wavelengths was found to be 100% effective in detecting fecal-contaminated chickens when tested on spectral images of chickens raised on a corn/soybean meal diet.

12.3.2   Multispectral imaging

Filter-wheel multispectral imaging
Based on the work of Chen and Massie (1993), Park and Chen (1994) developed an intensified multispectral imaging system using a single camera and six optical filters (at 542, 570, 641, 700, 720 and 847 nm) and neural-network classifiers for discrimination of wholesome poultry carcasses from unwholesome carcasses that included septicemia/toxemia and cadaver carcasses. This system acquired six-waveband multispectral images of the stationary birds by using each of the six filters in turn for imaging. The six images for each bird were input to the neural network and the classification accuracy for separating wholesome carcasses from unwholesome carcasses was 89.3%.

Dual-camera imaging
Since the previous work determined the 542 and 700 nm wavelengths to be significant for effective separation of wholesome and unwholesome birds, a dual-camera imaging system using 20 nm band-pass filters centered at 540 and 700 nm was developed for testing on a laboratory pilot-scale processing line (Park and Chen, 2000). Side-by-side black/white cameras were each fitted with one of the two filters and images were acquired for chickens with the pilot-scale processing line operating at a speed of 60 bpm. Offline, the two-waveband image intensity data for each bird were input to a neural network, which resulted in classification accuracies of 93.3% for the septicemia carcasses and 95.8% for cadaver carcasses (Chao *et al.*, 2000). Image acquisition was also performed on a 70 bpm commercial chicken evisceration line. Offline analysis of the images of 13 132 wholesome and 1459 unwholesome chicken carcasses from the commercial processing plant resulted in classification accuracies of 94% and 87% for wholesome and unwholesome carcasses, respectively (Chao *et al.*, 2002). Imaging at 70 bpm showed the operation of the dual-camera system near its upper limits regarding speed. This system was not truly multispectral in that the two-waveband images acquired for any one bird were not truly matched, given the side-by-side position of the two cameras; further increases in speed significantly increased image registration problems since the two images were not truly matched.

Common-aperture multispectral imaging
A common-aperture camera appears to be a single camera but allows for the broadband light entering the one objective lens to be separated via internal prism

assemblies into multiple optical channels that are directed to separate image detectors, allowing for spatially matched images at different wavebands. Park *et al.* (2004) developed a common-aperture multispectral imaging system for fecal contamination detection on chickens, using a common-aperture camera with three optical channels and filters for wavelengths selections at 514, 566 and 631 nm. A total of 300 birds with fecal contamination spots placed on the breast areas were imaged while stationary. The multispectral image classification algorithm achieved 96.8% overall accuracy and also encountered some problems with false positive detections.

Using the same common-aperture camera, Yang *et al.* (2005) developed a multispectral imaging system for chicken wholesomeness inspection using interference filters centered at 462, 542 and 700 nm. Multispectral image data acquisition was performed for 174 wholesome, 75 inflammatory process and 170 septicemia chicken carcasses on a pilot-scale processing line operating at 70 bpm, using optical channel integration times of 5, 10, and 18 ms for the 462, 542 and 700 nm channels, respectively. Offline image processing algorithms based on PCA and a customized spatial region of interest (ROI) were used as inputs to a decision tree classification model that accurately classified 89.6% of wholesome, 94.4% of septicemia and 92.3% of inflammatory process chicken carcasses. Despite individually adjustable settings for gain and integration time for the three optical channels, obtaining high-quality images simultaneously across all three channels was difficult as a result of multiple problems, such as the combination of tungsten halogen illumination and wavelength-specific CCD detector sensitivities that produced a tendency towards image saturation at 700 nm and yet inadequate signal at 460 nm. The limited dynamic range available from the 8-bit resolution of each channel was also a limiting factor to the maximum possible classification accuracies. Integration times for imaging of moving targets were difficult to optimize, as attempts to shorten the integration times to accommodate rapid imaging often resulted in blurry images.

12.4   Case studies for online line-scan poultry safety inspection

Development of automated methods to address specific issues of wholesomeness and fecal contamination for freshly slaughtered chickens was established as a research priority for the USDA Agricultural Research Service in the 1990s. This resulted initially from the 1996 implementation of HACCP standards by the USDA FSIS in all meat and poultry processing plants nationwide, and became even more important as FSIS began testing its HIMP in a small number of volunteer plants. FSIS inspectors employed on-site at poultry processing plants were already inspecting carcasses for a variety of conditions and problems. The new HACCP standards required plant personnel to develop and perform procedures based on verification of CCPs set through the entire length of the processing operations, to satisfy performance standards; the responsibility of monitoring HACCP performance was placed on the FSIS inspectors. HIMP testing sought to reduce the workload of FSIS inspectors by eliminating the task of bird-by-bird inspection by inspectors: plant personnel were given the responsibility of performing inspection, and executing HACCP procedures, with FSIS inspectors present to verify the acceptable condition of birds output at the end of the line and to observe and monitor the processing line operations overall.

Agricultural Research Service (ARS) sought to address two particular food safety standards, one for wholesomeness and the other for fecal contamination, relevant to HIMP in chicken processing plants. Wholesomeness inspection here refers to the detection of birds exhibiting symptoms of systemic disease, specifically those of septicemia or toxemia. Such systemically diseased birds occur infrequently, but are prohibited by law from being sold for human consumption and must be disposed of by processors. Fecal contamination can occur not only from obvious sources – inadequate washing of the birds as they are scalded and defeathered on the kill line – but also from automated evisceration line operations which draw out the internal organs to allow for examination of the organs and the internal body cavity of each bird. Contamination and cross-contamination of birds by fecal matter is of concern because pathogenic microorganisms may be present in fecal material from the digestive tract, and thus FSIS enforces a zero-tolerance standard for visible contamination on carcasses.

In recent years, ARS has made significant progress in its research to develop automated inspection methods for addressing (1) wholesomeness inspection to detect systemically diseased birds and (2) fecal contamination on carcasses. The following sections present overviews of this work, which is based on line-scan spectral imaging technologies. A prototype system has been developed as a result of the work described in Case Study 1, with regulatory approval granted for use in pre-sorting of chicken carcasses on high-speed commercial kill lines. The same platform is being used in Case Study 2 to implement line-scan imaging for fecal contamination detection – a real-time imaging algorithm has been developed and the research now targets incorporation of both inspection tasks (wholesomeness and fecal contamination) into a single 'box' system that allows flexibility in online installation and configuration.

12.4.1   Case study: wholesomeness inspection

Overview of wholesomeness inspection
Hyperspectral reflectance line-scan images of chickens were acquired online from a commercial processing line operated at 140 bpm and then compiled and analysed offline in the laboratory to develop algorithms for automated online inspection of chickens. Analysis included determination of an appropriate target area (ROI) across the bird breast area, as well as selection of spectral wavebands for differentiating the birds based on both reflectance intensity and intensity ratio. The multispectral algorithm was developed and its performance was evaluated through in-plant online testing.

Line-scan imaging system

The spectral line-scan imaging platform used for this research is based on an electron-multiplying charge-coupled-device (EMCCD) camera and an imaging spectrograph with a linear FOV created by a slit in front of the spectrograph. Two pairs of high-power, broad-spectrum white light-emitting-diode (LED) line lights (LL6212, Advanced Illumination, Inc., Rochester, VT, USA) were used for illumination, providing continuous broadband illumination from 380 to 750 nm. For each line-scan image, the ImSpector V10 spectrograph (Spectral Imaging Ltd, Oulu, Finland) disperses a collimated light beam from each pixel of the scanned line in order to obtain a spectrum for each pixel. This produces a 2-D image of reflectance intensity for each scanned line, with spatial position along one axis (spatial dimension) and spectral waveband along the other (spectral dimension). The 512 × 512 pixel detector array of the PhotonMax 512b EMCCD camera (Princeton Instruments, Roper Scientific, Inc., Trenton, NJ, USA) is thermoelectrically cooled to an operating temperature of −70°C, allowing for high-speed low-light image acquisition, and was used with a 10 MHz 16-bit digitizer. The imaging control software for continuous acquisition of line-scan images was built on a LabVIEW 8.2 (National Instruments, Austin, TX, USA) platform. For online imaging, the system was positioned on the poultry line with a lens-to-shackle distance of 914 mm. Using a reference target panel (Spectralon, Labsphere, North Sutton, NH, USA), the line lights were adjusted to maximize the reflectance intensity measurable for the linear FOV, which spanned a vertical distance of 178 mm from the shackles on which the chickens were hung.

The 178 mm linear distance of the FOV was captured by 512 spatial pixels in each individual line-scan image, for a spatial resolution of 0.35 mm × 0.35 mm per pixel. Without any binning, the 512 × 512 detector array acquires a single hyperspectral line-scan image consisting of 512 spectral data points for each of the 512 spatial points, to produce a total image size of 512 × 512 pixels. By binning the spectral dimension by four – that is, accumulating every four pixels along the spectral dimension as one pixel – the total line-scan image size was reduced to 512 × 128 pixels and the signal-to-noise ratio of the data was improved. With the LED line lights illuminating a 99% diffuse reflectance target, it was observed that for the first 19 spectral channels and the last 54 spectral channels (out of 128 total), the reflectance intensities from the LED lights were too low to be useful. These 73 channels were omitted from data acquisition (via software controls) and all hyperspectral imaging was performed using only the remaining 55 spectral channels (512 × 55 pixels). Spectral calibration was performed using six spectral peaks from the emission of a neon-mercury pencil light: 436 and 546 nm from mercury, and 614, 640, 703 and 724 nm from neon. The 55 spectral channels of each hyperspectral line-scan image spanned from 389 to 744 nm, respectively, with an average spectral bandwidth of 6.6 nm.

Calibration of the imaging system was performed prior to each session of online imaging. A white 99% diffuse reflectance panel was used as a reference target, positioned at a distance matching the chicken-to-lens distance used during online imaging. The line lights were adjusted to maximize reflectance intensity across all pixels of the line-scan images. White (W) and dark (D) reference images were then acquired: five line-scan images were acquired using the illuminated reference target and averaged to calculate the reference image W; with the lights off and the camera lens covered, another five line-scan images were acquired and averaged to calculate the reference image D. These reference images were used to convert the raw reflectance line-scan images (I_0) to relative reflectance line-scan images (I), according to the following equation: $I = (I_0 - D)/(W - D)$. In this study, all hyperspectral and multispectral line-scan images, including the reference images, were acquired using a camera exposure time of 0.1 ms with an electron-multiplying gain of 45.

The spectral line-scan system was first used to perform continuous online hyperspectral line-scan imaging for data collection. Following data analysis and algorithm development, the same system was used to conduct real-time continuous multispectral inspection – that is, after the acquisition of each line-scan image, selected waveband data for that line-scan image were immediately analyzed to produce decision information before the next line-scan image was acquired, and after a complete set of line-scan decision information was acquired for a single bird, a final decision for the bird was calculated before the next bird on the processing line enters the FOV. The significant advantage presented by this spectral imaging system is its capacity to continuously operate either in hyperspectral or in multispectral mode, with the ability to select a subset of any of the available spectral channels through the camera control software. The short 0.1 ms exposure time was critical to the effective performance of the system for online imaging inspection of chicken carcasses on the rapidly moving processing line.

Development of bird detection algorithm
Hyperspectral line-scan images were acquired for a total of 5309 chickens (5260 wholesome, 49 systemically diseased) on a 140 bpm kill line at a commercial poultry processing plant. These images were acquired across several 8 h shifts. A black panel was hung behind the processing line shackles, opposite the imaging system, to provide a dark background to facilitate image segmentation – that is, separation of bird from background pixels during later image analysis – and system calibration was performed at the start of each shift. The system was turned on to perform continuous hyperspectral line-scan imaging while chicken carcasses passed through the linear FOV. Simultaneously, during the imaging sessions, a veterinarian was posted beside the imaging system to observe the same birds so that unwholesome birds – that is, those systemically diseased birds exhibiting septicemia/toxemia symptoms that, while infrequent in occurrence, must be rejected from the processing line – could be identified in the data set.

Figure 12.1 shows an example of whole-bird image at 620 nm of two chickens, one wholesome and one systemically diseased, compiled from hyperspectral line-scan images. Analysis of the hyperspectral data found that the greatest difference in relative reflectance between chicken pixels and background pixels occurred at the 620 nm waveband. Therefore, a threshold value of 0.1 at 620 nm was selected as a method to classify pixels as being either carcass pixels (intensity ≥ 0.1) or background (intensity < 0.1).

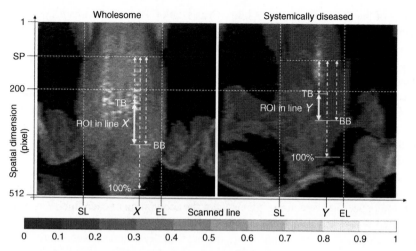

Fig. 12.1   Reference points defining the possible ROI areas for analysis: the SP, the SL, the EL, the TB and the BB.

A line-by-line algorithm to detect the entrance of a new bird in the FOV was developed based on the analysis of pixel intensities at 620 nm using the 0.1 threshold value. The bird detection algorithm examines only the upper 200 pixels in each line-scan image (512 pixels total). The uppermost 200 pixels correspond to the uppermost 69 mm of the FOV, where the legs would first appear for a properly positioned bird. This upper area of the line-scan image was termed the carcass detection length (CDL). If all CDL pixels show 620 nm intensity values below the 0.1 threshold, then the algorithm assumes that no bird is present in the FOV. The initial entry of a new bird is recognized when the 620 nm intensity increases above the 0.1 threshold for any single pixel within the 200 CDL pixels – that is, the detection of a non-background pixel. This method only examines the uppermost 200 pixels so as to disregard possible anomalies in the position of the wings. After the initial detection of a non-background pixel within the CDL, each subsequent line-scan image is monitored for the appearance of additional non-background pixels in the CDL. As a bird continues to move across the FOV, the bird's leg and body begin to fill in the CDL. Eventually, the algorithm finds a line-scan image containing only one remaining background pixel or several adjacent background pixels in the CDL, immediately followed by a line-scan image containing no background pixels. When this occurs, the spatial coordinate of the last background pixel is recorded, and the pixel located at this coordinate in the line-scan that contains no background pixels is identified as the starting point (SP) for the bird. This point is located along the leading edge of the chicken at the junction of the thigh and the side of the belly, as shown in Fig. 12.1. The line-scan image containing this SP pixel is identified as the starting line (SL) for the bird. The bird detection algorithm continues to monitor the pixels in subsequent line-scan images that are located at the same SP coordinate until those pixels indicate a turnover in 620 nm intensity from chicken (≥ 0.1) to background (< 0.1). This turnover indicates that the main body of the bird has passed through the FOV, and that the ending line (EL) of the bird has been found. The SP, SL and EL are marked on the two example bird images in Fig. 12.1.

Selection of ROI parameters and wavebands

To develop an algorithm for differentiating wholesome and systemically diseased birds, an effective area of analysis for each bird image needed to be determined. When considering the set of line-scan images that comprise a complete image of a single chicken, the SL and EL, respectively, as discussed above defined the peripheral limits to any selection for a ROI, eliminating the irregularities presented by the wing and leg features on the edges, for example. Analyzing all the pixels in every line-scan image between the SL and EL, however, would be undesirable due to the great variations in bird size and shape and the associated differences in shadow/edge effects and the number of background pixels present included within the line-scan images. An image area consisting only of chicken pixels selected from across the breast area would be most desirable; to accomplish this, it was necessary to define parameters that could select such an area customized to each chicken through the line-by-line imaging acquisition and analysis process even despite size and shape variations.

Beginning with the SP as a reference, it was determined that within each line-scan image, the pool of pixels available for selection can only include those chicken pixels located both below the SP coordinate and above any background pixels present (if any) at the lower end of the line-scan image. To define the area of possible ROIs, the relevant pixels in each line-scan image were counted, starting with 0% at the SP coordinate and proceeding down 100% at the lowermost chicken pixel (non-background pixel) as shown in Fig. 12.1. Possible ROIs to be evaluated were then defined by a top boundary (TB) and bottom boundary (BB) values which specify a percentage of the full 100% of pixels available for inclusion in an ROI. For example, for an ROI defined by TB-BB values of 30–70%, only the chicken pixels located between the 30% and 70% coordinates of the distance between the SP coordinate and the lowermost chicken pixel in each line-scan would be selected for analysis as ROI pixels.

Comparative analysis of potential ROIs defined by combinations of TB values at 10%, 20%, 30% and 40%, and BB values of 60%, 70%, 80% and 90%, was used to select ROI boundaries best suited for differentiating wholesome and systemically diseased chickens. For this analysis, the compiled hyperspectral line-scan images of 785 wholesome and nine systemically diseased chickens were used. For each TB/BB pairing, ROI pixels were extracted for the 785 wholesome bird images and the average spectrum for wholesome ROI pixels was calculated. Similarly, the average spectrum for systemically diseased ROI pixels was calculated using the nine systemically diseased chicken images. For each potential ROI – as defined by a TB/BB pairing – the difference spectrum was calculated between the average wholesome and average systemically diseased spectra. Figure 12.2 shows a comparison of the range of difference values (55 points per difference spectrum) calculated for each potential ROI. Overall, the 40–60% ROI boundary

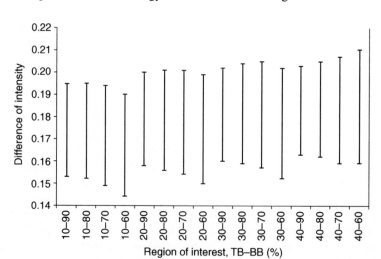

Fig. 12.2   The range of the 55-channel spectral differences between the average wholesome ROI pixel spectrum and the average systemically diseased ROI pixel spectrum for each of 16 possible ROIs.

limits produced greater spectral differences than the other ROI boundary combinations. Consequently, the 40–60% ROI boundary definition was the final selection for an ROI to use for developing a differentiation algorithm.

After selecting the ROI boundary parameters, 'key' wavelengths were selected for use in developing an algorithm to differentiate wholesome and systemically diseased birds. In Fig. 12.2, which shows for each potential ROI the range of difference values for the 55 spectral data points, the maximum difference value plotted for the 40–60% ROI is shown at 0.210. This value was the difference between the average wholesome ROI spectrum and average systemically diseased ROI spectrum at the 580 nm wavelength. For this reason, the 580 nm wavelength was selected as a key wavelength for differentiation based on relative reflectance intensity.

Figure 12.3 shows the average wholesome ROI spectrum and average systemically diseased ROI spectrum. Several wavelengths corresponding to the local maxima and minima exhibited by these spectra were evaluated for use in the differentiation algorithm as spectral ratios – that is, ratios of the relative intensities from two different wavelengths – in the following pairings: 435/455, 495/534 and 580/620. For each pairing, the ratio value was calculated for the average wholesome spectrum and for the average systemically diseased spectrum, and the difference between the two values was calculated. For the three pairings, 435/455, 495/534 and 580/620, the calculated ratio differences between wholesome and systemically diseased were 0.003, 0.039 and 0.121, respectively. The last pair clearly produced a much greater difference in ratio values; thus, the 580 and 620 nm wavebands were selected as key 'base' and 'peak' wavelengths to use in the differentiation algorithm.

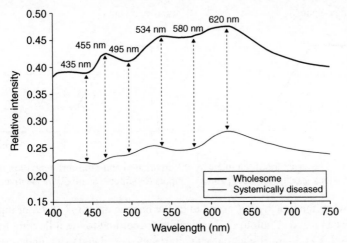

Fig. 12.3 The average spectrum for wholesome ROI pixels and average spectrum for unwholesome ROI pixels, for the 40–60% ROI.

Development of automated inspection algorithms

The multispectral differentiation algorithm was built around the line-by-line bird detection algorithm, and was developed to analyze ROI pixels for each bird by calculating an output value for every ROI pixel of a bird, and then averaging all the pixel outputs to obtain one final output to identify the condition of the whole bird. Using mapping functions to convert two input values (the pixel's intensity at 580 nm, I_A, and its ratio value, I_b/I_p) to two output values (intensity output, O_I, and ratio output, O_R) that are then averaged to produce one overall output value, O_{pixel}, each ROI pixel is classified as being either a wholesome pixel or a systemically diseased pixel. The accumulated values of O_{pixel}, for all ROI pixels are then averaged to calculate a final output value (O_f) that can be compared with a threshold value to classify the bird as being either wholesome or systemically diseased.

Figure 12.4 shows the two mapping functions used to calculate the O_I and O_R output values for each ROI pixel, where reference points a, b, c and d are defined as follows:

a = the average I_A for systemically diseased ROI pixels;
b = the difference between average I_A and its standard deviation for wholesome ROI pixels;
c = the average I_b/I_p for systemically diseased ROI pixels and
d = the difference between average I_b/I_p and its standard deviation for wholesome ROI pixels.

The intensity-based mapping function assigns the intensity output (O_I) to be 1 when the input $I_A \leq a$, and to be 0 when the input $I_A \geq b$. For input values between a and b, output O_I is calculated according to the following formula: $O_I = (b - I_A)/(b - a)$. Similarly, the ratio-based mapping function assigns the ratio output (O_R)

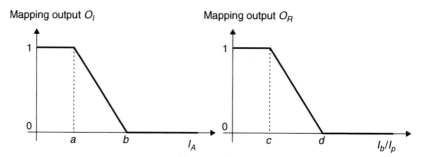

Fig. 12.4   The structure of the intensity-differentiation and ratio-differentiation mapping functions used to convert I_A and I_b/I_p inputs to outputs O_I and O_R, respectively.

to be 1 when the input $I_b/I_p \leq c$, and to be 0 when the input $I_b/I_p \geq d$. For input values between c and d, output O_R is calculated according to the following formula: $O_R = (d - I_b/I_p)/(d - c)$. Higher output values were correlated to a higher possibility of a pixel being indicative of systemic disease.

The multispectral differentiation algorithm begins for a single bird when the bird detection algorithm first detects the SP and the SL – that is, when the system recognizes that a bird is present and ready to be analyzed. Once the SL is detected, the ROI pixels within this line-scan image are immediately located. For each ROI pixel in this line-scan image, the I_A, I_p and I_b values are acquired, the I_b/I_p ratio is calculated, and the O_I and O_R values are immediately calculated and then averaged to produce one output value O_{pixel}, which is temporarily stored in memory as line-scan imaging continues. For each new line-scan image, this same process of immediately locating ROI pixels and calculating O_{pixel} for each pixel is repeated, until the bird detection algorithm detects the EL, which indicates that there are no more relevant line-scan images to be analyzed for the current bird. At this point, the algorithm has stored an O_{pixel} value for every ROI pixel for the bird, and now calculates the average O_{pixel} value for all the ROI pixels. This final averaged decision output O_f is compared with a pre-set threshold value T, to identify the bird as being either wholesome or systemically diseased: if $O_f \geq T$, then the bird is systemically diseased; if $O_f < T$, then the bird is wholesome. After the final classification of a bird as either systemically diseased or wholesome, the multispectral inspection algorithm begins again with the process of analyzing subsequent line-scan images to detect the SP of the next bird to be inspected.

Online evaluation
The performance of the spectral imaging system was evaluated based on continuous multispectral inspection of freshly slaughtered chickens on a 140 bpm kill line during two 8 h shifts in a commercial processing plant. Because immediate comparison between inspection system output and the plant's on-site FSIS inspector evaluations for individual birds was not feasible, the final counts of wholesome and systemically diseased birds produced by the imaging inspection system were compared with numbers from the FSIS tally sheets produced by the human inspectors who worked on the same processing line during those two 8 h shifts.

Additionally, several limited periods (30–40 min each) of inspection verification by direct comparison were performed by a veterinarian who observed individual wholesome and systemically diseased chickens on the line as they approached the inspection system FOV, to produce the data for verification of the inspection system's performance.

Due to logistical requirements determined by the plant management, the imaging inspection system was allowed to be set up on the line and turned on only after the processing line had already begun to move at the start of each shift, and the system was required to be disassembled and removed prior to final washdown of the processing line, which was immediately begun as the last bird passed through each stage on the processing line. Consequently, the system was tested during continuous periods of processing operation over most of the duration of each day shift, but was not able to inspect all the birds that were processed during the starting and ending hours of each shift.

During the first and second shifts, the inspection system identified 254 of 45456 chickens (0.56%) and 98 of 61020 chickens (0.16%), respectively, as being systemically diseased. FSIS tally sheets showed inspector counts of systemically diseased birds during those two shifts to be 84 of 53563 birds (0.16%) and 71 of 64971 birds (0.11%), respectively. The relative fractions of systemic birds identified by each side are similar, suggesting that the inspection system was operating in a consistent manner to identify systemically diseased birds. Figures 12.5 and 12.6 show the distribution of the O_f decision values for the first and second shifts, respectively, where all systemically diseased decision outputs are above the threshold value of 0.6. The bird detection and multispectral inspection algorithms detected and analyzed between 500 and 4000 ROI pixels per bird, depending on the size of the bird. Systemically diseased birds exhibited a general

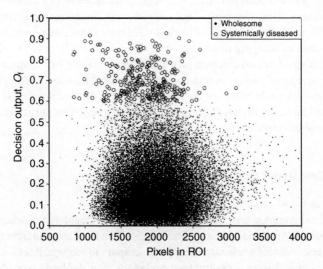

Fig. 12.5   The inspection system results for online multispectral line-scan inspection of chickens on the processing line during Shift 1.

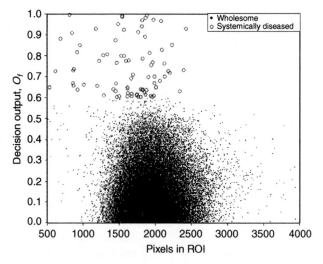

Fig. 12.6   The inspection system results for online multispectral line-scan inspection of chickens on the processing line during Shift 2.

tendency to be smaller than wholesome birds, but the distribution shows that the inspection system did not assume smaller birds to be systemically diseased, and that a sufficient number of ROI pixels were detected even for the smallest birds for effective inspection.

Added over all the short periods of direct system verification, the veterinarian observed a total of 43 878 birds as they approached the imaging inspection system. Of the 43 800 wholesome birds and 78 systemically diseased birds observed, the imaging system correctly identified 43 636 wholesome (99.6%) and 75 systemically diseased (96.1%) birds. These results, combined with the general performance of the system as considered against FSIS tally sheets, demonstrates that the spectral line-scan imaging inspection system can effectively classify wholesome and systemically diseased chickens on a 140 bpm processing line. Implemented for online pre-sorting of young chickens on high-speed kill lines in a HIMP processing plant, the system could allow for the small fraction of unwholesome and questionable birds to be easily diverted for rejection or reinspection by human personnel if desired for plant operations.

12.4.2   Case study: fecal contamination detection

Overview of fecal contamination detection
Chicken fecal samples were collected from freshly slaughtered chickens to provide contaminant material for experimental use. In a laboratory imaging environment, the contaminant material was applied in spots to the breast area of chicken carcasses obtained from a commercial processor, and the birds were hung on a pilot-scale poultry shackle line to simulate commercial line speeds needed to test

real-time multispectral line-scan image acquisition and analysis for fecal contamination detection.

Line-scan imaging system and software
The hyperspectral line-scan imaging system used for this research is based on an EMCCD camera and an imaging spectrograph with a linear FOV created by 40 μm slit between the object lens and the spectrograph. The object lens was a C-mount 1.4/23 mm lens (Schneider, Germany). Two pairs of high-power, broad-spectrum white LED line lights (LL6212, Advanced Illumination, Inc., Rochester, VT, USA) were used for illumination, providing continuous broadband illumination from spanning 380 to 750 nm. For each line-scan image, the Hyperspec-VNIR spectrograph (Headwall Photonics Inc., Fitchburg, MA) disperses a collimated light beam from each pixel of the scanned line in order to obtain a spectrum for each pixel. This produces a 2-D image of reflectance intensity for each scanned line, with spatial position along one axis (spatial dimension) and spectral waveband along the other (spectral dimension). The 1004 × 1002 pixel detector array of the Luca EMCCD camera (Andor Technology Inc., CT, USA) has an 8-μm pixel size and 13.5 MHz readout rate, and was used with a 14-bit digitizer board. Figure 12.7 shows the laboratory arrangement of the spectral line-scan imaging system on the 140 bpm pilot-scale shackle line.

The camera control and image acquisition/analysis software for the hyperspectral line-scan imaging system was developed in-house using a Microsoft Visual Basic (Version 6.0) platform. For spectral imaging, the spectral information of each pixel was collected in the vertical (traverse) direction of the EMCCD. For this study, multispectral images were acquired only at the 517 and 565 nm wavelengths rather than full-spectrum hyperspectral data, in order to test the

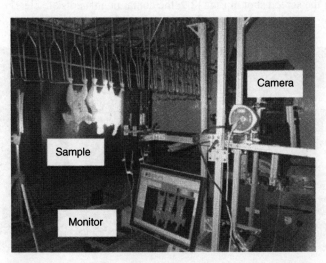

Fig. 12.7   Spectral line-scan imaging system set up on 140 bpm pilot-scale processing line.

ratio-based multispectral fecal detection algorithm developed during previous studies (Windham *et al.*, 2003).

Multispectral detection

A total of 50 chicken carcasses were used for testing the ratio-based real-time multispectral fecal detection algorithm. The contaminant fecal and ingesta material was applied to the breast area of the chickens, in eight spots per bird – four large and four small. The large and small spots each contained approximately 50 and 5 mg, respectively, of contaminant material.

Two wavebands, one centered at the 517 nm wavelength with approximately 11 nm bandwidth and the other centered at 565 nm with approximately 14 nm bandwidth, were used for multispectral imaging via random multitrack mode of the imaging system. No spatial binning was used for multispectral image acquisition, and the exposure time for each line-scan image was 0.5 ms. Real-time image processing was performed to calculate band ratio values using the ratio of these two wavebands and then the ratio value was compared with a threshold value of 1.42.

Figure 12.8 shows the images at 517 and 565 nm acquired by spectral line-scan imaging system for three representative birds; Fig. 12.9 shows the processed image produced during real-time imaging and analysis. The real-time multispectral inspection algorithm uses the relative reflectance values at 517 and 565 nm to remove the image background and to calculate a two-wavelength ratio value; the ratio value is compared with a threshold value in order to detect the contamination spots. Previous work using a different hyperspectral visible/near-infrared imaging system determined that this ratio method was effective with a threshold value of 1.05 for detecting contaminant spots (Park *et al.*, 2005). Using the equipment in this experiment, the ratio method was still effective, but, due to differences in illumination and calibration, the threshold for detection was adjusted to 1.42. As shown in the screen shot in Fig. 12.9, the contaminant spots are clearly identified by the image ratio algorithm with the new threshold value, although some false

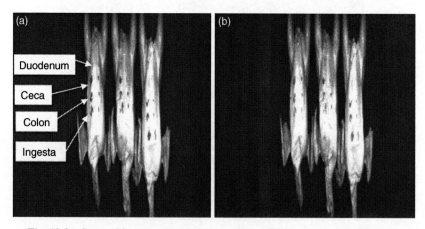

Fig. 12.8   Spectral images at (a) 517 nm and (b) 565 nm, captured by line-scan hyperspectral imaging system for three representative birds.

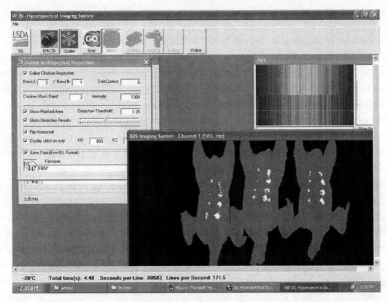

Fig. 12.9   Screen display of image processing results from spectral line-scan imaging system.

positives did occur due to residual feather cuticles. Acquired with the pilot-scale shackle line operating at 140 bpm (2.3 birds/s) and the imaging system operating at about 171 lines/s, the bird images consisted of approximately 74 lines per bird. These results demonstrate that the multispectral inspection method and the spectral line-scan imaging system in this study can be effectively used for detection of surface contamination spots for chickens on a 140 bpm processing line.

Further research concerns
This study demonstrated the effectiveness of the ratio-based multispectral imaging inspection method and the line-scan imaging system in a laboratory environment. Further research is needed to address concerns relevant to commercial implementation of the method, such as the occurrence of false positives due to residual feather cuticles. Previous and additional studies have found that the addition of a third wavelength – for example, 802 or 600 nm (Windham *et al.*, 2003; Park *et al.*, 2005) – can effectively remove false positives.

Additional issues will need to be addressed for truly feasible commercial implementation of the multispectral detection method for fecal contamination on chickens. For example, some processing plants operate kill lines at speeds as high as 180 birds per minute, a speed for which laboratory testing has not yet been demonstrated. The minimum size of contamination spots that must be detected also requires further study to determine the limits of the detection method, and detection will also need to address a means for implementation to be able to examine other areas such as chicken back, wings and vent areas.

12.5   Future trends

High-speed processing is essential to effective poultry processing operations. Growing awareness of food safety concerns has helped add food safety inspection to the list of tasks that automated computer vision can help address. Current research has demonstrated the feasibility of spectral line-scan imaging to detect unwholesome (systemically diseased) conditions and fecal contamination on high-speed bird processing lines. Successful commercialization of these technologies has great potential for benefits to multiple aspects of poultry processing.

The commercialization of online inspection methods will provide processors with an opportunity to streamline and reconfigure processing lines for a variety of purposes. For example, wholesomeness pre-sorting implemented early on the kill line will significantly reduce the presence of unwholesome birds downstream on the evisceration lines where the human inspectors are currently posted; the workload for bird-by-bird inspection can be reduced and inspectors can be reassigned for greater efficiency on tasks such as HACCP enforcement and end-of-line product verification. A pre-sorter can also be used to create a diverted line of 'questionable' birds requiring human eyes for closer examination, far fewer than requiring human inspection for every bird. For processors, the expense of 'paw harvesting' systems can be reduced by online pre-sorting as well. Current systems remove paws (chicken feet) from birds on the kill line, temporarily holding the feet in batches while the bodies proceed downstream toward inspection stations; whole batches of paws are discarded whenever an unwholesome bird is identified. Effective online fecal contamination detection will help prevent cross-contamination of processing equipment and also reduce the water volumes used for bird washing at specific CCPs.

Increased use of computer vision technologies in the future may also occur to improve processing and production management processes. For example, online inspection data can be recorded and tracked to better manage and adjust processing equipment, if damaged products are consistently detected during particular operation times, or to backtrack from high-frequency product problems to determine necessary adjustments to environmental controls in broilerhouse production and transportation practices before new birds are delivered to the processing plants.

12.6   Conclusions

Computer vision is becoming increasingly important in the poultry industry due to increasing use and speed of automation in processing operations. The implementation of computer vision technologies could greatly benefit poultry inspection applications, in particular by addressing some of the limitations of current inspection procedures that are performed by human inspectors. Research to develop suitable technology has included multiple approaches, including methodology based on spectroscopy and spectral imaging. The two case studies presented in this chapter focus on recent research targeting line-scan imaging for two food safety applications: automated wholesomeness inspection and fecal contamination detection for chicken carcasses. With further adaptation for additional inspection tasks, these line-scan imaging-based poultry inspection systems may prove even more useful to the poultry industry than their initial development has suggested.

12.7   References

Chao, K., Park, B., Chen, Y. R., Hruschka, W. R. and Wheaton, F. W. (2000) Design of a dual-camera system for poultry carcasses inspection. *Applied Engineering in Agriculture*, **16**, 581–7.

Chao, K., Chen, Y. R., Hruschka, W. R. and Gwozdz, F. B. (2002) On-line inspection of poultry carcasses by a dual-camera system. *Journal of Food Engineering*, **51**, 185–92.

Chen, Y. R. and Massie, D. R. (1993) Visible/near-infrared reflectance and interactance spectroscopy for detection of abnormal poultry carcasses. *Transactions of the ASAE*, **36**, 863–9.

Chen, Y. R., Huffman, R. W., Park, B. and Nguyen, M. (1995)A transportable spectropho-tometer system for on-line classification of poultry carcasses. *Applied Spectroscopy*, **50**, 910–16.

Chen, Y. R., Hruschka, W. R. and Early, H. (2000) On-line trials of a chicken carcass inspection system using visible/near-infrared reflectance. *Journal of Food Processing and Engineering*, **23**, 89–99.

Park, B. and Chen, Y. R. (1994) Intensified multi-spectral imaging system for poultry car-cass inspection. *Transactions of the ASAE*, **37**, 1983–8.

Park, B. and Chen, Y. R. (2000) Real-time dual-wavelength image processing for poultry safety inspection. *Journal of Food Processing and Engineering*, **23**, 329–51.

Park, B., Lawrence, K. C., Windham, W. R. and Smith, D. P. (2004) Multispectral imaging system for fecal and ingesta detection of poultry carcasses. *Journal of Food Processing and Engineering*,**27**, 311–27.

Park, B., Lawrence, K. C., Windham, W. R. and Smith, D. P. (2005) Detection of cecal contaminants in visceral cavity of broiler carcasses using hyperspectral imaging. *Transactions of the ASABE*, **21**, 627–35.

USDA (1996) Pathogen reduction: Hazard analysis and critical control point (HACCP) systems: final rule. *Federal Register*, **61**, 38805–989.

USDA (1997) HACCP-based inspection models project (HIMP): proposed rule. *Federal Register*, **62**, 31553–62.

Usher, C., Britton, D., Daley, W. and Stewart, J. (2005) Machine vision process monitoring on a poultry processing kill line: results from an implementation. *Proceedings of SPIE*, **5996**, 59960B1-8.

Windham, W. R., Smith, D. P., Park, B., Lawrence, K. C. and Feldner, P. W. (2003) Algorithm development with visible/near-infrared spectra for detection of poultry feces and ingesta. *Transactions of the ASAE*, **46**, 1733–8.

Yang, C. C., Chao, K. and Chen, Y. R. (2005) Development of multispectral imaging pro-cessing algorithms for food safety inspection on poultry carcasses. *Journal of Food Processing and Engineering*, **69**, 225–34.

13

Computer vision in the fish industry

J. R. Mathiassen, E. Misimi, S. O. Østvik and I. G. Aursand, Department of Processing Technology, SINTEF Fisheries and Aquaculture, Norway

Abstract: This chapter discusses the application of computer vision in the fish industry. Applications of computer vision are found in automated systems for sorting, grading and processing of fish and fish products. Computer vision is also used for understanding and optimization of practices related to fisheries, fish farming and fish processing. Based on the applications presented in this chapter, we outline the challenges and benefits related to the use of computer vision in the fish industry and point to some future trends.

Key words: automated sorting and grading of fish, automated fish processing, process understanding and optimization.

13.1   Introduction

In recent years, computer vision has been applied to several major areas of the fish industry. A large number of applications are in automation – including automated sorting, grading and processing of fish and fish products. Computer vision has also proven itself as a valuable tool in understanding and optimizing processes in the fish industry. This chapter presents some representative applications of computer vision in the fish industry, outlines the challenges related to the use of computer vision and points to some future trends. The fish industry has in recent years seen an increase in the number of computer vision applications, especially in large sectors in the fish industry, such as the Atlantic salmon industry.

Computer vision has proven itself capable of automated grading of whole Atlantic salmon according to their quality grade (Misimi *et al.*, 2006, 2008a). Automated weight estimation of Alaskan salmon (Balaban *et al.*, 2010b) and Alaskan pollock (Balaban *et al.*, 2010a) is possible with a high accuracy, thus enabling automation of the weight-based sorting of these species using 2-D computer vision. Weight estimation of whole herring has been done using 3-D computer vision (Mathiassen *et al.*, 2011a), and this research is ongoing – leading eventually to a replacement of manual labor in this high-throughput sector in the fish industry.

Automated processing is another major area where computer vision has benefited the fish industry. Robots and 3-D computer vision have been used in automatic slaughter lines for Atlantic salmon (Bondø *et al.*, 2011), substantially reducing the need for manual labor. Another successful application in processing is the use of computer vision and cutting manipulators for trimming of Atlantic salmon fillets, where several commercial solutions are available (Nordischer Maschinenbau Rud.Baader GmbH, Lübeck, Germany; Marel hf, Gardabaer Iceland). Pre-rigor removal of pin bones from Atlantic salmon fillets has been solved using computer vision (Thielemann *et al.*, 2007) and is now available in the commercial fish processing machinery (Trio Food Processing Machinery AS, Stavanger, Norway). De-heading of salmon (Buckingham and Davey, 1995; de Silva and Wickramarachchi, 1997; Buckingham *et al.*, 2001; Jain *et al.*, 2001) is another successful and established use of computer vision for fish processing.

Computer vision is used to understand or optimize processes, without necessarily automating the same processes. Process understanding and optimization involves shedding light on the effect of a change in the configuration of the process on the parameters of interest. Computer vision on magnetic resonance imaging (MRI) images has been used to understand and/or optimize many processes in the fish industry, including salting and desalting of cod (Erikson *et al.*, 2004; Foucat *et al.*, 2006; Gallart-Jornet *et al.*, 2007; Aursand *et al.*, 2010). An unusual and successful application of computer vision is in the computer-vision-assisted determination of the effect of fishing vessel water separator type on herring fillet quality (Aursand *et al.*, 2011). A more common application of computer vision is the determination of color attributes of fish muscle, as affected by irradiation (Yagiz *et al.*, 2010), high pressure and cooking (Yagiz *et al.*, 2009b).

Computer vision has, as demonstrated in the applications and discussion in this chapter, turned out to be a valuable technology for many and diverse applications in the fish industry. The general trend seems to point towards use of computer vision to increase cost-effectiveness through automation, and increase product quality through process understanding and optimization.

13.2   The need for computer vision in the fish industry

The fish industry is a global industry with a global market and competition arena. The development of the industry is forced by consumer trends driven through a continually concentrated retail industry. There are several dominant patterns of development that affect the retail industry in this respect – including the importance of food safety and the consumers' desire for healthy and easy-to-prepare meals. Another dominating trend in the food industry in general – that also applies to the fish industry – is the need to continuously offer new and innovative products in response to changing consumer habits and prefer-ences. Naturally, this requires great flexibility in the production process while maintaining both competitiveness and profitability – thus motivating the fish industry to move towards a high level of automation and high degree of process optimization.

Mechanized automation in the fish processing industry has a good track record, with a limited degree of adaptability to the fish or fish product being processed. The traditional Baader (Nordischer Maschinenbau Rud.Baader GmbH, Lübeck, Germany) filleting machines, introduced in the 1950s, are probably the best example of sophisticated machines that have had an enormous significance for the fish processing industry in the last century – substantially reducing the need for manual labor. Further advances in the automation of fish processing came in the early 1990s, with the introduction of the first 3-D measurement-based portion cutting machines using computer vision. Some years later, in the 2000s, the first computer-vision-based fillet-trimming machines were launched. The trend in the fish processing industry is that it wants to automate regular processes as much as possible, and computer vision is widely seen as an enabling technology for increased levels of advanced automation.

In the past decade, computer vision has been used in many applications to understand and optimize processes in fisheries, fish farming and fish processing. This recent development has its root in the need to increase cost-effectiveness while maintaining or increasing fish and fish product quality. Computer vision is a versatile tool in this respect – allowing an objective and high-speed measurement of many quality parameters in large data sets, and enabling rapid pre-process-ing and processing of large numbers of images. Computer vision is an enabling technology that, if properly applied, can address the need of the fish industry to implement more advanced automation and to understand and optimize processes throughout the value chain.

13.3   Automated sorting and grading

Grading and sorting are terms that are sometimes used interchangeably in the food industries (Hu *et al.*, 1998). Sorting usually involves more discrete categories of different object types, whereas grading may involve determining a continuous parameter of a single category of objects. Automated sorting and grading can also be defined as the physical separation process of moving objects to positions cor-responding to their respective categories. In this section we chose to define auto-mated sorting and grading as the process of automatically determining a discrete or continuous category or parameter of an object – in some cases also including physical separation of the objects based on their category or parameter. The fol-lowing subsections present the application of computer vision to quality grading of whole Atlantic salmon, weight estimation of whole herring, weight estimation of Alaskan pollock and salmon, and grading of fillets with respect to water and fat content.

13.3.1   Quality grading of whole Atlantic salmon

In the current salmon processing plants, whole Atlantic salmon is manually graded and sorted into the categories *superior*, *ordinary* and *production*. Superior salmon are free from significant defects and blemishes and ordinary salmon have minor defects and blemishes. With both the superior and ordinary grades, the entire fish is suitable for human consumption. In the case of production grade fish, there are substantial blemishes or defects, and limitations on the products that can be produced from the fish. In a typical salmon processing plant, between 90% and 97% of all salmon are superior grade and the remaining fraction is either ordinary or production grade. The external quality grading and sorting of whole salmon requires the manual labor of between two and four workers. Manual sorting is prone to error and relatively costly, and it is therefore desirable to automate this sorting.

A majority of the production grade fish have shape defects that distinguish them from superior and ordinary grades. Figure 13.1 illustrates the differences between production grade and the other two grades. Trained inspectors have attributed the distinguishing features of production grade to 'humpback' and 'short tail' (Misimi *et al.*, 2006). Based on the knowledge of trained inspectors, a set of four geometrical features was selected for discriminating this type of production grade fish from the two other grades. These four features were derived from the parameters illustrated on the silhouette image in Fig. 13.2, and were chosen for their ability to quantify the 'humpback' and 'short tail' properties.

The superior and ordinary grades can be discriminated using shape-based geometrical features (Misimi *et al.*, 2008a), although the differences between these two grades are more subtle. Trained inspectors attribute the distinguishing features of ordinary grade to slightly thicker and shorter tails and a more asymmetrical shape. The features used to discriminate superior from ordinary were extracted from geometrical parameters similar to those in Fig. 13.2, with the addition of more parameters in the posterior region.

Both above-mentioned references use a combination of a two-class linear discriminant analysis (LDA) classifier and a leave-one-out cross-validation (LOOCV) to evaluate the classifier performance. Two-class LDA is a simple and powerful classifier that reduces the dimensionality of multidimensional feature space to a single dimension wherein the two classes are optimally separated (Theodoridis and Koutroumbas, 1999). It is assumed that the two classes are normally distributed with mean vector μ_1 and covariance matrix $\sum_1$ for class 1, and similarly μ_2 and $\sum_2$ for class 2. The linear transformation of a feature vector x, to the single dimension that optimally separates the two classes, is

$$y(x) = w^T x,$$

where

$$w = (\Sigma_1 + \Sigma_2)^{-1}(\mu_2 - \mu_1).$$

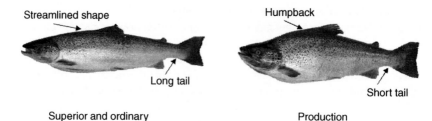

Fig. 13.1   Illustration of shape differences between superior/ordinary salmon and production salmon.

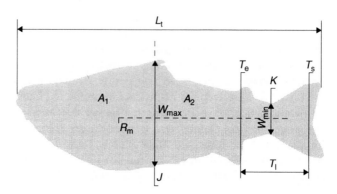

Fig. 13.2   Shape parameters for feature extraction.

The two-class classifier thus becomes a simple decision criterion on whether $y(x) < c$ for some threshold c. If the criterion is true, x belongs to class 2, otherwise it belongs to class 1. The threshold c can be varied to adjust the tradeoff between detection rate and false alarm rate.

The LOOCV method (Lachenbruch and Mickey, 1968; Theodoridis and Koutroumbas, 1999) is a method that can be used for evaluating the performance of a classifier. LOOCV provides an almost unbiased estimate of the expected classification error of a classifier, and is especially useful for evaluating classifiers that are susceptible to over-training due to a high dimensionality of the feature space. If a data set has n samples, the LOOCV trains the classifier on $n - 1$ samples and evaluates the classifier on the excluded sample. This is repeated n times, each time excluding a different sample.

Using two-class LDA classifiers and LOOCV to evaluate the classifier performance, the results of Misimi *et al.* (2006, 2008a) can be combined and summarized in the confusion matrix in Table 13.1. The superior grade salmon can be detected with a probability of 87%, and ordinary and production with a probability of 79% and 78%, respectively. The referenced research did not include a tradeoff analysis between the detection rate and false alarm rate. Despite this, the research shows that computer vision has the potential to reliably perform automated quality grading of whole Atlantic salmon.

Table 13.1 Confusion matrix for classifying salmon as superior, ordinary and production grades

	Estimated grade		
True grade	Superior	Ordinary	Production
Superior	87%	6%	7%
Ordinary	14%	79%	
Production	22%		78%

13.3.2   Weight estimation of whole herring

Weight is an important parameter by which the price of whole herring (*Clupeaharengus*) is determined. The price per kg of whole herring is determined by the individual round (ungutted) weight of the herring. The pricing is based on grouping the herring into a small number of standardized weight categories (e.g., from 125–99 g and 200–99 g). A weight estimation error of only a few grams on the category borders can result in an increase in price of up to 25% (Norges Sildesalgslag, 2011). Today's pelagic processing plants use mechanical increasing-gap graders, with manual post-sorting to correct major errors made by the mechanical graders. The throughput in a typical pelagic processing plant is 50 tons/h – equivalent to 200 000 fish per hour. Despite their capability for such a high throughput, the mechanical graders have some inaccuracies and the manual post-sorting of such large amounts of fish makes sorting errors and worker fatigue a potential problem. It is therefore desirable to explore alternative methods for high-speed weight estimation and sorting of whole herring.

Mathiassen *et al.* (2011a) describe a 3-D machine vision-based system for high-speed weight estimation of whole herring. Herring were singulated and imaged using a 3-D laser triangulation system mounted above a conveyor belt moving at 1000 mm/s. A 3-D image of a herring imaged with that system is shown in Fig. 13.3. Weight estimation was done using several 2-D and 3-D feature sets, and the performance of the weight estimators was evaluated using LOOCV.

The weight estimation performance for several feature sets is summarized in Table 13.2 (see Mathiassen *et al.*, 2011a for details). The root-mean-square error of cross-validation (RMSECV) is of the weight estimator that can be computed as

$$\text{RMSECV} = \sqrt{\frac{1}{n} \sum_{i=1}^{n} (\hat{y}_i - y_i)^2},$$

where $\hat{y}_i$ is the estimated weight of sample i, based on training the estimator on $n-1$ samples excluding sample i, and y_i is the true weight. By using this approach to evaluating estimator performance, a realistic estimate is obtained of the likely performance of the weight estimator, assuming that the data set is representative of real-life conditions. The results in Table 13.2 show that a combination of 2-D

(a)

(b)

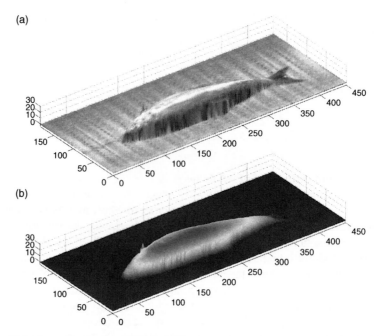

Fig. 13.3 Example of a geometrically calibrated intensity image (a) and 3-D thickness-map (b) obtained from a fresh herring. All axis units are in millimeters. Note the relaxed tail due to the freshness (no rigor mortis) of the fish. (Reprinted with permission from John Wiley and Sons.)

+ 3-D features gives the best weight estimation. Using only thickness gives relatively poor weight estimation. This latter result might explain the reason behind the sorting errors found in mechanical increasing-gap weight graders. These types of mechanical graders achieve selective grading of fish based primarily on their thickness. Figures 13.4 and 13.5 further illustrate the weight estimation performance. Using thickness alone results in a large spread, and 3-D projected volume has a similar RMSECV to using 2-D + 3-D features, but the 2-D + 3-D combination appears to significantly reduce the magnitude of the worst-case under- and overestimation of the weight. The results indicate that it is possible to undertake accurate high-speed weight estimation of whole herring using 3-D computer vision.

13.3.3   Other representative applications

Computer vision has been used for weight estimation of Alaskan pollock and several species of Alaskan salmon. Balaban *et al.* (2010a) estimated the weight of Alaskan pollock from the view area, using several linear and non-linear regression models. The most accurate estimation was obtained using a power fit with an exponent of approximately 1.5, with the resulting coefficient of determination being $R^2 = 0.993$. This power fit model has also been used to estimate the weight

Table 13.2 Weight prediction performance of different feature sets, evaluated using LOOCV

Feature set	RMSECV (g)	Largest negative deviation (g)	Largest positive deviation (g)	R^2
Thickness	20.6	−73.4	40.3	0.701
Area	10.3	−32.1	37.9	0.925
2D	8.1	−27.2	16.6	0.955
2D+	7.0	−23.0	16.4	0.967
3D	6.5	−20.5	19.8	0.970
2D + 3D	5.6	−13.9	14.0	0.980

Source: reprinted with permission from John Wiley and Sons.

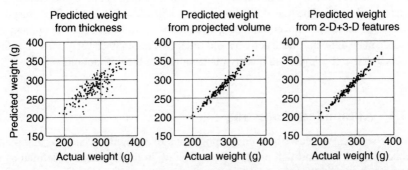

Fig. 13.4 Predicted weight using different feature sets, evaluated by LOOCV, of fresh herring. The feature sets are, from left to right: thickness, 3-D (consisting only of projected volume) and 2-D + 3-D. (Reprinted with permission from John Wiley and Sons.)

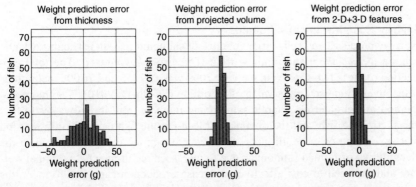

Fig. 13.5 Weight prediction error distribution, evaluated by LOOCV, of fresh herring. The feature sets are, from left to right: thickness, 3-D (consisting only of projected volume) and 2-D + 3-D. (Reprinted with permission from John Wiley and Sons.)

of several species of Alaskan salmon (Balaban *et al.*, 2010b), with an R^2 value of 0.987 for a data set consisting of four different species. These results show that computer vision can accurately estimate the weight of whole Alaskan pollock and salmon for sorting.

Multidimensional computer vision, based on multispectral imaging, and advanced statistical methods have been used to accurately determine the water and fat content of dried, salted and fresh fish fillets (ElMasry and Wold, 2008). Wold *et al.* (2006) used partial least squares (PLS) regression to estimate the water content in dried salted coalfish fillets from a high-speed near-infrared (NIR) imaging spectroscopy system in transflectance mode. The RMSECV of the estimator was assessed using LOOCV, and was 0.70%. These estimation errors were much smaller than those accomplished by manual graders, and the automated grading system meets the performance requirements for that application. A similar imaging setup and PLS regression algorithm was used to measure the water and fat content in fresh fillets of Atlantic halibut, catfish, Atlantic cod, mackerel, herring and saithe. The RMSECV for water and fat was 2.73% and 2.99%, respectively. This research shows that automatic estimation of water and fat content of fillets can be done non-destructively and reliably. This enables early sorting of the fillets and thereby improved quality management of the production.

13.4   Automated processing

In this section, we describe some applications of computer vision to automate processing of fish and fish products. Processing, in this context, is defined as a manipulation, handling or modification operation that involves interaction and modification of fish or fish products. The following subsections present the application of computer vision to slaughtering of salmonids, trimming of Atlantic salmon fillets, pre-rigor removal of pin bones from Atlantic salmon fillets, de-heading of fish and portioning of fish and fish products.

13.4.1   Slaughtering of salmonids

Norway produces and sells large volumes of salmonid fish, in particular Atlantic salmon (*Salmosalar*) and rainbow trout (*Oncorhynchus mykiss*). The total production of these two species in the year 2009 was 859 056 and 76 008 metric tons respectively, representing a value of $3 billion (Sandberg, 2010). The traditional method of slaughtering Atlantic salmon and rainbow trout in Norway is to first anesthetize the fish, and then manually bleed-cut the fish using a knife incision that severs one or more gill arcs. There is a strong motivation to automate the slaughtering of farmed salmonids, due to increased requirements for animal welfare and the need to cost-effectively increase production capacity. To meet this need, a new automated slaughter line, shown in Fig. 13.6, has been developed (Bondø *et al.*, 2011) and commercialized (Seaside AS, Stranda, Norway).The new slaughter line uses 3-D machine vision and bleed-cutting robots. Traditional

Fig. 13.6   A photograph showing the pilot installation of the slaughter line at Slakteriet AS. The in-feed is to the right and the operator station to the left of the image. (Reprinted with permission from Emerald Group.)

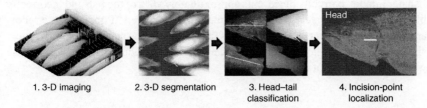

1. 3-D imaging   2. 3-D segmentation   3. Head–tail   4. Incision-point
                                                      classification   localization

Fig. 13.7   A flow chart showing the different steps in the 3-D machine vision system: (1) 3-D imaging, (2) 3-D segmentation, (3) head-tail classification and (4) incision point localization. (Reprinted with permission from Emerald Group.)

slaughter lines require four people to operate, whereas the new slaughter line reduces the number of operators to between one and two.

The fish are transported on a conveyor belt to an electro-stunner for anesthetization. Directly after being stunned, the fish are transferred to a second conveyor belt and imaged using a 3-D vision system, based on laser triangulation, placed above the conveyor belt. A flow chart of the 3-D machine vision system can be seen in Fig. 13.7. Step 1 shows the raw acquired 3-D image. This image is segmented into individual regions (corresponding to individual fish) in step 2. Based on the segmented 3-D image, the terminal ends of each region are detected and classified as being either a head or a tail in step 3. Detected heads are further processed in step 4 to compute the location of the incision point corresponding to the position of the gill arcs. The incision point coordinates are then sent to an array of four robots to perform the bleed-cutting operation that severs the gill arcs of the fish.

Fig. 13.8   Examples of high throughput (left image) and moving un-anesthetized fish (middle and right images). (Reprinted with permission from Emerald Group.)

Although 3-D machine vision based on laser triangulation is a well-known technology, the interpretation and analysis of 3-D images remains a challenge. Figure 13.8 shows some examples of 3-D images that are difficult to analyze. Partially overlapping fish and twitching fish are difficult to segment properly, and cause errors in the localization of the cutting position. These problems are exemplary of the types of challenges found in applying computer vision to the fish industry.

The automated salmonid slaughter line is capable of a throughput of up to 80 fish per minute, correctly bleed-cuts approximately 90% of the fish and incorrectly bleed-cuts approximately 0.5%. The remaining 9.5% are left uncut by the robots and bled-cut by the one or two operators that supervise the slaughter line. The new slaughter line as a whole reduces the need for manual labor, while increasing the welfare and quality of the fish.

13.4.2   Trimming of Atlantic salmon fillets

Atlantic salmon fillets are obtained by mechanical filleting, but, until recently, the fillets had to be trimmed manually to remove belly membrane, belly fat, tail fat, back fat, skin remnants and fin remnants. Computer vision has enabled this recent development of automated trimming of fillets, as exemplified in the Baader 988 trimming machine for salmon and sea trout (Nordischer Maschinenbau Rud.Baader GmbH, Lübeck, Germany) and the Marel ITM intelligent trimming machine (Marel hf, Gardabaer, Iceland). These trimming machines use 2-D computer vision to analyze the fillets, and to compute the trimming lines. The trimming line coordinates are sent to a system for cutting using knives oriented perpendicular to the conveyor belt. An alternative fillet-trimming system, based on 3-D computer vision, has been developed as a technology demonstrator for adaptive trimming of salmon fillets using a six-axis robot and a dual-use cutting tool (SINTEF Fisheries and Aquaculture AS, Trondheim, Norway). This latter trimming system, as an example of potential computer-vision-based solutions to automation of advanced processing tasks, will be described in more detail in this section.

The 3-D computer-vision-based trimming system is schematically illustrated in Fig. 13.9. The fillet is placed on a perforated pallet and held down by vacuum suction. A 3-D camera images the fillet in 3-D and in monochrome 2-D, and the

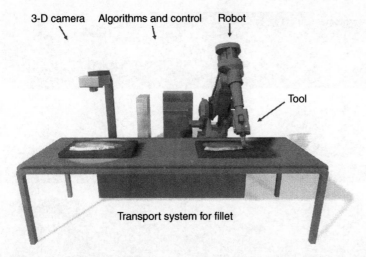

3-D camera   Algorithms and control   Robot

Tool

Transport system for fillet

Fig. 13.9   A simplified illustration of the proof-of-concept demonstrator for robotized automatic trimming of salmon fillets. The fillet is placed on a pallet and image with a 3-D camera (left), before being analyzed using machine vision algorithms. The machine vision system detects the trim objects and gives control signals to the robot (right), which uses a multi-purpose tool to remove the detected trim objects.

pallet is transported from the 3-D imaging station to the robotic trimming station. Based on the images, 3-D trimming paths are computed and sent to a six-axis for execution using a dual-use cutting tool. A more detailed description of the steps in the trimming system is shown in Fig. 13.10. In order to detect the trim objects (i.e., the objects, such as belly fat, to be removed in the trimming process) it was essential to acquire images that enhanced the visibility of the trim objects. The combined 3-D and monochrome 3-D images are seen in Fig. 13.11, together with regions of interest (ROIs) used to detect the trim objects. The monochrome image is obtained using blue LED illumination that enhances the image contrast between trim objects and fillet muscle.

The technology demonstrator was tested with 30 fillets of different sizes, and 100% accuracy was obtained for the detection of the trim objects, but with a cutting accuracy of only 80%. This shows the need for more research in adaptive robotic cutting of fish fillets using 3-D cutting paths. As a whole, the technology demonstrator illustrated how even difficult processing tasks can be automated. However, the cost-effectiveness, speed and accuracy of the approach were not yet on a par with that needed for an industrial solution. The Marel and Baader machines have demonstrated that automated trimming is possible in a cost-effective fashion, using a simplified approach as compared with a full six-axis robot and full 3-D cutting paths. The trimming machines and the technology demonstrator show that computer vision is already cost-effective for challenging fish processing tasks, and that there is future potential to perform such tasks more flexibly as soon as the research matures and the technology reaches a more cost-effective level.

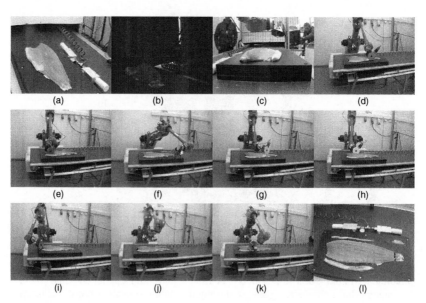

Fig. 13.10   An illustration of the entire automatic trimming process using the proof-of-concept demonstrator. (a) The fillet is placed on a perforated pallet with vacuum suction that keeps the fillet in place throughout the trimming. The pallet is mounted on the conveyor belt. (b) The pallet, with fillet, is conveyed into a closed camera housing and imaged using a 3-D camera. The camera acquires a blue-light monochrome image and a 3-D image. (c) The pallet, with fillet, is conveyed out of the closed housing, while the 3-D and monochrome images are being analyzed, and placed in the workspace of a robot. The robot has a multi-purpose oscillating knife tool. Once the analysis is complete, all trim objects will have been detected and located in the robot coordinate system, and 3-D cutting paths computed for each trim object. (d) and (e) The robot trims the belly fat off using the tool in 'vertical cut' mode. (f) The robot trims off the tail using the tool in 'vertical cut' mode. (g), (h) and (i) The robot trims the back fat using the tool in 'vertical cut' mode. (j) The robot changes tool mode to 'horizontal cut'. (k) The robot trims the belly membrane off using the tool in 'horizontal cut' mode. (l) The result after trimming. The belly membrane is seen in the upper-right corner of the pallet.

13.4.3   Pre-rigor removal of pin bones from Atlantic salmon fillet

Pin bones are relatively large bones that must be removed from Atlantic salmon fillets in order to produce boneless fillets. These bones adhere strongly to pre-rigor fresh fillets by means of tendons, and, therefore, it has been the norm to wait at least 3 days before filleting the fish, so as to allow the fillet to tenderize sufficiently to weaken the pin bone tendons and thus allow the pin bones to be mechanically pulled out. Although this process achieves boneless fillets, the downside is a reduction in freshness and shelf-life. This has motivated the development of an automated solution that allows pre-rigor removal of the pin bones.

A computer vision system was developed for successfully estimating the position of pin bones with a high accuracy (Thielemann *et al.*, 2007). Using this computer vision system, an automated line for pre-rigor pin bone removal has been

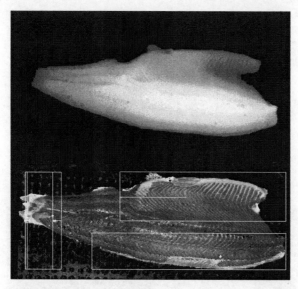

Fig. 13.11   The top image shows the 3-D image acquired using the z-Snapper 3-D camera. The bottom image shows the monochrome image acquired by the z-Snapper 3-D camera, using blue LED illumination. In this bottom image, the trim objects are detected and located in 3-D before sending cutting path coordinates to the robot. The left rectangle outlines the area of the fillet in which to search for tail fat, with the detected tail fat region delimited by the vertical lines. The upper rectangle outlines the area of the fillet in which to search for belly fat, with the horizontal line indicating the horizontal cutting path. The lower rectangle outlines the area of the fillet in which to search for the back fat.

commercialized (Trio Food Processing Machinery AS, Stavanger, Norway). This automated line is based on cutting the pin bone attachment tendons before pulling the pin bones out. In order to minimize waste, and maximize fillet quality, the cutting must be done precisely. Figure 13.12 shows a drawing of the pin bone removal machine, with inset photos illustrating tendon cutting and pin bone removal. Detecting the pin bones is a difficult computer vision problem, due to the pin bones ends often being partially or completely embedded in the flesh. Analysis of MRI images and color images, as seen in Fig. 13.13, showed that the tendon attachment position could be predicted from the location of the epaxial septum. The epaxial septum can be detected using a RANSAC/MLESAC algorithm with some prior constraints on the ROI (Thielemann *et al.*, 2007). The precision of the epaxial septum detection algorithm was within 3 mm for 78% of all fish.

Automated pre-rigor pin bone removal has enabled a great increase in freshness and shelf-life of Atlantic salmon fillets, by removing the previously required tenderizing time of at least 3 days. The success of automated pre-rigor pin bone removal can be attributed to computer vision being a cost-effective method for accurately locating the pin bone tendon attachments.

Fig. 13.12   Schematic of the Trio FDS 2000 PR-PB-4 pre-rigor pin-boning and skinning machine: (1) The machine has a vision system and cutting knife and (2) pin-boning and skinning mechanism. Copyright Trio Food Processing Machinery AS.

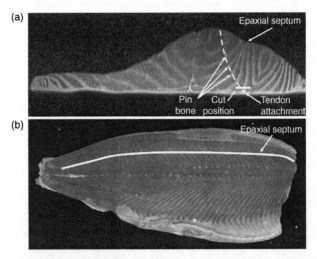

Fig. 13.13   Nomenclature for describing fish anatomy. (a) MRI-scan of a salmon fillet. The scan shows a single cut through the fillet, perpendicular to the length axis of the fish. The tendon attachment can be seen almost directly below the epaxial septum. White horizontal line indicates cut position. Stapled line indicates pin bone. (b) Photo of fillet seen from above. White line indicates the epaxial septum that is to be identified. (Adapted with permission from Thielemann et al., 2007.)

13.4.4   Other representative applications

Purely mechanical de-heading of salmon is known to be inefficient and wasteful. To increase the efficiency of the de-heading, a system was developed that uses computer vision to localize the collar bone position and adjust the de-header accordingly (de Silva and Wickramarachchi, 1997). Data fusion between computer vision and ultrasonic displacement sensors results in further improvements (Jain *et al.*, 2001). Other approaches to automated fish de-heading have also been investigated (Buckingham and Davey, 1995; Buckingham *et al.*, 2001).

Portioning of fish fillets or whole fish is a common application of 3-D computer vision to fish processing. Using analysis of 3-D images, it is possible to compute the slices that result in pieces with equal volume or weight. Several available industrial solutions are available for 3-D-based portion cutting, including the knife-based Marel I-Cut line of products (Marel hf, Gardabaer, Iceland) and the waterjet-based DSI Waterjet Portioning Systems (JBT FoodTech, IL, USA).

13.5   Process understanding and optimization

Computer vision is used to understand or optimize processes, without necessarily automating the same processes. Process understanding and optimization involves shedding light on the effect of a change in the configuration of the process on the parameters of interest. Computer vision is often used to partially or completely automate the measurement of these parameters, often for large data sets and with multiple process configurations. The following subsections present the application of computer vision to understanding the effects of first handling conditions on process parameters, choice of fishing vessel water separator type on the quality of herring fillets, and measuring the effect of irradiation, high-pressure processing and cooking on fillet appearance.

13.5.1   Effect of first handling conditions on optimal processing
parameters and end product quality

Magnetic resonance imaging (MRI) is a powerful imaging modality that can produce high-quality cross-section images of biological systems and, thus, can be used for studying the chemical and physical properties, anatomical structure and dynamic processes in foods (Ruan and Chen, 2001; Martinez *et al.*, 2003). Proton (^{1}H) MR imaging can provide information about spatial distribution of water and fat. This, in addition to its non-destructiveness, make it an excellent tool for establishing knowledge about effects of raw material properties, storage and processing on fish product quality. Moreover, sodium (^{23}Na) MRI can be used to visualize variations in the sodium distribution of similar raw materials salted or processed by different methods and, thus, ^{23}Na can be used as a tool in understanding and optimizing processes that affect salt content. Furthermore, theoretical sodium transport models can in turn be used to interpret the images. For example, for the aquaculture industry, MRI studies may be used to study the effect of diet and different feeding regimes on fat contents and distribution in fish. In fish processing, MRI can be helpful for optimization of various unit operations such as freezing, thawing, salting, smoking, etc. A detailed review of various MRI applications in food science is written by Hills (1998). However, due to the large footprint of MRI instrumentation, high investment costs and the necessary related infrastructure, MRI cannot presently be considered as a standard analytical tool in the food processing industry and is not suited for online applications today.

[23]Na MRI has been used to study salting and desalting of cod (Erikson et al., 2004), and the salting process of cod (Gallart-Jornet et al., 2007) and salmon (Foucat et al., 2006; Gallart-Jornet et al., 2007; Aursand et al., 2010). In the latter study, the effect of first handling conditions on end product quality was investigated by ^{1}H and [23]Na MRI. A combination of the two MR imaging methods gives information about salt uptake and distribution as affected by both fat content and perimortem handling stress. Figure 13.14 shows the ^{1}H and [23]Na MR images of the same slice of anesthetized and exhausted fillet pieces. The study showed that salt uptake is dependent on fat distribution. In the ^{1}H MR images, the subcutaneous fat layer was evident (top of fillet piece). These features were most clearly visible in column A of Fig. 13.14, where a [23]Na MR image of a thinner slice (2 mm) was acquired with considerably longer acquisition time. When comparing these thin slice (column A) [23]Na and the ^{1}H MR images (top and bottom rows, respectively), it is evident that the high-fat areas (seen in the ^{1}H images as high-intensity stripes) were lower in sodium content, which can be seen as low-intensity stripes in the corresponding [23]Na MR image, indicating low salt areas. The [23]Na MR images showed that the highest salt content was observed near the surface of the fillet with a gradual decrease inwards, due to the presence of the skin acting as a diffusion barrier (Gallart-Jornet et al., 2007). Furthermore, the study revealed that there was an evident relationship between pre-slaughter conditions and salt uptake. The salt had penetrated deeper into the exhausted fillets than into the anesthetized fillets. The fact that the [23]Na MR images revealed a deeper salt penetration into the exhausted fillets, compared with the anesthetized fillets, indicates that stronger rigor contractions in exhausted fish had an effect on the muscle microstructure. This assumption was also supported by an earlier study (Aursand et al., 2009), where pre-rigor, post-rigor and frozen-thawed salmon flesh was compared.

Based on the [23]Na intensities in the MR images, simple computer vision algorithms can be used to compute sodium concentration profiles. Figure 13.15 shows examples of such profiles of the salmon fillets along with a reference solution containing 4% NaCl. Considering the absolute values of NaCl, it should be realized that the MRI approach used here (spin-echo) detected mainly free sodium. A fraction of bound sodium is not detected (Veliyulin and Aursand, 2007). [23]Na MRI revealed salt distribution differences between anesthetized and exhausted fillets, where the salt penetration in the latter fillets was more pronounced.

MR imaging and computer vision was used for optimizing the processing conditions of two different Atlantic salmon raw material qualities (Aursand et al., 2010). The study gave conclusive results showing that sodium uptake is affected by both perimortem handling stress and fat distribution, and thus showing that

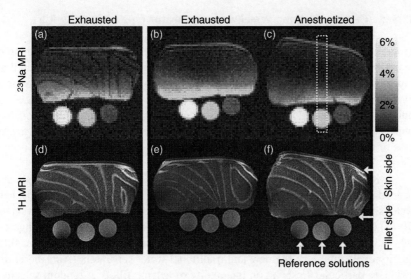

Fig. 13.14   Sodium (^{23}Na) and proton (^{1}H) magnetic resonance images from the same cross-sectional slice of brine salted anesthetized and exhausted Atlantic salmon fillet pieces. Reference solutions (2%, 4% and 6% NaCl) are shown as circles (10 mm tube cross-sections) in the lower part of each image. Lighter areas in the ^{1}H MR images correspond to fat-rich areas. The image slice thicknesses were 2 mm for images (a) and (b), and 10 mm for images (c)–(f). The selected ROI indicates the area used for calculation of profiles in Fig. 13.19. (Reprinted from Aursand *et al., Journal of Food Chemistry*, 2010, 120, Copyright (2010), with permission from Elsevier.)

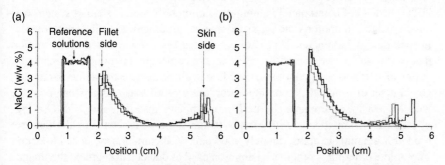

Fig. 13.15 ^{23}Na MRI concentration profiles through brine salted anesthetized (a) and exhausted (b) Atlantic salmon fillets after brine salting in 7.8% NaCl for 3 days. The skin side of the fillet pieces was in all cases placed opposite to the reference glass container with a 4% NaCl solution (to the right in the panel). The selected region of interest (ROI) for calculation of the profiles is shown in Fig. 13.18. (Reprinted from Aursand *et al., Journal of Food Chemistry*, 2010, 120, Copyright (2010), with permission from Elsevier.)

industrial salting procedures need to be optimized for different raw material qualities. Although MR imaging is still not suited for online applications today, the use of this imaging technique is shown to give useful information in industrial process optimization.

13.5.2   Effects of fishing vessel water separator type on herring fillet quality

Proper design of deck equipment is important for pelagic fishing vessels, in order to optimize the quality of the fish throughout the catching and onboard handling processes. Deck equipment, for large pelagic fishing vessels, has been developed by MMC Tendos (MMC Tendos AS, Fosnavåg, Norway) for pumping the fish onboard from the net. A large amount of sea water is pumped onboard together with the fish, and this water is removed by means of a water separator. The process of separating the water from the fish must be done gently in order to ensure a high fish quality.

MMC Tendos developed a new onboard handling system for pelagic fish, and needed an objective method of determining the fish quality obtained with the new water separator compared with the traditional water separator. Figure 13.16 shows the (1) traditional and (2) new water separators installed on the same fishing vessel. Computer vision was chosen as a tool to assist in obtaining objective fillet quality evaluations for the new and traditional water separators (Aursand et al., 2011), in particular focusing on the quality of herring fillets. A random selection of 1040 pairs of fillets was made – after landing and machine-filleting the fish – where 537 and 503 fillet pairs originated from herring that had passed through the traditional and new water separators, respectively. The fillet quality was evaluated with respect to the following three quality parameters: red tail, red base color and blood spots.

As each quality parameter had large variations in how they manifested themselves in the fillets, it was concluded that completely automated computer vision algorithms were not yet suited for quantifying these parameters (Aursand et al., 2011). Instead of a completely automated computer vision system, a system was developed for computer-vision-assisted manual grading of the fillets with respect to each quality parameter. This semi-automated system was based on the notion that pair-wise comparison of two objects with respect to a parameter is manually easier to perform than continuous grading with respect to the same parameter. Given a set of n objects, a continuous ordering of all these objects can be done with respect to a parameter, by performing binary insertion sort with $O(n \log_2 n)$ pair-wise comparisons with respect to the same parameter. For the data set used, this amounts to approximately 9000 pair-wise comparison needed to obtain a complete ordering. Computer vision was used to detect and segment the images of the fillets, and these where then scaled to the same size and presented randomly and pair-wise on a neutral background, as illustrated in Fig. 13.17. The pair-wise comparison is done by the manual inspector clicking on the fillet with the highest degree of the quality parameter being evaluated. The images were presented blindly with respect to the type of water separator that was used in order to obtain an unbiased quality evaluation of the water separators.

The result of the computer-vision-assisted quality evaluation, as applied to the 'red tail' quality parameter, is illustrated in Fig. 13.18, showing the fillets with varying degrees of 'red tail'. For simplicity in discussing the results, the complete ordering of all fillet images is remapped to a quantized scale from 0 (normal tail)

Fig. 13.16 Photograph of the traditional water separator (1) and new water separator (2) mounted on the same fishing vessel. The flow direction of fish is indicated with arrows.

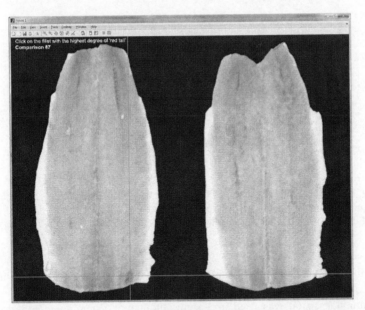

Fig. 13.17 Graphical user interface for computer-vision-assisted quality grading of herring fillets.

to 10 (very red tail). A comparison of the two water separators, with respect to the quantized 'red tail' scale, is shown in Fig. 13.19. The results show a reduction in the fraction of fillets with the reddest tails (degree 8.0–10.0 on the quantized scale) by almost 30% when comparing the new with the traditional water

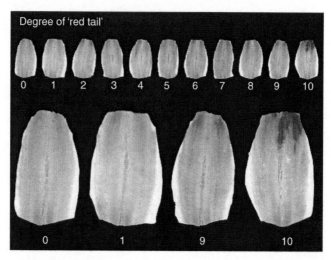

Fig. 13.18   Illustration of the grading scale (with respect to 'red tail') obtained from the computer-vision-assisted grading of herring fillets. A grade of 0 corresponds to normal fillets, whereas a grade of 10 corresponds to very red tail.

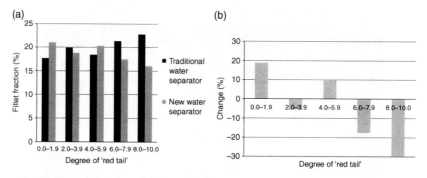

Fig. 13.19   (a) Fraction of fillets with different degrees of 'red tail'. (b) Percentage change in number of fillets with different degrees of 'red tail' when using new water separator instead of traditional water separator. Degree of 'red tail': 0, normal; 10, very red.

separator. The same trend, but somewhat lower in magnitude, is seen for the 'red base color' and 'blood spots' parameters.

Computer vision was used in a semi-automated system for comparing the effect of water separator type on the herring fillet quality. Experiments on a large data set gave conclusive results showing that the new water separator was gentler to the fish than the traditional water separator, and thus markedly increased the quality of the resulting fillets. Although completely automated computer vision is not yet suited for this task, the use of computer vision enabled an effective semi-automatic method for evaluating the quality of the herring fillets as affected by choice of water separator onboard the fishing vessel.

13.5.3   Other representative applications

In general, seafood is highly perishable, with a cold storage shelf-life of approximately 14 days for fresh or thawed products. Beyond 7 days, the product is usually considered to be of a lower grade (Yagiz *et al.*, 2009b). Processing techniques that can extend the shelf-life of seafood past 14 days, while maintaining the sensory attributes of a fresh product, are demanded by the consumers. Color is a sensory attribute that plays an important role in consumer perception of seafood quality. Thus, it is desirable to find processing techniques that both preserve shelf-life and also preserve the visual appearance of the fresh product. Computer vision has been used to measure the color of Atlantic salmon fillets, as affected by storage on ice after varying irradiation doses (Yagiz *et al.*, 2010). Medium radiation doses preserved the appearance, whereas high radiation doses had an adverse effect on appearance, resulting in reduced redness. This effect was attributed to the degradation of the pigment astaxanthin. Alternative treatments that may preserve shelf-life are high-pressure processing and cooking. The effect of high-pressure processing and cooking on the quality of Atlantic salmon has been investigated using computer vision (Yagiz *et al.*, 2009b). The results show that pressure treatment at 300 MPa substantially reduces the microbial count, and thus extends shelf-life freshness. However, computer vision analysis shows that the lightness (L*) and yellowness (b*) increases while redness (a*) decreases. This change is indicative of a cooked appearance due to protein denaturation. Computer vision has proven itself as a useful tool in the analysis of fish color as affected by different shelf-life-extending treatments, and research also shows that it can provide color measurements that are closer to the average real color than measurements from the Minolta Chroma Meter handheld colorimeter (Yagiz *et al.*, 2009a).

13.6   Challenges in applying computer vision in the fish industry

We would like to point out some challenges that are found in computer vision applications in the fish industry:

- Despite advances in computer vision, one of the greatest technological problems is still coping with natural biological variation of raw material manifested in shape, size, color, texture, freshness and other properties. Even with a good and representative sample selection, the biological origin implies that the samples will not be identical in their properties. This is in contrast to some other industrial applications of computer vision, where products have the less variability in their properties.
- Given that most specific physical fish properties relevant for inspection can be sensed by online imaging technologies, the next challenge is the interpretation and use of these data. Therefore, an advance and enlarged scope in classification and recognition algorithms and general machine learning is a precondition

for a successful application of computer vision in sensing and detection of more agglomerated/subjective properties such as fish quality.

- In-feed of fish samples to the imaging module, as well as in other handling/ processing stations, is still a challenging problem in most fish applications. Since many applications deal with live or anesthetized fish, issues such as twitching fish in the field of view, overlapping and non-singulated fish on the conveyor belt pose great challenges during, for instance, the image acquisition and image analysis stages.

- When talking about automation and how computer vision and imaging technology facilitates automation of fish applications and processes, the choice of suitable end-effectors (fixtures, grippers, cutting tools, etc.), as system elements in contact with the raw material, is also of great importance for the success of the computer-vision-based automated inspection, handling and processing of fish. Due to their high perishability, fish have to be handled gently, in online conditions, and this presently sets a number of limitations as to which end-effectors can be used, and which specific applications can fully be automated.

- Flexible automation is seen as a way to address high natural biological variation of fish-specific applications. The mechanical, semi-automated and fixed automated solutions based on the existing technology are simply unable to perform a higher degree of automation in handling, processing and higher raw material utilization. Therefore, in recent years, there has been an increase in demand from fish processors for automated integrated systems involving robots, intelligent sensing and end-effectors. Such a flexible automation, involving several processing operations, will naturally imply a challenge in creating reconfigurable and adaptive image acquisition and image analysis systems – something that is significantly more challenging than detecting specific properties in a single specific species as presented to the imaging system in a single processing stage.

13.7   Future trends

The general trend seems to point towards use of computer vision to increase production cost-effectiveness through automation and increase product quality through process understanding and optimization. Based on the referenced works and the observed trends therefrom, we point to several potential future directions in the application of computer vision in the fish industry:

- Increasing numbers of applications not currently automated will become the focus of this technology. As computer vision and application-specific knowledge is developed, more applications will lie within the realm of cost-effective automation.

- Cost-effective solutions to challenging problems of the future will make use of integration of computer vision using multiple imaging technologies, imaging modes and non-imaging sensors. Smart combinations of several cost-effective technologies can have the capabilities of a single and more expensive technology.

- As the objective assessment of fish products, with respect to various properties, becomes available, there will be a use for this assessment in order to increase product homogeneity and enable a greater market and product differentiation.
- There will be a transition towards online internal and external 3-D imaging. The limits today are expensive technology (CT, MRI) and challenging processing and analysis of images. Internal and external 3-D imaging has the potential to enable a complete understanding and utilization of products – providing detailed information to automated sorting, grading and processing stations, but also enabling online process understanding and optimization of fish processing systems and factories.

As more processes become automated, a tighter/seamless integration is possible between automated sorting, grading and processing stations. This is a transition from automation of single grading and processing stations to a more coherent and potentially more flexible automation of entire processing plants. As processing plants become completely automated – with greater requirements on flexible production – multi-use systems for inspection, handling and processing can play an important role in the fish factory of the future.

13.8   Further information

Computer vision has been used extensively, together with several imaging technologies, to inspect fish and fish products. For a review of the state-of-the-art, see Mathiassen *et al.* (2011b). Recently, Luten *et al.* (2006), Rehbein and Oehlenschlager (2009) and Nollet and Toldra (2010) have discussed tools, including computer vision, for analysis of seafood for quality and process control. Multisensor approaches for fish quality and freshness determination, including electronic noses, VIS imaging and spectroscopy, have been reviewed by Olafsdottir *et al.* (2004). Interested readers may wish to consult the above-mentioned reviews and books for more information on the applications of computer vision in the fish industry.

Computer vision is being used by more and more research groups in the area of seafood research as a tool for process optimization and understanding, and for automation purposes. Based on the frequency of recent publications, there are a number of the research groups that have become more visible when it comes to the use of computer vision and imaging technologies for muscle food applications. In the USA, Professor Balaban's group, from the Fishery Industrial Technology Center (www.sfos.uaf.edu/fitc) at Kodiak, Alaska, has been steadily publishing on the use of computer vision. One of the recent papers of the group is regarding the use of computer vision in quantification of gaping in fillets. In Latin America, Professor Aguilera's group (www.iufost.org/isfe/jaguilera.htm), from the Pontifica Universidad Catolica de Chilehas, has also been active in exploiting the computer vision for seafood applications. In Europe, Cemagref in Rennes (France) (www.cemagref.fr) has published some interesting papers regarding lipid content and correlation with computer vision parameters. Computer vision for muscle food analysis is also a prominent research topic with the Food Refrigeration and Computerized Food Technology (FRCFT) group at University College Dublin (Ireland) (www.ucd.ie/refrig). DTU (www.food.dtu.dk) in Denmark has recently published some papers regarding the use of multispectral imaging for determination of astaxanthin in salmonids. In Norway, the Nofima Research Organization (www.nofima.no) and SINTEF (www.sintef.no) are very active in the use of computer vision and imaging technologies for muscle food analysis. SINTEF is the largest research institute in Scandinavia and offers a leading competence along the entire value chain – meaning that computer vision is used for analysis, process understanding and optimization, and for handling, manipulation and automation of muscle food product processing lines. For more information on these research groups, please see the web addresses given.

13.9   Conclusions

Computer vision has been applied to a broad set of applications in the fish industry – including automated sorting and grading, automated processing and process understanding and optimization. The general trend seems to point towards use of computer vision to increase production cost-effectiveness through automation and increase product quality through process understanding and optimization. Despite the many successful applications, there is still great untapped potential in the use of computer vision in the fish industry. Many applications are still too challenging or costly to solve. However, as computer vision technology and application-specific knowledge is developed, more applications will lie within the realm of cost-effective automation.

13.10   References

Aursand, I. G., Erikson, U. and Veliyulin, E. (2010) Water properties and salt uptake in Atlantic salmon fillets as affected by ante-mortem stress, rigor mortis and brine salting: A low-field 1H NMR and 1H/23Na MRI study. *Food Chemistry*, **120(2)**, 482–9.

Aursand, I. G., Mathiassen, J. R., Bondø, M. and Toldnes, B. (2011) Technology for optimal on board handling of pelagic fish. Design of water separator and sampling for automatic weight estimation of single fish. SINTEF Report. F18949. Confidential.

Aursand, I. G., Veliyulin, E., Böcker, U., Rustad, T. and Erikson, U. (2009) Water and salt distribution in Atlantic salmon (*Salmo salar*) studied by low-field 1H NMR, 1H and 23Na MRI and light microscopy – effects of raw material quality and brine salting. *Journal of Agricultural and Food Chemistry*, **57**, 46–54.

Balaban, M., Chombeau, M., Cırban, D. and Gümüş, B. (2010a) Prediction of the weight of Alaskan pollock using image analysis. *Journal of Food Science*, **75(8)**, E552–E556.

Balaban, M., Gulgun, F., Soriano, M. and Ruiz, E. G. (2010b) Using image analysis to predict the weight of Alaskan salmon of different species. *Journal of Food Science*, **75(3)**, E157–E162.

Bondø, M., Mathiassen, J. R., Vebenstad, P., Misimi E., Skjøndal, E. M., Toldnes, B. and Østvik, S. O. (2011) An automated salmonid slaughter line using machine vision. *Industrial Robot: An International Journal*, **38(4)**, 399–405.

Buckingham, R. and Davey P. (1995) This robot's gone fishing. *Industrial Robot: An International Journal*, **22(5)**, 12–14.

Buckingham, R., Graham, A., Arnarson, H., Snaeland, P. and Davey, P. (2001) Robotics for de-heading fish – a case study. *Industrial Robot: An International Journal*, **28(4)**, 302–9.

de Silva, C. W., and Wickramarachchi, N. (1997) An innovative machine for automated cutting of fish. *IEEE Transactions on Mechatronics*, **2(2)**, 86–98.

Hu, B.-G., Gosine, R. G., Cao, L. X. and de Silva, C. W. (1998) Application of a fuzzy classification technique in computer grading of fish products. *IEEE Transactions on Fuzzy Systems*, **6(1)**, 144–52.

ElMasry, G. and Wold, J. P. (2008) High-speed assessment of fat and water content distribution in fish fillets using online imaging spectroscopy. *Journal of Agricultural and Food Chemistry*, **56(17)**, 7672–7.

Erikson, U., Veliyulin, E., Singstad, T. E. and Aursand, M. (2004) Salting and desalting of fresh and frozen-thawed cod (*Gadus morhua*) fillets: a comparative study using 23Na NMR, 23Na MRI, low-field 1H NMR, and physicochemical analytical methods. *Journal of Food Science*, **69**, FEP107–14.

Foucat, L., Ofstad, R., Renou, J.-P. (2006), How is the fish meat affected by technological processes? In G. A. Webb (ed.), *Modern Magnetic Resonance*. The Netherlands: Springer, 957–61.

Gallart-Jornet, L., Barat, J. M., Rustad, T., Erikson, U., Escriche, I. and Fito, P. (2007) A comparative study of brine salting of Atlantic cod (Gadusmorhua) and Atlantic salmon (Salmosalar). *Journal of Food Engineering*, **79**, 261–70.

Hills, B. (1998) *Magnetic Resonance Imaging in Food Science*. New York: John Wiley and Sons , 96.

Jain, A., de Silva, C. W. and Wu, Q. M. J. (2001) Intelligent fusion of sensor data for product quality assessment in a fish cutting machine. *Proceedings of Joint IFSA World Congress and 20th NAFIPS International Conference*, **1**, 316–21.

Lachenbruch, P. A. and Mickey, R. M. (1968) Estimation of error rates in discriminant analysis. *Technometrics*, **10**, 1–11.

Luten, J. B., Jacobsen, C., Bekaert, K., Sæbø, A. and Oehlenschlager, J. (2006) *Seafood Research from Fish to Dish*. Netherlands: Wageningen Academic.

Martinez, I., Aursand, M., Erikson, U., Singstad, T. E., Veliyulin, E. and van der Zwaag, C. (2003) Destructive and non-destructive analytical techniques for authentication and composition analyses of foodstuffs. *Trends in Food Science and Technology*, **14**, 489–98.

Mathiassen, J. R., Misimi, E., Toldnes, B., Bondø, M., and Østvik, S. O. (2011a) High-speed weight estimation of whole herring (Clupeaharengus) using 3-D machine vision. *Journal of Food Science*, **76(6)**, E458–64.

Mathiassen, J. R., Misimi, E., Bondø, M., Veliyulin, E., and Østvik, S. O. (2011b) Trends in application of imaging technologies to inspection of fish and fish products. *Trends in Food Science and Technology*, **22**, 257–75.

Misimi, E., Erikson, U. and Skavhaug, A. (2008a) Quality grading of Atlantic salmon (Salmosalar) by computer vision. *Journal of Food Science*, **73(5)**, E211–E217.

Misimi, E., Erikson, U., Digre, H., Skavhaug, A. and Mathiassen, J. R. (2008b) Computer vision-based evaluation of pre- and postrigor changes in size and shape of Atlantic cod (Gadusmorhua) and Atlantic salmon (Salmosalar) fillets during rigor mortis and ice storage: Effects of perimortem handling stress. *Journal of Food Science*, **73(2)**, E57–E68.

Misimi, E., Mathiassen, J. R., and Erikson, U. (2006) Computer vision-based sorting of Atlantic salmon (SalmoSalar) according to size and shape. *Proceedings of International*

Conference on Computer Vision Theory and Applications – VISAPP 2006 (1). Setubal, Portugal, 265–70.

Nollet, L. M. L., and Toldra, F. (2010) *Handbook of Seafood Products Analysis*. Boca Raton, FL: CRC Press.

Norges Sildesalgslag Minimum prices for herring [internet] (2011) Available from http://www.sildelaget.no/Fishery.aspx?Fish=NVG-Sild (accessed 25 January 2011).

Olafsdottir, G., Nesvadba, P., Di Natale, C., Careche, M., Oehlenschlager, J., Tryggvadottir, S., *et al.* (2004). Multisensor for fish quality determination. *Trends in Food Science & Technology*, **15**, 86–93.

Rehbein, H. and Oehlenschlager, J. (2009) *Fishery Products: Quality, Safety, Authenticity*. Oxford, UK: Wiley-Blackwell.

Ruan, R. R. and Chen, P. L. (2001) Nuclear magnetic resonance techniques and their application in food quality analysis. In Gunasekaran, S. (ed.), *Nondestructive Food Evaluation*. New York: Marcel Dekker, 165–216.

Sandberg, P. (2010). *Nøkkeltall fra norsk havbruksnæring* (Key figures from aquaculture industry). Fiskeridirektoratet, Bergen, Norway.

SINTEF Fisheries and Aquaculture AS (2007) Prototype salmon fillet-trimming robot. Available from http://www.sintef.no/fish.

Theodoridis, S. and Koutroumbas, K. (1999) *Pattern Recognition*. San Diego, CA: Academic Press.

Thielemann, J., Kirkhus, T., Kavli, T., Schumann-Olsen, H., Haugland, O. and Westavik, H. (2007) System for estimation of pin bone positions in pre-rigor salmon. *Lecture Notes in Computer Science*, **4678**, 888–96.

Veliyulin, E. and Aursand, I. G. (2007) ^{1}H and ^{23}Na MRI studies of Atlantic salmon (Salmosalar) and Atlantic cod (Gadusmorhua) fillet pieces salted in different brine concentrations. *Journal of the Science of Food and Agriculture*, **87**, 2676–83.

Wold, J. P., Johansen, I.-R., Haugholt, K. H., Tschudi, J., Thielemann, J., Segtnan, V. H., Narum, B. and Wold, E. (2006) Non-contact transflectance near infrared imaging for representative on-line sampling of dried salted coalfish (bacalao). *Journal of Near Infrared Spectroscopy*, **14**, 59–66.

Yagiz, Y., Balaban, M. O., Kristinsson, H., Welt, B. A. and Marshall, M. R. (2009a) Comparison of Minolta colorimeter and machine vision system in measuring colour of irradiated Atlantic salmon. *Journal of the Science of Food and Agriculture*, **89(4)**, 728–30.

Yagiz, Y., Kristinsson, H., Welt, B. A., Balaban, M. O., Ralat, M. and Marshall, M. R. (2009b) Effect of high pressure processing and cooking treatment on the quality of Atlantic salmon. *Food Chemistry*, **116(4)**, 828–35.

Yagiz, Y., Kristinsson, H., Welt, B. A., Balaban, M. O. and Marshall, M. R. (2010) Correlation between astaxanthin amount and a* value in fresh Atlantic salmon (Salmosalar) muscle during different irradiation doses. *Food Chemistry*, **120(1)**, 121–7.

14

Fruit, vegetable and nut quality evaluation and control using computer vision

J. Blasco, IVIA – Instituto Valenciano de Investigaciones Agrarias, Centro de Agroingeniería, Spain, N. Aleixos, I3BH, Universitat Politècnica de València, Spain and S. Cubero and D. Lorente, IVIA – Instituto Valenciano de Investigaciones Agrarias, Centro de Agroingeniería, Spain

Abstract: Machine vision has become a key technology in the area of quality control of horticultural products due to the increasing quality demands of manufacturers and customers, providing a fast and reliable method of inspecting features such as ripeness, size, weight, shape, colour, the presence of blemishes and diseases, the presence or absence of fruit stems, the presence of seeds, sugar content and so forth. This chapter provides a comprehensive review of recent advances in computer vision applied to the inspection of nuts, fruits and vegetables and outlines potential applications in the industry.

Key words: external quality, internal quality, automatic inspection, in-line inspection, computer vision, machine vision.

14.1 Introduction

Machine vision has become a key technology in the area of quality control of horticultural products, due to the increasing quality demands of manufacturers and customers. This technology is based on industrial image processing through the use of cameras mounted over sorting lines in order to automatically inspect products in real-time without operator intervention. A machine vision system can consist of a number of cameras, all capturing, interpreting and sorting products individually using a control system programmed to some pre-determined tolerance or requirement.

The use of machine vision in the agricultural industry has continually grown over the last 15 years. This is due to the increasing capability of processors, which has allowed the technology to be employed in a wide number of applications. This technology allows the estimation of quality analysis over the whole production line to be far more accurate than it would be if it were carried out manually. One area in which this technology has been widely used is in the inspection of agri-food commodities (Sun, 2007), particularly in the automatic inspection of fruits and vegetables, since the technology is more reliable, more objective and faster than human inspection. The quality of a particular fresh or processed piece of fruit or vegetable is defined by a series of physicochemical characteristics which make it more or less attractive to the consumer, such as its ripeness, size, weight, shape, colour, the presence of blemishes and diseases, the presence or absence of fruit stems, the presence of seeds, its sugar content and so forth (Cubero *et al.*, 2011).

Because the decisions made by operators are affected by psychological factors, such as fatigue or acquired habits, there is a high risk of human error in classification processes and this is one of the most significant drawbacks that can be prevented by using automated inspection systems based on computer vision. Nevertheless, there are certain particularities and problems associated with the automated inspection of agricultural produce, which are due to the biological nature of the produce. While manufactured products often present similar colours, shapes, sizes and other external features, fruits and vegetables are not uniform and may show very different characteristics from one item to another. One single fruit can have a different colour, size and shape from another one belonging to the same cultivar. Moreover, it is essential that the presence of stem-ends, leaves, shells, stones, dirt or any other extraneous material is identified and not confused with other natural features or skin defects. In this sense, the capability of machine vision-based systems to inspect beyond the visible spectrum means that they have the potential to detect defects or damage that is not visible to human operators (Sun, 2010). On the other hand, markets demand very fast image processing and for this reason a compromise between speed and accuracy must be achieved.

The aim of this chapter is to provide a comprehensive review of recent advances in computer vision and potential applications with regard to the inspection and quality control of nuts, fruits and vegetables. Some of the technologies used to acquire images are described, paying particular attention to the arrangement of the scene, lighting and acquisition technology. Representative works are then discussed, covering the application of computer vision technology to the inspection of some of the most important features that define the quality of the products from a commercial point of view. The focus then moves to in-line real-time systems for automatic inspection, as they represent the state of the art of applications of this technology to the industry. Finally, future trends and conclusions are given. In the case of future trends, these focus on non-standard systems for image acquisition, such as magnetic resonance imaging (MRI) or thermography and the potential applications developed for the assessment of internal quality.

14.2   Basics of machine vision systems for fruit, vegetable and nut quality evaluation and control

The sections below consider some of the fundamental requirements for the successful functioning of this technology in the field of fruit, vegetable and nut assessment and processing.

14.2.1   The need for singulation

The in-line application of machine vision systems in the food industry normally requires a previous singulation of the products, usually when large objects have to be physically separated. If this process is not carried out, the system may perceive several touching objects as a single product, making the analysis invalid. There are different ways to divide up the products, depending on their size and shape. The main method employed in the inspection of fruits and vegetables is the use of singulators composed of angled conveyor belts. These are usually installed before the feeding area of an electronic sorter, allowing the individual supply of fruits from a random reception. The machine consists of two inclined conveyor belts, arranged in a V formation, which move at different speeds in the same way. Upon entering the fruits in the singulator, the difference in speed of the tapes gives a slight rotation to the fruits as they advance, which forces them to separate from each other, exiting the machine one at a time.

However, when a high throughput is required, the mechanisms that separate the fruits do not work efficiently, thus allowing several fruits to travel together in the same place while passing under the inspection system. Moreover, a large fruit could touch one or both of its neighbours. In these cases, a conventional algorithm designed to determine the size or shape of a fruit from its boundaries could misinterpret the image if it is only able to locate the pixels that lie between the fruit and the background. In this way, the fruits that are in contact could be identified as one large fruit.

These errors occasionally occur in the three situations shown in Fig. 14.1: (a) when a large fruit is in contact with a neighbouring fruit, (b) when an extra fruit travels between two fruits that are correctly positioned on their cups, or (c) when two or more small fruit travel in the same cup. In all these cases, the calculation of size from the apparent boundaries can lead to an over-estimation of the fruit size. To solve this problem, Aleixos *et al.* (2002) developed a specific algorithm for separating the individual fruit in the image and then determining if one of the above situations has occurred. The algorithm detects the contact points between fruits by locating unexpected changes in the tangent of the extracted boundary. Sudden changes of the tangent in the X direction are used to detect cases (a) and (b), while large changes in the Y direction are used to detect case (c). In a similar way, Al-Mallahi *et al.* (2010a) developed an algorithm designed to detect touching potatoes travelling on a conveyor belt. This algorithm detects in-line potato tubers without singulation when they are mixed with clods, scanning images from left to right until it encounters a cluster of tubers and clods. Once encountered, a scan

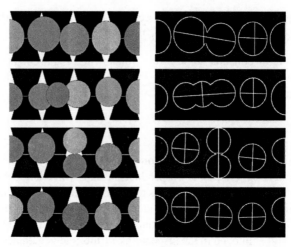

Fig. 14.1   Typical errors caused by a bad singulation during fruit transport.

around the periphery of the cluster is initiated, finishing when the scan returns to the starting point. At the end of the scan, the peripheral pixels are erased and the new width of the cluster is compared with the previous width. Contouring of the periphery continues until a sudden shrink in the width of the cluster occurs, which indicates the detection of a contact point. The left-to-right scanning resumes from the contact point until it reaches the right-hand side of the image and the periphery scanning is repeated every time a cluster is encountered. The algorithm was tested by running scans of 380 contact points between tubers and 112 contact points between tubers and clods. The results showed that approximately 99% of the contact points were detected successfully.

In some cases, due to the small size of the products, the singulation has to be done in a different way, for instance using vibration platforms as in the case of pomegranate arils (Blasco *et al.*, 2009a) or satsuma segments (Blasco *et al.*, 2009b). In both cases the product falls to a platform that vibrates at a specific frequency and has an inclination that forces the product onto high-speed conveyor belts, which facilitates the separation of the product. However, the products cannot always be singulated, which is particularly difficult in the case of small fruits or nuts. In these cases, the position of the object has to be calculated from the images, and each object tracked individually until it reaches a separating system, where the product is ejected by air nozzles. Attempts to inspect fruit without previous singulation have been carried out by Reyes and Chiang (2009). They used a robotic arm that located and separated each piece of fruit in the image analysis after the real-time inspection. Although the fruit is only partially inspected (just one half) and classified by colour, the main benefit of this approach is that it uses a robotic arm to sort the fruit physically instead of a large machine with dozens of outlets.

Once the products have been singulated, it is necessary to inspect the whole surface of the product at high speed. Many current solutions use mechanical systems to rotate and move each product under the camera (Leemans and Destain, 2004), some use mirrors (Reese *et al.*, 2010) or robotic arms (Blasco *et al.*, 2003) and others use several cameras that acquire different views of the product (Xiao-bo *et al.*, 2010).

14.2.2   The importance of good lighting

The success of an application for the inspection of fruits or vegetables depends on the quality of the images captured. This quality is defined by several factors. The quality of the camera and the illumination of the scene are considered to be two of the most important aspects. A good quality lighting system should provide uniform radiation throughout the scene, avoiding the presence of glare or shadows. It must be spectrally uniform and stable over time. If the scene is adequately illuminated for a particular purpose, image pre-processing is less necessary, thus saving processing time and obtaining more reliable results. If the objects are not properly illuminated, more time will have to be spent on removing noise, shadows or glares from the images. For instance, Gómez-Sanchis *et al.* (2008a) proposed a methodology for correcting the effect of bad illumination of spherical fruits. They proposed the construction of an elevation model of the fruit which can be used to estimate the theoretical height of a pixel and the angle of incidence of the light for each pixel. This made it possible to estimate what corrections may be required to reduce distorting effects of the illumination. However, this process is time consuming and time is a valuable resource when the products have to be inspected in the industry in real-time.

When designing a lighting system for these applications, it is important to keep in mind that this technology is to be used at an industrial level in packing-houses, using machines based on computer vision (inspection lines) that transport the products at very high speeds. In order to acquire images with sufficient quality, cameras can freeze the movement by using a progressive scanning mode and high-speed electronic shutters. The shutters regulate the time during which light excites the sensor of the camera. For instance, a standard camera that acquires 25 images per second will need 40 ms to capture each image. Any object travelling at a speed of 0.5 m/s will move 20 mm during the image acquisition, resulting in a moved image. However, as the shutter speed increases, the intensity of lighting must also increase to avoid underexposure.

Ideally, lamps directed at an angle of 45° to the flat object are normally effective. By doing this, specular reflection is reduced, thereby avoiding unwanted glare. This configuration was used by Fernández *et al.* (2005) to illuminate apple slices in order to control and evaluate their dehydration by tracking their colour and shape. A similar system was later used by Pedreschi *et al.* (2006) to illuminate potato chips in a system designed to measure the kinetics of colour changes under different frying temperatures. However, this method does not work with spherical products like apples, citrus fruits, peaches, etc. In these cases, the reflection of the light can cause specular bright spots that reach the camera and can cause the omission of important information about the product from the image captured. To avoid this problem, it is common in commercial in-line sorters to use polarizing filters to avoid these reflections by means of cross-polarization (Ludwig, 1973).

When the object to be inspected is translucent or when its contour has to be defined in the image with precision, backlighting can be used, which consists of placing the object between the light source and the camera. This solution was adopted by Blasco *et al.* (2009b) in an industrial prototype for in-line inspection of mandarin segments travelling over semi-transparent conveyor belts. By illuminating the segments from behind, the segments in the images showed up in sharp contrast with the background, thus making it easier to estimate their size and shape. At the same time, seeds appeared darker than the rest of the segment, facilitating their detection.

For adequate image acquisition, the emission spectrum of the source is crucial. Fluorescent tubes have been traditionally used in commercial sorters because they are suitable for applications in which visible light is important (i.e., colour sorting). They are also used to illuminate relatively large scenes, which allows for several products to be captured in one image. This speeds up the process of automatic inspection as the products can be captured and inspected one by one. However, they produce a characteristic flickering that is generally corrected by using high-frequency electronic ballasts. One of the main disadvantages of fluorescent tubes is that they produce less radiation in the infrared region in comparison, for instance, with incandescent lamps. These emit more infrared radiation, but also generate a lot of heat, can be considered directional sources and have a low colour temperature, which are significant drawbacks for applications requiring colour analysis. However, both types of lamps can be combined. Near-infrared (NIR) images can be used to segment the images more easily since there is a great contrast between the objects of interest and the background in this kind of image. Colour images can then be used to extract features.

An alternative that is beginning to be used in the industry is based on LEDs (light emitting diodes). An LED is a semiconductor light source. The use of LEDs in the lighting field is expected to increase in the next few years, as their benefits are greater than the incandescent and fluorescent lamps from various points of view. Such benefits include reliability, greater energy efficiency, greater vibration resistance, lower power dissipation, less environmental risk, ability to operate in strobe or continuous mode, etc. On the other hand, they are very directional and illumination power is still limited.

14.2.3   Technology for image acquisition

The selection of the camera for a vision system is normally a technical decision based on the particular needs of each inspection system. Cameras used in these systems convert the light that they receive from the scene into electronic signals. Different types of camera are available which capture the light and transfer the

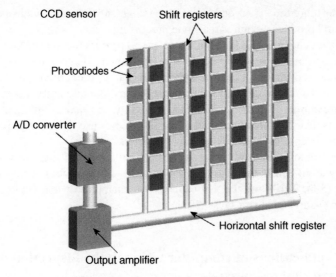

Fig. 14.2   Structure of a CCD sensor.

information to a computer, or any other decision unit, in different ways. The most common industrial cameras used in machines for the inspection of nuts, fruits and vegetables are based on a matrix CCD (charge-coupled device), which consists of an array of sensors (pixels), each of which is a photocell and a capacitor (Peterson, 2001). The load acquired by the capacitor depends on the amount of light received by the photocell. These charges are converted into a voltage and subsequently into a video signal. In applications where the inspection of objects needs to be continuous, like in the textile industry, these cameras can be based on a linear CCD array, composed of a one-dimensional array of cells which acquire a narrow strip of the scene. These cameras, known as line-scan cameras, are suitable for applications where the object moves below the camera so that the complete image of its surface is gradually acquired, line by line while the object advances.

Matrix cameras are the most widely used for commercial applications. They acquire a scene by using a two-dimensional (2-D) CCD array. First, the cameras capture colour images using a single colour CCD array, the sensors of which are sensitive to the red, green and blue (RGB) colours using a Bayer filter (Fig. 14.2). This provides interlaced images of sizes and frequencies compatible with the PAL (phase alternating line) or NTSC (National Television System Committee) standards. Currently, modern sophisticated cameras split the captured light using a dichoric prism towards three or even more CCDs. Each of them is sensitive to a particular wavelength (normally corresponding to R, G and B). The cameras are capable of acquiring high-resolution images with a high number of megapixels per inch. The images are then transferred to the computer using fast protocols like USB3 or GigaEthernet. However, it is important to bear in mind that the larger the image, the more time is needed for its processing, and the industrial inspection of agricultural produces requires real-time processing.

The latest cameras also have the capability to process the images they capture. The microprocessor is incorporated into the camera, which can operate like a small computer. Image processing applications can be programmed and executed in cameras that also have communication ports and digital input/outputs to interact with external devices or machines. Currently, the main limitation of these cameras is their low computational power, since they use old and relatively slow processors in comparison with modern computers.

Other technologies used for image acquisition are promising, but still have a long way to go before being used in the industry. Multispectral or hyperspectral cameras (Sun, 2010) and thermal cameras are examples of the new technology available (Vadivambal and Jayas, 2011).

14.3   Applications of computer vision in the inspection of external features

The ultimate purpose of industrial inspection systems based on computer vision is to analyse one or more features of the product of interest and compare them with the quality standards required. In most of these applications, image analysis is employed to classify the products by assessing features like colour, size and presence of damage. Other features, like shape and texture, are of less importance for commercial purposes.

14.3.1   Use of colour information

Colour is one of the most important attributes for biological products, since consumers may be influenced to choose or reject a particular fruit by its colour. Producers therefore strive to prevent products with defective colorations from reaching the market, as well as ensuring that individual products are packed in batches of a similar colour. Colour is often measured using colorimeters (Hoffman, 2011). However, these devices are limited to the measurement of small regions or in applications where the integration of the colour all over the sample is of interest. They are not well suited to measuring objects with a heterogeneous colour (Gardner, 2007). When it is necessary to measure the colour of entire samples (fruit, vegetable, nut, etc.), it is more appropriate to use devices that provide images in which the colours of the individual pixels that comprise the whole area can be measured (Mendoza *et al.*, 2006). The most widely used colour spaces are Hunter *Lab* or CIE 1976 L^*,a^*,b^* (CIELAB), which can be basically provided by colorimeters and are most used in science, and RGB which is the natural colour model for computers and cameras. However, RGB is not a perceptually uniform colour model. This means that the same Euclidian distance between two points in different regions of these spaces does not produce the same difference of perception in a standard observer. In these cases, a CIELAB is more suitable for colour analysis or comparison (León *et al.*, 2006).

Simple algorithms based on a single CIELAB coordinate are also used for fruit classification. The a^* coordinate was used by Liming and Yanchao (2010) to grade strawberries into three-colour categories. Compared to the human sorting, the system based on image analysis achieved an 89% success rate.

Hue angle and chroma are colour features derived from the above-mentioned uniform spaces. Kang et al. (2008) quantified the effect of curvature on the calculation of hue angle and chroma. They demonstrated that the hue angle provided a valuable quantitative description of the colour and colour changes of individual and batches of heterogeneously coloured mangoes. Sometimes fruits or vegetables belonging to the same variety have a high rate of texture and colour variability that complicates image analysis. López-García et al. (2010) proposed a method that combines colour and texture information for the detection of skin defects in four cultivars of citrus. Arzate-Vázquez (2012) used the L^*, a^*, b^* colour coordinates to evaluate the ripeness of avocados. They found that the a^* coordinate can be used to classify avocados by maturity with a success rate of 80%.

14.3.2   Morphological parameters

Size is particularly important in industry as a means of classifying objects into different commercial categories. The price of many agricultural products is directly related to their size. Measurement of this attribute in spherical or quasi-spherical objects is relatively easy, but it becomes more and more complex in irregularly shaped fruits and vegetables. The features most commonly used to estimate size are area, perimeter, length and width. However, due to the natural irregularities in the shape of agricultural produce, these measurements frequently depend on the orientation of the object with respect to the camera. For this reason, many authors combine size information obtained from images taken from different relative angles between the object and the camera. For instance, Blasco et al. (2003) estimated the size of Golden Delicious apples from four images of the fruit. The size was calculated from the view in which the stem was located nearest to the centroid of the object. Throop et al. (2005) measured the size of 14 cultivars of apple on fruits travelling on rollers by adjusting their translational and rotational speeds in such a way that images of one complete revolution of each fruit were captured regardless of their size. In this work, equatorial diameter and area of the apple were first calculated, then the apple was modelled as an ellipse and height was estimated from its major axis. This sizing was used later for orienting the fruit and surface analysis.

In terms of shape, Fourier descriptors have been widely used to estimate and classify fruits and nuts in different commercial categories. Ghazanfari et al. (1997) used the projected area of pistachio nuts to obtain Fourier descriptor, in order to classify the nuts into four different categories. Also the outline of the projected area was used by Currie et al. (2000) to estimate the shape of apples. Abdullah et al. (2006) also used Fourier descriptors to classify fresh cut star fruits into three different shape categories. Elliptical Fourier descriptors were used by Menesatti et al. (2008) to determine different cultivars of hazelnuts by shape.

14.3.3 Detection of external defects

Detection of skin defects or damage is one of the most widely extended applications of image analysis for the inspection of fruits, vegetables or nuts. The presence of external damage is a clear sign of poor product quality. Many applications aimed at such detection have been described. One difficulty which is common to most of these applications is that of distinguishing defective areas of the fruit or vegetable from natural organs like calyxes or stems. Although some defects simply depreciate the commercial value of the fruit and do not evolve, others may prevent the fruit from being sold, may grow and appear during transport or storage, or may contaminate other fruits and vegetables that were initially sound (i.e., fungal infestations). The industry is particularly interested in the early detection of this second group, to avoid the spread of rottenness. However, sometimes the appearance of the damage is very similar to the appearance of sound skin, thus making it difficult to detect using standard current commercial systems (Fig. 14.3). Detection of blemishes alone is not enough, individual identification of the different types of damage is necessary for adequate post-processing. Figure 14.3 illustrates how different regions in images of citrus of the same cultivar (cv. Valencia) have similar colour and textural properties, even though they belong to sound and damaged areas. For this reason, alternative techniques have to be considered. Unay and Gosselin (2006) compared several methods to segment different types of defects in multispectral images of apples with the aim of creating a real-time inspection system. Blasco *et al.* (2007) used a region-growing algorithm to separate contrasted regions of citrus fruits. These regions were subsequently merged by using colour distances in the hue, saturation and intensity (HSI) colour space with the aid of an unsupervised algorithm. This work allowed 11 different types of defects to be detected, but did not identify them. A later work combined images with different spectral information to identify these defects, the results were clearly improved by adding morphological parameters and decision algorithms (Blasco *et al.*, 2009c).

When the product to be inspected is of small size, the automatic methods become particularly important. For instance, the evaluation of peeling efficiency of hazelnuts after toasting is generally conducted by trained operators. By taking into account the size of the product and the potential defects, Pallottino *et al.* (2010) developed an alternative technique to evaluate the quality of the peeling based on image analysis. In this case, the solution adopted was to use a flat-bed scanner instead of traditional video cameras.

14.4   Real-time automatic inspection systems

Packinghouses require machines capable of inspecting large quantities of product in short time. Sometimes the harvest season is concentrated into a short period of time and there is a need to process all the products in that period, since the possibility of storing the fruits or vegetables is limited. Therefore, machines for automatic inspection must be capable of inspecting several tons of products per hour,

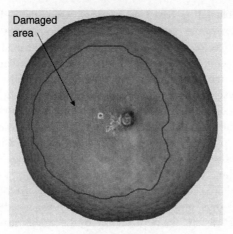

Fig. 14.3 Decay in citrus caused by fungus infection. The damaged area (inside the marked region) is similar to the sound skin.

which means inspecting each particular piece at a very high speed. Moreover, to maintain their competitiveness and position in the complex agri-food market, there is demand from growers for more sophisticated systems capable of inspecting beyond the traditional parameters of size and colour, allowing for more detailed analysis of the products. However, the processing of more information implies more processing time. Thus, it is very important to continue investigating and improving real-time algorithms capable of adapting to the growers' and machinery makers' requirements.

In order to achieve real-time operation, image processing must be very fast. For this reason, some authors have used specific hardware to minimize processing time. For instance, Aleixos *et al.* (2002) developed a multispectral camera capable of acquiring two images simultaneously (one standard RGB image and one monochromatic NIR image) from the same scene in a citrus packing line. Image processing algorithms were parallelized and implemented on two DSP (digital signal processors) for simultaneous analysis. Fruit was then graded by means of a non-linear discriminant analysis procedure that depended on colour, size and defect features. Aleixos *et al.* (2002) demonstrated that the system was capable of correctly classifying lemons and mandarins at a rate of 10 fruits per second with the technology available at the end of the 1990s. Currently, modern computers allow for processing of images much faster than before – hence, more complex algorithms can be implemented. Some recent works that deal with the problem of working in real-time are described below.

The grading of apples into quality classes is a complex task involving different stages. Leemans and Destain (2004) presented a real-time computer vision-based system to classify cv. Jonagold apples with external defects. The acquisition of the images was performed by CCD cameras during the motion of the fruit on an adapted commercial machine. After segmentation of the image, the fruits were detected using a clustering based on k-means. Once the defects were located, they were characterized by a set of features including colour, shape and texture descriptors, as well as the distance of the defects to the nearest calyx or stem end. These data were accumulated for each fruit and summarized in order to transform the dynamic table into a static table. The grading was performed using quadratic discriminant analysis. The correct classification rate was 73%.

Xiao-bo *et al.* (2010) also describe the use of three-colour cameras to inspect rotating apples and classify them into two categories (accepted or rejected), depending on the presence of defects. The three cameras work simultaneously to capture three different views of the fruit. Al-Mallahi *et al.* (2010b) developed an automatic machine vision system for the in-line sorting of potatoes using UV imaging. The goal was to differentiate sound potatoes from undesirable material (e.g., stones) – thus, only one image per potato was needed. They processed one image every 94 ms and the performance of the system reached a success rate of 98%.

In the following section, a case study on the development of a real-time inspection prototype for the in-line sorting of pomegranate arils is presented. Pomegranates (*Punica granatum*) are composed of small seeds (arils) ranging in colour from white to deep red or purple. The development of a whole prototype for the automatic sorting of pomegranate arils into commercial categories using in-line machine vision is described here as an example of a typical inspection machine, such as those that operate in packinghouses. Apart from the adaptations to the handling of this product, the prototype includes all the basic components of a machine used for the inspection of other products, such as mechanisms for the singulation of fruit, computer vision algorithms for the inspection and sorting of individual objects, mechanisms for controlling the prototype and for synchronizing the inspection unit with the output system (Fig. 14.1).

Due to the small size of the arils, it is not feasible to use mechanical elements that singulate them one by one as with other fresh fruits and vegetables. In order to singulate the arils, they fall down into a vibrating plate with an inclination that makes the arils advance upwards towards the end part, where they are spilt onto the conveyor belts. As they fall, the high speed of the conveyor belts separates them from the forthcoming arils. The width of the conveyor belt is only 30 mm to ensure the alignment of the product and facilitate the further sorting in the outlets.

The prototype uses two progressive scan cameras to acquire images in RGB connected to a main computer. The scene is illuminated by fluorescent tubes located above the conveyor belts, which are powered by means of high-frequency electronic ballast to avoid the flicker effect. The presence of bright spots in the wet product caused by directional lighting is minimized using cross-polarizing filters placed on the tubes and the camera lens. The scene captured by each camera has a size of approximately 360 mm along the direction of the advance and includes three conveyor belts.

To get real-time specifications, the image acquired by the camera is processed at the same time as the next image is being acquired from the other camera. The result is an overlap in time between the processing of one image and the acquisition of the next, thus saving time and optimizing the operation. The acquisition of the images is governed by the pulses received from an optical encoder attached to the carrier roller and connected to the computer. The application has been programmed to trigger the camera every time the conveyor belts advance 350 mm. The application is programmed using different threads. One of these threads is aimed at counting the encoder pulses and triggers the camera when a particular value is reached, thus making the image acquisition independent of the speed of the conveyor belt, ensuring that there is never any overlapping, or gaps between consecutive images.

An algorithm for image processing segments the regions of interest from the background, and then determines which parts are actual arils and what corresponds to undesirable material, such as skin pieces or internal pulp membranes. Arils are small; therefore, after first analysis, those objects with large size are classified as undesired material and labelled for refusing. The colour of the remaining objects is then analysed. In this case, the categories are determined by the colour of each aril with the aim of refusing those with defective colours (immature, decayed). The good ones are then separated into lots of different shades (pink, red, purple). To achieve this, the averaged R and G colour coordinates of each aril are measured to obtain the R/G ratio, which is the feature used to determine the colour category of the arils. The user interface allows the operator to select the thresholds between four different categories. The information of the category is used to determine from which air ejector the aril should be ejected.

After the inspection chamber, the sorting area is determined. Three outlets are located on one side of each conveyor belt. In front of each outlet, at the other side of each conveyor belt, air ejectors are placed to expel the product. When any object is detected in the image, its position is also recorded. A metallic piece of the machine, which has a predefined distance from the different outlets, is captured in all the images serving as a location reference. By measuring the distance between each object and this reference point in the images, it is possible to estimate the position of each aril. This is used to estimate the number of pulses from the position of the arils to the air ejector. The control computer stores different pulse counts for each aril in its memory, decreasing all counts each time when it receives a pulse. When a count reaches the zero value, the aril is passed just in front of the air jet and the computer opens the correct electro-valve to shoot the aril. Since all the synchronization is based on the encoder pulses, it depends totally on the advance of the conveyor belts, independent of their speed. Having three pulses per each millimetre of the conveyor belt advance, this mechanism is very precise if the arils do not move from the acquisition of the image to the moment of ejection.

This prototype, which is shown in Fig. 14.4, has been tested in the food inspection industry, working specifically on the classification of pomegranate arils (Blasco et al., 2009a). It has also been adapted for the classification of satsuma segments (Blasco et al., 2009b). In this case, the main adaptation consisted of the

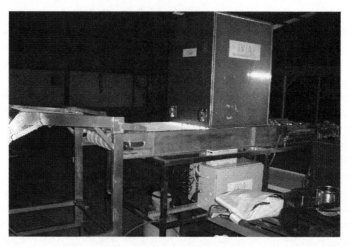

Fig. 14.4   Prototype developed for the automatic sorting of pomegranate arils.

use of semi-transparent conveyor belts in order to light the fruit from below, thus enhancing the contrast between the segments and the background to facilitate the contour analysis. The vision system was capable of working at more than 20 images per second, but mechanical limitations due to difficulty in handling the segments reduced the working speed of the system to six images per second.

14.5   Future trends

14.5.1   Non-standard computer vision systems

Hyperspectral imaging systems are starting to be used as a scientific tool for food quality assessment (Lorente *et al.*, 2012). However, the high amount of information given by these systems makes their incorporation into an industry that demands high-speed processes more complicated, such as in the case of automatic or non-destructive inspection. A typical hyperspectral image is composed of dozens of correlative monochromatic images that normally contain redundant information, or may exhibit a high degree of correlation. In addition, the computational complexity of the classifier can become prohibitive for high-dimensional data sets.

A hyperspectral image is composed of a set of relatively wide-ranging continuous wavelengths, while a multispectral image is composed of a few wavelengths that do not necessarily have to be continuous. These are, therefore, more suited to the industrial needs. Multispectral systems are an alternative that go beyond traditional RGB imaging, allowing for the detection of features that are usually difficult to detect with standard systems. The main advantages are the relatively low cost of the systems in comparison with hyperspectral systems and the fact that they can be more specific for real applications. However, they are still undergoing further development and they are not yet standardized. Many of the works related to the inspection of fruit and vegetables that use hyperspectral systems to acquire the images are aimed to find out particular wavelengths that allow developing systems capable of obtaining similar results using simpler technology or that could be used for in-line applications.

Gómez-Sanchis *et al.* (2008b) analysed the feasibility of detecting rottenness in citrus fruits caused by *Penicillium digitatum* in the early stages of infection using a hyperspectral computer vision system. The aim was to find a reduced set of optimally selected bands in visible and NIR in order to develop a system capable of substituting the current manual inspection based on dangerous UV-induced fluorescent lighting. ElMasry *et al.* (2008) determined some important wavelengths for detecting bruises in apples cv. McIntosh using this technique over hyperspectral images. Three wavelengths in the NIR region (750, 820, 960 nm) were found that could potentially be implemented in multispectral imaging systems. Gowen *et al.* (2009) studied spectral bands related with water to investigate the spectral behaviour of white mushrooms under different levels of mechanical vibration, using a hyperspectral imaging system, based on a spectrograph and operating in the NIR wavelength of 950–1700 nm. Karimi *et al.* (2012) used this technology for the classification of avocados.

Once hyperspectral data have been analysed and a few wavelengths obtained, simpler systems can be developed. A multispectral vision system was developed by Unay and Gosselin (2006) to study several thresholding and classification-based techniques for pixel-wise segmentation of multispectral images of cv. Jonagold apples using a multilayer perceptron-based method, in order to develop an automatic prototype for the detection of surface defects. This work was later enhanced by extracting several features from defective skins aimed at classifying the fruit into different categories (Unay *et al.*, 2011).

14.5.2   Internal quality assessment

Current technology allows scientists to acquire detailed images of the internal parts of fruits and vegetables that could be used for industrial quality assessment. However, the technologies are still very expensive and have particular problems that prevent them from being transferred to use in industry. Promising results allow us to believe that they are a good alternative for the future. Milczarek *et al.* (2009) developed an in-line method to detect damaged pericarp tissue of tomatoes using multivariate analysis of magnetic resonance (MR) images. The technique proved to be effective for predicting the conductivity score of pericarp tissue in tomatoes, with a processing time between 400 ms and 1200 ms per image. Barreiro *et al.* (2008) studied different MR sequences to detect the presence of seeds in mandarins, with 100% accuracy using a radial–spiral sequence. Internal browning of pears was also detected using MR images by Hernández-Sánchez *et al.* (2007), who compared MR relaxometry with MRI.

Thermography is still expensive because of the high price of the cameras, but it can be an alternative to assess the internal quality of fruits and vegetables. It was employed by Baranowski *et al.* (2008) to detect apple watercore. The time derivative of apple temperature per unit of mass was a good parameter to distinguish between apples with and without affected internal tissues.

Hyperspectral imaging has also been widely used to measure internal quality attributes, such as sugar or SSC, flesh and skin colour, firmness, acidity and starch index, of fruits; most of these tests were conducted on apples. With regards to internal quality, maturity is extremely important in harvest time determination and from a marketing context. Several parameters are normally used to assess fruit maturity, such as firmness. ElMasry *et al.* (2007) determined MC, SSC and acidity in strawberries by means of a visible-NIR hyperspectral imaging system (400–1000 nm), showing a good prediction performance for moisture content ($r = 0.91$), SSC ($r = 0.80$) and pH ($r = 0.94$). Fernandes *et al.* (2011) reported a system based on neural networks for the estimation of grape anthocyanin concentration using hyperspectral technology. The internal quality of whole pickles was studied by Ariana and Lu (2010) using transmittance (675–1000 nm) and reflectance (400–675 nm) hyperspectral imaging systems. Images of pickles were obtained using a prototype of in-line hyperspectral imaging system. Principal component analysis was applied to the hyperspectral images of the pickle samples for bloater damage detection. Transmittance images were much more effective for internal defect detection compared with reflectance ones. An overall defect classification accuracy of 86% was achieved, compared with an accuracy of 70% by the human inspectors.

Although, at this time, these technologies are still not developed enough for industrial use in the agricultural field, many groups are working on them to complete applied systems that are economically attainable and whose results are beneficial.

14.6   Conclusions

This chapter has summarized the current state of the art applications of computer vision for the inspection of fruits, vegetables and nuts. The industry has several requirements which are a priority, such as product singulation, real-time processing and the need to inspect the whole surface of the products. In the development of automatic systems, the latter is extremely important, otherwise, problems related with the illumination or the position of the products during their transport may result in a decrease in the performance. Previously, the capacity of processors and memory of the computers limited the size of the images captured, since the processing time is directly related with the size of the image to be analysed. Current processors allow for the analysis of large images with high resolution that permits the inspection of colour with great reliability and the detection of the smallest defects. Hence, this technology can be used to analyse a large variety of different types and species of fruits, vegetables or nuts. In the case of nuts, however, it is used less, probably due to mechanical limitations related with their small size and the difficulty of handling. A case study of a real inspection machine has been described, which involves the development of all the subsystems necessary to implement an automatic in-line inspection of a small processed fruit.

The increasing interdisciplinary nature of research groups offers the possibility of combining genetic, biological and physiological knowledge with physics and computer vision research to take an important step towards integrated solutions for the fruit and vegetable industry. However, there are still challenges in this topic that have to be overcome by researchers. Some future trends that are currently under research have been discussed. Although, nowadays systems to analyse internal quality are probably more suited to laboratory developments, many current works try to provide the industry with important practical solutions. In the future, the inspection of the internal quality of products will most likely be incorporated into the packing lines using either of the available technologies.

This chapter has discussed the technology that is currently used in the industry for the inspection of external quality of products. The future developments for the inspection of internal quality of many horticultural products have also been reviewed. This development is mainly due to the constant reduction in price of the components for machinery and the increase in computational capacity of modern computers. New products and new consumer demands mean that research in this field must be constantly increased in order to discover the full potential of computer vision as a non-destructive analytical tool.

14.7   Sources of further information

There have been some key books published directly related to this topic; among them, the most recent are those edited by Sun (2007, 2009, 2010). These books deal with the application of standard and non-standard computer vision systems to the analysis of food, some chapters being dedicated to the automatic inspection of different fruits, vegetables and nuts. A recent and exhaustive review about this topic has been done by Cubero *et al.* (2011), which describes from laboratory developments to the real applications of computer vision to the inspection of fruit and vegetables, and by Lorente *et al.* (in-press), who have revised most recent applications of hyperspectral imaging applied to the internal and external inspection of fruits and vegetables. A list of the main manufacturers of commercial electronic sorters for agricultural product inspection based on computer vision is provided below.

- Anhui Hongshi Hi-Tech Co., Ltd. http://www.ahhongshi.com. http://www. radixsystems.co.uk/. Inspection of nuts and small products.
- Aweta. http://www.autolinesorters.com. Inspection of external features of fruit and some vegetables using visible and NIR.
- Compac Sorting Equipment. http://www.compacsort.com. Inspection of external and internal features of fruit and some vegetables using visible and NIR.
- FOMESA Machinery. http://www.fomesa.com. Inspection of external features of fruit and some vegetables using visible and NIR.
- Frumac. http://www.frumac.com. Inspection of external features of fruit and vegetable using images in the visible range.

- Greefa. http://www.greefa.nl. Inspection of external and internal features of fruit and some vegetables using visible and NIR.
- MAF RODA. http://www.maf-roda.com. Inspection of external features of fruit and some vegetables using visible and NIR.
- MaxFrut. http://www.maxfrut.com. Inspection of external features of fruit and some vegetables using visible and NIR.
- Multiscan Technologies. http://www.multiscantechnologies.com. Inspection of external features of fruit using visible and NIR, and internal quality using X-rays.
- Radix Systems Limited. http://www.radixsystems.co.uk/. Inspection of nuts and small products.
- SEA s.r.l. http://www.seasort.com/en. Inspection of nuts and small products.
- Tecnofruta Levante, S.L. http://www.tecnofruta.com. Inspection of external features of fruit and using visible.
- Unitec S.p.A. http://www.unitec-group.com. Inspection of external features of fruit and some vegetables using visible and NIR.

14.8   Acknowledgements

This work was partially funded by the Instituto Nacional de Investigación y Tecnologia Agraria y Alimentaria de España (INIA) through research project RTA2009-00118-C02-01 with the support of European FEDER funds.

14.9   References

Abdullah, M. Z., Mohamad-Saleh, J., Fathinul-Syahir, A. S. and Mohd-Azemi, B. M. N. (2006) Discrimination and classification of fresh-cut starfruits (*Averrhoa carambola* L.) using automated machine vision system. *Journal of Food Engineering*, **76**, 506–23.

Aleixos, N., Blasco, J., Navarrón, F. and Moltó, E. (2002) Multispectral inspection of citrus in real-time using machine vision and digital signal processors. *Computers and Electronics in Agriculture*, **33(2)**, 121–37.

Al-Mallahi, A., Kataoka, T., Okamoto, H. and Shibata, Y. (2010a) An image processing algorithm for detecting in-line potato tubers without singulation. *Computers and Electronics in Agriculture*, **70**, 239–44.

Al-Mallahi, A., Kataoka, T., Okamoto, H. and Shibata, Y. (2010b) Detection of potato tubers using an ultraviolet imaging-based machine vision system. *Biosystems Engineering*, **105(2)**, 257–65.

Ariana, D. P. and Lu, R. (2010) Hyperspectral waveband selection for internal defect detection of pickling cucumbers and whole pickles. *Computers and Electronics in Agriculture*, **74(1)**, 137–44.

Arzate-Vázquez, I., Chanona-Pérez, J. J., Perea-Flores, M. J., Calderón-Domínguez, G., Moreno-Armendáriz, M. A., Calvo, H., Godoy-Calderón, S., Quevedo, R. and Gutiérrez-López, G. (2012) Image processing applied to classification of avocado variety Hass (*Persea americana* Mill.) during the ripening process. *Food and Bioprocess Technology*, **5(2)**, 474–84.

Baranowski, P., Lipecki, J., Mazurek, W. and Walczak, R. T. (2008) Detection of watercore in 'Gloster' apples using thermography. *Postharvest Biology and Technology*, **47**, 358–66.

Barreiro, P., Zheng, C., Sun, D.-W., Hernández-Sánchez, N., Pérez-Sánchez, J. M. and Ruiz-Cabello, J. (2008) Non-destructive seed detection in mandarins: comparison of automatic threshold methods in FLASH and COMSPIRA MRIs. *Postharvest Biology and Technology*, **47**, 189–98.

Blasco, J., Aleixos, N. and Moltó, E. (2003) Machine vision system for automatic quality grading of fruit. *Biosystems Engineering*, **85(4)**, 415–23.

Blasco, J., Aleixos, N. and Moltó, E. (2007) Computer vision detection of peel defects in citrus by means of a region oriented segmentation algorithm. *Journal of Food Engineering*, **81**, 535–43.

Blasco, J., Cubero, S., Gómez-Sanchis, J., Mira, P. and Moltó, E. (2009a) Development of a machine for the automatic sorting of pomegranate (*Punica granatum* L) arils based on computer vision. *Journal of Food Engineering*, **90**, 27–34.

Blasco, J., Aleixos, N., Cubero, S., Gómez-Sanchis, J. and Moltó, E. (2009b) Automatic sorting of satsuma (*Citrus unshiu*) segments using computer vision and morphological features. *Computers and Electronics in Agriculture*, **66**, 1–8.

Blasco, J., Aleixos, N., Gómez-Sanchis, J. and Moltó, E. (2009c) Recognition and classification of external skin damage in citrus fruits using multispectral data and morphological features. *Biosystems Engineering*, **103**, 137–45.

Cubero, S., Aleixos, N., Moltó, E., Gómez-Sanchis, J. and Blasco, J. (2011) Advances in machine vision applications for automatic inspection and quality evaluation of fruits and vegetables. *Food and Bioprocess Technology*, **4(4)**, 487–504.

Currie, A. J., Ganeshanandam, S., Noiton, D. A., Garrick, D., Shelbourne, C. J. A. and Oraguzie, N. (2000) Quantitative evaluation of apple (*Malus x domestica* Borkh.) fruit shape by principal component analysis and Fourier descriptors. *Euphytica*, **111**, 219–27.

ElMasry, G., Wang, N., ElSayed, A. and Ngadi, M. (2007) Hyperspectral imaging for non-destructive determination of some quality attributes for strawberry. *Journal of Food Engineering*, **81**, 98–107.

ElMasry, G., Wang, N., Vigneault, C., Qiao, J. and ElSayed, A. (2008) Early detection of apple bruises on different background colors using hyperspectral imaging. *LWT*, **41**, 337–45.

Fernandes, A. M., Oliveira, P., Moura, J. P., Oliveira, A. A., Falco, V., Correia, M. J. and Melo-Pinto, P. (2011) Determination of anthocyanin concentration in whole grape skins using hyperspectral imaging and adaptive boosting neural networks. *Journal of Food Engineering*, **105(2)**, 216–26.

Fernández, L., Castillero, C. and Aguilera, J. M. (2005) An application of image analysis to dehydration of apple discs. *Journal of Food Engineering*, **67**, 185–93.

Gardner, J. L. (2007) Comparison of calibration methods for tristimulus colorimeters. *Journal of Research of the National Institute of Standards and Technology*, **112**, 129–38.

Ghazanfari, A., Irudayaraj, J., Kusalik, A. and Romaniuk, M. (1997) Machine vision grading of pistachio nuts using Fourier descriptors. *Journal of Agricultural Engineering Research*, **68(3)**, 247–52.

Gómez-Sanchis, J., Gómez-Chova, L., Aleixos, N., Camps-Valls, G., Montesinos-Herrero, C., Moltó, E. and Blasco, J. (2008b) Hyperspectral system for early detection of rottenness caused by *Penicillium digitatum* in mandarins. *Journal of Food Engineering*, **89(1)**, 80–6.

Gómez-Sanchis, J., Moltó, E., Camps-Valls, G., Gómez-Chova, L., Aleixos, N. and Blasco, J. (2008a) Automatic correction of the effects of the light source on spherical objects. An application to the analysis of hyperspectral images of citrus fruits. *Journal of Food Engineering*, **85(2)**, 191–200.

Gowen, A. A., Tsenkova, R., Esquerre, C., Downey, G. and O'Donnell, P. D. (2009) Use of near infrared hyperspectral imaging to identify water matrix co-ordinates in mushrooms

(*Agaricus bisporus*) subjected to mechanical vibration. *Journal of Near Infrared Spectroscopy*, **17(6)**, 363–71.

Hernández-Sánchez, N., Hills, B., Barreiro, P. and Marigheto, N. (2007) A NMR study on internal browning in pears. *Postharvest Biology and Technology*, **44**, 260–70.

Hoffman, G. (2011) CIE colour space. Available from http://www.fho-emden.de/~hoffmann/ciexyz29082000.pdf (accessed June 2011).

Kang, S. P., East, A. R. and Trujillo, F. J. (2008) Colour vision system evaluation of bicolour fruit: A case study with 'B74' mango. *Postharvest Biology and Technology*, **49**, 77–85.

Karimi, Y., Maftoonazad, N., Ramaswamy, H. S., Prasher, S. O. and Marcotte, M. (2012) Application of hyperspectral technique for color classification avocados subjected to different treatments. *Food and Bioprocess Technology*, **5(1)**, 252–64.

Leemans, V. and Destain, M.-F. (2004) A real-time grading method of apples based on features extracted from defects. *Journal of Food Engineering*, **6**, 83–9.

León, K., Domingo, M., Pedreschi, F. and León, J. (2006) Color measurement in L*a*b* units from RGB digital images. *Food Research International*, **39**, 1084–91.

Liming, X. and Yanchao, Z. (2010) Automated strawberry grading system based on image processing. *Computers and Electronics in Agriculture*, **71(S1)**, S32–S39.

López-García, F., Andreu-García, A., Blasco, J., Aleixos, N. and Valiente, J. M. (2010) Automatic detection of skin defects in citrus fruits using a multivariate image analysis approach. *Computers and Electronics in Agriculture*, **71**, 189–97.

Lorente, D., Aleixos, N., Gómez-Sanchis, J., Cubero, S., García-Navarrete, O. L. and Blasco, J. (2012). Recent advances and applications of hyperspectral imaging for fruit and vegetable quality assessment. *Food and Bioprocess Technology*, **5(4)**, 1121–42.

Lorente, D., Aleixos, N., Gómez-Sanchis, J., Cubero, S. and Blasco, J. (in-press) Selection of optimal wavelength features for decay detection in citrus fruit using the ROC curve and neural networks. DOI: 10.1007/s11947-011-0737-x.

Ludwig, A.. (1973) The definition of cross polarization. *IEEE Transactions on Antennas and Propagation*, **1**, 116–19.

Mendoza, F., Dejmek, P. and Aguilera, J. M. (2006) Calibrated color measurements of agricultural foods using image analysis. *Postharvest Biology and Technology*, **41**, 285–95.

Menesatti, P., Costa, C., Paglia, G., Pallottino, F., D'Andrea, S., Rimatori, V. and Aguzzi, J. (2008) Shape-based methodology for multivariate discrimination among Italian hazelnut cultivars *Biosystems Engineering*, **101**, 417–24.

Milczarek, R. R., Saltveit, M. E., Garvey, T. C. and McCarthy, M. J. (2009) Assessment of tomato pericarp mechanical damage using multivariate analysis of magnetic resonance images. *Postharvest Biology and Technology*, **52**, 189–95.

Pallottino, F., Menesatti, C., Costa, C., Paglia, G., De Salvador, F. R. and Lolletti, D. (2010) Image analysis techniques for automated hazelnut peeling determination. *Food and Bioprocess Technology*, **3(1)**, 8, 155–9.

Pedreschi, F., León, J., Mery, D. and Moyano, P. (2006) Development of a computer vision system to measure the color of potato chips. *Food Research International*, **39**, 1092–8.

Peterson, C. (2001) How it works: the charged-coupled device or CCD. *Journal of Young Investigators*. Available from http://www.jyi.org/volumes/volume3/issue1/features/peterson.html (accessed June 2011).

Reese, D., Lefcourt, A. M., Kim, M. S. and Lo, Y. M. (2010) Using parabolic mirrors for complete imaging of apple surfaces. *Bioresource Technology*, **100**, 4499–506.

Reyes, J. F. and Chiang, L. E. (2009) Location and classification of moving fruits in realtime with a single color camera. *Chilean Journal of Agricultural Research*, **69(2)**, 179–87.

Sun, D.-W. (ed.) (2007) *Computer Vision Technology for Food Quality Evaluation*. London: Academic Press, Elsevier Science .

Sun, D.-W. (ed.) (2009) *Infrared Spectroscopy for Food Quality Analysis and Control*. London: Academic Press, Elsevier Science .

Sun, D.-W. (ed.) (2010) *Hyperspectral Imaging for Food Quality Analysis and Control.* London: Academic Press, Elsevier Science .

Throop, J. A., Aneshansley, D. J., Anger, W. C. and Peterson, D. L. (2005) Quality evaluation of apples based on surface defects: Development of an automated inspection system. *Postharvest Biology and Technology,* **36**, 281–90.

Unay, D. and Gosselin, B. (2006) Automatic defect segmentation of 'Jonagold' apples on multi-spectral images: A comparative study. *Postharvest Biology and Technology,* **42**, 271–9.

Unay, D., Gosselin, B., Kleynen, O., Leemans, V., Destain, M. F. and Debeir, O. (2011) Automatic grading of Bi-colored apples by multispectral machine vision. *Computers and Electronics in Agriculture,* **75(1)**, 204–12.

Vadivambal, R. and Jayas, D. S. (2011) Applications of thermal imaging in agriculture and food industry – a review. *Food and Bioprocess Technology,* **4(2)**, 186–99.

Xiao-bo, Z., Jie-wen, Z., Yanxiao, L. and Holmes, M. (2010) In-line detection of apple defects using three color cameras system. *Computers and Electronics in Agriculture,* **70**, 129–34.

15

Grain quality evaluation by computer vision

D. S. Jayas and C. B. Singh, University of Manitoba, Canada

Abstract: Grain quality is defined by several factors such as physical (moisture content, bulk density, kernel size, kernel hardness, vitreousness, kernel density and damaged kernels), safety (fungal infection, mycotoxins, insects and mites and their fragments, foreign material odour and dust) and compositional factors (milling yield, oil content, protein content, starch content and viability). This chapter discusses several computer vision technologies such as colour imaging, hyperspectral imaging, X-ray imaging and thermal imaging and reviews their applications in grain quality evaluation based on the above described grain quality factors.

Key words: grain quality, colour imaging, hyperspectral imaging, X-ray imaging, thermal imaging.

15.1   Introduction

Due to increasing consumer awareness for high quality and healthy food, grain merchandisers and food processors are paying more attention to the quality of grain purchased from farms and grain elevators (grain handling facilities) with zero-tolerance to grain contamination. Most of the grains produced globally are cereals (wheat, paddy, barley, corn, oats and rye), oilseeds (canola, soybeans, safflower, sunflower, flaxseed) and pulses (peas, beans, chickpeas and lentils), and are consumed in different forms. Most of the grain-exporting countries have introduced strict regulations to determine grain quality and grade with very low tolerance to contaminations to remain competitive in the international grain market. Grain quality is evaluated by several parameters such as physical factors (moisture content, bulk density, kernel size, kernel hardness, vitreousness, kernel density and damaged kernels), sanitary factors (fungal infection, mycotoxins, insects and mites and their fragments, foreign material, odour and dust) and intrinsic factors (milling yield, oil content, protein content, starch content and viability) (Muir, 2001). Most of the countries use many of these factors as the main criteria in grading and quality evaluation of grains. For example, Canadian wheat is graded based on the bulk density (test weight), varietal purity, vitreousness, soundness and foreign material content (CIGI, 1993).

At present, manual inspection and/or chemical analyses methods are used for grain quality evaluation at the grain handling/processing facilities, which are time consuming. Manual inspection methods are subjective and less efficient and wet chemical analysis is destructive. Therefore, a rapid, objective and accurate auto-mated technique is needed for grain quality estimation in real-time which meets the standards of the international grain market.

Machine vision systems have shown promise for real-time automated inspec-tion of grains and other agri-food products. A machine vision system, includes a detector (camera), electromagnetic radiation/light source (e.g. visible, infrared, near-infrared, X-rays), frame grabber board, interface (cable) and image acqui-sition and processing software. The schematic of a continuous line-scan imag-ing system for inspecting the grain on a conveyer belt is shown in Fig. 15.1. Electromagnetic radiation covers energy waves from gamma rays, X-rays, ultra-violet (UV), visible (VIS), near-infrared (NIR), infrared (IR), microwaves, to radiowaves in descending order of energy. Based on the radiation source used, the machine vision systems are classified into colour imaging (visible light source), NIR hyperspectral imaging (NIR radiation source), thermal imaging (IR radia-tion) and X-ray imaging (X-ray source). There are other imaging systems with different radiation sources (e.g., IR light, UV, radiowaves); however, they are not covered in this chapter due to their limited applications in automated grain inspection.

Colour machine vision systems have been used in grain quality analyses for discriminating different grain types, classes, varieties, impurities, fungal-infected and insect-damaged kernels (Ahmad et al., 1999; Choudhary et al., 2008; Paliwal et al., 2003; Zayas and Flinn, 1998). Colour images of individual kernels are described by colour, textural and morphological features and are used in qual-ity assessment of grains. Colour cameras are less expensive and thus are widely used in machine vision systems in many industrial inspection applications. Near-infrared hyperspectral imaging (NIR HSI) provides spectral information in a spa-tially resolved manner. Spatial information is useful in the grain quality monitoring as it can provide chemical mapping of the sample from the analysis of hyperspec-tral data (hypercube). The mappings can also be visually examined without any significant subjectivity. Hyperspectral imaging (HSI) has been used for decades in remote sensing for earth resource monitoring and is finding new applications in medicine, agriculture, manufacturing and forensics (Gat, 2000). Though rela-tively new in the agri-food industry, HSI has shown potential for quality evalua-tion and inspection of several agricultural products (Gowen et al., 2007). Thermal imaging systems, which use thermal IR radiation detectors (long-wave), have also

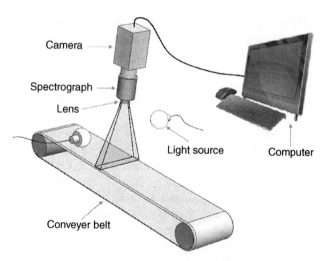

Fig. 15.1   Schematic of a line-scan machine vision system.

been investigated for their applications in quality and safety inspection of agricultural and food products (Gowen *et al.*, 2010; Vadivambal and Jayas, 2011). The IR imaging systems using IR light sources and interferograms are different from thermal imaging systems and are not covered in this chapter due to their limitation in real-time automated inspection. Soft X-rays imaging systems using a fluoroscope (produces soft X-rays of 1–100 nm wavelengths) have also been used in detecting damaged grains.

In this chapter, various machine vision systems such as colour imaging, NIR HSI, X-ray imaging and thermal imaging, their hardware components and working principles are described and their applications in grain quality analysis are reviewed.

15.2   Colour imaging

Colour plays a significant role in evaluating the quality of grain. The sections below consider the practical implications of this within the technology, and how these can be practically applied.

15.2.1   Colour imaging systems

A colour imaging system mainly consists of light source, colour camera (analogue or digital camera), digital converter for analogue camera, frame grabber board, zoom lens (mounted in front of camera for focusing the object) and a computer for image acquisition, storage and processing. An area-scan colour imaging system is shown in Fig. 15.2. Light emitting diodes, halogen lamps, incandescent lamps

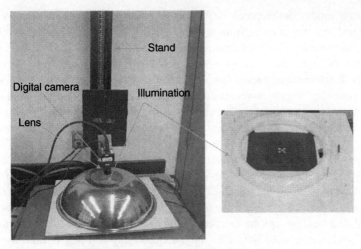

Fig. 15.2   An area-scan colour imaging system.

and fluorescent lamps are generally used as light sources to illuminate the field of view (FOV) of the camera. Uniform light should be delivered to the FOV to obtain high quality images. To keep the light intensity at the same level, efficient illumination lamp controllers equipped with photodiode light sensors are used. Illumination standardization and spatial calibration are also conducted prior to the imaging.

The charge-coupled device (CCD) and complementary metal oxide semiconductor (CMOS) are two main types of cameras available for colour imaging. The sensors in the colour cameras are sensitive to the visible (400–700 nm) part of electromagnetic spectrum. The multispectral images at three wavebands of red (R), green (G) and blue (B) produce colour images. For line-scan imaging, the cameras consist of a linear array of detectors. For area-scan imaging, cameras consist of a two-dimensional (2-D) array of detectors.

FireWire (IEEE 1394) and CAMERA Link are standard communication interfaces used for transferring digital image data from camera to a computer. CAMERA Link cable with a standard channel link chip (interfaces camera, and a frame grabber board attached to computer) provides higher data transfer speed than FireWire. The GiGE VISION (Egri, 2006) is a newly developed interface, which provides very high data transfer rate but has a limited bandwidth. FireWire communication interfaces do not require a frame grabber board and are directly connected to the computer. The length of cables in GiGE VISION can be as long as 100 m, against the CAMERA Link cable length of approximately 10 m maximum.

15.2.2   Colour image features
Colour images are quantitatively described by morphological, colour, textural and wavelet features and these features are used in objective classification and

prediction model development (Choudhary *et al.*, 2008). The common morphological features derived from a colour image are area, perimeter, maximum radius, minimum radius, mean radius, major axis length, minor axis length, shape moments, radial Fourier descriptors and boundary Fourier descriptors. The colour features that can be extracted from the colour image of an object (sample) are mean, variance, ranges, moments and histograms of the RGB colours, and mean, intensity and range of hue (H), saturation (S) and intensity (I) values. Textural features extracted from an image describe the object patterns by roughness, smoothness or repetitiveness (Gonzalez and Woods, 2002). Textural features are derived from grey-level co-occurrence matrix (GLCM) and grey-level run length matrix (GLRM). The GLCM provides information regarding the distribution of grey-level intensities in an image and GLRM gives the information about the occurrence of collinear and consecutive pixels of similar grey levels in the image. The textural features that can be extracted from GLCM are mean, variance, uniformity, correlation, cluster shade, entropy, homogeneity and inertia from the grey, red, blue and green bands. The textural features extracted from GLRM are short run, long run, colour non-uniformity, run length non-uniformity, entropy and run percentage from the grey, red, blue and green bands. Wavelet analysis of an image that decomposes it into low- and high-frequency components (up to a given level of decomposition) is also used for textural analysis. Energy and entropy determined from wavelet coefficients at each level of decomposition (or each resolution) describe the object texture and are used for discrimination. These extracted colour image feature sets (colour, texture, morphological, wavelets) still form a large data size for training of classifiers and prediction models. The dimension of an extracted feature set can be further reduced by multivariate analysis and most important features contributing to classification and prediction models can be selected. The features that do not significantly contribute to the model can be eliminated from the training set of classification models. Using top features from combined morphological, colour, textural and wavelet features instead of each individual feature tends to give better classification and prediction accuracy (Choudhary *et al.*, 2008).

15.2.3   Applications of colour imaging

Colour imaging has been extensively used in the agriculture and food industry for quality and safety inspection, damage detection, sorting, grading and monitoring of food processing systems. Colour image features extracted from grains quantitatively describe colour, shape, size and texture of the grains and these features when used in developing supervised classification/prediction algorithm give high accuracy in grain quality analyses. Steenhoek *et al.* (2001) used colour imaging to segment corn kernels surface damage into blue-eye mould, germ damage, sound germ, shadow in sound germ, hard starch and soft starch. RGB colour intensities representing colour features were extracted from the images and were given as input to a probabilistic neural network (PNN) classifier. A histogram threshold was applied to remove the background. The PNN classifier gave up to 75% of classification accuracy; misclassification occurred due to significant feature over-lap between similar categories. The improved performance of 95% accuracy was achieved by grouping pixels into only three categories, namely damage (blue-eye mould and germ damage), germ (sound germ and shadow in sound germ) and starch (soft and hard starch). A colour imaging-based grader was developed for corn grading (Ni *et al.*, 1998). Morphological image features (kernel length, width, thickness and projected area) were extracted after removing background using fixed thresholding. The ratio of the kernel thickness to width was used in a tree-structured linear discriminant classifier to separate flat kernels from round-shaped kernels. Kernel length, width and projected area were used in classifying kernels based on their size (large, medium and small). Moderate to high classifi-cation accuracy of 74–90% was achieved. Ng *et al.* (1998) discriminated mechan-ical and mould-damaged corn kernels using colour machine vision. Mechanically damaged kernels were dyed (green) before imaging to obtain higher values of G (green) component compared with the R (red) component of the damaged por-tions. Using G/R ratio as the thresholding cut-off, damaged pixels were discrim-inated from undamaged pixels on the surface of corn kernels. Fungal-damaged corn kernels could not be dyed as dye adhered to only the starchy area of damaged kernels. The RGB intensity values were extracted from the corn kernel images and used in developing a back propagation neural network (BPNN) classifier. The model correctly classified more than 99.5% mechanical damaged kernels and 98.7% mould-damaged corn kernels. Although higher accuracy was achieved in classifying mechanical damaged kernels, dyeing is not realistic for real-time grain sorting at grain handling facilities.

Classification algorithms were also developed using digital colour images to sort rice into sound, cracked, chalky, immature, dead, broken, damaged and off-type kernels (Wan *et al.*, 2002). The image processing and feature extraction program developed in C programming language removed the background using histogram thresholding and extracted 16 kernel features. Logical 'and' was used for the classification and rice kernels were categorized. This system correctly classified more than 90% kernels on average.

Zayas and Flinn (1998) detected the insects in bulk wheat samples using a colour machine vision system. Thirty-five image features, such as mean, standard deviation, skewness and kurtosis, and derived features from histograms of red, green, blue and hue, saturation and lightness were extracted and used as input in discriminant classifiers. The developed two-class (insects versus non-insects) and three-class (insects, dark grass seeds and others) models gave 93.8% and 92.6% average accuracy, respectively. Kernel orientation (ventral or dorsal side), par-ticles clinging to the insect and dockage with similar colour to insects affected the performance of the model. Hidden insects and early development stages were not detected in this study. Other imaging techniques such as X-rays can potentially detect hidden insects.

Majumdar and Jayas (2000a, 2000b, 2000c, 2000d) developed image process-ing algorithms for classification of cereal grains, namely Canada western red spring (CWRS) wheat, Canada western amber durum (CWAD) wheat, barley, oats and rye. They extracted colour, textural and morphological features from colour images of individual kernels and developed classification algorithms. The model achieved the highest average classification accuracy of 99.7% and 99.8%, respectively, for CWRS and CWAD wheat when textural, colour and morphological features were combined and 20 most significant features were given as input to discriminant classifiers. In this study clean grain samples were used; however, grain received at grain handling facilities may contain some impurities and foreign materials such as broken grain, other variety grain, chaff, straw and dust. Paliwal *et al.* (2003) developed a model to classify cereal grains (CWRS wheat, CWAD wheat, barley, oats and rye) and detect dockage (broken wheat kernels, chaff, buckwheat, wheat spikelets and canola seeds) present in the grain. Morphological, colour and textural features were extracted from the colour images and the top 20 significant features each from texture, colour and morphological features and combined features were selected by statistical analysis and used in training and testing of BPNN algorithms. The developed model correctly classified most of the cereals; however, dockage particles with irregular shape and undefined features gave lower classification accuracy (90%). Proper cleaning and removal of dockage is essential for improving classification accuracy.

Choudhary *et al.* (2008) used a colour imaging system to classify cereals, namely CWRS wheat, CWAD wheat, barley, oats and rye. Wavelet, colour, morphological and textural features were extracted from individual kernels and used in developing linear discriminant analysis (LDA) and quadratic discriminant analysis (QDA) classification algorithms. Linear discriminant classifiers gave the highest classification accuracy (89.4–99.4%) using combined wavelet, colour, morphological and textural image features in training.

15.3   Hyperspectral imaging

The following sections look at the way HSI systems work, their differences with other systems and their suitability for various applications.

15.3.1   Hyperspectral imaging systems

Similar to other machine vision devices, the HSI system also consists of a detector, radiation source (light), image acquisition software and an integrated computer for data acquisition and storage. In addition, HSI must have a wavelength selection or filtering device to obtain NIR radiation at desired wavelengths and remove out-of-band radiation. Silicon detector-based CCD cameras are used for only 700–1100 nm wavelength region of NIR spectral range (700–2500 nm). The CCD cameras are cost effective but are not suitable for longer wavelength scanning application beyond 1100 nm as silicon becomes transparent and needs a coating material that reduces the quantum efficiency of the detector. The most common type of camera used in HSI systems is indium–galium–arsenide (InGaAs) detector with high sensitivity in 900–1700 nm wavelength range (available up to 2500 nm) and low thermal noise. The InGaAs-detector-based cameras have high quantum efficiency (>85%), wider spectral range and fast response, which produces very low dark current, and are operational at room temperature by thermoelectric cooling. Indium–antimonide (InSb), mercury–cadmium–telluride (HgCdTe), and quantum well infrared photodetectors (QWIPs) are other common detectors used in multispectral/hyperspectral imaging (Tran, 2003).

Based on sample presentation techniques, there are mainly two types of HSI systems available: line-scan (pushbroom) and area-scan. The selection of a wavelength-filtering device is highly dependent on the type of HSI system. In line-scan HSI systems (moving objects on conveyer), a dispersing device such as spectrograph is used for wavelength selection. In line-scan HSI systems, which are more suitable for on-line inspections, full spectral information is collected for each pixel in spatial dimension (width) and successive line scans are combined to form a three-dimensional (3-D) hypercube. In area-scan imaging system, the spatial information for each pixel is simultaneously collected at each wavelength sequentially to form a 3-D hypercube. This hypercube (from both types of systems) contains the spatial features (pixel information) in the first two dimensions along x and y axis and spectral features (wavelength) in a third dimension (z axis) (Fig. 15.3). Since objects (samples) are stationary in area-scan imaging, sequential wavelength selection devices are used that pass the light at a selected wavelength to the detector after reflecting from the sample. Two types of electronically tunable filters (ETFs), namely acousto-optical tunable filter (AOTF) and liquid crystal tunable filter (LCTF), are used in area-scan HSI systems. These filtering devices have large optical apertures, high spectral resolution, wide spectral range and randomly accessible tuning wavelengths. An area-scan NIR HSI system is shown in Fig. 15.4.

Most of the detectors suffer from thermal noise, termed as 'dark current', which is caused by the flow of charge in the detectors due to the thermal energy at ambient temperature when there is no radiation (light reflected from the sample) reaching the detector. Thermoelectric cooling or liquid nitrogen cooling systems in advanced NIR cameras minimize dark current, but hyperspectral image data must be corrected for dark current effect. Dark current should be recorded prior to imaging of samples and then subtracted from subsequent acquired hyperspectral images. Dark current should be collected at regular intervals.

Hyperspectral images produced by HSI systems also suffer from geometric distortions and need to be corrected. Due to moving parts in a spectrograph, pushbroom HSI systems cause two registration problems, namely 'smile' and 'keystones'. Smile is the curvature distortion of horizontal lines and keystones are the distortion of rectangles into trapezoids; these can be corrected by applying geometric control points (GCP) calibration (Lawrence et al., 2003; Polder et al., 2003). Use of sequential filtering, in the area-scan HSI systems, changes wavelengths sequentially but focus does not change, which produces optical errors called lateral chromatic aberration (LCA) and axial chromatic aberration (ACA). The LCA is the geometric image distortion produced by different magnifications at each scanning wavelength and ACA is blurring of images at specific wavelengths

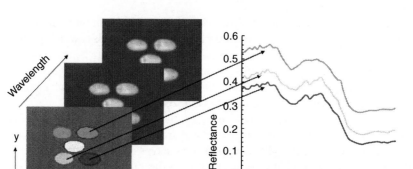

Fig. 15.3   Three-dimensional (3-D) hyperspectral data representation in a hypercube.

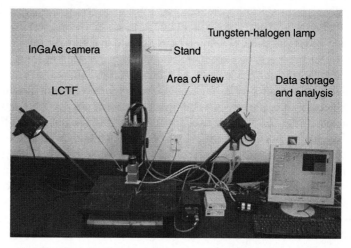

Fig. 15.4   Area-scan NIR hyperspectral imaging system.

caused by defocusing. These optical errors also need to be corrected (Wang and Paliwal, 2006).

15.3.2   Hyperspectral image features

The large number of hyperspectral data collected in a 3-D hypercube require an appropriate data reduction/feature extraction method. HSI data can be analysed as an image at each waveband using imaging methods described in previous sections and/or as a spectrum at each pixel by applying a wide range of chemometric methods used in spectroscopy. Multivariate image (MVI) analysis is a common method used to reduce the large size of hyperspectral data (Geladi and Grahn, 1996) and to identify and select significant wavebands. In MVI, the 3-D hyperspectral data of a hypercube are reshaped into a 2-D array by arranging all pixel intensity values of an NIR image at each waveband into a column (each column represents a variable wavelength) and then principal component analysis (PCA) is applied. Principal component (PC) loadings are used to identify significant wavelengths for multispectral imaging. Pseudo colour maps (also called chemical maps) formed by PC scores are used to identify heterogeneity/uniformity within and/or among samples. A number of other methods have also been reported in HSI for dimension reduction, wavelength selection, classification and prediction, including band (wavelength) ratios, band differences, partial least squares (PLS), PCA-Fisher discriminant analysis (FDA), multiple linear regression (MLR), artificial neural networks (ANN), decision trees, spectral angle mappers, hybrid PC-ANN, fuzzy logic, support vector machines (SVM), image thresholding, independent component analysis (ICA), wavelet analysis and discriminant analysis (Singh et al., 2010a).

15.3.3   Applications of hyperspectral imaging

Recent studies have demonstrated the potential of this technique for grain quality analysis such as moisture, protein and oil content determination, damage detection and classification. Near-infrared HSI has been used to evaluate vitreousness of durum wheat (Gorretta et al., 2006). Vitreousness, defined as glossy or shiny appearance of the wheat kernel, is considered an important characteristic that affects the milling performance of durum wheat and quality of products made from semolina. The wheat kernels were scanned using an NIR HSI system (650–1100 nm wavelength range) and averaged reflectance spectra were obtained by averaging spectral reflectance of all pixels within a kernel. Wheat kernels were classified by developing a PLS-FDA algorithm using reflectance spectral data. Maximum classification accuracy of 94% was achieved. There are significant absorption bands (e.g., protein, starch) in the 1100–1700 nm wavelength region that may further improve the performance of classifiers.

Hyperspectral imaging has been applied in quantitative analysis of corn kernels to determine moisture and oil content (Cogdill et al., 2004). Single corn kernels were scanned in transmittance mode in 750–1090 nm wavelength range and the size of hypercube was reduced by averaging spectra of all pixels within the kernel. Reference oil and protein content data were collected and calibration models were developed using PCR and PLS to predict the moisture and oil content of the corn. Averaging of spectra improves the signal-to-noise ratio but spatial information is lost, therefore, averaging cannot be used to investigate the uniformity/heterogeneity within a sample.

A short-wave NIR imaging system was used to classify wheat into colour classes by Archibald et al. (1998). Six wheat classes, three each from hard red spring (HRS) and hard white winter (HDWW) wheat were first scanned in an NIR spectrometer to obtain spectral characteristics and then significant wavelengths were selected to determine the percentage of red colour of a sample using MLR. Mixtures of HRS and HDWW wheat in 50:50 proportions were scanned as bulk samples using the spectral imaging system at 11 selected wavelengths. Hyperspectral data were analysed by PCA and eight PC scores were examined and PC scores were mapped into pseudo colour images. The score identified the differences between red and white wheat types, but spectroscopic analysis gave better results. Several factors such as non-uniform illumination, saturated pixels and white pixels might have reduced the performance. Development of a supervised classification algorithm could possibly improve the performance as analysis of scores was subjective.

Near-infrared HSI was used to group western Canadian wheat into eight commercial classes (Mahesh *et al.*, 2008). Bulk wheat samples were scanned in 960–1700 nm wavelength range at 75 wavebands in 10 nm intervals. The hyperspectral image data were analysed by cropping an area of 200 × 200 pixels around the central pixels, averaging the spectra and normalizing using standard 99% reflectance panel spectra. The averaging reduced the size in spatial dimension but kept the original spectral dimension. Thus, the size of spectral dimension was reduced by applying PROC STEPDISC (SAS, 2002) procedure to the averaged spectra and significant wavelengths were selected using the criteria of partial R-square and average squared canonical correlation (ASCC). Linear and quadratic discriminant analyses and BPNN (standard and Wardnet) classifiers were used to develop classification algorithms. The BPNN classifier correctly classified more than 90% of wheat classes. Linear discriminant classifier gave the best classification accuracy, with correct classification of 94–100%.

Choudhary *et al.* (2009) reduced the size of hypercube and extracted features by wavelet texture analysis for identification of wheat classes. A 256 × 256 pixel area around the central image pixel was cropped and Daubechies-4 wavelet transform was applied up to five levels of resolution to each of 75 waveband images. Textural features (energy and entropy) were extracted from each image band at each level in the horizontal, vertical and diagonal directions. At each resolution level an additional rotationally invariant feature was extracted by adding the above three features (horizontal, vertical and diagonal) resulting in 40 features (8 × 5 = 40) per image band and 3000 features (40 × 75) per hypercube. The extracted features still formed a large data set (3000 variables per sample). Therefore, significant features (top 10–100) were extracted from a total of 3000 features by statistical analysis and linear and quadratic discriminant and neural network classification models were developed for wheat classification. The multivariate analysis was also used to reduce the dimensionality and PCA was applied to the normalized and reshaped (from 3-D to 2-D) hyperspectral data. The first three PCs explained more than 99% of the variation of original data. These PC scores were mapped into pseudo colour images and the same wavelet features described earlier were extracted, resulting in 120 features (40 × 3 = 120). These wavelet features from each of the three score images, combined 120 features, and top selected features from 120 features were used to develop classification model for comparison with previous analysis. Wavelet texture features extracted from NIR images gave better classification compared with PC score images. Linear discriminant classifiers classified more than 99% of wheat samples using the top

90 features from NIR images. It was observed that the features at coarser reso-lutions at 4th and 5th level did not significantly contribute to the classification. Wavelet energy features contributed more to the classification than entropy fea-tures. Orientation invariant features contributed more to classification than each of individual orientation features.

Berman *et al.* (2007) classified wheat kernels as sound or stained (discoloured) using HSI. Hyperspectral images were generated by a point-based scanning spec-trometer (grid mapping) in the 350–2500 nm wavelength range at 1 nm interval (2151 points), which took approximately 10 h to scan 300 kernels placed in a grid. The hyperspectral data were pre-processed by eliminating the first 70 wave-lengths due to high noise that would cause blurring in the images. The remaining spectral points were co-averaged (10 points), reducing the spectral points to 208 from 2151. The spectral data were analysed using 420–2500 nm (all 208 wave-bands), 420–1000 nm (58 wavebands) and 420–700 nm (28 wavebands) spectral ranges. Wheat kernels were classified as sound or stained by a penalized dis-criminant analysis (PDA) classifier with more than 95% classification accuracy. Classification results were similar, using full as well as reduced spectral ranges. Use of 28 wavebands in classification is still large in comparison to real-time mul-tispectral imaging (two to ten wavelengths); therefore, fewer wavelengths (opti-mally, two or three) should be selected using multivariate analysis.

Recently, long-wave NIR (1000–1600 nm) (LWNIR) and short-wave NIR (700–1100 nm) (SWNIR) hyperspectral and colour machine vision systems were investigated to detect damaged wheat kernels such as insect damaged (Singh *et al.*, 2009a, 2010b), fungal damaged (Singh *et al.*, 2007, 2012) and sprouted and midge damaged (Singh *et al.*, 2009b, 2010c). Hyperspectral data of scanned damaged and control (healthy) individual wheat kernels were analysed using MVI analysis and significant wavebands were identified. Statistical and histo-gram features were extracted from selected waveband NIR images and used in classification. Top colour, textural and morphological features were selected from colour images of damaged and healthy kernels. Classification algorithms (lin-ear, quadratic, Mahalanobis) were developed using features from each system independently. The SWNIR and colour imaging feature classification suffered from high false positive errors in some cases – that is, healthy (control) ker-nels were misclassified as damaged. The LWNIR gave the lowest false positive errors. Combining SWNIR image features with top colour image features gave the best classification accuracy, with minimum false positive error of only up to 4%. However, the LWNIR region has several absorption bands associated with protein, starch and water, which are critical for quantitative analysis.

15.4   X-ray imaging

X-ray technology has been available for some time. The following sections study its use and disadvantages (including possible damage to products and equipment, and safety concerns), then looking at its actual and potential applications.

15.4.1   X-ray imaging systems

An X-ray imaging system consists of an X-ray radiation source, monochrome camera, frame grabber board and a computer. X-ray imaging systems are divided into soft and hard X-ray imaging systems based on energy of X-rays. Hard X-rays, with high energy photons of 0.1–1 nm wavelength and higher penetration power, are used for scanning dense materials such as metals. In agricultural and food applications, mostly soft X-rays with low energy of 1–100 nm wavelengths and low penetrating power are used. A soft X-ray imaging system is shown in Fig. 15.5. A fluoroscope is used as soft X-ray energy source. For soft X-ray imaging, front illuminated (3–10 keV) and back thinned CCD (2–20 keV) detectors are used. High energy X-rays can damage the sensors in CCD that will cause reduction in sensitivity and increase in dark current. Care must be taken to protect the camera from damage by X-rays. In principle, X-ray imaging provides density-based differences among various scanned samples. When an X-ray beam (incident energy) passes through a sample, a portion of it is absorbed and/or scattered (known as attenuation) and the remaining energy is transmitted through material and recorded by the detector or camera. The absorption of X-ray energy by the sample depends on its thickness and density. Samples with higher density and/or thickness will absorb more incident energy, resulting in lower transmitted energy reaching the X-ray detector. In a biological tissue, different types of damages cause a change in density, which can be used for discrimination by X-ray imaging. At present, continuous X-ray systems are available but there is not yet a mechanism to separate the kernels into a single layer in a non-touching manner. Grains can be, first, separated mechanically and then by software programs after the X-ray images are acquired. Software programs are also required to extract the kernel features and discriminate the kernels in real-time at high throughput. Due to potential risk of X-ray radiation hazards, installation of these machines would require federal/provincial approval and should comply with safety guidelines established by government legislations/laws. The safety guidelines would restrict the access to X-ray systems and may require sealing the area, making such systems complicated for use in grain handling facilities.

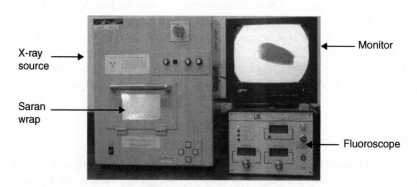

Fig. 15.5 X-ray imaging system.

15.4.2   X-ray image features

The images acquired from an X-ray imaging system are grey-scale images. For single kernel analysis, the background, usually a Saran wrap on which samples are placed, must be removed prior to feature extraction. The background is removed and objects are segmented by converting grey-scale (also colour) images into binary images and then applying fixed or automatic thresholding. The textural features from segmented (background removed) single kernel images are extracted as described in the colour image feature section. Normalized histogram features (binned into different groups) and histogram and shape moments can also be extracted (Karunakaran *et al.*, 2004a, 2004b) and used in developing classification models. Normalized histogram features are position- and size-invariant. The histogram and shape moments derived from binary and grey-scale images describe shape and grey-level distribution of an object (Gonzalez and Woods, 2002).

15.4.3   Applications of X-ray imaging

X-ray imaging is used as a quality control tool in the food processing industry to detect defects and contaminants (Haff and Toyofuku, 2008). Kumar and Bal (2007) detected the cracks in unhulled rice grain (called paddy). During rice milling, the cover (hull) is removed by pressing the paddy between counter-rotating pneumatic rubber rolls of a dehusker and then bran is removed by whitener. Due to the pressure applied in the dehusker, the cracked kernels will either break during the dehulling or whitening operation, thus significantly reducing milling yield; therefore, they need to be separated before milling. The single paddy kernels were imaged using an X-ray system with setting of 12 keV energy, 8 mA current and 10 s exposure time. Image pre-processing was applied, kernels were segmented and Hough transform was used to automatically determine the cracks in the final segmented image. The developed algorithm detected horizontal, vertical and total number of cracks in paddy kernels with over 97% accuracy. Despite the high accuracy, separation of touching kernels in single layer on a continuously moving conveyer belt by efficient programming and feature extraction for real-time classification is still a challenging task.

In Canada, there is a zero-tolerance policy for live insects in grains (Canada Grain Act, 1975) and grain needs to be disinfested if any live insects are found before shipping. This delays the shipment of grain and incurs additional cost. Therefore, timely and accurate detection is needed to detect insect infestation. Visual examination is subjective and cannot detect internal insect infestation. Berlese funnel, a method adopted by the Canadian Grain Commission, is slow and its detection efficiency and accuracy depend on the grain moisture, temperature and insect developmental stages (Smith, 1977). Karunakaran *et al.* (2004a, 2004b) investigated the potential of soft X-ray imaging to detect insect infestation in wheat by most common insect species of *Rhyzopertha dominica* (lesser grain borer) and *Tribolium castaneum* (red flour beetle). Artificially insect infested and control wheat kernels were scanned using a soft X-ray machine equipped with a monochrome CCD camera. Wheat kernels were manually placed on Saran wrap on the platform between X-ray tube (energy source) and detector (CCD camera) and kernel images were acquired at 15 kV potential and 65 mA current. The captured images were digitized into 8-bit grey-scale images at a resolution of 60 pixels/mm by Dazzle digital video creator (Dazzle Multimedia Inc., Fremont, CA) and saved in a personal computer for analysis. Background (Saran wrap) was removed by fixed thresholding process (known grey value of background) and 57 image features (histogram features, textural features and histogram and shape moments) were extracted by developing an algorithm. Classification models were developed using BPNN and statistical classifiers. The BPNN correctly identified all the healthy (control) and pupae-adult infested kernels and more than 99% of kernels infested by *R. dominica* larvae. However, classification of *T. castaneum* was lower and only 73% and 86% control and kernels infested by four larval instars, respectively, were correctly identified.

The potential of soft X-ray imaging was also investigated to detect fungal infection in wheat (Narvankar *et al.*, 2009). Healthy and infected wheat kernels with the common storage fungi namely *Aspergillus niger, A. glaucus* group and *Penicillium spp.* were imaged using a soft X-ray imaging system. Algorithms were developed to extract a total of 34 image features (maximum, minimum, mean, median, variance, standard deviation and 28 GLCM features). These features were used in developing statistical discriminant classifiers (linear, quadratic and Mahalanobis) and BPNN classifier. Statistical classifiers gave better classification accuracies than the BPNN classifier. Quadratic and Mahalanobis classifiers gave better detection rates of infected kernels but had high false positive errors (up to 28%). Linear classifiers detected 82.8% healthy and 89–92.8% fungal-infected kernels in two-way classification (healthy versus infected).

In the studies described in this section, single kernels were manually placed in a sample chamber in a small X-ray machine. Real-time applications of X-ray imaging would require continuous scanning of grain at high throughput. Haff and Slaughter (2004) built a line-scan X-ray imaging system for sorting of grain and detection of insect infestation by *Sitophilus granaries* (L.) (granary weevil). The X-ray system consisted of an X-ray tube with a beryllium window, a high-voltage source, an image intensifier with a beryllium window, optical coupling, a CCD camera, a frame grabber board and a computer. Images were acquired at 12 keV energy and 99 mA current. Usually, lower energy gives better contrast and higher current improves signal-to-noise ratio. The FOV of the system was sufficient (6 cm^2) to image approximately 350 kernels of grain in a single frame. With an exposure time of 149 ms (per frame), the system would yield a maximum throughput rate of around 2500 kernels per second (by calculation only). Healthy (control) and wheat kernels infected by granary weevil with different development stages (eggs, larvae, pupae and adults) were imaged using this system. X-ray films of these kernels were also produced and compared with digital X-ray images acquired by this system. Visual examination by human subjects gave correct recognition rate of 84.4% and 90.2%, respectively, for the X-ray digital images and X-ray films. Human observation of X-ray images would not be suitable for high-throughput screening of grain as it is subjective, slow and tedious. The supervised classification by using X-ray image features would give more accurate and consistent results for high-throughput screening. There is further need to develop a mechanism to separate damaged kernels after they are detected.

15.5   Thermal imaging

Thermal imaging is another technology employed in gathering grain quality information using machine vision. The sections below describe how this works and various applications for this technology.

15.5.1   Thermal imaging systems

Thermal imaging system is different from other types of imaging systems discussed previously; it does not require a radiation light source and the thermal camera detects the heat dissipated from the sample. A thermal camera converts the emitted thermal radiation into electric signals in a similar way to that in which a colour camera converts the reflected light from an object into digital form. A thermal IR heating source (9000–14 000 nm) is used as a radiation source for heating the material. Upon heating, the material first absorbs the IR heat and then emits it. According to the black body radiation law, all objects emit IR radiation above absolute zero temperatures. The amount of radiation emitted by an object is proportional to the body temperature. The thermal camera detects the emitted radiation from a sample and maps it as a pseudo colour image, also known as a thermogram. Thermograms are analysed to see the temperature distribution using image processing techniques. Since cooling rates of different biological materials will be different due to different thermal capacities or thermal conductivities, the analysis of thermograms of different biological materials (e.g., healthy versus fungal or insect-damaged grain) will provide discriminatory information. The cameras used in IR thermal imaging are of a focal plane array (FPA) type, namely InSb, InGaAs, HgCdTe and QWIP, with HgCdTe being the most common. Thermal cameras can be cooled at room temperature by cryogenic cooling to reduce dark current and increase sensitivity; however, cryogenic cooling increases the cost of the thermal camera. Uncooled thermal cameras are relatively less expensive, compact and can stabilize at room temperature. Thermal cameras would be most suitable for applications at grain handling facilities, considering the dusty and rough environment at these facilities. The need for a light source in other types of imaging systems makes continuous and consistent operation in such environment quite difficult. A thermal IR radiation source would be ideal for such applications since a thermal heating source can be kept at a significant distance away from the conveyer belt and thermograms can be obtained for analysing the quality of grain based on temperature distribution. Bulk grain in multilayers, flowing in conveyer belt, can be analysed using a thermal camera in real-time. An area-scan thermal camera is shown in Fig. 15.6. At present, commercial thermal imaging systems for the grain industry are not available, but the research has shown promising results.

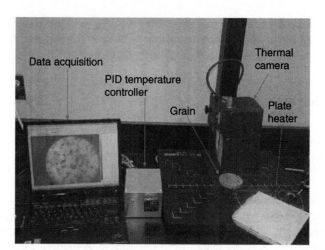

Fig. 15.6 Infrared thermal imaging system.

15.5.2   Thermal image features

Thermal images, which are pseudo colour maps, indicate temperature distribution within an object. Extracted features should represent this distribution pattern in quantitative form. Statistical features such as maximum, minimum, mean, median, standard deviation, variance and maximum and minimum difference are extracted (Manickavasagan *et al.*, 2008). The whole intensity (temperature) range can be binned into different groups and then features can be extracted locally within that group to identify any significant changes between radiation emitted from different objects. Supervised classification models can be developed for discrimination of damaged grain (insect and fungal damaged) from healthy grain.

15.5.3   Applications of thermal imaging

Though very promising for real-time grain quality analysis at grain handling facilities, thermal imaging is relatively new, with fewer applications compared with other imaging techniques. The technique has been investigated for use in detecting damage in wheat such as insect damage, fungal damage and incidence of sprouting. Thermal imaging was used to detect hidden insect infestation by *Cryptolestes ferrugineus* (rusty grain beetle) in wheat kernels (Manickavasagan *et al.*, 2008). It is challenging to detect internal/hidden insect infestation by visual examination and colour imaging systems. Thermal imaging was investigated to detect six insect development stages (four larval instars, pupae and adults) by artificially infesting wheat kernels by Manickavasagan *et al.* (2008). The infested kernels were removed from the incubation room (30°C), refrigerated at 5°C for 60 s, maintained at ambient conditions for 20 s and then imaged using a thermal camera. The thermal distribution of the infested and control kernels (placed in similar environment) were mapped as colour images. It was noticed that the means of the highest 5% and 10% temperature values on the surface of the grain were significantly higher for wheat kernels with young larvae inside compared with wheat kernels with pupae inside. The temperature distribution analyses of the infested (with different development stages) kernels' surface showed a high correlation with the respiration rate of each developmental stage. The classification models (linear and quadratic) developed using extracted thermal image features gave moderate average accuracy of 77.6–83.5% in two-way classification. Developed stages of pupae and adult gave higher accuracy of above 90%. In two-way discrimination, however, early developmental stages instar 1 and instar 2 gave only 80% and 70% classification accuracies. High false positive errors (19–25%) and pre-cooling requirements are constraints in industrial applications. Grains usually dissipate heat at a slower rate but grain moves on conveyer belts for a significant amount of time, which allows the grain to heat or cool prior to imaging. This study used insect species of very small size – therefore, infestation by larger size insect species would give better classification as they would most likely have higher respiration rate.

In another study, Chelladurai *et al.* (2010) used IR thermal imaging to detect wheat grain infected by fungal species of *Aspergillus glaucus, A. niger* and *Penicillium*. The bulk wheat grain infected by fungi as well as control wheat samples were heated for about 180 s with a plate heater placed 10 mm above the grain (plate maintained at 90°C) and then cooled in ambient air for about 30 s. Thermal images were then acquired and image features were extracted to develop classification models. The classification models gave high accuracy in discriminating infected kernels from control wheat samples but failed to correctly identify infection by each individual species.

Sprout damage in wheat, caused by germination due to excessive moisture absorption, adversely affects the baking quality of wheat flour for bread making, due to high alpha amylase concentration in sprouted kernels. In Canada, all commercial wheat classes are visually examined for sprout damage. In laboratories, falling number, an indirect method of monitoring alpha amylase activity, is commonly adopted. Vadivambal *et al.* (2010) used thermal imaging to detect sprout-damaged wheat kernels objectively and non-destructively (visual inspection is subjective and falling number is a destructive method). Artificially sprouted and control (healthy) wheat single kernels were imaged using an IR thermal camera and images were stored. Thermal image features were extracted and supervised classification algorithms were developed using LDA, QDA and an artificial neural network. Statistical classifiers gave false positive errors of nearly 12% and detected 95–98% damaged wheat kernels. Neural networks correctly identified all the healthy kernels and more than 91% damaged wheat kernels. Despite high accuracy, the classification rate by thermal imaging was lower compared with HSI classification of artificially sprouted and healthy wheat kernels (98–100%) by Singh *et al.* (2009b).

15.6   Conclusions

The image processing techniques (colour, hyperspectral, X-ray and thermal imaging) discussed in this chapter have demonstrated high potential for real-time automated grain quality monitoring. Colour machine vision systems are the least expensive of these techniques and are already used in colour-based commercial grain sorters for sorting of grains based on surface characteristics (colour, shape and size). However, colour imaging does not provide information about chemical composition and its distribution in the kernel as it relies on external surface features of the object under investigation. NIR hyperspectral technique provides spectral information along with spatial labelling – hence, compositional distribution can be obtained. NIR region (900–1700 nm) can be used for rapid and multi-constituent analysis (e.g., moisture, fat, oil, protein, starch, fibre and solid contents) simultaneously with almost no sample preparation. Multispectral systems from NIR hyperspectral analyses can be built for rapid quantitative and qualitative analysis and sorting of grain. X-ray imaging has also shown its potential for application in the grain industry – particularly, detection of internal insect infestation. Thermal imaging has also been investigated for detection of damage in grains. However, X-ray and thermal imaging do not provide any compositional information about the sample, but thermal imaging would be least affected by environmental conditions (e.g., day and night lighting conditions; intensity, fluctuation and life of light source; dust) inside the grain processing/handling facilities due to the use of thermal detectors that detect the heat dissipated from a sample. For real-time grain sorting and single kernel analyses, kernels from acquired images ideally containing single layer of kernels, should be separated, the dimension of acquired imaging data should be reduced after pre-processing and relevant information should be extracted in quantitative form for objective classification and calibration model development. Most of the studies reported used laboratory-sized imaging equipment – hence, there is a need to develop a mechanism to spread the grain into single layer prior to imaging and then to develop algorithms to separate touching kernels in the grain layer. Once the kernels are separated, individual kernel features can then be extracted and used as input to a classifier for classification, and a classification decision can finally be passed to the automatic separation mechanism to accept or reject the grain. Building a system for completing these tasks in real-time is challenging. With new developments in optical instrumentation, computer hardware, optical sensors and fast computational methods, machine vision techniques are expected to become essential components of grain quality and safety monitoring tools.

15.7   Acknowledgements

The authors thank the Canada Research Chairs programme, the Natural Sciences and Engineering Research Council of Canada, Canada Foundation for Innovation and the Government of Manitoba for partial funding support.

15.8   References

Ahmad, I. S., Reid, J. F. and Paulsen, M. R. (1999) Color classifier for symptomatic soybean seeds using image processing. *Plant Disease*, **83**, 320–7.

Archibald, D. D., Thai, C. N. and Dowell, F. E. (1998) Development of short-wavelength near-infrared spectral imaging for grain color classification. *Proceedings of SPIE*, **3543**, 189–98.

Berman, A., Connor, P. M., Whitbourn, L. B., Coward, D. A., Osborne, B. G. and Southan, M. D. (2007) Classification of sound and stained wheat grains using visible and near infrared hyperspectral image analysis. *Journal of Near Infrared Spectroscopy*, **15**, 351–8.

Canada Grain Act (1975) Canada grain regulations. In *Canada Gazette*, **Part II 109 (14)**, 1708–839.

Chelladurai, V., Jayas, D. S. and White, N. D. G. (2010) Thermal imaging for detecting fungal infection in stored wheat. *Journal of Stored Products*, **46**, 174–9.

Choudhary, R., Paliwal, J. and Jayas, D. S. (2008) Classification of cereal grains using wavelet, morphological, colour, and textural features of non-touching kernel images. *Biosystems Engineering*, **99(3)**, 330–7.

Choudhary, R., Mahesh, S., Paliwal, J. and Jayas, D. S. (2009) Identification of wheat classes using wavelet features from near infrared hyperspectral images of bulk samples. *Biosystems Engineering*, **102(2)**, 115–27.

CIGI (1993) *Grains & Oilseeds: Handling, Marketing, Processing*, 2nd edn. Winnipeg: Canadian International Grains Institute.

Cogdill, R. P., Hurburgh, C. R., Rippke, G. R., Bajic, S. J., Jones, R. W., McClelland, J. F., Jensen, T. C. and Liu, J. (2004) Single kernel maize analysis by near-infrared hyperspectral imaging. *Transactions of the ASAE*, **47(1)**, 311–20.

Egri, J. (2006) Two high-speed buses make it easier for test engineers to interface a digital camera to a host computer. *Test & Measurement World*. Available at http://www.tmworld.com/article/320005-Ethernet_vs_Camera_Link.php (accessed 8 January 2010).

Gat, N. (2000) Imaging spectroscopy using tunable filters: a review. *Proceedings of SPIE*, **4056**, 51–64.

Geladi, P. and Grahn, H. (1996), *Multivariate Image Analysis*. Chichester: John Wiley and Sons.

Gonzalez, R. C. and Woods, R. E. (2002), *Digital Image Processing*. Singapore: Pearson Education.

Gorretta, N., Roger, J. M., Aubert, M., Bellon-Maurel, V., Campan, F. and Roumet, P. (2006) Determining vitreousness of durum wheat kernels using near infrared hyperspectral imaging. *Journal of Near Infrared Spectroscopy*, **14(4)**, 231–9.

Gowen, A. A., O'Donnell, C. P., Cullen, P. J., Downey, G. and Frias, J. M. (2007) Hyperspectral imaging: an emerging process analytical tool for food quality and safety control. *Trends in Food Science & Technology*, **18**, 590–8.

Gowen, A. A., Tiwari, B. K., Cullen, P. J., McDonnell, K. and O'Donnell, C. P. (2010) Applications of thermal imaging in food quality and safety assessment. *Trends in Food Science & Technology*, **21**, 190–200.

Haff, R. P. and Slaughter, D. C. (2004) Real-time X-ray inspection of wheat for infestation by the granary weevil, *Sitophilus granarius* (L.). *Transactions of the ASAE*, **47**, 531–7.

Haff, R. P. and Toyofuku, N. (2008) X-ray detection of defects and contaminants in the food industry. *Sensing and Instrumentation for Food Quality*, **2**, 262–73.

Karunakaran, C., Jayas, D. S. and White, N. D. G. (2004a) Detection of internal wheat seed infestation by *Rhyzopertha dominica* using X-ray imaging. *Journal of Stored Products*, **40**, 507–16.

Karunakaran, C., Jayas, D. S. and White, N. D. G. (2004b) Identification of wheat kernels damaged by the red flour beetle using x-ray images. *Biosystems Engineering*, **87**, 267–74.

Kumar, P. A. and Bal, S. (2007) Automatic unhulled rice grain crack detection by X-ray imaging. *Transactions of the ASABE*, **50(5)**, 1907–11.

Lawrence, K. C., Park, B., Windham, W. R. and Mao, C. (2003) Calibration of a push-broom hyperspectral imaging system for agricultural inspection. *Transactions of the ASABE*, **46(2)**, 513–21.

Mahesh, S., Manickavasagan, A., Jayas, D. S., Paliwal, J. and White, N. D. G. (2008) Feasibility of near-infrared hyperspectral imaging to differentiate Canadian wheat classes. *Biosystems Engineering*, **101**, 50–7.

Majumdar, S. and Jayas, D. S. (2000a) Classification of cereal grains using machine vision: I. Morphology models. *Transactions of the ASABE*, **43(6)**, 1669–75.

Majumdar, S. and Jayas, D. S. (2000b) Classification of cereal grains using machine vision: II. Color models. *Transactions of the ASABE*, **43(6)**, 1677–80.

Majumdar, S. and Jayas, D. S. (2000c) Classification of cereal grains using machine vision: III. Texture models. *Transactions of the ASABE*, **43(6)**, 1681–7.

Majumdar, S. and Jayas, D. S. (2000d) Classification of cereal grains using machine vision: IV. Morphology, color, and texture models. *Transactions of the ASABE*, **43(6)**, 1689–94.

Manickavasagan, A., Jayas, D. S. and White, N. D. G. (2008) Thermal imaging to detect infestation by *Cryptolestes ferrugineus* inside wheat kernels. *Journal of Stored Products*, **44**, 186–92.

Muir, W. E. (2001) Production and marketing of cereal and oil seed crops.. In Muir, W. E. (ed.), *Grain Preservation Biosystems*, unpublished manual, Winnipeg: Department of Biosystems Engineering, University of Manitoba.

Narvankar, D. S., Singh, C. B., Jayas, D. S. and White, N. D. G. (2009) Assessment of soft x-ray imaging for detection of fungal infection in wheat. *Biosystems Engineering*, **103**, 49–56.

Ng, H. F., Wilcke, W. F., Morey, R. V. and Lang, J. P. (1998) Machine vision evaluation of corn kernel mechanical and mold damage. *Transactions of the ASABE*, **41(2)**, 415–20.

Ni, B., Paulsen, M. R. and Reid, J. F. (1998) Size grading of corn kernels with machine vision. *Applied Engineering in Agriculture*, **14(5)**, 567–71.

Paliwal, J., Visen, N. S., Jayas, D. S. and White, N. D. G. (2003) Cereal grains and dockage identification using machine vision. *Biosystems Engineering* , **85(1)**, 51–7.

Polder, G., Van-der Heijden, G. W. A. M., Keizer, L. C. P. and Young, I. T. (2003) Calibration and characterization of imaging spectrographs. *Journal of Near Infrared Spectroscopy*, **11(3)**, 193–210.

SAS (2002) *SAS User's Guide*. Cary, NC: Statistical Analysis Systems Institute, Inc.

Singh, C. B., Jayas, D. S., Paliwal, J. and White, N. D. G. (2007) Fungal detection in wheat using near-infrared hyperspectral imaging. *Transactions of the ASABE*, **50(6)**, 2171–6.

Singh, C. B., Jayas, D. S., Paliwal, J. and White, N. D. G. (2009a) Detection of insect-damaged wheat kernels using near-infrared hyperspectral imaging. *Journal of Stored Products*, **45(3)**, 151–8.

Singh, C. B., Jayas, D. S., Paliwal, J. and White, N. D. G. (2009b) Detection of sprouted and midge-damaged wheat kernels using near-infrared hyperspectral imaging. *Cereal Chemistry*, **86(3)**, 256–60.

Singh, C. B., Jayas, D. S., Paliwal, J. and White, N. D. G. (2010a) Near-infrared hyper-spectral imaging for quality analysis of agricultural and food products. *Proceedings of SPIE*, **7676**, 767603.

Singh, C. B., Jayas, D. S., Paliwal, J. and White, N. D. G. (2010b) Identification of insect-damaged wheat kernels using short-wave near-infrared hyperspectral and digital colour imaging. *Computers and Electronics in Agriculture*, **73(2)**, 118–25.

Singh, C. B., Jayas, D. S., Paliwal, J. and White, N. D. G. (2010c) Detection of midge-damaged wheat kernels using short-wave near-infrared hyperspectral and digital colour imaging. *Biosystems Engineering*, **105(3)**, 380–7.

Singh, C. B., Jayas, D. S., Paliwal, J. and White, N. D. G. (2012) Fungal damage detection in wheat using short-wave near-infrared hyperspectral and digital colour imaging. *International Journal of Food Properties*, **15(1)**, 11–24.

Smith, L. B. (1977) Efficiency of Berlese-Tullgren funnels for removal of the rustygrain beetle, *Cryptolestes ferrugineus*, from wheat samples. *Canadian Entom*, **109**, 503–9.

Steenhoek, L. W., Misra, M. K., Batchelor, W. D. and Davidson, J. L. (2001) Probabilistic neural networks for segmentation of features in corn kernel images. *Applied Engineering in Agriculture*, **17(2)**, 225–34.

Tran, C. D. (2003) Infrared multispectral imaging: principle and instrumentation. *Applied Spectroscopy Reviews*, **38(2)**, 133–53.

Vadivambal, R., Chelladurai, V., Jayas, D. S. and White, N. D. G. (2010) Detection of sprout-damaged wheat using thermal imaging. *Applied Engineering in Agriculture*, **26(6)**, 999–1004.

Vadivambal, R. and Jayas, D. S. (2011) Applications of thermal imaging in agriculture and food industry – a review. *Food Biotechnology*, **4(2)**, 186–99.

Wan, Y. N., Lin, C. M. and Chiou, J. F. (2002) Rice quality classification using an automatic grain quality inspection system. *Transactions of the ASABE*, **45(2)**, 379–87.

Wang, W. and Paliwal, J. (2006) Calibration and correction for non-uniform pixel sensitivity in digital NIR imaging. CSBE, Winnipeg, Paper no. MBSK06108.

Zayas, I. Y. and Flinn, P. W. (1998) Detection of insects in bulk wheat samples with machine vision. *Transactions of the ASABE*, **41(3)**, 883–8.

16

Computer vision in the bakery industry

C.-J. Du, University of Warwick, UK, Q. Cheng, University of Reading, UK and D.-W. Sun, University College Dublin, Ireland

Abstract: Computer vision techniques used for quality evaluation of bakery products are highly application dependent. The development of segmentation method, the choice of colour space, the extraction of size, shape and texture features, and the selection of classification technique have to be tailored to the task to be performed. In this chapter, techniques employed for different bakery products are summarized in separate sections. The colour, size and shape inspection techniques for bread, muffins, biscuits and pizza bases, along with others, are described, respectively. In addition, the methods for texture inspection of crumb grain are explained, and the segmentation and classification methods developed for different bakery products are elucidated.

Key words: computer vision, colour, size and shape, image texture, bakery products.

16.1 Introduction

Bakery is one of the most important sectors in the food industry. The global bakery industry is forecast to be worth $410 billion by 2015, according to a comprehensive global report by market research firm Global Industry Analysts (Anon., 2011). Driven by the increasingly fast pace of modern consumer lifestyle, demand for industrial bakery products is set to significantly increase in the long term. Additionally, sustained demand from developing countries for bakery products is set to further energize the market. Bread remains a staple part of the Western diet, and biscuits and cakes are increasingly regarded as regular snacking items rather than occasional treats. However, apart from these opportunities, the global baking industry is currently facing a range of difficult challenges due to a high level of competition. Following the 2009 global recession, consumers are increasingly focused on purchasing products with the best quality/price ratio, so it is essential for producers to cater to this need.

Product appearance plays a vital role in making a purchase decision, with colour, size and shape significantly influencing impulse purchases of bakery products.

Colour enhances the appearance of food, giving it aesthetic value (Abdullah, 2008) while also influencing the anticipated oral and olfactory sensations by triggering memories of previous eating experiences (Dubose and Cardello, 1980). Size and shape are also important indications of a product's overall quality; a broken or deformed product may taste the same as a whole one, but customers are reluctant to buy seemingly damaged goods (Abdullah, 2008). As such, quality assurance is an important aspect of production. Traditionally, quality assurance tasks are performed by trained inspectors. However, the assessment methods are tedious, laborious, costly and time consuming and can be easily influenced by physiological factors, inducing subjective and inconsistent evaluation results.

The need for consistent quality assurance has led to the recent rapid development of computer vision techniques, which can quantitatively characterize complex size, shape, colour and texture properties of foods. Computer vision systems are playing an increasingly important role in food quality evaluation, as they maintain accuracy and consistency while eliminating the subjectivity of manual inspections. They also offer flexibility in application and are reasonable substitutes for the human visual decision-making process in the quality evaluation of bakery products. The quality assurance of bakery products as a whole sector using computer vision has been reviewed by Abdullah (2008). However, there is no general computer vision system that can be applied to every bakery product for quality evaluation. Although typical functions are found in many computer vision systems, the image processing techniques used are highly application dependent. Most systems have to be tailored to the task to be performed, and different methods have to be developed for different bakery products according to their specific characteristics. The main focus of this chapter is to summarize advances in the application of image processing techniques for the quality evaluation of a variety of bakery products.

16.2   Computer vision applications for analysing bread

The following sections investigate how colour, size and shape and texture in bread are analysed using these technological applications.

16.2.1   Colour

The formation of bread colour during baking can be measured using a variety of techniques, including chemical methods, colorimetry and computer vision. The chemical methods are mostly intended to measure the concentration of browning reactions of the products, or alternatively the consumption of reactants. However, these chemical methods can be destructive, expensive and time-consuming. This is in contrast to both colorimeter and computer vision techniques, which quantify the amount of reflected light from the surface of the food by using colour sensors. Computer vision is an automated, non-destructive, cost-effective technique, and has various advantages over colorimeter use. A colorimeter will often give an inaccurate colour estimate due to the small capture area, whereas a whole bread surface could be measured by means of computer vision. Furthermore, computer vision does not imply any contact with samples for measurement, which is essential when dealing with deformable materials such as bread (Purlis, 2010). In addition, while a colorimeter can only evaluate the colour of the bread, computer vision techniques can also evaluate other important quality properties, including size, shape and texture features.

Before measuring the bread colour, it is essential to carry out a calibration procedure of the computer vision system in order to ensure the validity of results. This can be done using the calibrating plates of a conventional colorimeter and adjusting the camera parameters. Nashat *et al.* (2011) calibrated the computer vision system using four colour samples manufactured by Labsphere Inc., USA, which are SCS-RD-010, SCS-GN-010, SCS-BL-010 and SCS-YW-010, corresponding to red, green, blue and yellow colour standards, respectively. Figure 16.1a shows sample images of the bread during baking, which clearly exhibit the development of browning on the bread surface at different temperatures.

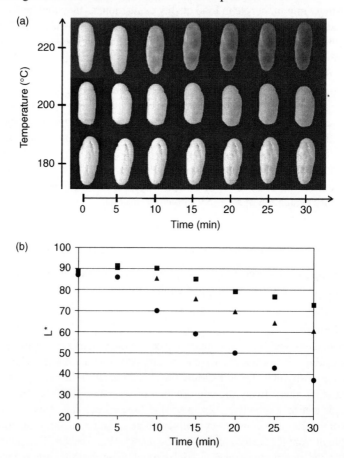

Fig. 16.1   (a) Bread sample images showing development of browning at bread surface during baking at different temperature. (b) Variation of lightness of bread surface during baking at 180°C (squares), 200°C (triangles) and 220°C (circles) (Purlis, 2010).

As a cheaper method of image acquisition, CCD cameras are frequently employed in computer vision systems for bread quality evaluation. The image acquired is generally in RGB (red, green and blue) colour space. However, evaluation of the literature on this subject suggests that bread is mostly evaluated using the CIE L*a*b* colour space, due to its ability to perceptually linearize colour differences. The Euclidean distance between two colour points in the CIE L*a*b* colour space corresponds to the perceptual difference between the two colours in the human vision system. This property of CIE L*a*b* has made it particularly attractive and useful in food science.

RGB colour space involves red, green and blue light being added together in various ways to reproduce a broad array of colours. In contrast, the three parameters of CIE L*a*b* represent the lightness of colour (L*), which ranges from 0 to 100 (black to white), its position between red and green (a*, values between −120 and +120) and its position between yellow and blue (b*, values between −120 and +120). In order to transform RGB to CIE L*a*b*, it is necessary to convert RGB to CIE XYZ tristimulus values, first, by:

$$\begin{cases} X = 0.412453R + 0.357580G + 0.180423B \\ Y = 0.212671R + 0.715160G + 0.072169B \\ Z = 0.019334R + 0.119193G + 0.950227B \end{cases} \qquad [16.1]$$

There is no simple formula for conversion between XYZ and L*a*b*, because RGB is device dependent. A white reference has to be acquired before conversion. If it is obtained by taking a perfectly reflecting diffuser under CIE standard D_{65} illumination, the three colour components of CIE L*a*b* can be obtained as follows:

$$\begin{cases} L^* = 116 f\left(Y/Y_w\right) - 16 \\ a^* = 500\left[f\left(X/X_w\right) - f\left(Y/Y_w\right)\right] \\ b^* = 200\left[f\left(Y/Y_w\right) - f\left(Z/Z_w\right)\right] \end{cases} \qquad [16.2]$$

where

$$f(v) = \begin{cases} v^{1/3} & \text{if } v > 0.008856 \\ 7.787v + 16/116 & \text{otherwise} \end{cases} \qquad [16.3]$$

X_w, Y_w and Z_w represent the white reference, which is defined through its trichromatic values.

After the conversion, the mean, standard deviation and histogram of each colour component can be calculated. Figure 16.1b demonstrates the variation of lightness of bread surfaces during baking at temperatures of 180°C, 200°C and 220°C, respectively. If colour values $(L^*_{ref}, a^*_{ref}, b^*_{ref})$ of the crust of the control bread are taken as the reference, the colour difference can be determined as

$$\text{Diff}^* = \sqrt{\left(L^* - L^*_{ref}\right)^2 + \left(a^* - a^*_{ref}\right)^2 + \left(b^* - b^*_{ref}\right)^2}$$ [16.4]

16.2.2   Size and shape

The internal structure of yeast-leavened bread when sliced is commonly referred to as crumb grain. It can be described as a complex of interconnected cells in a heat-set glutinous starch matrix (Kamman, 1970). Crumb grain is an important element of bread quality, reflecting flour characteristics, dough formulation and processing (Scanlon and Zghal, 2001). It largely governs the sensory properties of the bread (Baardseth *et al.*, 2000) and so influences consumer purchase (Pyler, 1988). The quality of crumb grain is generally evaluated visually based on the cell size, cell shape and cell wall thickness, which can be characterized using computer vision techniques. An image of a slice of bread acquired using a CCD camera is shown in Fig. 16.2a.

Cells should, first, be segmented for the characterization of crumb grain structure. The techniques of image segmentation developed for food quality evaluation

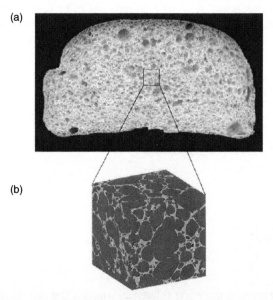

Fig. 16.2   (a) A slice of bread acquired by using a CCD camera (14.8 × 8.7 cm²). (b) A crumb sub-volume (the square area in (a)) captured by X-ray tomography at ESRF (European Radiation Synchrotron Facility, F-38 Grenoble, beamline ID19) (Lassoued *et al.*, 2007).

using computer vision can be divided into four different approaches as follows: thresholding-based, region-based, gradient-based and classification-based segmentation (Du and Sun, 2004a). Thresholding-based and classification-based methods are used in most applications for crumb grain segmentation.

Thresholding-based segmentation is a particularly effective technique for scenes containing solid objects resting upon a contrasting background, which distinguishes the object from the remaining part of an image with an optimal value. The key to this method is the selection of the threshold value. There is no universal methodology for threshold selection that works on all kinds of images, and a variety of techniques have been proposed to set the threshold value under different circumstances. The suitability of six thresholding methods (isodata, Otsu, minimum error, moment-preserving, Pun and fuzzy) to consistently segment bread crumb images was investigated by Gonzales-Barron and Butler (2006). They reported that the fuzzy, Otsu, isodata and moment-preserving methods are able to yield good and consistent binary images. The two most cited methods (isodata and Otsu) will now be briefly described.

The isodata algorithm proposed by Ridler and Calvard (1978) is an iterative threshold selection technique. A threshold value T^0 is initialized as midway between the maximum and minimum grey level. Next the means μ_F^0 and μ_B^0 of the foreground and background pixels are calculated, respectively. A new threshold value $T^1 = \mu_F^0 + \mu_B^0/2$ is then obtained. This process is repeated until the threshold value no longer changes.

This problem can also be explored by considering the values in the two regions as two classes. The optimum threshold separates those two classes in a way that their combined spread (within-class variance) is minimal (Otsu, 1979). The within-class variance is defined as the weighted sum of the variances of each class:

$$\sigma_w^2(T) = S_F(T)\sigma_F^2(T) + S_B(T)\sigma_B^2(T) \qquad [16.5]$$

where $S_F(T) = \Sigma_{i=0}^{T-1} p(i)$ and $S_B(T) = \Sigma_{i=T}^{N-1} p(i)$ are the probabilities of the foreground and background respectively and $\sigma_F^2(T)$ and $\sigma_B^2(T)$ are their variances. It was shown that minimizing the within-class variance is the same as maximizing the between-class variance, which is:

$$\sigma_b^2(T) = \sigma^2 - \sigma_w^2(T) = S_F(T)S_B(T)\left[\mu_F(T) - \mu_B(T)\right]^2 \qquad [16.6]$$

Thus, the optimal threshold can be obtained by iteratively updating the class probabilities S_i and class means μ_i in turn.

Classification-based methods attempt to assign each pixel to different objects based on classification techniques, including supervised and unsupervised methods. Supervised methods require a training dataset to be specified from which a classifier can be trained, while unsupervised methods learn a classification directly from the data, with no training data required. Unsupervised techniques –

use of the k-means algorithm, for example – appear to be most efficient in reliably and consistently separating crumb grain image into parts of interest without human intervention. For example, Sapirstein *et al.* (1994) employed the k-means algorithm to classify monochrome images of bread slices into cells and background.

Due to its sound theory, simplicity and speed, k-means is one of the most commonly used methods for crumb grain image segmentation. Let $X \subseteq R^3$ be the given pixels represented by RGB values. The objective of k-means is to partition K clusters $\overline{X}_1, \ldots, \overline{X}_K$ and assign each point in every cluster $\overline{X}_i$ to a centre σ_i, so as to locally minimize the potential cost $\Sigma_{i=1}^{K} \Sigma_{x \in \overline{X}_i} \|x - \delta_i\|^2$. The most popular approach used in k-means problem solving is the Lloyd-style method (Lloyd, 1982), which employs an iterative procedure. At each iteration, it clusters all of the data in the Voronoi region of a centre together, and then recalculates the centre as the centroid of its cluster. The procedure stops when the potential cost stabilizes.

However, the traditional Lloyd method and its variant converge only to local optima and are sensitive to the choice of the initial centres (Milligan, 1980). To produce a good solution, it is usually rerun many times with different initializations, which works well only when the number of clusters is small and chances are high that at least one random initialization is close to a good solution (Frey and Dueck, 2007). By augmenting k-means with a non-uniform sampling process for centre initialization technique, an algorithm that is $\Theta(\log K)$ competitive with the optimal clustering could be obtained (Arthur and Vassilvitskii, 2007). At the beginning, an initial centre δ_1 is uniformly chosen at random from X. The next centre is picked as $\delta_i = x' \in \overline{X}$ with probability $D(x')^2 / \Sigma_{x \in X} D(x)^2$, where $D(x)$ denotes the shortest distance from a pixel x to the closest centre that has been chosen. The above procedure is repeated until a total of K centres have been chosen. In this way, the starting centres of the partitions are chosen with probability proportional to their contribution to the overall potential. Based on the initialized centres, the algorithm proceeds, as with the Lloyd method, until convergence occurs.

Based on the segmented binary cell image, morphological features of crumb grain such as mean cell area, cell density (number of cells/cm^2), cell shape distribution, cell uniformity, mean cell wall thickness and void fraction can be computed. In pixel-based representation this is the number of pixels within the segmented area, which can be straightforwardly determined by counting. Cell shape can be quantified using a shape factor $SF = 4\pi \times$ cell area $\times$ cell perimeter^2 (Brescia *et al.*, 2007). This measurement calculates the circularity of an object; a perfect circle has a shape factor of one, while a line has a shape factor approaching zero. Cell shape distribution analysis is performed by counting the percentages of cells that fall into four predefined shape factor classes: $0 < SF \leq 0.2$, $0.2 < SF \leq 0.5$, $0.5 < SF \leq 0.8$ and $0.8 < SF \leq 1$. Cell uniformity is indicated by the ratio between the number of cells lower than 5 mm^2 to the number of cells higher than 5 mm^2. The cell wall thickness is calculated as the average value of a distance measured between the boundaries of a given cell and the boundaries of cells detected at regular angular intervals in its neighbourhood (Sapirstein *et al.*, 1994). 'Void fraction' is the ratio between the total area corresponding to the bread pores and the total area of bread.

The analysis of bread cellular structure from 2-D images is limited because cells are truncated due to bread sectioning (Lassoued *et al.*, 2007). Breads are three-dimensional (3-D) entities and most activities occur in 3-D space. Thus, 3-D imaging techniques such as magnetic resonance imaging (MRI) and X-ray tomography are needed to visualize bread structure in 3-D. Naito *et al.* (2003) applied MRI to non-frozen and frozen dough and breads to evaluate the grain structures of baked breads. Falcone *et al.* (2005) analysed crumb samples using X-ray tomography to capture high quality 3-D images of cellular structure. Using X-ray tomography, Babin *et al.* (2005) linked mechanical properties of breads to their crumb cellular structure by finite element analysis. In the work of Babin *et al.* (2006), the potential of X-ray tomography for the dynamic follow-up of dough during proofing and baking was demonstrated by determining bubble growth kinetics and evidence of coalescence phenomena.

Due to the good contrast of the 3-D X-ray images, as shown in Fig. 16.2b, the cells can be easily segmented using a simple threshold method (Lassoued *et al.*, 2007). Volumic distributions of cell size can then be determined by 3-D morphological granulometry, performed by iterations of closing (successive application of dilation and erosion of the same size) using octahedral structuring elements of increasing size. Similarly, volumic distributions of cell wall size can be obtained by openings (successive application of erosion and dilation) of increasing size. The volumic means of cell size, cell wall thickness and crumb porosity (cell volumic fraction) are derived from these distributions. The 3-D structure allows for analysis of the connectivity between cells, which is calculated as the ratio of the largest cell volume to the total void volume.

The 2-D and 3-D imaging techniques are applied at different scales, with a whole cut section being analysed in 2-D, whereas only a sub-volume of crumb was analysed in 3-D (as shown in Fig. 16.2b). Both 2-D and 3-D images allow the assessment of useful crumb characteristics. Lassoued *et al.* (2007) compared 2-D results with those obtained by 3-D X-ray tomography. They reported that 2-D images can be applied to the rapid control of crumb grain and quantify the cellular structure for the calculation of mechanical properties.

16.2.3   Texture

Image texture is an attribute representing the spatial arrangement of the grey levels of the pixels in a region (Anon., 1990). As a useful feature for area description, image texture can quantify some characteristics of the grey-level variation within an object, such as fineness, coarseness, smoothness and graininess. Crumb grain images with small cells produce frequent grey-level changes and therefore a fine texture, while those with large cells generate a coarser texture. It is important to note that, in the context of computer vision, the concept of texture is completely different to the one generally understood and used in food science. Food texture refers to the manner in which the food behaves in the mouth and is characterized by parameters such as hardness, cohesiveness, viscosity, elasticity, adhesiveness, brittleness, chewiness and gumminess.

Several image texture techniques have been used to quantify various crumb grain structures, including first-order statistical measures (Zayas, 1993), the grey-level co-occurrence matrix method (GLCM) (Gonzales-Barron and Butler, 2008), transform-based methods such as Fourier transform (Rogers *et al.*, 1995), Haar transform (Bertrand *et al.*, 1992) and fractal dimension (FD) (Gonzales-Barron and Butler, 2007a). The first-order grey-level statistics can be derived from the normalized grey-level histogram $\{h_i\}$. Besides mean and variance, skewness and kurtosis can be computed for texture characterization as follows:

$$GLS = \frac{1}{\sigma^3} \sum_{i=0}^{N_g-1} (i-\mu)^3 h_i \qquad [16.7]$$

$$GLK = \frac{1}{\sigma^4} \sum_{i=0}^{N_g-1} (i-\mu)^4 h_i \qquad [16.8]$$

where N_g is the number of grey levels, and μ and σ are the mean and standard deviation of grey level respectively.

One of the most frequently cited methods for texture analysis is GLCM, which is a general procedure presented by Haralick *et al.* (1973) for extracting image texture information in the spatial domain. To obtain the spatial relationship contained in the image, a spatial-dependence GLCM is, first, constructed by estimating the second-order joint conditional probability density functions of pixel intensity. Each element (i, j) of GLCM represents the probability that two pixels with the grey level i and j co-occur in the image separated by a distance d in direction θ. Theoretically, a variety of GLCMs could be constructed from the image with different values of direction and distance. For the direction θ, the four angles $0°$, $45°$, $90°$, and $135°$ are commonly used to achieve rotation invariance by averaging the results of each angle. For the distance d, values other than one are rarely used. A set of features can be derived from each GLCM to describe the texture characteristics within the image, such as the homogeneity, contrast, presence of organized structure and the complexity and nature of grey-level transitions occurring.

A wide range of texture descriptions can be extracted from the Fourier frequency domain (Liu and Jernigan, 1990). $P(u,v)$ denotes the power spectrum and $\phi(u,v)$ the phase spectrum obtained from a local region of bread. The normalized power spectrum is computed as

$$NP(u,v) = \frac{P(u,v)}{\sum_{u,v \neq 0} P(u,v)} \qquad [16.9]$$

Twenty eight spatial frequency domain features can be derived to capture relevant characteristics of texture manifest in their spectra (Liu and Jernigan, 1990). To facilitate understanding, a few are listed as follows:

$$\text{Energy in major peak} = NP(u_1, v_1) \times 100 \qquad [16.10]$$

where u_1, v_1 are the frequency coordinates of the maximum peak of the power spectrum.

$$\text{Laplacian of major peak} = \nabla^2 P(u_1, v_1) \qquad [16.11]$$

$$\text{Laplacian of major peak phase} = \nabla^2 \phi(u_1, v_1) \qquad [16.12]$$

The Haar transform is the simplest of the wavelet transforms. The Haar function is defined as

$$\psi(x) = \begin{cases} 1 & 0 \le x < 1/2 \\ -1 & 1/2 \le x < 1 \\ 0 & \text{otherwise} \end{cases} \qquad [16.13]$$

Its scaling function can be described as

$$\phi(x) = \begin{cases} 1 & 0 \le x < 1 \\ 0 & \text{otherwise} \end{cases} \qquad [16.14]$$

By iteratively dilating (scaling) a factor of two and translating (shifting) its width at any given scale j, it can generate an orthonormal wavelet basis functions:

$$\psi_{j,k}(x) = 2^{j/2} \psi(2^j x - k) \qquad [16.15]$$

and orthonormal scaling basis functions:

$$\phi_{j,k}(x) = 2^{j/2} \phi(2^j x - k) \qquad [16.16]$$

where j, k are non-negative integers and $0 \le k < 2^j$. The Haar transform is derived from the Haar matrix. An example of a 4×4 Haar transformation matrix is shown below:

$$H_4 = \frac{1}{\sqrt{4}} \begin{bmatrix} 1 & 1 & 1 & 1 \\ 1 & 1 & -1 & -1 \\ \sqrt{2} & -\sqrt{2} & 0 & 0 \\ 0 & 0 & \sqrt{2} & -\sqrt{2} \end{bmatrix} \qquad [16.17]$$

The image can be viewed as a hilly terrain where height from the ground is proportional to the grey value of the image (Peleg *et al.*, 1984). The image texture features such as roughness, smoothness and graininess can be described by the FD of this surface, which is invariant under translation, rotation and certain scale transformation. Several approaches exist to estimate the FD of an image, such as the ε-blanket method, which calculates FD by using dilatation and erosion of an image (Peleg *et al.*, 1984), the frequency domain method, which determines FD from the Fourier power spectrum of the image data (Quevedo *et al.*, 2002), and the differential box counting (DBC) method (Sarkar and Chaudhuri, 1994). The fractal texture determined by the box counting method varies monotonically with image FD true values, and can be a useful descriptor in texture recognition. Furthermore, compared with other approaches in terms of computer complexity and accuracy, DBC is a simple, accurate and computationally efficient method. The image is divided into N_{sr} distinct self-similar pieces, each of which is scaled down by a ratio sr in all dimensions, and becomes statistically identical to the original one. FD of the scaled image can be computed by

$$FD = \frac{\log(N_{sr})}{\log(1/sr)} \qquad [16.18]$$

Using the DBC method, a number of values of N_{sr} can be estimated for different values of sr. FD of the bread image can then be estimated by the slope of the least-squares linear fit of the plot of $\log(N_{sr})$ versus $\log(1/sr)$.

16.3   Computer vision applications for analysing muffins

The following sections investigate how colour, size and shape in muffins are analysed using these technological applications.

16.3.1   Colour

Colour plays a very significant role in the determination of muffin quality, as the variation in muffin colour indicates the degree of cooking and hence the product quality (Abdullah, 2008). Figure 16.3 shows four blueberry muffins from a single batch with different degrees of 'doneness', including undercooked, optimum, slightly overcooked and substantially overcooked. An efficient approach for muffin colour classification was presented by McConnel and Blau (1995). This approach uses a reference histogram based on the colour distribution of each class of interest, and a test histogram, representing the colour distribution of the object to be classified. The dissimilarity of distributions for a variety of likely reference colour distributions was measured and the reference class with the smallest measured dissimilarity was chosen.

The use of hue, saturation, intensity (HSI) colour space to grade a brown muffin was investigated by Abdullah *et al.* (2000, 2002). The choice of HSI colour space is not only because of its close representation of the human perception of

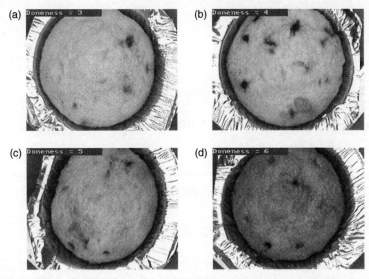

Fig. 16.3 Blueberry muffin images with different degrees of doneness from a single batch: (a) undercooked, (b) optimum, (c) slightly overcooked and (d) substantially over-cooked. (Adapted from McConnel and Blau, 1995.)

colour, but also, and more importantly, because it helps to compress information for easier colour discrimination. HSI separates colour into three components – that is, hue, saturation and value. H (hue) distinguishes among the perceived colours, such as red, yellow, green and blue. I (intensity) represents the total amount of light passing through a particular area and S (saturation) refers to how much white light is mixed with a hue. As I increases, the corresponding colour becomes increasingly bright.

To convert RGB to HSI, the first step is to obtain the normalized values of the red, green and blue components as follows:

$$nr = \frac{R}{(R+G+B)} \qquad [16.19]$$

$$ng = \frac{G}{(R+G+B)} \qquad [16.20]$$

$$nb = \frac{B}{(R+G+B)} \qquad [16.21]$$

where nr, ng and nb are the normalized values between 0 and 1, which give that nr + ng + nb = 1. Then the transformation to HSI can be achieved via the following equations:

$$S = 1 - 3 \times \min(nr, ng, nb) \qquad [16.22]$$

$$I = \frac{(R+G+B)}{(3 \times 255)} \tag{16.23}$$

$$H = \begin{cases} \cos^{-1} \left\{ \dfrac{0.5 \times \left[(nr-ng)+(nr-nb) \right]}{\left[(nr-ng)^2 + (nr-nb)(ng-nb) \right]^{1/2}} \right\} & h \in [0, \pi] \text{ for } nb \le ng \\[4ex] 2\pi - \cos^{-1} \left\{ \dfrac{0.5 \times \left[(nr-ng)+(nr-nb) \right]}{\left[(nr-ng)^2 + (nr-nb)(ng-nb) \right]^{1/2}} \right\} & h \in [\pi, 2\pi] \text{ for } nb > ng \end{cases}$$

$$\tag{16.24}$$

In this space, only the hue component (H) was analysed by Abdullah *et al.* (2000), since this attribute directly characterizes the colour properties of an object and thus the degree of doneness of muffins.

The hue distributions for different groups of muffins overlap each other, making it impossible to use the method based on the direct thresholding technique for classification. Abdullah *et al.* (2002) developed a classification strategy by first treating all hue values in the set ranging from 10° to 60° as input features, before invoking the Wilks's lambda analysis (Everitt and Dunn, 1991) to produce a subset containing all principal features and, finally, applying discriminant analysis to establish classification.

16.3.2   Size and shape

The techniques for bread size and shape analysis described in Section 16.2.2 can also be used in muffin analysis. A variety of 2-D quantitative information, including air cell count, mean cell size, total cell area and relative cell area, can be extracted from the reconstructed 2-D horizontal slice images captured by X-ray micro-computed tomography (Lim and Barigou, 2004). Similar to bread, a number of parameters including spatial cell size distribution, cell wall-thickness distribution, index of connectivity and degree of anisotropy can be extracted for 3-D muffin cellular structure analysis. Figure 16.4 exhibits a rendered 3-D model of a chocolate muffin, which is numerically cut to reveal its inner structure. Instead of using the morphological granulometry employed by Lassoued *et al.* (2007) for 3-D bread analysis, Lim and Barigou (2004) extracted the above parameters by combining image analysis with a stereological technique. They used an established stereological method, the Saltykov area analysis method (Xu and Pitot, 2003), to determine the cell size distribution.

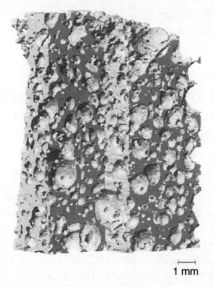

1 mm

Fig. 16.4 A rendered 3-D model of a chocolate muffin, which can be numerically sliced at any level and in any direction in order to view the inner structure (Lim and Barigou, 2004).

The index of connectivity is calculated by comparing the solid area and perimeter before and after an image dilation

$$IC = \frac{P_b - P_a}{A_b - A_a}$$
[16.25]

where P and A indicate perimeter and solid area, respectively. It is a measure of relative convexity or concavity of the total solid surface. The degree of anisotropy is a measure of 3-D structural symmetry, and can be obtained by using the mean intercept length and eigen analysis (Harrigan and Mann, 1984). The anisotropy analysis is based on the stereological calculus. First, some direction-dependent measurements are obtained by probing the segmented images and sending multiple lines over the full range of 3-D angles. Many parallel test lines are sent covering the whole test volume for every angle, and the mean intercept length for each angle is calculated as an average for all these lines. A multivariable linear least square fitting technique is then used to fit an ellipsoid to the mean intercept length data. Finally, a mean intercept length tensor based on the ellipsoid coefficients is defined, and their eigenvalues and eigenvectors are determined, leading to the following definition:

$$\text{Degree of anisotropy} = 1 - \frac{\min(\lambda)}{\max(\lambda)}$$
[16.26]

where λ represents the eigenvalues. The index of connectivity and the degree of anisotropy are some of the most important determinants of the mechanical strength of a cellular structure (Odgaard, 1997).

16.4   Computer vision applications for analysing biscuits

The following sections investigate how colour, size and shape in biscuits are analysed using these technological applications.

16.4.1   Colour

Surface colour of biscuits is controlled by many factors, such as water content, pH, amino acids and baking conditions. The baking process is quite complex, with biochemical reactions and physical transformations giving rise to biscuits with different shades of colour (Moore, 1991). Consequently, biscuit colour is an important indication of quality. Figure 16.5 shows eight biscuits from different groups, baked at 200°C in an industrial baking oven for 8, 10, 12, 14, 16, 18, 20 and 24 min, respectively. It demonstrates that, during biscuit manufacture, colour inspection plays a key role at all stages of the process, from raw dough to the finished product.

To separate a biscuit image from the background, Nashat *et al.* (2011) proposed a combination of the auto thresholding method with the watershed transformation technique. The auto thresholding method (Otsu, 1979) is used to distinguish between untouched and touched biscuits. In the former case, further image

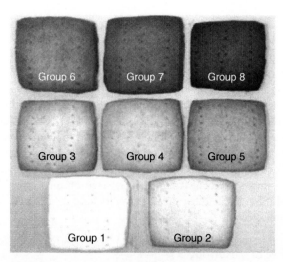

Fig. 16.5   Biscuit images baked at 200°C in an industrial baking oven for different times: Group 1 (8 min), Group 2 (10 min), Group 3 (12 min), Group 4 (14 min), Group 5 (16 min), Group 6 (18 min), Group 7 (20 min) and Group 8 (24 min) (Nashat *et al.*, 2011).

processing only involves smoothing and the removal of artefacts or noise. For the latter, watershed transformation is applied to separate the touching biscuits.

The concept of watershed transformation was originally proposed by Beucher and Lantuejoul (1979) to solve the problem of image segmentation, which simulates a flooding process over the image surface. The biscuit image to be segmented is considered as a topographic surface, in which the altitude of a position is equal to the intensity of the corresponding pixel in the image. Since the watershed lines are the highest crests separating the regional minima, it seems natural to compute the watersheds of the image gradient (Vincent, 1993). Unfortunately, as the biscuit images usually feature much noise and many local irregularities, there are a great number of minima generated in the image gradient. Therefore, a major problem with the watershed algorithm is that it may over-segment the biscuit image, yielding incorrect results.

To overcome the problem of over-segmentation, Meyer and Beucher (1990) proposed a method called marker-controlled watershed. The foreground markers are connected pixels within each of the objects, while the background markers are pixels out of any object. The design of robust marker detection techniques involves using the knowledge specific to the series of images under study (Vincent, 1993). For example, if the biscuits are circular, the Laplacian of Gaussian (LoG) filter (Wang et al., 2008) can be used to obtain the markers. This is defined as

$$\nabla^2 I(x,y) = \left(\frac{\partial^2}{\partial x^2} + \frac{\partial^2}{\partial y^2} \right) G_\sigma(x,y) \qquad [16.27]$$

where G_σ is the Gaussian function with variance σ

$$G_\sigma(x,y) = \frac{1}{\sqrt{2\pi}\sigma} \exp\left(-\frac{x^2+y^2}{2\sigma^2} \right) \qquad [16.28]$$

A LoG operation results in strong positive responses for dark blobs of extent σ and strong negative responses for bright blobs of similar size, upon which the biscuit marker can be obtained via a simple thresholding method. After marker extraction, the gradient image is modified, where it only has regional minima at the locations of biscuits and background markers. Based on the modified gradient image, the watershed transform can then be used to obtain the desired segmentation results.

From the segmented image, Hamey et al. (1998) used RGB colour space to monitor colour development throughout the baking process of commercial biscuit samples. However, the RGB colour space of the segmented images has no accurate definition for a human observer, where the proximity of colours in the space does not indicate colour similarity in perception. Therefore, it is difficult to distinguish directly among images with RGB colour space. Nashat et al. (2011) transform RGB information sensed by the machine vision to HSL (hue, saturation and lightness) colour space. HSL space was selected as it not only defines colour in the sense of perceptual uniformity, but more significantly, it matches to the human perception of colour (Camastra and Vinciarelli, 2008). Only the hue component (*H*) was analysed by Nashat *et al.* (2011), and the calculation of the value of *H* is the same as equation [16.24]. Moreover, the CIE L*a* b* colour space was also used to describe colour of corn biscuits by Lara *et al.* (2011).

Based on the extracted colour information, Hamey *et al.* (1998) employed the feed-forward neural network (FFNN) with back-propagation learning algorithm to recognize biscuits and categorize them into three groups (undercooked, correctly cooked and overcooked). In order to overcome the speed limitations associated with colour-based machine inspection, the self-organizing map (SOM) was used to reduce the input dimension of RGB space. SOM is a Kahonen-type neural network which is characterized by its ability to create a topological feature map modelling the probability distribution function of training samples (Hiotis, 1993). In another work, Nashat *et al.* (2011) used the Wilks's lambda analysis for dimension reduction, which is also known as the stepwise discriminant analysis (Rencher, 2002). Only those hue values with the highest discriminant powers are kept for classification using the support vector machine (SVM) technique. SVM is a state-of-the-art learning algorithm, which has a good theoretical foundation in statistical learning theory (Vapnik, 1995). More details about SVM can be found in the subsequent Section 16.5.2.

16.4.2   Size and shape

Size and shape are frequently used for the quality evaluation of biscuits. Smolarz *et al.* (1989) presented a method for processing the visual image of the transverse cut of a biscuit. The mathematical morphology (Serra, 1982) was employed to define structural elements of extruded biscuits. Structural parameters related to individual cells and the overall biscuit were determined to discriminate biscuit types.

To distinguish between whole and broken crackers, a computer vision system was proposed by Gunasekaran and Ding (1994). Sample images of the lion-shaped whole and broken crackers are shown in Fig. 16.6a and 16.6b, respectively. The test images were aligned with the reference image to obtain the best-fit edge contour. The size and shape features of area, aspect ratio, curvature, continuity and radius were then calculated. The classification was performed using a three-layer back-propagation neural network, with five input nodes, five hidden nodes and two output nodes.

In inspecting the shape of cookies, Nicchiotti and Ottaviani (1997) plotted the histogram of the radii distribution as a contour signature. Based on the histogram obtained, the shape of the object can be established by comparing the histogram signature of the tested image with the reference one, allowing analysis, for example, of the sum-square distances between object and template signatures. A simple rule-based method can then be employed to classify cookies based on their shapes.

(a)

(b)

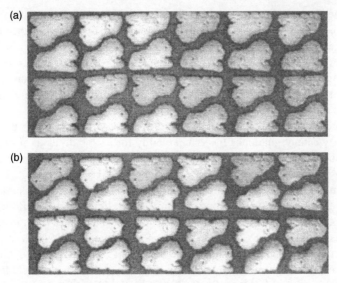

Fig. 16.6   Sample images of the lion-shaped crackers: (a) whole crackers and (b) broken crackers. (Adapted from Gunasekaran and Ding, 1994.)

16.5   Computer vision applications for analysing pizza bases

The following sections investigate how colour, size and shape in pizza bases are analysed using these technological applications.

16.5.1   Colour

Colour information can be extracted from an image of a pizza base by examining every pixel within the pizza base boundaries. For pizza base production, the determination of the colour information of total surface area after heat processing, especially the degree of browning, is important because excessive heating of proteins produces charring and results in irreversible lysine loss, caused by deamination or decarboxylation (Anglemier and Montgomery, 1976). As an objective measurement method, a computer vision approach has been developed by Unklesbay *et al.* (1983) for determining the relative degree of browning of pizza shells, prepared with and without soy flour. The percentage of the total area of each brown intensity level was obtained by constructing a histogram of the grey level representing the image of pizza base. In their study, the usefulness of colour information in the prediction of available lysine content was also confirmed. It was shown that the developed technique is very promising for use in cases where a rapid, non-destructive test for available lysine in baked dough is needed.

16.5.2   Size and shape

The most basic convenient measurement for the size of pizza base is area. For pixel-based representation of pizza base image, the area can be straightforwardly determined by counting the number of assigned pixels. Another way to obtain the area of the pizza is, first, to determine the Feret diameters (the distance between two tangents on opposite sides of the area perpendicular to a given direction) at 1° intervals to a maximum of 180°. The area of pizza can then be calculated using the average Feret diameter. The advantage of this method is that it can obtain the areas of the smallest and largest rounds of pizza for further shape analysis, as shown in the work of Sun and Brosnan (2003).

In practice, to produce a perfect-round pizza base is difficult as the present technology cannot guarantee the visual quality for each individual of each batch in the massive-scale manufacturing process. Generally, there are three types of pizza base defect: flowing base, poor alignment and poor pressing. A flowing base is simply where one side of the dough has spread out (flowed) more on the tray, increasing the diameter in one direction. Poor alignment occurs if the dough-ball is not centred when being pressed out, and hence the pizza base, instead of a round shape, may feature a straight side, or form an uneven circle. Poor pressing is characterized by an out-of-shape base that is not completely round, but is still of an acceptable quality. Figure 16.7 shows four illustrated images including three unacceptable quality levels – that is, flowing base (Fig. 16.7a), poor alignment (Fig. 16.7b) and poor pressing (Fig. 16.7c) – and one acceptable quality level – that is, standard (Fig. 16.7d).

Three steps of image processing algorithm could be developed to extract pizza base contours from digital images: noise reduction by median filter, thresholding-based segmentation and edge detection. In order to improve the quality of the captured image, operations need to be performed to remove or decrease degradations

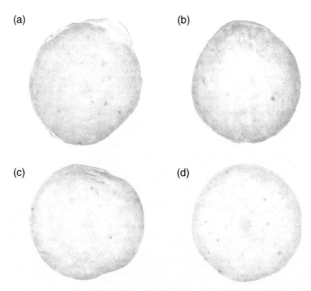

Fig. 16.7   Sample images of pizza base: (a) flowing base, (b) poor alignment, (c) poor pressing and (d) standard.

suffered in its acquisition. A median filter is a non-linear filtering technique, which allows the edges to be preserved while filtering out the unwanted noise, and is suitable for the removal of possible noises within the pizza base image. A thresholding-based image segmentation method can then be applied to separate the pizza base from the background. The segmentation step is necessary to obtain a closed and continuous boundary of the pizza base, which is difficult to obtain directly using traditional edge detection technology. Thus, the boundary becomes that set of interior points, each of which has at least one neighbour outside the pizza base. From the segmented image, the shape of the pizza base can be detected by the Canny edge detector (Canny, 1986) with selected low and high hysteresis thresholds. From the extracted shape and the segmented image of the pizza base, the shape feature of circularity can be computed by:

$$\text{Circularity} = \frac{\text{Perimeter}^2}{\text{Area}} \qquad [16.29]$$

Considering that the circularity alone was not enough for adequate shape analysis, two further criteria were introduced (Sun and Brosnan, 2003):

Spatial ratio I (SRI) = pizza base area/area of the smallest round which can hold the pizza base.
Spatial ratio II (SRII) = area of the biggest round which can be fit into by the pizza base/pizza base area.

According to the results of the four indices – that is, area, SRI, SRII and circularity – Sun and Brosnan (2003) classified pizza bases into acceptable and unacceptable levels. First, evaluation of the standard samples was performed to allow comparison. From the results, the lowest value of the area, SRI, SRII and the poorest circularity in the analysed samples were considered as the basis for classification into acceptable and unacceptable quality levels. Any pizza base with results smaller than its corresponding limit area, SRI and SRII or larger circularity than the limit was considered defective. It was found that, after studying the results based on these limits, an inaccuracy of 13% was obtained when using a combination of these indexes (Sun and Brosnan, 2003). In practice, as high-speed processing is important, the order of the four indices used when running the programme has to be determined in order to produce an efficient and effective quality analysis system. The most effective classification was attained by use of the circularity index. However, even though SRI and SRII are much less efficient, they should still be used for the classification of some samples. Hence, the best order to apply the indexes to determine acceptable and defective pizza bases is circularity, area, SRII and SRI.

Besides the above shape features with their combination of size measurement, a shape description technique that is independent of size measurement, the Fourier descriptor, can be applied to describe the boundary of pizza base. The Fourier transform of one cycle of the boundary function is an alternative representation of the associated object's shape, which could characterize the magnitude of changes of the shape frequency in the spatial domain. Using Fourier transform, the boundary function spectrum can be low-pass (only allowing low-frequency impulses to pass) filtered without destroying the characteristic shape of the object. Only the amplitudes and phases of the low-frequency impulses in the spectrum – that is, the low-order Fourier coefficients – are required to characterize the basic shape of the object. For example, Du and Sun (2004b) found that seven coefficients of the Fourier transform contained most of the information about the shape of pizza base, and were adequate for representing its characteristics. These values are candidates for shape descriptors.

Using the Fourier coefficients of each image as input, SVMs have shown promise as tools for pizza base classification (Du and Sun, 2004b). The classification of pizza bases into acceptable and unacceptable quality levels can be examined as a binary categorization problem. Suppose there are l pizza base samples in the training data. Each sample is denoted by a seven-dimensional vector $x_i \in R^7$ – that is, the first seven coefficients of the Fourier transform of the pizza base shape. Pizza base classification can be described as the task of finding a classification decision function $f: x_i \rightarrow y_i, y_i \in \{-1, +1\}$ using training data. Subsequently, the classification decision function f is used to correctly classify the unseen test data. If $f(x) > 0$, the input vector x is assigned to the class $y = +1$ – that is, the acceptable quality level – otherwise to the class $y = -1$ – that is, the unacceptable quality level.

For the linearly separable training vectors x_i, the classification function has the following form:

$$f(x) = \text{sgn}\left(w^T x + b\right)$$

[16.30]

where w is normal to the hyperplane and b is a bias term, which should satisfy the following conditions:

$$y_i\left(w^T x_i + b\right) \geq 1, \ i = 1, 2, \ldots, l$$

[16.31]

For the linearly non-separable case, the constraints in equation [16.31] are relaxed by introducing a new set of non-negative slack variables $\{\xi_i | i = 1, 2, \ldots, l\}$ as the measurement of violation of the constraints (Vapnik, 1995) as follows:

$$y_i\left(w^T x_i + b\right) \geq 1 - \xi_i, \ i = 1, 2, \ldots, l$$

[16.32]

The optimal hyperplane is the one that minimizes the following formula:

$$\frac{1}{2} w^T w + C \sum_{i=1}^{l} \xi_i$$

[16.33]

where C is a parameter used to penalize variables ξ_i, subject to constraints in equation [16.32].

For a non-linearly separable case, the training vectors x_i can be mapped into a high-dimensional feature space $\mathcal{H}$ by a non-linear transformation $\phi(\cdot)$. The training vectors become linearly separable in the feature space $\mathcal{H}$, and are then separated by the optimal hyperplane as described earlier. In many cases, the dimension of $\mathcal{H}$ is infinite, which makes it difficult to work with $\Phi(\cdot)$ explicitly. Since the training algorithm only involves inner products in $\mathcal{H}$, a kernel function $k(x_i, x_j)$ is used to solve the problem, which define the inner product in $\mathcal{H}$:

$$k\left(x_i, x_j\right) = \left\langle \phi(x_i), \phi(x_j) \right\rangle$$

[16.34]

Besides a linear kernel, polynomial kernels and Gaussian radial basis function (RBF) kernels are usually applied in practice. These are defined as:

$$k\left(x_i, x_j\right) = \left(x_i x_j + b\right)^d$$

[16.35]

$$k\left(x_i, x_j\right) = \exp\left(-\frac{\left\|x_i - x_j\right\|^2}{2\sigma^2}\right)$$

[16.36]

where b is the bias term and d is the degree of polynomial kernels. The classification function then has the following form in terms of kernels:

$$f(x) = \mathrm{sgn}\left[\sum_{i=1}^{l} y_i \alpha_i k\left(x_i, x\right) + b\right]$$

[16.37]

where α_i can be obtained by solving a convex quadratic programming problem subject to linear constraints. The support vectors are those x_i with $\alpha_i > 0$ in equation [16.37].

For multi-classification of pizza base samples into four shape quality levels (flowing base, poor alignment, poor pressing and standard), the construction and combination of several binary categorization problems are necessary. The earliest approach for multi-classification using SVM is one-versus-all. The classification of pizza base with this method can be described as the task of constructing four binary SVMs. The ith SVM is trained using the samples from the ith class as positive and the samples from all the other classes as negative. Four classification decision functions can be found:

$$f^i(x) = \sum_{j=1}^{l} y_j^i \alpha_j^i k\left(x_j^i, x\right) + b^i, i = 1, \ldots, 4$$

[16.38]

where $y_j^i \in \{+1, -1\}$, k is a kernel function, b_i is a bias term, and α_j^i is the coefficient obtained by solving a convex quadratic programming problem. Given an unknown sample of pizza base, the input vector x is assigned to the class that has the largest value of the decision function in equation [16.38].

Another approach which uses a combination of several binary SVM classifiers is called the one-versus-one method. The classification of pizza base with this method involves the construction of six binary SVMs, one classifier C_{ij} for every pair of distinct classes – that is, the ith class and the jth class, where $i \neq j$, $i = 1,\dots,4$; $j=1,\dots,4$. Each classifier C^{ij} is trained with the samples in the ith class with positive labels, and the samples in the jth class with negative labels. Six classification decision functions can be constructed as detailed below:

$$f^{ij}(x) = \sum_{n=1}^{\text{sum}} y_n^{ij} \alpha_n^{ij} k\left(x_n^{ij}, x\right) + b^{ij}, i \neq j, i = 1,\dots,4; j = 1,\dots,4 \qquad [16.39]$$

where sum is the total number of the ith and jth classes from the training data, $y_n^{ij} \in \{+1,-1\}$, k is a kernel function, b^{ij} is a bias term and α_n^{ij} is the coefficient obtained by solving a convex quadratic programming problem. Given an unknown sample of pizza base (denoted by x), if the decision function in equation [16.39] states that the input vector x is in the ith class, the classifier C^{ij} casts one vote for the ith class, otherwise the vote for the jth class was added by one. When all the votes from the six classifiers are obtained, the unknown sample x is assigned to the class with the most votes.

The third approach is the directed acyclic graph (DAG) SVM, which is a learning algorithm combining many two-class classifiers into one multi-class classifier using a decision graph. The training phase of the classification of pizza base is the same as the one-versus-one method – that is, it constructs six binary classifiers. However, in the test phase, it utilizes a new multi-class learning architecture called the decision directed acyclic graph (DDAG). Each node of the DDAG associates with a one-versus-one classifier. Given an unknown sample x, first, the binary decision function at the root node is evaluated. Then, if the value of the binary decision function is -1, the node exits via the left edge; otherwise, if the value is $+1$, it exits via the right edge. The binary decision function of the next internal node is then evaluated in the same manner. The class of x is the one associated with the final leaf node.

16.6   Computer vision applications for analysing other bakery products

It is feasible to use computer vision techniques for the inspection of other bakery products. Computer vision methods were developed by Bergaz *et al.* (2010) for quality control during production, including colour, shape and topping defects detection, of napolitana (a traditional Spanish fermented bakery product). For colour inspection, they used the histogram of the red plane and evaluated the percentage of red component. To decide whether it follows a regular shape or not, the blue plane was used to segment the napolitana and the brightness plane for edge detection. For topping quality evaluation, the napolitana's area was divided into six ROIs (2×3 regions of interest). The percentage of chocolate topping was then calculated for each ROI. A napolitana was rejected if at least one of ROIs had less than 12% of chocolate topping or presented topping excess more than 45%. In another work, the internal structure of cake was examined by computer vision (Sapirstein, 1995). The brightness, cell density, cell area and uniformity were analysed, and the results indicated that even the most minor deviations from the required specifications were obvious to the developed system, which allows corrective measures in the bakery to be taken more quickly.

16.7   Future trends and further information

There are still a few issues which need to be addressed regarding computer vision application in bakery product assurance. Due to the non-uniform sizes, shapes, surfaces and colours of bakery products, the image segmentation might not produce a high enough degree of accuracy, due to the complex and non-uniform cellular structure of crumb grain. In the work of Zghal *et al.* (1999), they showed that the interconnection of cells and the complexity of the cellular structure of breads were probably responsible for the over-estimation of cell characteristics using the cell segmentation method. Lassoued *et al.* (2007) also reported that, although a satisfactory characterization of images based on crumb cells and cell walls was obtained, the procedure tended to slightly underestimate the void fraction and overestimate cell wall thickness. Gonzales-Barron and Butler (2006) evaluated the suitability of six thresholding techniques for bread segmentation and found substantial variations in crumb features such as cell uniformity and void fraction according to threshold variations. Robust and accurate segmentation of such images still remains a challenge.

Physical models could be developed for simulation of food processing via integrating image-derived quantitative information. For example, Tokumoto and Hirai (2002) used shape signature in studying the relationship between shape deformation of food dough and control parameters during the forming process. A key issue here is the difficulty in selecting image features that are valuable, and issues of accuracy, repeatability and variability must be considered in the long run. Furthermore, the effective use of these models will require new knowledge from a variety of areas.

As image sizes grow larger and algorithms become more complex, the speed of analysis will be reduced and will not be able to satisfy the high speed requirements of real-time systems. Parallel algorithms should be implemented for those time-consuming tasks, which can be executed one piece at a time on many different processing devices, and then put back together again at the end to produce the necessary result. Nashat *et al.* (2011) presented an in-depth analysis of the multicore implementation of SVM-based artificial classifier for real-time inspection of biscuits. However, while parallel algorithms are valuable because of substantial improvements they offer in multiprocessing system, some problems are impossible or very difficult to parallelize.

For further reading in image processing, *The Image Processing Handbook* by Russ (1998) is an excellent elementary textbook. Another book is *Digital Image Processing* by Burger and Burge (2007), which provides a modern, self-contained introduction to digital image processing. Numerous complete Java implementations are provided, all of which work within the open source software of ImageJ, the programmer extensible imaging system developed, maintained and distributed by Wayne Rasband of the National Institutes of Health. Classical textbooks, including Gonzalez and Woods (1992) and Pratt (1991) are also worthwhile reading.

For further information about computer vision and its applications for food quality evaluation, the reader should refer to the handbook of Sun (2008), and specific reviews (Brosnan and Sun, 2004; Du and Sun 2004a; Gunasekaran, 1996).

16.8   Conclusions

This chapter has focused on describing the relevant techniques of computer vision for quality evaluation of bread, muffin, biscuit and pizza base, along with others. The CIE L*a*b* colour space is mainly used for bread colour feature extraction, while thresholding-based and classification-based methods are used in most applications for crumb grain segmentation. From the segmented image, a number of morphological features of crumb grain can be computed, including mean cell area, cell density, cell shape distribution, cell uniformity, mean cell wall thickness and void fraction. For 3-D bread analysis, the volumic means of cell size, cell wall thickness and crumb porosity (cell volumic fraction) can be obtained. Several image texture techniques have been applied for quantifying various crumb grain structures, including first-order statistical measures, GLCM methods, transform-based methods, such as Fourier transform, Haar transform and FD.

The use of HSI colour space to grade a brown muffin has been investigated. A number of parameters including spatial cell size distribution, cell wall-thickness distribution, index of connectivity and degree of anisotropy can be extracted for 3-D muffin cellular structure analysis.

RGB, HSL and L*a* b* colour spaces can be used to monitor the colour development of biscuit, and the classification of biscuit according to colour can be performed by FFNN and SVM. A variety of techniques have been employed for size and shape inspection, including mathematical morphology, neural network and histogram of the radii distribution.

The relative degree of browning of a pizza base can be determined by constructing a histogram of the grey level representing the image. Various features can be extracted for size and shape inspection of pizza base, including area, circularity, spatial ratios and Fourier descriptors. With the extracted features of each image as input, SVM can differentiate acceptable pizza bases from defective ones, and can classify these as resulting from flowing base, poor alignment or poor pressing. It is also feasible to use computer vision techniques for inspecting other bakery products, such as napolitana and cake.

16.9   References

Abdullah, M. Z. (2008) Quality evaluation of bakery products. In D.-W. Sun (ed.), *Computer Vision Technology for Food Quality Evaluation*. London: Elsevier, 481–522.

Abdullah, M. Z., Abdul-Aziz, S. and Dos-Mohamed, A. M. (2000) Quality inspection of bakery products using color-based machine vision system. *Journal of Food Quality*, **23**, 39–50.

Abdullah, M. Z., Abdul-Aziz, S. and Dos-Mohamed, A. M. (2002) Machine vision system for online inspection of traditionally baked Malaysian muffins. *Journal of Food Science Technology*, **39(4)**, 359–66.

Anglemier, A. F. and Montgomery, M. W. (1976) Amino acids, peptides, and proteins. In Fennema, O. R. (ed.), *Principles of Food Science. Part 1. Food Chemistry*. New York: Marcel Dekkar, 205–84.

Anon. (1990) *IEEE Standard 610.4-1990 in IEEE Standard Glossary of Image Processing and Pattern Recognition Terminology*. New York: IEEE Press.

Anon. (2011) Bakery products: A global strategic business report [Internet], Global Industry Analysts, Inc. 6150 Hellyer Ave., San Jose, CA 95138, USA. Available from: http://www.strategyr.com/Bakery_Products_Market_Report.asp (accessed 25 May 2011).

Arthur, D. and Vassilvitskii, S. (2007) K-means++: The advantages of careful seeding. *Proceedings of the Eighteenth Annual ACM-SIAM Symposium on Discrete Algorithms*, 1027–35.

Baardseth, P., Kvaal, K., Lea, P., Ellekjær, M. R. and Færgestad, E. M. (2000) The effects of bread making process and wheat quality on French baguettes. *Journal of Cereal Science*, **32**, 73–87.

Babin, P., Della Valle, G., Dendievel, R., Lassoued, N. and Salvo, L. (2005) Mechanical properties of bread crumbs from tomography based finite element simulations. *Journal of Material Science*, **40**, 5867–73.

Babin, P., Della Valle, G., Chiron, H., Cloetens, P., Hoszowska, J., Pernot, P., Réguerre, A. L., Salvo, L. and Dendievel, R. (2006) Fast X-ray tomography analysis of bubble growth and foam setting during breadmaking. *Journal of Cereal Science*, **43**, 393–7.

Bergaz, L. P., Ruiz, G. R., Gracia, L. M. N., Guimaraes, A. C. and Gil, J. G. (2010) Bakery products quality control using computer vision: Napolitana case. *CIGR Workshop on Image Analysis in Agriculture*, 26–27 August 2010, Budapest.

Bertrand, D., Le Guernevé, C., Marion, D., Devaux, M. and Robert, P. (1992) Description of the textural appearance of bread crumb by video image analysis. *Cereal Chemistry*, **69**, 257–61.

Beucher, S. and Lantuejoul, C. (1979) Use of watershed in contour detection. *International Workshop on Image Processing, Real-Time Edge and Motion Detection/Estimation*, France, 12–21.

Brescia, M., Sacco, D., Sgaramella, A., Pasqualone, A., Simeone, R., Peri, G. and Sacco, A. (2007) Characterisation of different typical Italian breads by means of traditional, spectroscopic and image analyses. *Food Chemistry*, **104**, 429–38.

Brosnan, T. and Sun, D.-W. (2004) Improving quality inspection of food products by computer vision – a review. *Journal of Food Engineering*, **61(1)**, 3–16.

Burger, W. and Burge, M. J. (2007) *Digital Image Processing: An Algorithmic Approach Using Java*. New York: Springer-Verlag.

Camastra, F. and Vinciarelli, A. (eds), (2008) *Machine Learning for Audio Image and Video Analysis Theory and Applications*. London: Springer.

Canny, J. (1986) A computational approach to edge detection. *IEEE Transactions on Pattern Analysis and Machine Intelligence*, **8(6)**, 679–98.

Du, C.-J. and Sun, D.-W. (2004a) Recent developments in the applications of image processing techniques for food quality evaluation. *Trends in Food Science & Technology*, **15(5)**, 230249.

Du, C.-J. and Sun, D.-W. (2004b) Shape extraction and classification of pizza base using computer vision. *Journal of Food Engineering*, **64(4)**, 489–96.

Dubose, C. N. and Cardello, A. V. (1980) Effect of colorants and flavorants on identification, perceived flavor intensity, and hedonic quality of fruit flavored beverages and cakes. *Journal of Food Science*, **45**, 1393–415.

Everitt, B. S. and Dunn, G. (1991) *Applied Multivariate Data Analysis*. London: Edward Arnold.

Falcone, P. M., Baiano, A., Zanini, F., Mancini, L., Tromba, G., Dreossi, D., Montanari1, F., Scuor, N. and Del Nobile, M. A. (2005) 3D quantitative analysis of bread crumb by X-ray tomography. *Journal of Food Science*, **70**, 265–72.

Frey, B. J. and Dueck, D. (2007) Clustering by passing messages between data points. *Science*, **315(5814)**, 972–6.

Gonzales-Barron, U. and Butler, F. (2006) A comparison of seven thresholding techniques with the k-means clustering algorithm for measurement of bread-crumb features by digital image analysis. *Journal of Food Engineering*, **74**, 268–78.

Gonzales-Barron, U. and Butler, F. (2007a) Fractal texture analysis of bread crumb digital images. *European Food Research and Technology*, **226(4)**, 721–9.

Gonzales-Barron, U. and Butler, F. (2007b) Prediction of panellists' perception of bread crumb appearance using fractal and visual textural features. *European Food Research and Technology*, **226(4)**, 779–85.

Gonzales-Barron, U. and Butler, F. (2008) Discrimination of crumb grain visual appearance of organic and non-organic bread loaves by image texture analysis. *Journal of Food Engineering*, **84**, 480–8.

Gonzalez, R. C. and Woods, R. E. (1992) *Digital Image Processing*. Reading, MA: Addison-Wesley.

Gunasekaran, S. (1996) Computer vision technology for food quality assurance. *Trends in Food Science and Technology*, **7**, 245–56.

Gunasekaran, S. and Ding, K. (1994) Using computer vision for food quality evaluation. *Food Technology*, **48(6)**, 151–4.

Hamey, L. G. C., Yeh, J. C. H., Westcott, T. and Sung, K. Y. (1998) Pre-processing color images with self organising map: Baking curve identification and bake image segmentation. *Proceedings of the 14th International Conference on Pattern Recognition*, Brisbane, Queensland, Australia, 16.

Haralick, R. M., Shanmugan, K. and Dinstein, I. (1973) Textural features for image classification. *IEEE Transactions on Systems, Man, and Cybernatics*, **3(6)**, 610–21.

Harrigan, T. P. and Mann, R. W. (1984) Characterisation of microstructural anisotropy in orthotropic materials using a second rank tensor. *Journal of Materials Science*, **19**, 761–7.

Hiotis, A. (1993) Inside a self organising map. *Artificial Intelligence Expert*, June, 38–43.

Kamman, P. W. (1970) Factors affecting the grain and texture of white bread. *The Bakers Digest*, **44**, 34–8.

Lara, E., Cortés, P., Briones, V. and Perez, M. (2011) Structural and physical modifications of corn biscuits during baking process. *LWT—Food Science and Technology*, **44**, 622–30.

Lassoued, N., Babin, P., Valle, G. D., Devaux, M. F. and Reguerre, A. L. (2007) Granulometry of bread crumb grain contributions of 2D and 3D image analysis at different scales. *Food Research International*, **40(8)**, 1087–97.

Lim, K. S. and Barigou, M. (2004) X-ray micro-computed tomography of cellular food products. *Food Research International*, **37**, 1001–12.

Liu, S. S. and Jernigan, M. E. (1990) Texture analysis and discrimination in additive noise. *Computer Vision, Graphics, and Image Processing*, **49**, 52–67.

Lloyd, S. P. (1982) Least-squares quantization in PCM. *IEEE Transactions on Information Theory*, **28(2)**, 129–37.

McConnel, R. K. and Blau, H. H. (1995) Color classification of non-uniform baked and roasted foods. *Proceedings of the Food Processing Automation Conference IV*, American Society of Agricultural Engineers, St Joseph, MI, 40.

Meyer, F. and Beucher, S. (1990) Morphological segmentation. *Journal of Visual Communication and Image Representation*, **1(1)**, 21–46.

Milligan, G. W. (1980) An examination of the effect of six types of error perturbation on fifteen clustering algorithms. *Psychometrica*, **45**, 325–42.

Moore, C. A. (ed.), (1991) *Automation in the Food Industry*. Glasgow, New York: Blackie.

Naito, S., Ishida, N., Takano, H., Koizumi, M. and Kano, H. (2003) Routine evaluation of the grain structures of baked breads by MRI. *Food Science and Technology Research*, **9**, 155–61.

Nashat, S., Abdullah, A., Aramvith, S. and Abdullah, M. Z. (2011) Support vector machine approach to real-time inspection of biscuits on moving conveyor belt. *Computers and Electronics in Agriculture*, **75**, 147–58.

Nicchiotti, G. and Ottaviani, E. (1997) A simple rotation invariant shape signature. *Proceedings of the 6th International Conference on Image Processing and Its Applications*, Dublin, Ireland, 722.

Odgaard, A. (1997) Three-dimensional methods for quantification of cancellous bone architecture. *Bone*, **20**, 315–28.

Otsu, N. (1979) A threshold selection method from gray level histograms. *IEEE Transactions on Systems, Man and Cybernetics*, **9**, 62–6.

Peleg, S., Naor, J., Hartley, R. and Avnir, D. (1984) Multiple resolution texture analysis and classification. *IEEE Transactions on Pattern Analytical and Machine Intelligence*, **6**, 518–23.

Pratt, W. (1991) *Digital Image Processing*, 2nd edn. New York: Wiley.

Purlis, E. (2010) Browning development in bakery products – A review. *Journal of Food Engineering*, **99**, 239–49.

Pyler, E. J. (1988) *Baking Science and Technology*, 2nd edn. Kansas City, MO: Sosland.

Quevedo, R., Carlos, L. G., Aguilera, J. M. and Cadoche, L. (2002) Description of food surfaces and microstructural changes using fractal image texture analysis. *Journal of Food Engineering*, **53(4)**, 361–71.

Rencher, A. C. (ed.), (2002) *Methods of Multivariate Analysis*. Canada: John Wiley and Sons.

Ridler, T. W. and Calvard, S. (1978) Picture thresholding using an iterative selection method. *IEEE Transactions on Systems, Man and Cybernetics*, **8**, 630–2.

Rogers, D. E., Day, D. D. and Olewnik, M. C. (1995) Development of an objective crumb grain measurement. *Cereal Foods World*, **40(7)**, 498–501.

Russ, J. C. (1998) *The Image Processing Handbook*, 3rd edn. Boca Raton, FL: CRC.

Sapirstein, H. D. (1995) Quality control in commercial baking: machine vision inspection of crumb grain in bread and cake products. *Food Processing Automation IV Proceedings of the FPAC Conference*, ASAE, St Joseph, Michigan.

Sapirstein, H. D., Roller, R. and Bushuk, W. (1994) Instrumental measurement of bread crumb grain by digital image analysis. *Cereal Chemistry*, **71(4)**, 383–91.

Sarkar, N. and Chaudhuri, B. (1994) An efficient differential box counting approach to compute fractal dimension of images. *IEEE Transactions on Systems, Man, and Cybernatics*, **SMC-24**, 115–20.

Scanlon, M. G. and Zghal, M. C. (2001) Bread properties and crumb structure. *Food Research International*, **34**, 841–64.

Serra, J. (1982) *Image Analysis and Mathematical Morphology*. London: Academic Press.

Smolarz, A., van Hecke, E. and Bouvier, J. M. (1989) Computerized image analysis and texture of extruded biscuits. *Journal of Texture Studies*, **20**, 223–34.

Sun, D.-W. (ed.) (2008) *Computer Vision Technology for Food Quality Evaluation*. London: Elsevier.

Sun, D.-W. and Brosnan, T. (2003) Pizza quality evaluation using computer vision – part 1 Pizza base and sauce spread. *Journal of Food Engineering*, **57**, 81–9.

Tokumoto, S. and Hirai, S. (2002) Deformation control of rheological food dough using a forming process model. *Proceedings 2002 IEEE International Conference on Robotics and Automation*, Washington, DC, 1457.

Unklesbay, K., Unklesbay, N., Keller, J. and Grandcolas, J. (1983) Computerised image analysis of surface browning of pizza shells. *Journal of Food Science*, **48**, 1119–23.

Vapnik, V. (1995) *The Nature of Statistical Learning Theory*. New York: Springer.

Vincent, L. (1993) Morphological grayscale reconstruction in image analysis: Applications and efficient algorithms. *IEEE Transactions on Image Processing*, **2(2)**, 176–201.

Wang, Y.-P., Wu, Q. and Castleman, K. R. (2008) Image enhancement. In Wu, Q. Merchant, F. and Castleman, K. (eds), *Microscope Image Processing*. New York: Academic Press.

Xu, Y. H. and Pitot, H. C. (2003) An improved stereologic method for three-dimensional estimation of particle size distribution from observations in two dimensions and its application. *Computer Methods and Programs in Biomedicine*, **72(1)**, 1–20.

Zayas, I. Y. (1993) Digital image texture analysis for bread crumb grain evaluation. *Cereal Food World*, **38**, 760–6.

Zghal, M. C., Scanlon, M. G. and Sapirstein, H. D. (1999) Prediction of bread crumb density by digital image analysis. *Cereal Chemistry*, **76(5)**, 734–42.

17

Development of multispectral imaging systems for quality evaluation of cereal grains and grain products

M. A. Shahin, D. W. Hatcher and S. J. Symons, Canadian Grain Commission, Canada*

Abstract: The commercial value of small grains such as wheat and barley is determined by their overall quality. Grading systems established by the major countries exporting these grains establish maximum tolerances for various contaminants and damage factors to the grain kernels. Grading factors are mainly visual characteristics. Collectively, the grading factors describe grain quality and safety as affected by growing conditions, handling and storage practices. Human visual inspection is the current method of grain quality assessment, which can be subjective and inconsistent. Alternative instrumental approaches to quality assurance are constantly being researched. Hyperspectral imaging has been used effectively for the detection and quantification of grain damage due to various grading factors such as mildew, fusarium, sprouting and green immature seeds.

Hyperspectral systems, while effective, are expensive research tools. Multispectral systems using a few wavelengths provide practically viable, less expensive and simpler imaging solutions. A multispectral system using three band-pass filters successfully detected and scored green barley kernels. Development of fast, accurate and low-cost multispectral systems is expected to have a profound effect on the acceptance of instrumental methods of grain grading by the grain industry.

Key words: cereal, grain, wheat, barley, imaging, hyperspectral, multispectral, grain grading, quality, mildew, fusarium, sprouting, green immature, flour colour.

17.1   Introduction

Appearance is the single most important evaluating factor that determines the value of grains. A number of grading factors (or degrading factors to be more precise) adversely affect the appearance of cereal grains. The adverse effects of these grading factors on the end use quality of both the common wheat and amber durum wheat have been well understood and documented (Dexter and Edwards, 1998a, 1998b). The purpose of grain grading, regardless of the type of commodity, is to ensure that all parties involved, including sellers and buyers, are receiving/paying the appropriate value for that specific lot of grain. Within North America, grading factors are associated with the commodity's food safety as well as the final end-product performance. While end-product performance tradition-ally was the key motivator, there is increasing concern to ensure that the current grading systems also adequately address the food safety issue. Environmental conditions throughout the grain's development and maturity is the prime com-ponent influencing the nature and severity of the grading factor that affects both safety and quality.

Mildew damage is a serious quality defect that has a negative impact on process-ing quality and commercial value of wheat (Dexter and Edwards, 1998a). Mildew, caused by the fungus Cladosporium, is normally associated with poor harvesting conditions characterized by high amounts of rainfall, often predicting or coincid-ing with sprout damage. There are no known reports linking mildew with food safety issues, specifically the production of mycotoxins. Mildew infection is asso-ciated with grey tufts caused by the presence of the spores. As the mildew infection becomes more severe, kernels are characterized as being blackened. While mildew damage offers no toxicology hazard in wheat, its presence can be detrimental to the end-products made from wheat flour and therefore significantly impacts the price paid for a specific level of mildew damage. Isolating mildew's detrimental role on the overall quality of wheat is difficult as the causal factor – that is, poor weather conditions – also directly impacts the kernels' development, leading to a lower wheat test weight and falling number value (Dexter and Edwards, 1998a). Due to its grey to blackening of the wheat kernel, the presence of mildew has a negative impact on the flour colour, producing a duller less bright product of both lower value and quality. As the vast majority of flour buyers purchase on the basis of flour colour, increasing mildew content requires a decrease in the extraction or flour yield that a miller can achieve for a specific level of mildew. Wheat mill-ing performance decreases with increasing levels of mildewed kernels of common wheat (Everts et al., 2001), but no impact of mildew damage on semolina yield has been established (Dexter and Matsuo, 1982). Unfortunately, assessing the level or impact of mildew damage is extremely difficult as it is currently based on a trained inspector's subjective visual assessment. Although this has been sufficient in the past, increased focus on ensuring true representative value for a specific lot of wheat has many clamouring for an objective and quantifiable assessment for the grain industry.

Mildew is normally a key first indicator of wet grain. While it impacts end-product quality, specifically flour colour, sprout damage is a more significant problem that results from excessive exposure to moisture and resulting in premature kernel sprouting. Premature sprouting generates a series of enzymes that are deleterious to a variety of different end products. It has been demonstrated that sprout damage is extremely detrimental to end-product quality, specifically in wheat and rye, but also in canola and malting barley. Considerable research has been done on identifying and describing the deleterious effects of the enzyme α-amylase in wheat due to the extended processing time required for making bread or Asian steam buns. Identified problems in bread-making, for example, include reduced water-holding capacity, stickier doughs, coarser crumb structure and frequent gummy crumb. All of these issues have a strong economic impact on the manufacturer. Lower water-holding capacity reduces the number of loaves produced per unit weight of flour, while stickier doughs are harder to handle mechanically. Gummy crumbs not only result in considerable product 'loss' as loaves are damaged by sticking to slicing blades, they can also clog automated slicing equipment leading to frequent production line stoppages. Visually, due to the presence of increased free sugars from the action of α-amylase, the crust is often darkened due to caramelization during the bake cycle, which reduces visual appeal and deters the consumers from purchasing the product. If the consumer does buy the product, the coarser and more open interior crumb structure is not appealing thereby negating any further purchases, which is extremely detrimental to brand recognition.

Although the Hagberg falling number test is used commercially throughout the world for contractual/commerce purposes, the test itself is only an indirect measurement of sprout damage. Due to its time and laboratory requirements, the falling number test is not used regularly at the initial site of wheat arrival – that is, a primary elevator in Canada, but is increasingly used in the USA and Australia. The primary means of assessing sprout damage at a receival point still remains visual, using the expertise of a trained inspector's subjective visual assessment of a kernel. Unfortunately, the definition of what constitutes sprouted versus severely sprouted grain is open for interpretation, which can lead to dispute between buyer and seller. Additionally, because the levels of amylase in severely sprouted grain can range over 1000-fold between single kernels, trying to assign 'proper value' to a sample of grain as a predictor of processing capacity using the visual approach is extremely difficult and not without significant economic risk.

Fusarium head blight (FHB), also known as scab in the USA, is primarily found on cereals and is a result of the growth of existing spores present naturally on the surface of the grain, as well as specific environmental conditions at anthesis that facilitate spore growth. Unlike mildew, however, FHB poses a food safety issue due to the production of a trichothecene mycotoxin known as deoxynivalenol (commonly referred to as DON or vomitoxin). Unfortunately, a number of studies (Dexter, 1997; Dexter et al., 1996a, 1996b; Dexter and Edwards, 1998a, 1998b) have demonstrated that the mycotoxin remains active and present during both the milling process and subsequent end-product processing. In wheat milling, it is partitioned throughout various resulting millstreams and has been shown not to be destroyed during processing or baking. Consequently, DON in grains leads to the presence of the mycotoxin in the end-products.

Aside from food safety issues FHB has been clearly shown to impact end-product quality with numerous studies of its detrimental influence on wheat flour products. Economically, in addition to DON there is often an associated reduction in flour yield with high levels of FHB and poorer flour colour, and, due to its hydrolytic enzymes, a decline in protein strength parameters which is reflected in the final end-products' performance.

Another economically significant, but traditionally difficult wheat quality parameter to measure both rapidly and quantifiably has been the presence of hard vitreous kernels (HVK). While there is discussion on the degree of HVK impact on common wheat product performance, and wheat protein content that serves as the basis for segregation is easily determined via near-infrared measurement, there is no ambivalence with respect to its impact on milling yield of semolina from amber durum wheat. Factors such as weather conditions, heredity and, especially, soil fertility have all been linked with kernel vitreousness. While a visual assessment by trained inspectors is common, the nature of what constitutes loss of vitreousness remains variable across countries. Within Canada, a non-vitreous kernel is the one that has any portion, no matter how small, of white/yellowish regions within the kernel which can be confirmed by slicing open the kernel with a scalpel. It is these starchy regions that contribute to the lowering of semolina yield during milling causing significant economic loss due to the production of undesirable, poor colour quality durum flour, which is excluded from high quality pasta production.

Presence of green immature kernels downgrades grain shipments because of their negative impact on the appearance and processing quality of the grain. During the early stages of development, most seeds have photosynthetic capacity and contain chlorophyll for this function. When conditions during maturation shorten or change the process, chlorophyll residues remain in the seed imparting a green colour to the immature seed. Normal mature seeds of cereals contain only residues of photosynthetic pigments, while other seeds retain their green seed coats – for example, green laird type lentils or green peas as grown in Canada. The inclusion of green seeds in the mature harvest of cereals is, therefore, an indicator that the maturation process has been terminated early, either by a short growing season whereby the maturation period was too short for the process to complete or due to environmental factors such as frost which prematurely halt the maturation process. Green seeds in a mature crop are often a sign of problems to be encountered later on during processing. As such, the amount of green seeds in grain shipments is limited according to specifications set by the seller, buyer or both. In our example, for Canadian grain, the Canadian Grain Grading Guide specifies the amount of green seeds allowed in each class of each grain (CGC, 2011). The inspection process is typically a visual assessment either by count or appearance as defined for various grains in the Canadian Grain Grading Guide (CGC, 2011).

The effect of green seeds on quality differs according to the grain type, but is always negative. In canola, green seeds lead to an off-colour in the pressed oils. In malting barley, green seeds reduce the uniformity of germination during the malting process, lowering both the yield and quality of the malt. Uniformity of germination during malting is desirable to create the best malt. In wheat, green seeds are an indicator of immaturity, with its associated reduced storage protein production and higher α-amylase content, or frost damage. The darkened seed coat layers impart dark specks into flour during milling, creating a less desirable product which commands a lower price.

The Canadian Grain Commission (CGC) is the Canadian federal agency responsible for establishing tolerances for the various grades of grains grown in Canada and has carried out extensive research to ensure that the levels of kernel damage, due to various grading factors within each grade are appropriate and reflect the intrinsic value of the grain (CGC, 2011). Current grading systems are based on a relatively slow and subjective human visual inspection by trained inspectors, whereby the degree of damage to kernels is assessed by comparison to standard samples to give an overall visual assessment of severity of damage. However, it remains difficult to quantify the amount of damage through visual inspection, especially for samples with light to moderate degrees of damage. It is important to determine the extent of damage as the severity of damage is inversely related to the quality of grain and grain products. Fast and accurate objective methods are required to assess grain quality to cope with high-volume inspection demanded by the grain industry to meet the needs of national as well as international trade. Identification of damaged regions of individual kernels by imaging appears to be a logical solution; however, variations in the discoloration of the damaged regions and the colour of underlying kernels make this impossible to achieve using a traditional imaging platform. Hyperspectral imaging (HSI) can provide a workable solution in this situation.

Hyperspectral imaging has been used in a wide variety of fields, including remote sensing, pharmaceutical, medical and agricultural industries. Recently, this technology has emerged as a research tool for food quality and safety control (Gowen et al., 2007). In the agro-food industry, HSI applications have been reported for quality assessment of fruits (Kim et al., 2002; Lu, 2003), vegetables (Cheng et al., 2004; Gowen et al., 2008), poultry (Park et al., 2002), beef steaks (Naganathan et al., 2008) and cereal grains (Cogdill et al., 2004; Goretta et al., 2006; Lin et al., 2006; Shahin and Symons, 2008). Recent studies have shown that hyperspectral imaging could distinguish sprout-damaged (Singh et al., 2009; Xing et al., 2009) as well as stained and fungal-infected (Berman et al., 2007; Singh et al., 2007) wheat kernels from sound kernels. Shahin and Symons (2007) investigated the use of HSI to study spectral characteristics of wheat kernels as affected by various grading factors. Spectra of damaged kernels were reported to be significantly different from those of sound undamaged kernels. These spectral differences can be utilized to determine the extent of damage in wheat samples.

A brief outline of the chapter is as follows. Description of hyperspectral imaging and methods of spectral data/image analysis are described in Section 17.2. The use of HSI for detection of damaged kernels due to various grain grading

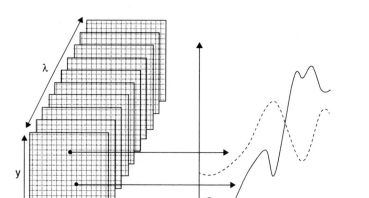

Fig. 17.1   A schematic of the HSI hypercube showing spatial (x, y) and spectral (λ) dimensions; spectra on the right correspond to single pixel locations in the image.

factors is presented in Sections 17.3–17.6, followed by quality evaluation of grain products in Section 17.7. Steps involved in the development of a multispectral system are described in Section 17.8.

17.2   Hyperspectral imaging

Hyperspectral imaging (HSI) is a relatively new technology that combines conventional imaging and spectroscopy technologies to provide both spatial and spectral information with high resolution. There are two major types of HSI systems based on the mode of data acquisition: (a) area imagers capture a series of images each at a specific wavelength using tuneable filters and (b) line scanners capture spectral data for one line in the scene at a time. Regardless of the mode of operation, HSI systems produce three-dimensional (3-D) data structures commonly known as hyperspectral images or hypercubes, where associated with each pixel is a spectrum that can be used as a fingerprint to characterize the composition of that particular pixel (Fig. 17.1). Hyperspectral images typically contain hundreds of spectral bands acquired at contiguous wavelengths. The precise spectral information contained in an HSI image enables better characterization and identification of targets.

17.2.1   Image acquisition and calibration
Sample illumination plays a critical role in acquiring high quality hyperspectral images. Therefore, due consideration must be given to designing a proper lighting system. The design of illumination systems is based on a wide range of diverse requirements, but often the most important requirements are maximizing intensity and/or uniformity of light distribution. Regulated light sources with power controllers are often used to ensure consistent illumination over time. In most imaging applications, appropriately positioned lamps constitute the illumination system. However, some applications require sophisticated arrangements to guide

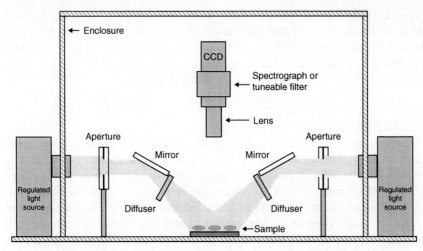

Fig. 17.2 A generalized schematic of a hyperspectral imaging system in a light-tight enclosure along with lighting arrangements for sample illumination.

light beams on to the test object. This is particularly the case when a hyperspectral camera system is enclosed in a light-tight cabin to prevent erroneous peaks due to stray ambient light. The light sources often generate significant heat that can lead to overheating the cabin interior. This can cause serious performance issues for some camera systems that have heat-sensitive detectors. Excessive heat can also damage the samples. Such applications would require the lamp-housings to be kept outside the cabin. The task then is to guide the light beam to the sample through optical lenses, mirrors, light guides or fibre bundles. Light guides are more suitable for applications requiring illumination of relatively smaller areas whereas optical components such as mirrors, lenses, prisms, diffusers, etc. are typically used in applications that require illumination over relatively larger areas. A typical in-cabin camera set-up, along with the lighting arrangement that minimizes shadowing effect, is shown in Fig. 17.2.

Illumination light intensity level and camera exposure time are set so that a strong spectral signal is obtained without saturating the detector array. If permitted by the application, spatial and/or spectral binning should be used as binning leads to cleaner low-noise spectra as well as data reduction, which is often desired in spectroscopic applications. Reference images are collected to calibrate spectra at each pixel in terms of a percentage reflectance value. Standard panels with known reflectance properties (Spectralon, Labsphere, USA) are used to collect reference images. Dark current response images are collected with the lamp off and a cap covering the focusing lens. Calibrated reflectance images (R) can be calculated by using a simple mathematical expression such as the one given below:

$$R = \frac{I_{raw} - I_{dark}}{I_{white} - I_{dark}}$$

[17.1]

where I_{raw} is the non-calibrated original image of a sample, I_{white} is the image of the white reference and I_{dark} is the dark current image. Sophisticated algorithms such as Savitzky and Golay (1964) are often used for noise removal while maintaining the features of the spectral curve. Wavelength calibration of the system can be performed using mercury–argon and krypton pencil lights.

17.2.2   Methods of spectral analyses

Hyperspectral images usually consist of 100 or more contiguous spectral bands resulting in huge spectral datasets to be analysed where collinearity exists among the original variables (wavelengths/bands). Large datasets usually contain large amounts of information, much of which is redundant to the problem under question. Important information may be partly hidden because the data are too complex to be easily interpreted. Some means of data compression is required prior to performing any multivariate analyses such as multiple linear regression (MLR) and/or linear discriminant analysis (LDA). Principal component analysis (PCA) is one of several possible bilinear modelling (BLM) approaches for data compression in situations where the original variables are mutually correlated. Common information in the original variables is used to build new latent variables that reflect the underlying structure. The latent variables are estimated as linear functions of both the original variables and the observations, thereby the name 'bilinear'. BLM techniques also include principal component regression (PCR), partial least squares (PLS) regression and multivariate curve resolution (MCR). These are all based upon the fundamental PCA procedure; therefore, a solid understanding of PCA is absolutely necessary for researchers involved in multivariate spectral data analyses. Detailed description of PCA can be found in Esbensen (2002), Jackson (1991) and Mardia *et al.* (1979).

In the spectroscopy community, PLS regression and PCR are commonly used methods for quantitative analyses. For classification, PLS discriminant analysis (PLS-DA), support vector machine (SVM) and soft independent modelling of class analogy (SIMCA) are frequently used. Pixel-by-pixel image classification is more popular among the researchers interested in spatial aspect of the information in hypercubes. Classification methods, such as minimum distance, Mahalanobis distance, maximum likelihood, spectral angle mapper (SAM), SVM and neural networks are frequently used to classify each pixel based on its spectral characteristics compared with a set of pre-defined target spectra. The performance of regression models is evaluated based on coefficient-of-determination (R^2) and root-mean-squared-error (RMSE), as well as the slope and offset of the trend line that fits the reference versus predicted values. For discriminant models, overall as well as class-wise accuracy of classification is the preferred measure of performance. A number of commercial software packages are available to perform chemometrics (WinISI, FOSS, Denmark; Unscrambler, CAMO Software, Oslo; SAS, SAS Institute Inc., Cary, NC) and spectral image analyses (ENVI-IDL, ITT Visual Information Solutions, Boulder, CO; MATLAB, Mathworks Inc., Natick, MA) to identify some products available today.

A number of studies conducted at the Grain Research Laboratory (GRL/CGC) explored the use of hyperspectral imaging for assessing grain quality, as well as the effect of grain damage on the colour of flour. Brief descriptions of these studies follow.

17.3   Detection of mildew damage in wheat

Mildew damage is a serious grading factor in Canadian wheat, especially in the eastern wheat classes. Luo *et al.* (1999) used conventional image analysis to detect mildew and five other types of damage factors on CWRS (Canada western red spring) wheat kernels. Visually obvious severely damaged mildewed kernels were detected highly accurately (97%); however, detection of visually challenging slight or moderate levels of mildew damage was not attempted. Berman *et al.* (2007) investigated hyperspectral imaging over the full spectral range (420–2500 nm) for fungal detection in wheat grains. Accuracy of fungal detection for visible-NIR (VNIR) range (420–1000 nm) was comparable to that for the full spectral range (420–2500 nm). In an unpublished study, hyperspectral imaging was used to detect and quantify the extent of mildew damage in wheat samples. Spectral characteristics of mildewed kernel regions were found to be significantly different from those of sound kernel regions (Fig. 17.3). Image classification based on these spectral differences, using spectral angle mapper or any other classification algorithm, could identify mildewed and non-mildewed (sound) regions of kernels in an image (Fig. 17.4). Image classification-based approach, to determine the extent of mildew damage, appears to be an ideal solution; however, this method does not coincide well with the current grading system where only one inspector score for the entire image (sample) is available for comparison. True classification of each pixel is never known to validate the method. Another key point to remember, which plays a crucial role in the accuracy of classification, is that the target spectra used for classification must be true representations of the target classes.

To be consistent with the current grading system, Shahin *et al.* (2010) used hyperspectral imaging in the VNIR spectral range (400–1000 nm) to quantify and predict visually assessed inspector scores, indicating various mildew levels in wheat samples. Samples (65) of Canada eastern soft red winter (CESRW) wheat with a wide range of mildew damage were collected from different locations in eastern Canada. These samples represented nine levels of mildew damage over three grades, with three levels in each grade. The samples were visually inspected by a trained inspector who assigned a score from 1 to 9 based on the extent of mildew damage: a lower score meant a lower mildew damage (1 means top of grade 1; high quality grain) while a higher score meant a higher level of mildew damage (9 means bottom of grade 3; low quality grain). Hyperspectral images of bulk samples were acquired in reflectance mode. Two spectra were measured from each image representing a sample – image mean spectrum and standard deviation spectrum. PLS calibrations were developed to predict mildew levels based on these spectra as input to the models. A PLS regression model based on image mean spectra predicted the

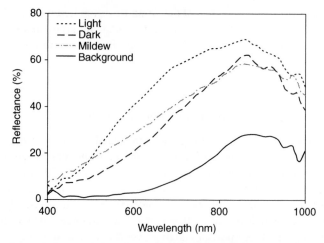

Fig. 17.3   Spectral characteristics of normal wheat kernels (light and dark coloured), mildewed regions of kernels and image background or voids.

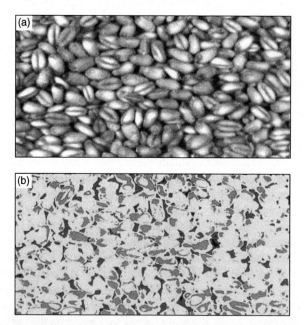

Fig. 17.4   (a) Original image of wheat and (b) classification image, showing normal kernel regions, mildew regions, image background or voids and unclassified pixels.

inspector scores for 9 mildew levels with an R^2 approaching 0.82 and RMSE of 0.89 on an independent test set. A PLS regression model based on image standard deviation spectra performed better with an R^2 exceeding 0.86 and an RMSE of 0.76 on the test set. Combining the two spectra did not improve the model performance

Table 17.1   Performance of PLS regression models for mildew detection based on different spectra as model input

PLS model input data	Number of PLS factors	Calibration		Validation	
		R^2	RMSE	R^2	RMSE
Image mean spectra	6	0.860	0.77	0.815	0.89
Image standard deviation spectra	4	0.866	0.75	0.861	0.77
Combined image mean and standard deviation spectra	4	0.869	0.75	0.865	0.76

(Table 17.1). Accuracy of classification into 9 mildew levels with a PLS-DA classifier was low (48–50%); however, most misclassifications were within one level. Accuracy of predicting mildew damage within ±1 level approached 91%, whereas accuracy of predicting three visually assessed inspector grades was 84%.

When tested on a second test set comprising of samples collected from an additional crop year, the model based on image mean spectra worked better than the one based on image standard deviation spectra as reported in Shahin *et al.* (2010). This highlights the importance of a totally independent test set consisting of samples from different crop years in developing robust calibrations capable of generalizing to unknown samples. The model based on image standard deviation apparently was an overfit as it failed to predict unknown samples. With the image mean spectra, accuracy of mildew predictions within ±1 level was 78%, which improved significantly to 93% after bias correction, leading to an accuracy of visual grade predictions approaching 89% for the second test set (Fig. 17.5). Six wavelengths spreading over the entire spectrum (450–950 nm) were needed to achieve this classification. These results have shown that hyperspectral imaging in the VNIR wavelength range offers the ability to detect and discern varying degrees of severity of mildew damage in commercial CESRW wheat samples.

17.4   Detection of fusarium damage in wheat

Fusarium head blight is a fungal disease that may infect a number of small grain cereals such as wheat, barley, corn and oats (Gallenberg, 2002). FHB is favoured by wet weather at anthesis; however, *Fusarium* ssp. can infect the grain until harvest given suitable conditions for infection. Infection at the early stages of seed development causes the greatest physical damage to the seed and the highest

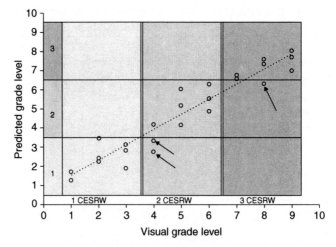

Fig. 17.5   Inspector vs predicted grade levels for CESRW wheat samples from 2009 crop after bias correction of the model initially developed with samples from 2008 crop. Arrows point to the samples misclassified to adjacent grade.

levels of mycotoxin production (McMullen *et al.*, 1997; Windels, 2000). The principal causal agent of wheat FHB in North America is *Fusarium graminearum* Schwabe (Goswami and Kistler, 2004). In Canada, *F. graminearum* is only one of many species of fusarium, but it is considered the most important one because of the impact that it has on yield and grain quality, its ability to produce several different toxins and its abundance in eastern Canada and the western prairies. Fusarium-damaged kernels (FDK) may contain mycotoxins such as DON, historically referred to as vomitoxin, which can cause serious health problems. Fusarium infection has detrimental effects on flour colour, ash content and baking performance (Dexter *et al.*, 1996), as well as other quality and safety issues (Dexter and Nowicki, 2003). A positive correlation between fusarium damage and DON levels has been found (Teich *et al.*, 1987).

Fusarium-damaged wheat is typically characterized by thin or shrunken chalk-like kernels. In the current Canadian grading system, a representative sample is visually inspected for kernels showing any evidence of infection by *Fusarium* spp. Infected kernels are characterized by a white or pinkish mould or fibrous growth. This visual process is slow when only slight damage is apparent as inspectors have to use a 10× magnifying lens to examine each suspect seed for the degree of mould growth. This contrasts with the USDA definition of FDK, which considers only those kernels that are severely infected by *Fusarium* spp. appearing as chalk-like or tombstone kernels. Severely damaged wheat FDK can be easily detected by human inspectors. However, visual detection of fusarium damage at an early stage is challenging. Fast and accurate instrumental methods are required to meet the needs of the grain industry for FDK detection, even at low levels of infection.

Several laboratory methods are available for the detection and measurement of moulds and mycotoxins in cereal grains and flour samples, including high-performance liquid chromatography (Campagnoli et al., 2011; Fredlund et al., 2009), gas chromatography (Tacke and Casper, 1996), mass spectrometry (Campagnoli et al., 2011; Scott et al., 1981) and enzyme-linked immunosorbent assay (Casale et al., 1988). However, these laboratory-based chemical detection methods are not suitable for rapid online inspection and quality assurance protocols, requiring expert technical use and clean non-commercial environments. Rapid inspection or sorting methods for grain are typically based on kernel density using a gravity table (Tkachuk et al., 1991) or optical properties (Ruan et al., 1998). In a recent preliminary research, Takenaka et al. (2009) have reported encouraging results on DON decontamination using a combination of gravity separation and optical sorting. A recent study claimed that metal-oxide-semiconductor sensors, commonly known as electronic noses, could potentially be used as a screening tool to distinguish between DON contaminated and non-contaminated durum wheat samples (Campagnoli et al., 2011). Williams (1997) used near-infrared (NIR) spectroscopy to model DON levels in whole grain bulk samples with only moderate success. Dowell et al. (1999) investigated NIR reflectance to measure DON concentration in single kernels. NIR spectroscopy of individual kernels segregated sound and FDK with high accuracy (95–97%) under controlled laboratory conditions (Delwiche and Hareland, 2004). In contrast, test results using this technique on commercial samples under commercial sorting operational conditions had a much lower accuracy (50%; Delwiche, 2008). Peiris et al. (2009) examined NIR absorbance characteristics of various concentrations of DON as well as of sound and fusarium-damaged single wheat kernels. This study indicated that NIR spectrometry in the 1000–2100 nm range could estimate DON levels in kernels having more than 60 ppm DON. However, further investigations were required to detect DON in asymptomatic kernels.

Hyperspectral imaging in the shortwave infrared (SWIR) range (1000–2500 nm) has shown potential for detecting fungal contamination in wheat (Singh et al., 2007; Zhang et al., 2007) and fusarium damage in maize corn (Williams et al., 2010) and wheat (Polder et al., 2005). The extremely high cost of cameras sensitive in the SWIR range has been a limiting factor in the development of commercially viable application systems in this waveband range. Berman et al. (2007) has shown that accuracy of fungal detection in wheat grain with hyperspectral imaging over the 420–1000 nm wavelength range could be just as good as over the full spectral range 420–2500 nm. The use of high-power bichromatic emitting diodes (LEDs) emitting at red and green wavelengths have been reported to achieve moderate levels of overall accuracies (50–85%) for detection of fusarium-damaged wheat kernels (Delwiche, 2008; Yang et al., 2009).

Recently, Delwiche et al. (2011) used two hyperspectral imaging systems in the NIR (1000–1700 nm) and VNIR (400–1000 nm) spectral ranges for recognition of FDK in wheat. Exhaustive searches were performed on the 144 and 125 wavelength pair images that, respectively, comprised the NIR and visible systems to determine accuracy of classification using an LDA classifier. On a limited set of wheat samples with the best wavelength pairs, either with visible or NIR wavelengths, were able to discriminate FDK from sound kernels at an average accuracy of approximately 95%. Accuracy dropped off substantially when the visual contrast between the two kernel conditions became imperceptible. Shahin and Symons (2011) used hyperspectral imaging in the VNIR range to detect FDK in CWRS wheat. Using a PCA-based approach, an FDK detection rate of 92% was achieved with 8% false positives using six wavelengths within 450–950 nm. Accuracy of distinguishing between severely damaged and mildly damaged sub-categories of FDK kernels ranged from 84–86% on the average.

As an extension to the earlier work of Shahin and Symons (2011), HSI was used for the detection of FDK in multiple classes of Canadian wheat that vary significantly in the physical characteristics of their kernels. More than 5200 kernels, representing seven major classes of Canadian wheat, with varying degrees of infection with symptoms ranging from sound through mild to severe were imaged in the VNIR (400–1000 nm) wavelength range. The classes of wheat were CWRS, Canada western amber durum (CWAD), Canada western red winter (CWRW), Canada eastern red spring (CERS), CESRW, Canada eastern hard red winter (CEHRW) and Canada eastern white winter (CEWW). Figure 17.6 shows normalized spectra of sound and FDK kernels for four different classes of wheat. The class labels are purposely omitted as they are not relevant. What is important to note is that sound and FDK kernels have different spectral characteristics for all four classes shown. Based on these spectral difference sound and damaged kernels can be identified.

A PLS regression model using kernel mean-normalized spectra at four selected wavebands (494, 578, 639 and 678 nm) that were chosen based on regression coefficient vector, was comparable to that based on the full spectrum (450–950 nm)

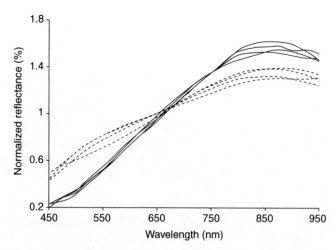

Fig. 17.6   Mean-normalized spectra of sound (solid lines) and fusarium-damaged (dashed lines) kernels for four different classes of wheat.

and resulted in highly accurate kernel classification based on fusarium damage. A partial least squares discriminant analysis (PLS-DA) model was used to segregate kernels into sound and damaged (FDK) categories. A single PLS-DA model based on four selected wavelengths detected FDK in all seven classes with an overall accuracy of 90% with 9% false positives. Class-wise accuracy of FDK detection ranged from 85% for CWAD to 95% for CERS (Table 17.2). These results have demonstrated that FDK kernels with slight to severe symptoms of infection in multiple wheat classes can be detected with four selected wavebands. The fact that a few specific wavebands are required for the detection of FDK in multiple classes of wheat suggests that it may be possible to solve this problem using a low-cost imaging system built around a monochrome digital camera and a set of optical filters in a motorized filter wheel. For industrial uptake, the lower cost of such a multispectral approach would be appealing. Such a system would initially have the potential to identify samples requiring further chemical analysis for toxin contamination.

17.5   Sprout damage in wheat

Mature cereal grains that are rained upon while in the head, either as a standing crop or in the swath, start to germinate as a result of the hydration. The resulting so called 'pre-harvest sprouting' causes increasing damage with time to the cereal kernel's biochemical and physical characteristics, which consequently reduces the value of the crop. To ensure that the intrinsic value of exported grain is maintained, Canadian wheat grading has tolerances for both sprouted and severely sprouted kernels based upon visual assessment by trained inspectors.

Sprouted grains contain degraded starch and protein components that reduce milling yield and lower flour or semolina quality, resulting in poor quality endproducts. While grains at early stages of sprouting exhibit no visible symptoms, the level of α-amylase, an enzyme activated during the germination process, may be sufficient to cause product quality degradation. A bulk wheat sample containing

Table 17.2   Kernel classification results for a universal PLS-DA model based on four wavelengths/bands (494, 578, 639 and 678 nm)

Wheat class	Classification (%)		
	Total	FDK	FP
CEHRW	93	92	5
CERS	90	95	18
CESRW	87	89	15
CEWW	93	93	6
CWAD	86	85	12
CWRS	90	87	6
CWRW	94	94	6
Overall	91	90	9

only a few severely sprouted kernels mixed with sound kernels may exhibit significant α-amylase activity on a ground sample basis. Because of this heterogeneity in enzyme activity between kernels, starch pasting tests on bulk ground samples do not give any indication of the variation in α-amylase activity among the individual kernels, given that a few kernels with high activity can have significant impact on overall sample quality.

To improve the grading system, which provides only a count of sprouted kernels, and to reduce the dependence on time-consuming starch pasting methods, a system that both identifies individual sprouted kernels and determines their individual level of sprout damage is desirable. The feasibility of using hyperspectral imaging and FT-NIR spectroscopy was explored at the Grain Research Laboratory for predicting the α-amylase activity of individual wheat kernels in order to be able to detect early stage sprout damage.

17.5.1   Prediction of the α-amylase activity for individual wheat kernels

Xing *et al.* (2009) reported the application of hyperspectral imaging in the SWIR range (1000–2500 nm) to predict α-amylase activity levels in CWAD and CWRS wheat samples collected from across western Canada. In total, 288 CWAD and 264 CWRS kernels exhibiting different degrees of sprout damage were imaged using the SWIR hyperspectral system. The α-amylase activity was individually determined for each kernel using an α-amylase substrate-hydrolysis-based assay. For both classes of wheat, α-amylase activity was not linked to any single wavelength (Fig. 17.7). PLS models were built to predict α-amylase activity based on the wavelength region between 1255 and 2300 nm. R^2 values of 0.54 and 0.73 were obtained for CWAD and CWRS, respectively. Higher R^2 value for the CWRS might be due to larger enzyme activity variations among CWRS kernels than those of CWAD. The numbers of kernels were distributed more evenly across the different α-amylase activity ranges in the CWRS samples and this could account for the better performance of prediction model for CWRS as well. In addition, these two classes of wheat have different translucency due to significantly different hardness and bran pigmentation, which would result in different light-path lengths in kernels and different surface reflectivity. While the prediction accuracy obtained from this study is limited, it does, however, suggest potential to develop a high-speed optical sorter for predicting the α-amylase activity of individual wheat kernels, especially for CWRS.

FT-NIR spectroscopy and SWIR hyperspectral imaging were compared for predicting α-amylase activity in individual CWRS kernels based on spectral information in the 1235–2450 nm wavelength range (Xing *et al.*, 2011). The FT-NIR spectra looked similar to those from the SWIR system in terms of the characteristic wavelength regions; however, FT-NIR spectra were smoother than the SWIR spectra. The PLS model predictions of α-amylase for both systems were comparable in terms of R^2 and RMSE values. However, by using log-transformation of the α-amylase activity level as the target value, the prediction accuracy improved significantly, more so for the HSI system than the FT-NIR system. The improved

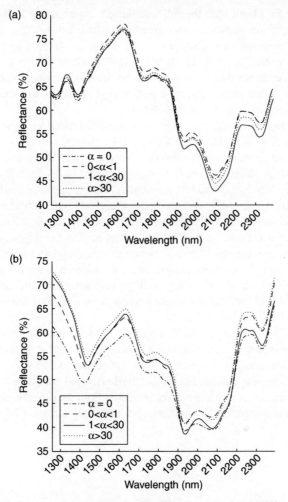

Fig. 17.7   Mean spectra of kernels with different α-amylase activity levels:
(a) CWAD and (b) CWRS.

R^2 values for the SWIR hyperspectral imaging system and the FT-NIR instrument
were 0.88 and 0.82, respectively.

17.5.2   Detection of sprouted wheat kernels
Reflectance spectra (1250–2390 nm) of two classes of western Canadian wheat
(CWAD and CWRS) were studied to detect sprouting damage in individual kernels at the early stages of germination (Xing *et al.*, 2010a). α-amylase activity
levels were used as an indicator for the sprouting stage. PLS-DA was utilized to
build classification models based upon α-amylase activity levels. Classification

models were first built with the full wavelength region using PLS-DA and then with the significant variables determined according to the VIP score plots. The selected wavelengths were centred at 1290, 1397, 1634, 2076 and 2165 nm for CWAD and centred at 1334, 1397, 1634, 1899, 2076 and 2165 nm for CWRS. This technique shows that α-amylase can be detected in wheat kernels at levels below that of visual grain inspection with a total classification accuracy of over 91% in CWRS and around 86% in CWAD.

Sprouted kernels usually exhibit some exterior/visual evidence that distinguishes them from sound kernels (CGC, 2011). Hence, a VNIR imaging system may be able to capture some additional morphometric information that can be useful for kernel classification. Individual kernels (144) of CWRS wheat were scanned with a VNIR hyperspectral imaging system (Xing *et al.*, 2010b). Average spectra of sprouted and severely sprouted kernels had higher reflectance responses compared with sound kernels in the wavelength region above 720 nm (Fig. 17.8). The ratio of the reflectance at 878 nm to that at 728 nm was used to identify sprouted from non-sprouted kernels. However, this waveband ratio did not separate the severely sprouted kernels from either the sound or sprouted groups. While only using this parameter, all of the sound kernels were correctly classified; however, about 14% (12/97) of not-sound kernels were misclassified into the sound kernel group.

Also in the study (Xing *et al.*, 2010b), PCA loadings plot identified four wavelengths, centred at 475, 525, 670 and 875 nm, which contributed to distinguishing the different quality of wheat kernels. Using the morphological features of smoothness, size and value range of the third principal component score image, severely sprouted kernels were clearly separated from the sound kernels. Utilizing a classification procedure that combines both spectral and spatial features, 100%

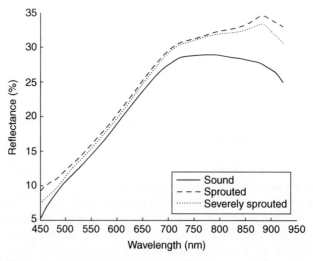

Fig. 17.8   Mean reflectance spectra of CWRS from the VNIR system.

of the sound kernels, 94% of the sprouted kernels and 98% of severely sprouted kernels were correctly classified. Since only four wavelengths were required, this method shows promise for the future development of a low-cost multispectral system for the identification of sprout-damaged wheat kernels. The recognition of individual sprouted kernels, and their subsequent removal, will vastly improve the value of any lot of grain.

17.6   Determination of green immature kernels in cereal grains

The evaluation of greenness is important to grain quality, value and price. It can be hypothesized that automated systems, which can determine green seeds in samples or create a score for sample greenness, would add overall value to the grain handling and processing system and provide a means for future technology to upgrade samples by removal of the seeds most damaging to quality. Such demands cannot be met by visual inspection. Spectroscopy has been shown to be able to predict chlorophyll content in canola (Daun, 2003), but is limited to an overall sample assessment. While this can relate to crushing, it is not a model for sample upgrading and has not proven to be successful in other grains. Imaging has the potential to identify individual seeds and determine a green score for both the sample and individual seeds.

In wheat and barley, it is an assessment of the whole sample appearance that determines the impact of green immature seeds on the quality of flour and malt, respectively. Given that a sample of wheat or barley is essentially of the same grain type, variation within the sample is inherent. The sample may contain several registered varieties from several locations across the western Canadian prairies due to the bulk handling procedures of Canadian cereal grains. The damaged kernels within a sample can vary from a few to many and from lightly to severely damaged. A human inspector cannot achieve a full analysis of such variability in the limited time allowed to inspect and grade a sample. In contrast, objective measurement systems based upon vision technology have the potential to close this gap and provide an evaluation of the sample as individuals in a population and to eventually determine the best possible evaluation of grain value, both financially and in terms of processing outcomes.

Since greenness is a direct result of chlorophyll content and its absorption maxima is in the red and blue regions of the spectrum, measurements made in the visible region of the spectrum logically provide a measure of this factor as the two types of chlorophyll known as a and b, respectively. Chlorophyll a absorbs electromagnetic radiation with peaks at 665 and 465 nm, while chlorophyll b absorbs with peaks at 453 and 642 nm. Thus, greenness should be measurable with both RGB devices such as a camera or scanner and spectral devices such as hyperspectral camera.

Desktop scanners form an ideal image capture system for grain samples. They have been used to grade lentils based upon green colour (Shahin and Symons,

2001; Shahin *et al.*, 2004), but remain difficult to control, as the TWAIN interface (www.twain.org) allows for both common and proprietary communication functions. Proprietary commands limit interaction with non-vendor-specific imaging software. Scanning also takes several seconds to complete, which does not fit well into the operational inspection environment. Our results show that a scanner-based system consisting of KS400 software (Carl Zeiss Imaging, Halbergmoos, Germany) and a flatbed scanner (Scanmaker 4, Microtek USA), when properly calibrated (Shahin and Symons, 2003), can create a seed count that relates to the number of green seeds added to high quality barley samples where known amounts of green seeds were added (Fig. 17.9). While the precision required for grading was not reached and similar results were found in two additional cultivars, this approach shows the potential for machine vision for scoring green seeds in cereals. The lack of precision is probably related to the broad band filters used in such RGB devices and their lack of specificity for the chlorophyll pigments.

Hyperspectral imaging provides a platform where several hundred narrow wavebands can be collected across the area surface of a sample. As chlorophyll has absorption peaks in the range of 640–660 nm, a VNIR spectroscope with the wavelength range of 400–1000 nm can be used to detect chlorophyll in green seeds. With the high spectral resolution of such an instrument, it can be seen that regions of green barley and wheat kernels have an absorption peak around the 670 nm region in contrast to fully matured kernels (Fig. 17.10). Using waveband ratios, green kernels can be identified with accuracy in both wheat and barley samples (Fig. 17.11). While this approach to measurement gives excellent green kernel detection and a score for greenness, it remains far from a practical solution due to cost, datacube size and operating requirements that require a dust-free, clean laboratory for operation. This research methodology is, however, very useful for identifying key wavebands that can be used in a simpler multispectral system to emulate this analysis.

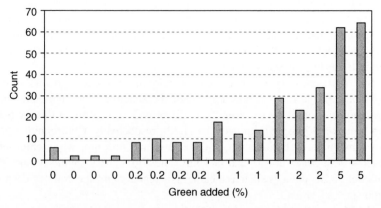

Fig. 17.9   The seed count for green barley kernels added to clean barley samples of cv Metcalfe at levels of 0.2%, 1%, 2% and 5% ($R = 0.89$).

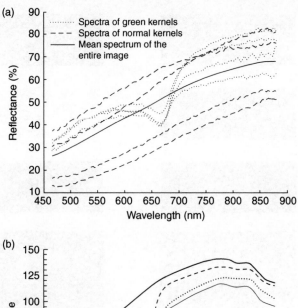

Fig. 17.10   Reflectance spectra for normal and green barley (a) as well as wheat kernels as captured with a calibrated VNIR hyperspectral system (b); green kernels absorb at about 670 nm.

17.7   Effect of mildew on the quality of end-products

Mildew damage is a serious quality defect that has a negative impact on process-ing quality and commercial value of wheat. Mildew imparts a grey discoloration to the entire kernel as damage increases. Wheat milling performance decreases with increasing levels of mildewed kernels due to a decrease or darkening of flour colour. Internal reports (unpublished) indicate that CEWW displayed a significant shift in the resulting flour colour, decreased flour yield straight grade flour from −1.6 Kent Jones (KJ) units for a No.1 grade to −0.3 KJ units for a No. 3 grade. The impact of mildew on durum wheat products is primarily a significant increase (2–3 fold) in the speckiness of the resulting spaghetti (Dexter and Matsuo, 1982).

As described in Section 17.3, Shahin *et al.* (2010) analysed 65 samples of CESRW wheat for mildew damage with a HSI system in the 400–1000 nm

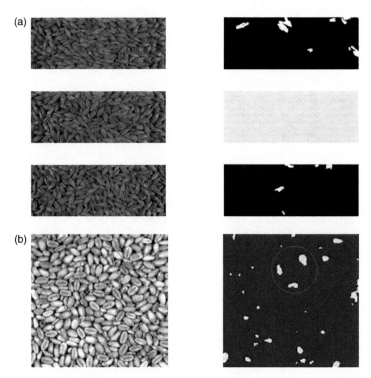

Fig. 17.11 (a) Detection of green seeds in barley. Left: composite pseudo-colour image composed from image planes at 560, 668 and 737 nm. Right: green kernels segmented in wavebands at 560 nm (top), 668 nm (middle) and 737 nm (bottom) with a score based upon intensity in the grey level image. (b) Detection of green seeds in wheat. Left: composite true colour image composed of RGB channels at 450, 550 and 650 nm. Right: green kernels detected with (560/668 nm) band ratio.

wavelength range. These samples represented a wide range of mildew damage over three grades (1–3). Based on the severity of mildew damage, samples were graded into nine levels (three levels in each grade). After imaging, the samples were milled at a constant extraction rate (72%) for flour colour measurements (KJ and Agtron) and other laboratory analyses. PLS regression models were developed to predict flour colour based on spectral measurements of whole grain samples of CESRW. Predictions of Agtron units from the PLS model were highly correlated with the measured values with R^2 ranging between 0.754 and 0.844 and RMSE between 2.87 and 3.66 (Table 17.3; Fig. 17.12). Relatively lower correlation between measured and predicted flour KJ colour was observed for the PLS model (Table 17.3). The measured and predicted Agtron flour colour values closely matched for all grade levels investigated, although no strong relationship existed between the visually assessed mildew level and flour colour (Fig. 17.13).

These results demonstrate that, in addition to mildew level assessment, prediction of flour colour from spectral measurements of grain samples is possible.

Being able to predict flour colour without milling offers huge labour and cost benefits. It eliminates the need for a separate instrument for colour measurements. What is not known at this moment is whether the quality of end-products such as noodles can be predicted from spectral measurement of grain samples. If that turns out to be true, then the quality of grain, flour and noodles can be assessed/ predicted by just measuring the whole grain samples.

17.8   Development of multispectral imaging systems

The following sections consider the way this technology is honed to cater for the specific needs of the grain distribution industry.

17.8.1   Wavelength selection

Hyperspectral imaging is a great research tool due to its ability to detect, distinguish and quantify multiple target constituents. However, high equipment cost

Table 17.3 Performance summary of PLS models to predict Agtron and KJ flour colour based on HSI measurements of CESRW whole grain samples

Flour colour	Dataset	Slope	Offset	RMSE	R^2
Agtron	Calibration	0.8438	8.6933	2.8714	0.8438
	Validation	0.7792	12.3987	3.6630	0.7537
Kent-John	Calibration	0.7295	0.1918	0.4850	0.7295
	Validation	0.6173	0.2568	0.6501	0.5289

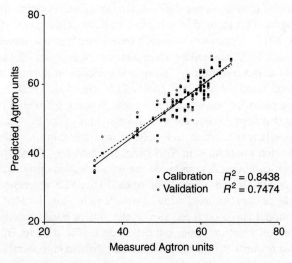

Fig. 17.12 Prediction of flour colour based on spectral measurements of grain samples for calibration and validation datasets along with the target line.

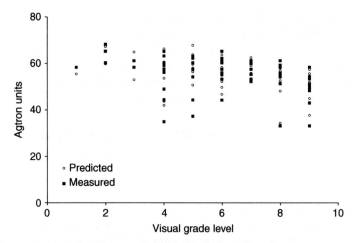

Fig. 17.13   Mildew level grade of grain sample versus
flour colour (measured and predicted).

and slow processing speed due to huge amount of data to be analysed are limiting factors in its acceptance as a general purpose tool for sample analysis. Wavelength selection is often required for developing relatively inexpensive and fast instruments for inspection. Wavelength selection is straightforward for spectra exhibiting distinct spectral features linked with known constituents, such as chlorophyll absorption in green seeds (Fig. 17.10). In most cases, however, the wavelength-constituent link is not readily obvious, hence requiring some 'digging' to find suitable combinations of wavelengths. Exhaustive search is the most commonly used method reported to select a pre-defined number of wavelengths from all available wavelengths. This method of selection, however, tends to overfit the training data (Fearn, 2011). Optimization search techniques, such as genetic algorithms (Brandye *et al.*, 2000) and independent component analysis (Comon, 1994), are often used to avoid overfitting. Through experience in spectral data analyses, we have found that PCA loadings and/or PLS regression coefficients provide a great starting point for wavelength selection. Starting off with a reduced set of wavelengths that approximate the overall behaviour of the PCA loadings or PLS regression coefficient function, selection refinements can be achieved by using variable selection procedures in SAS (SAS Institute Inc., Cary, NC): proc step-disc for classification models and proc reg with selection option for quantitative/regression models. In our example of wheat FDK, PLS regression vector (Fig. 17.14a) could be approximated with 13 wavelengths (Fig. 17.14b). Performance of the PLS model based on these 13 wavelengths or bands closely matched the one based on full spectrum. Four out of 13 bands (Fig. 17.14c; B2–B5) contributed the most towards the model performance without considerable reduction in the performance. Hence, these four wavelengths (494, 578, 639, 678 nm) were selected for the development of a multispectral system for wheat FDK.

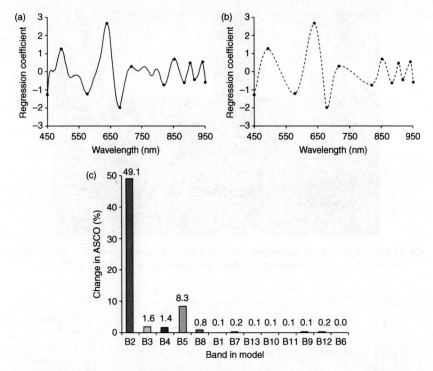

Fig. 17.14 (a) Regression coefficients plot of the FDK-PLS model using full spectrum, (b) approximation with 13 selected wavelengths/bands and (c) contribution of bands in the model.

17.8.2   Multispectral imaging for detecting green seeds

Multispectral imaging typically uses ten or fewer wavebands in the imaging system and therefore is considerably less expensive to implement, has a more rugged architecture and the data stream created is significantly lower than a hyperspectral system. There are several different approaches to the construction of a multispectral system: one is to use multiple filters and cameras (Kise *et al.*, 2010) enabling capture of the image simultaneously at all wavelengths; another uses a liquid crystal tuneable filter where wavebands can be programmed to occur in quick succession. We chose to construct a multispectral system with a single camera and a high-speed filter wheel to switch wavebands for collection (Fig. 17.15).

Using off-the-shelf available components, a GigE camera (Basler Germany, Model SCA 1600) was chosen as it presents the opportunity for operation from a long distance over the local area network. One inch diameter band-pass filters (Edmond Optics, USA) were selected as close to the wavebands identified by the research hyperspectral system, namely 580, 671 and 730 nm, and mounted into a filter wheel assembly (ASI Imaging, USA). In contrast to the narrow bandwidth of hyperspectral bands, typically around 1 nm in our work, the filters have a half-

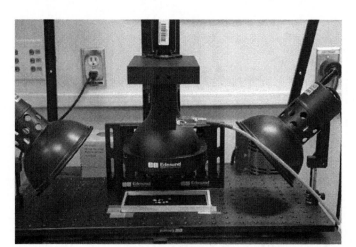

Fig. 17.15 Multispectral system consisting of a monochrome GigE camera, a filter wheel housing 1 in diameter band-pass filters, halogen illumination and a VNIR lens.

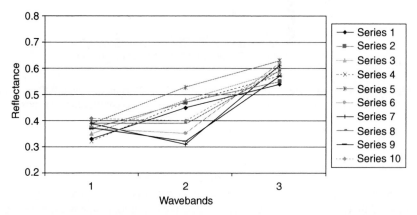

Fig. 17.16 Ten kernels, five green and five normal, from a commercial barley sample, with measurements at three wavelengths (580, 671, 730 nm) using a multispectral system.

height bandwidth of around 10 nm. The increased bandwidth was of concern as it reduces the specificity of the measurement system; however, off-the-shelf filters could not be matched to the exact wavebands desired, so those chosen represented the best compromise for constructing a prototype system. Experiments measuring the three wavebands on several samples of barley kernels indicated that the multi-spectral system could indeed identify chlorophyll in barley kernels (Fig. 17.16).

Several models predicting seed greenness were built using various combinations of the three wavelengths. A simple ratio of 671–730 nm was found to be the most effective in scoring the greenness of individual kernels. Through experimentation, it was established that a threshold ratio of 671/730 = 75% marked the transition to green kernels that related to the detection level by visual inspection.

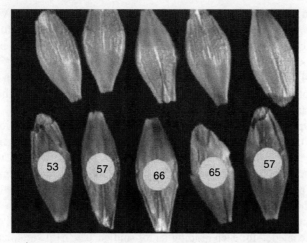

Fig. 17.17   Ten individual barley kernels, five sound and five green, scored by a multi-spectral imaging system using two wavebands at 671 and 730 nm. (The green kernels are in the bottom row.)

Kernels with a ratio lower than 75% were green. The lower the ratio value, the greener the kernel (Fig. 17.17) is. To achieve these results, the normalization of the data from each filter proved critical since the transmission characteristics varied from filter to filter.

The next step in our development is to use the waveband ratio to identify and score kernels in bulk barley samples with a flatted surface. Our aim is to count and score individual kernels. Once this can be achieved in a reliable manner, comparison of this data to malting parameters appears the obvious step. Can our approach to acquiring data be used to predict the malting performance of barley? This intriguing question remains to be answered.

17.9   Conclusions

Hyperspectral imaging was used for detection and quantification of damage to cereal grains due to various degrading factors including mildew, fusarium, sprouting and green immature kernels. Alpha-amylase activity in sprouted wheat kernels can be predicted with reasonable accuracy using hyperspectral imaging in the SWIR (1000–2500 nm) wavelength range. Accuracy of sprout detection based on alpha-amylase activity was 91% in CWRS and around 86% in CWAD wheat using five wavelengths in the SWIR range. Sprout detection with the VNIR spectra (400–1000 nm) was more accurate – 100%, 94% and 98%, respectively, for sound, sprouted and severely sprouted CWRS wheat. Fusarium damage in CWRS wheat can be detected accurately (92%) with a VNIR system. A single PLS-DA model using four wavelengths can detect FDK in multiple classes of Canadian wheat with accuracy between 85% and 95%. Mildew levels in CESRW wheat can be predicted with a PLS-DA model based on six wavelengths in the VNIR range with accuracy between 91% and 93% for mildew level and 84–89% for inspector grade predictions. Flour colour as affected by mildew damage can be assessed from the spectral measurements of the whole grain samples with R^2 values of 0.754–0.844 and RMSE of 2.87–3.66 Agtron units. Green kernels in wheat and barley samples can be detected and scored for greenness using three bands in the visible spectral range around chlorophyll absorption at 670 nm.

While hyperspectral systems can identify kernel damage due to various grading factors, multispectral systems can be much simpler and far less expensive. A multispectral system using three band-pass filters was demonstrated to successfully detect and score green barley kernels. The simple multispectral system proved as accurate in prediction at a fraction of the cost of the more expensive hyperspectral unit. The multispectral approach is now our preferred platform for further development of prototype operational imaging systems. A few wavelengths (six or fewer) are needed for each grading factor studied. The hardware platform used for green seed detection can be adapted for different grading factors by plugging in a suitable set of filters. The simple model will also be tested with other cereal grains to determine if the platform has varied applications in grain grading/handling. Development of fast, accurate and low-cost multispectral systems is expected to have a profound effect on the acceptance of instrumental methods of grain grading by the grain industry.

17.10   Acknowledgements

The authors thank the Inspection Division of the Canadian Grain Commission (CGC) for providing inspected samples for this research. They would also like to thank Lisa Van Schepdael, Loni Powell and Janice Bamforth of the Image Analysis & Spectroscopy Program (CGC) for scanning the samples. Special thanks to Juan Xing for her support in wheat sprout detection studies.

17.11   References

Berman, M., Connor, P. M., Whitbourn, L. B., Coward, D. A., Osborne, B. G. and Southan, M.D. (2007) Classification of sound and stained wheat grains using visible and near-infrared hyperspectral image analysis. *Journal of NearInfrared Spectroscopy*, **15(6)**, 351–8.

Brandye, M. Smith, P. and Gemperline, J. (2000) Wavelength selection and optimization of pattern recognition methods using the genetic algorithm. *Analytica Chimica Acta*, **423**, 167–77.

Campagnoli, A., Cheli, F., Polidori, C., Zaninelli, M., Zecca, O., Savoini, G., Pinotti, L. and Dell'Orto, V. (2011) Use of the electronic nose as a screening tool for the recognition of durum wheat naturally contaminated by deoxynivalenol: A preliminary approach. *Sensors*, **11**, 4899–916.

Casale, W. L., Pestka, J. J. and Hart, L. P. (1988) Enzyme-linked immunosorbent assay employing monoclonal antibody specific for deoxynivalenol (vomitoxin) and several analogues. *Journal of Agricultural and Food Chemistry*, **35**, 663–8.

CGC (2011) Official Grain Grading Guide 1 July 2011. Available from:http://www.grain-scanada.gc.ca/oggg-gocg/ggg-gcg-eng.htm

Cheng, X., Chen, Y. R., Tao, Y., Wang, C. Y., Kim, M. S. and Lefcourt, A. M. (2004) A novel integrated PCA and FLD method on hyperspectral image feature extraction for cucumber chilling inspection. *Transactions of the ASAE*, **47(4)**, 1313–20.

Cogdill, R. P., Hurdburgh, C. R. and Rippke, G. R. (2004) Single kernel maize analysis by near-infrared hyperspectral imaging. *Transactions of the ASAE*, **47**, 311–20.

Comon, P. (1994) Independent component analysis, a new concept? *Signal Processing*, **36(3)**, 287–314.

Daun, J. K. (2003) How green is green? Long-term relationships between green seeds and chlorophyll in canola grading. *Journal of the American Oil Chemists' Society*, **80**, 119–22.

Delwiche, S. R. (2008) High-speed bichromatic inspection of wheat kernels for mold and color class using high-power pulsed LEDs. *Sensing and Instrumentation for Food Quality and Safety*, **1(2)**, 103–10.

Delwiche, S. R. and Hareland, G. A. (2004) Detection of scab-damaged hard red spring wheat kernels by near-infrared reflectance. *Cereal Chemistry*, **8(15)**, 643–9.

Delwiche, S. R., Kim, M. S. and Dong, Y. (2011) Fusarium damage assessment in wheat kernels by Vis/NIR hyperspectral imaging. *Sensing and Instrumentation for Food Quality and Safety*, **5(2)**, 63–71.

Dexter, J. E. (1997) Making the grades: Implications of frequently encountered wheat grading factors. *Forum on Stored Grain Products*, **12**, 8–9.

Dexter, J. E. and Edwards, N. M. (1998a) The implications of frequently encountered grading factors on the processing quality of common wheat. *Association of Operative Millers – Bulletin*, June, 7115)

Dexter, J. E. and Edwards, N. M. (1998b) The implications of frequently encountered grading factors on the processing quality of Durum wheat. *Association of Operative Millers – Bulletin*, May, 7165)

Dexter, J. E. and Matsuo, R. R. (1982) Effect of smudge and blackpoint, mildewed kernels, and ergot on durum wheat quality. *Cereal Chemistry*, **59**, 63–9.

Dexter, J. E. and Nowicki, T. W. (2003) Safety assurance and quality assurance issues associated with Fusarium head blight in wheat. In Leonard, K. J. and Bushnell, W. R. (eds), *Fusarium Head Blight of Wheat and Barley* . St Paul, MN: APS, 420–60.

Dexter, J. E., Clear, R. M. and Preston, K. R. (1996a) Fusarium head blight: Effect on the milling and baking of some Canadian wheats. *Cereal Chemistry*, **73**, 695–701.

Dexter, J. E., Clear, R. M., Preston, K. R., Marchylo, B. A. and Clarke, J. M. (1996b) Fusarium head blight: Wheat processing implications. *Proceedings of the 1996 Regional Fusarium/Scab Forum*, October 1996, Winnipeg, Manitoba, Canada.

Dowell, F. E., Ram, M. S. and Seitz, L. M. (1999) Predicting scab, vomitoxin, and ergosterol in single wheat kernels using near-infrared spectroscopy. *Cereal Chemistry*, **76**, 573–6.

Esbensen, K. (2002) *Multivariate Data Analysis – in Practice*, 5th edn, CAMO Process AS, Oslo.

Everts, K. L., Leath, S. and Finney, P. L. (2001) Impact of powdery mildew and leaf rust on milling and baking quality of soft red winter wheat. *Plant Disease*, **85**, 423–9.

Fearn, T. (2011) Partial least squares regression versus multiple linear regression. *NIR News*, **22(1)**, 15–16.

Fredlund, E., Thim, A. M., Gidlund, A., Brostedt, S., Nyberg, M. and Olsen, M. (2009) Moulds and mycotoxins in rice from the Swedish retail market. *Food Additives and Contaminants—Part A Chemistry, Analysis, Control, Exposure and Risk Assessment*, **26(4)**, 507–11.

Gallenberg, D. J. (2002) Wheat scab. Extension Extra, ExEx 8097. Cooperative Extension Service, SDSU/USDA. Available from: http://agbiopubs.sdstate.edu/articles/ExEx8097.pdf

Goswami, R. S. and Kistler, H. C. (2004) Heading for disaster: Fusarium graminearum on cereal crops. *Molecular Plant Pathology*, **5(6)**, 515–25.

Goretta, N., Roger, J. M., Aubert, M., Bellon-Maurel, V., Campan, F. and Roumet, P. (2006) Determining vitreouness of durum wheat kernels using near-infrared hyperspectral imaging. *Journal of NearInfrared Spectroscopy*, **14**, 231–239.

Gowen, A. A., O'Donnell, C. P., Cullen, P. J., Downy, G. and Frias, J. M. (2007) Hyperspectral imaging – an emerging process analytical tool for food quality and safety. *Trends in Food Science & Technology*, **18**, 590–8.

Gowen, A. A., O'Donnell, C. P., Taghizadeh, M., Gaston, E., O'Gorman, A., Cullen, P.J ., Frias, J. M., Esquerre, C. and Downey, G. (2008) Hyperspectral imaging for the investigation of quality deterioration in sliced mushrooms (*Agaricus bisporus*) during storage. *Sensing and Instrumentation for Food Quality and Safety*, **2(3)**, 133–43.

Jackson, J. E. (1991) *A Users Guide to Principal Components*. New York: Wiley & Sons.

Kim, M. S., Lefcourt, A. M., Chao, K., Chen, Y. R., Kim, I. and Chan, D. E. (2002) Multispectral detection of fecal contamination on apples based on hyperspectral imagery. *Transactions of the ASAE*, **45(6)**, 2027–37.

Kise, M., Park, B., Heitschmidt, G. W., Lawrence, K. C. and Windham, W. R. (2010) Multispectral imaging system with interchangeable filter design. *Computers and Electronics in Agriculture*, **72**, 61–8.

Lin, L. H., Lu, F. M. and Chang, Y. C. (2006) Development of a near-infrared imaging system for determination of rice moisture. *Cereal Chemistry*, **83**, 498–504.

Lu, R. (2003) Detection of bruise on apples using near-infrared hyperspectral imaging. *Transactions of the ASAE*, **46**, 523–30.

Luo, X., Jayas, D. S. and Symons, S. J. (1999) Comparison of statistical and neural network methods for classifying cereal grains using machine vision. *Transactions of the ASAE*, **42(2)**, 413–19.

Mardia, K. V., Kent, J. T. and Bibby, J. M. (1979) *Multivariate Analysis*. London: Academic Press.

McMullen, M., Jones, R. and Gallenberg, D. (1997) Scab of wheat and barley: a re-emerging disease of devastating impact. *Plant Disease*, **81**, 1340–8.

Naganathan, G. K., Grimes, L. M., Subbiah, J., Calkins, C. R., Samal, A. and Meyer, G. E. (2008) Partial least squares analysis of near-infrared hyperspectral images for beef tenderness prediction. *Sensing and Instrumentation for Food Quality and Safety*, **2(3)**, 178–88.

Park, B., Lawrence, K. C., Windham, W. R. and Buhr, R. J. (2002) Hyperspectral imaging for detecting fecal and ingesta contaminants on poultry carcasses. *Transactions of the ASAE*, **45**, 2017–26.

Peiris, K. H. S., Pumphrey, M. O. and Dowell, F. E. (2009) NIR absorbance characteristics of deoxynivalenol and of sound and fusarium-damaged wheat kernels. *Journal of NearInfrared Spectroscopy*, **17**, 213–21.

Polder, G., Van Der Heijden, G. W. A. M., Waalwijk, C. and Young, I. T. (2005) Detection of fusarium in single wheat kernels using spectral imaging. *Seed Science and Technology*, **33**, 655.

Ruan, R., Ning, S., Song, A., Ning, A., Jones, R. and Chen, P. (1998) Estimation of fusarium scab in wheat using machine vision and a neural network. *Cereal Chemistry*, **75**, 455–9.

Savitzky, A. and Golay, M. J. E. (1964) Smoothing and differentiation of data by simplified least squares procedures. *Analytical Chemistry*, **36(8)**, 1627–39. doi:10.1021/ac60214a047.

Scott, P. M., Lau, P.-Y. and Kanhere, S. R. (1981) Gas chromatography with electron capture and mass spectrometric detection of deoxynivalenol in wheat and other grains. *Journal of Association of Official Analytical Chemists*, **64**, 1364–71.

Shahin, M. A. and Symons, S. J. (2001) A machine vision system for grading lentils. *Canadian Biosystems Engineering*, **43**, 7.7–7.14.

Shahin, M. A. and Symons, S. J. (2003) Color calibration of scanners for scanner-independent grain grading. *Cereal Chemistry*, **80(3)**, 285–9.

Shahin, M. A. and Symons, S. J. (2007) The use of hyperspectral imaging to characterize wheat grading factors. *Proceedings of the 13th International Conference on NIR*, June 2007, Umea, Sweden.

Shahin, M. A. and Symons, S. J. (2008) Detection of hard vitreous and starchy kernels in amber durum wheat samples using hyperspectral imaging. *NIR News*, **19(5)**, 16–18.

Shahin, M. A. and Symons, S. J. (2011) Detection of fusarium damaged kernels in Canada Western Red Spring wheat using visible/near-infrared hyperspectral imaging and principal component analysis. *Computers and Electronics in Agriculture*, **75**, 107–12.

Shahin, M. A., Hatcher, D., Symons, S. J. (2010) Assessment of mildew levels in wheat samples based on spectral characteristics of bulk grains. *Quality Assurance and Safety of Crops & Foods*, **2**, 133–40.

Shahin, M. A., Meng, A., Symons, S. J., Dorrian, E. and Manivannan, U. (2004) Determination of short-term sequential scan variability in a scanner-based vision system for grain grading. *Journal of Electronic Imaging*, **13(4)**, 673–9.

Singh, C. B., Jayas, D. S., Paliwal, J. and White, N. D. G. (2007) Fungal detection in wheat using near-infrared hyperspectral imaging. *Transactions of the ASABE*, **50**, 2171–6.

Singh, C. B., Jayas, D. S., Paliwal, J. and White, N. D. G. (2009) Detection of sprouted and midge-damaged wheat kernels using near-infrared hyperspectral imaging. *Cereal Chemistry*, **86(3)**, 256–60.

Tacke, B. K. and Casper, H. H. (1996) Determination of deoxynivalenol in wheat, barley, and malt by column cleanup and gas chromatography with electron capture detection. *Journal of Association of Official Analytical Chemists International*, **79**, 472–5.

Takenaka, H., Kawamura, S., Sumino, A. and Yano, Y. (2009) New combination use of gravity separator and optical sorter for decontamination deoxynivalenol of wheat. *Proceedings of the 5th CIGR International Technical Symposium on Food Processing, Monitoring Technology in Bioprocesses and Food Quality Management*, CD-ROM, ISBN: 978-3-000-028811-1 , Potsdam, Germany, 1036–9.

Teich, A. H., Shugar, L. and Smid, A. (1987) Soft white winter wheat cultivar field-resistance to scab and deoxynivalenol accumulation. *Cereal Research Communication*, **15**, 109–14.

Tkachuk, R., Dexter, J. E., Tipples, K. H. and Nowicki, T. W. (1991) Removal by specific gravity table of tombstone kernels and associated trichothecenes from wheat infected with Fusarium head blight. *Cereal Chemistry*, **68**, 428–31.

Williams, P. C. (1997) Recent advances in near-infrared applications for the agriculture and food industries. In Steele, J. L. and Chung, O. K. (eds), *Proceedings of International Wheat Quality Conference*. Manhattan, KS: Grain Industry Alliance, 109–28.

Williams, P., Manley, M., Fox, G. and Geladi, P. (2010) Indirect detection of fusarium in maize (Zea mays L.) kernels by near-infrared hyperspectral imaging. *Journal of NearInfrared Spectroscopy*, **18**, 49–58.

Windels, C. E. (2000) Economic and social impact of Fusarium head blight: Changing farms and rural communities in the northern Great Plains. *Phytopathology*, **90**, 17–21.

Xing, J., Hung, P., Symons, S., Shahin, M. and Hatcher, D. (2009) Using a SWIR hyperspectral imaging system to predict alpha amylase activity in individual Canadian Western wheat kernels. *Sensing and Instrumentation for Food Quality and Safety*, **3**, 211–18.

Xing, J., Symons, S., Shahin, M. and Hatcher, D. (2010a) Sprouting detection at early stages in individual CWAD and CWRS wheat kernels using SWIR spectroscopy. *Sensing and Instrumentation for Food Quality and Safety*, **4**, 95–100.

Xing, J., Symons, S., Shahin, M. and Hatcher, D. (2010b) Detection of sprout damage in Canada Western Red Spring wheat with multiple wavebands using visible/near-infrared hyperspectral imaging. *Biosystems Engineering*, **106(2)**, 188–94.

Xing, J., Symons, S., Shahin, M. and Hatcher, D. (2011) Comparison of short-wavelength infrared (SWIR) hyperspectral imaging system with an FT-NIR spectrophotometer for predicting alpha-amylase activities in individual Canadian Western Red Spring (CWRS) wheat kernels. *Biosystems Engineering*, **108**, 303–10.

Yang, I.-C., Delwiche, S. R., Chen, S. and Lo, Y. M. (2009) Enhancement of Fusarium head blight detection in free-falling wheat kernels using bichromatic pulsed LED design. *Optical Engineering*, February, **48(2)**, 023602.

Zhang, H., Paliwal, J., Jayas, D. S. and White, N. D. G. (2007) Classification of fungal infected wheat kernels using near-infrared reflectance hyperspectral imaging and support vector machine. *Transactions of the ASABE*, **50**, 1779.

Index

α-amylase activity, 466–7
  kernels mean spectra, 467
A-mode, 281
acousto-optic tunable filter (AOTF), 34, 35, 407
Adaboost, 213
adjacent method, 73
algorithm development, 194–5
American Society for Testing and Materials (ASTM) International, 43
animal sorting, 311–12
artificial intelligence, 212
artificial neural networks (ANN), 115
Atlantic salmon
  fillets trimming, 362–4
  pre-rigor removal of pin bone from fillet, 364–6
  quality grading, 355–7
automated cutting, 206–31
  application examples, 213–28
    backlit image for yield determinations, 220
    backlit image of a frame, 218
    breast meat harvesting, 217–20
    clavicle region location, 219
    correlation thickness and intensity, 221
    corresponding skeleton, 216
    fan bone region location, 219

    indicators of search region and the found body point, 218
    larger animal applications, 226–8
    positive fan bone sample region, 220
    poultry cutting applications, 215–17
    poultry slaughter and production operations, 213–15
    sample CAT scan model, 216
    sample non-fan bone regions, 220
    sequence of operations in poultry slaughter facility, 214
    SIFT feature location at a fixed scale and orientation, 217
    steps in beef slaughter process, 226
  feature selection, extraction and analysis, 211–12
    Haar-like features, 212
    histogram of oriented gradients, 212
    scale invariant feature transform, 211–12
  future trends, 228–30
    Kinect device and processing to locate WOGs, 229
    ultrasonic holographic image of shoulder joint, 229
    ultrasonic holographic image of wing joint, 229
  machine and computer vision, 208–11

automated cutting (*Cont.*)
    human compared to machine vision,
        209
    machine vision techniques, 210–11
    machine learning algorithms, 212–13
        Adaboost, 213
        K-nearest neighbour, 213
        support vector machine, 213
    time-of-flight (TOF) sensor for
        automatic chicken tracking, 221–6,
        228
        estimating the belt frame, 222–4, 223
        experiment, 226
        extracting without giblets (WOG)
        point cloud, 224–5
        intensity image, extracted point
        cloud, polynomial approximation,
        principal components, 225
        maximum principal curvature,
        minimum principal curvature,
        mean curvature and average mean
        curvature, 225
        orientation of WOG, 225
        set-up, 222
        TOF camera with visible imaging to
        locate position and orientation of
        WOGs, 228
        tracked WOG orientations, 226
        transformation of range data from
        camera co-ordinate system to belt
        co-ordinate system, 222
automated grading, 354–60
    quality of whole Atlantic salmon, 355–7
    representative applications, 358–60
automated inspection algorithm, 343–4
    structure of the intensity-differentiation
        and ratio-differentiation mapping,
        344
automated processing, 360–7
    Atlantic salmon fillets trimming, 362–4
    pre-rigor removal of pin bone from
        Atlantic salmon fillet, 364–6
    representative applications, 367
    salmonids slaughtering, 360–2
automated sorting, 354–60
    representative applications, 358–60
    weight estimation of whole herring,
        357–8
automatic sorting, 150–77

    advanced techniques and their
        application, 160–72
    architecture of complete contaminant
        detection system, 171
    biscuit sorting, 164–5
    cake sorting, 162
    chocolate cakelets and mode of
        illumination, 162
    cream biscuit inspection, 165
    ergot location among wheat grains,
        172
    erosion and dilation, 167
    locating contaminants in cereals,
        169–72
    morphological analysis, 166–9
    object location based on point
        features, 163–5
    receiver operating characteristic, 167
    robust object location, 160–3
    statistical pattern recognition,
        165–6
    use of Hough transform, 165
    use of Hough transform for biscuit
        location, 161
    use of morphology for scratch
        detection, 168
    use of morphology to recover rat
        droppings, 270
alternative image modalities, 172–3
basic techniques and their application,
    152–60
    basic properties of the masks, 153
    basic texture detection using busy-
        ness concept, 157
    centroidal profile method for object
        shape analysis, 159
    insect location, 156
    locating insects in cereals, 155
    principle of Hough transform, 160
    result of basic algorithm for
        connected component analysis,
        158
    shape analysis, 157
    skeleton concept, 160
    texture analysis, 157
    thresholding and edge detection, 152
    thresholding and feature detection,
        152–5
future trends, 176–7

recent advances, 175–6
special real-time hardware for food
    sorting, 173–5
axial chromatic aberration (ACA), 407–8

B-mode, 281
B-scan, 16
back propagation neural network (BPNN)
    classifier, 405
bakery industry
    applications for analysing biscuits,
        436–9
    applications for analysing bread, 423–32
    applications for analysing muffins,
        432–6
    applications for analysing other bakery
        products, 444–5
    applications for analysing pizza bases,
        439–44
    computer vision, 422–47
    future trends, 445–6
band sequential (BSQ), 29
base image, 45
Bayes decision theory, 114
beam-splitting devices, 38–9
beef, 263, 266–9
    applications, 268–9
        muscle images corrected for shading,
            268
    carcass grade prediction, 267–8
    palatability prediction, 266–7
biconical antennas, 83
bilinear modelling (BLM), 458
bird detection algorithm, 339–41
    reference points defining ROI, 340
biscuits
    computer vision applications for
        analysis, 436–9
        colour, 436–8
        lion-shaped crackers, 439
        size and shape, 438–9
black body, 18
blob, 141
bread
    colour, 423–6
        browning at bread surface during
            baking, 424
    computer vision applications for
        analysis, 423–32

bread slice acquired by CCD camera
    and X-ray tomography, 426
size and shape, 426–9
texture, 429–32

C/C++, 54
C-scan, 16
calibration, 139
CAMERA Link, 403
Canada eastern hard red winter (CEHRW),
    464
Canada eastern red spring (CERS), 464
Canada western amber durum (CWAD),
    464, 466
Canada western red spring (CWRS), 459,
    464, 466
Canada western red winter (CWRW),
    464
Canadian Grain Commission (CGC), 455
Canadian Grain Grading Guide, 454–5
Canny edge detector, 441
carcass
    grade prediction of beef, 267–8
    production system and market
        characteristics, 310–13
    RTU for composition and meat
        traits prediction in large animals,
        285–93
    correlation coefficients and
        confidence interval between SFD
        and LMA, 287
    in vivo carcass traits prediction study
        for sheep and goat, 289–90
    RTU for composition and meat traits
        prediction in small animals and
        fish, 293–303
    breast measurements and body weight
        with broilers, 295–6
    cross-sectional view of broiler
        chicken breast muscle, 298
    fillet cross-section, 303
    fish carcass traits prediction, 301–2
    image acquisition procedure with a
        linear probe, 299
    ultrasound evaluation for composition
        and meat traits, 278–82
    milestones for ultrasound
        development and application in
        animal science, 279–80

carcass detection length (CDL), 340
cattle, 285–8
central processing units (CPU), 122
centroidal profile method, 159
cereal grains
    green immature kernels determination,
        469–71
    hyperspectral imaging, 456–9
    mildew effect on quality of end-
        products, 471–3
    multispectral imaging systems
        development, 451–78, 473–7
        green seeds detection, 475–7
        wavelength selection, 473–5
    wheat fusarium damage detection,
        461–5
    wheat mildew damage detection,
        459–61
    wheat sprout damage, 465–9
charge-coupled device (CCD), 5, 137, 260,
    385, 389–90, 403, 406, 425
chemical maps, 409
Chi-square, 109
circularity measure, 158
code reading, 141
collimator, 66
colour, 386–7
colour calibration, 139
colour features, 257, 258
colour imaging, 402–6
    applications, 404–6
    features, 403–4
    systems, 402–3
        area-scan colour imaging system, 403
colourspace, 258
common-aperture multispectral imaging,
    335–6
complementary metal oxide semiconductor
    (CMOS), 137, 403
complementary metal oxide silicon
    (CMOS) cameras, 11
component analysis, 211
computed tomography (CT), 65, 66–7,
    259, 294, 298
computer-assisted tomography (CAT), 65
computer vision, 120, 133
    automated cutting in food industry,
        206–31
        application examples, 213–28

feature selection, extraction and
    analysis, 211–12
future trends, 228–30
machine learning algorithms, 212–13
machine vision and computer vision,
    208–11
automatic sorting in the food industry,
    150–77
    advanced techniques and their
        application, 160–72
    alternative image modalities, 172–3
    basic techniques and their application,
        152–60
    future trends, 176–7
    recent advances, 175–6
    special real-time hardware for food
        sorting, 173–5
bakery industry, 422–47
    applications for analysing biscuits,
        436–9
    applications for analysing bread,
        423–32
    applications for analysing muffins,
        432–6
    applications for analysing other
        bakery products, 444–5
    applications for analysing pizza
        bases, 439–44
    future trends, 445–6
fish industry, 352–76
    applications, 373–4
    automated processing, 360–7
    automated sorting and grading,
        354–60
    future trends, 374–5
    needs, 353–4
    process understanding and
        optimisation, 367–73
food processing, 133–48
    3-D image of cookies, 144
    application examples, 143–8
    bread shape and baking
        degree, 146
    chocolate and cookie inspection, 143
    fill level in soup can and fill level
        measurement of margarine
        packages, 147
    foreign object detection, 145–6
    freshness watchdog, 144–5

hamburger bun 3-D image, 146
meat slaughtering and portioning, 147
package fill level, 146–7
quality grading of seafood, 143–4
robotised sorting and bun picking, 148
foreign body detection and removal in the food industry, 181–200
fundamentals of X-ray inspection, 188–95
optical inspection, 183–8
X-ray inspection of food products, 195–200
fresh and processed meat, 255–72
application and implementation, 260–9
application and implementation for lamb and pork, 269–71
image features, 256–60
fruit, vegetable and nut quality evaluation and control, 379–95
applications in inspection of external features, 386–8
machine vision systems, 381–6
real-time automatic inspection system, 388–92
grain quality evaluation, 400–18
colour imaging, 402–6
hyperspectral imaging, 406–11
overview, 400–2
thermal imaging, 415–18
X-ray imaging, 411–15
image analysis methods selection, 139–43
3-D smart camera scans bacon, 142
blob analysis, 141
calibration, 139
chicken breasts 3-D image, 142
code reading, 141
edge finding, 141
example of pattern matching, 141
filters, 139–40
image rectification through removal of lens and perspective distortion, 139
inspection of logotype and reading of bar code, 140
optical character recognition text reading, 143

pattern matching, 140
pixel counting, 140–1
region of interest, 140
special mechanical considerations, 143
image processing techniques in the food and beverage industries, 97–123
classification, 113–19
digital image analysis techniques, 99–113
relevance, impact and trends, 119–22
infrared techniques for image acquisition in the food and beverage industries, 3–26
acquisition systems, 7–24
electromagnetic spectrum, 5–7
nomenclature and abbreviation, 26
poultry industry, 330–51
future trends, 350
online line-scan safety inspection, 336–49
processing applications, 331–2
spectral imaging inspection, 333–6
standard sensors or vision systems, 134
technology selection, 136–9
choice of conveyor colour to create contrast, 138
imager types, 137–9
laser triangulation set-up, 138
vision system types, 136–7
tomographic techniques in the food and beverages industries, 64–96
applications, 89–93
electrical impedance, 69–86
image reconstruction, 86–9
nuclear tomography, 65–9
vision system typically operates in four-step cycles, 134
vs humans, 134
connected components analysis, 158
connection weights, 115
contrast, 192
correspondence problem, 210
cost function, 87
critical control points (CCP), 332
cycloidal arcs, 242

DataMatrix codes, 141
decision directed acyclic graph (DDAG), 444

Dewar, 21
difference of Gaussian (DoG) filters, 212
differential box counting (DBC), 432
diffuse illuminator, 11
digital demodulator, 78
digital imaging
    quality control applications, 234–7
        adjusted count based on image and
            object dimension, 236
        bubbles rising in beer, 237
        distribution of the length values for
            the rice grains, 236
        measured dispersed rice grains length,
            234
        measuring length as a projected
            distance, 235
digital signal processor, 78, 122, 174, 389
digital video creator, 414
directed acyclic graph (DAG), 444
dual-band optimisation technique, 187
dual-camera imaging, 335
dual-energy detection (DEXA), 173, 190
dynamic range, 192

EAN13, 141
edge, 141
edge-based segmentation, 100, 101
edge finding, 141
electrical capacitance tomography (ECT),
    71
electrical resistance tomography (ERT), 71
electron-multiplying charge-coupled-
    device (EMCCD), 338
electronic noses, 463
electronically tunable filters, 34–8, 407
elliptical Fourier descriptors, 387
embedded techniques, 109
Environment for Visualising Images
    (ENVI), 53
equivalence principle, 88
3-D ERT P2000 system, 89
European CCIR (Comite
    Consultatif International des
    Radiocommunication), 13
extreme ultraviolet (EUV), 6

fan beam geometry, 67
far ultraviolet (FUV), 6
farm to table approach, 182

fast Fourier transform (FFT), 78, 174
feature extraction, 106
feature selection, 106
fecal contamination detection, 346–9
    further research, 349
    line-scanning imaging system and
        software, 347–8
    multispectral detection, 348–9
    overview, 346–7
Federal Communications Committee
    (FCC), 82
feedforward neural network, 117, 438
feedlot feeding, 311–12
Feret diameter, 440
field programmable gate arrays (FPGA),
    122, 174
fillet trimming
    Atlantic salmon, 362–4
        3-D image acquired using the
            z-Snapper 3-D camera, 365
        proof-of-concept demonstrator, 363
filleting machines, 354
filter techniques, 108
filter-wheel multispectral imaging, 335
filters, 139–40
FireWire, 403
fish, 300–3
fish industry
    automated processing, 360–7
    automated sorting and grading, 354–60
    computer vision, 352–76
    computer vision applications challenges,
        373–4
    future trends, 374–5
    needs for computer vision, 353–4
    process understanding and optimisation,
        367–73
        first handling condition parameters
            and end product quality, 367–9
        fishing vessel water separator effect
            on herring fillet quality, 370–2
        magnetic resonance images of
            Atlantic salmon fillet, 369
        representative applications, 373
flat-field correction, 50
floating voltage-to-current converter, 75
fluid bed drying, 89
focal plane array (FPA), 415
focal spot size, 189

food and beverage industries
  computer vision and infrared techniques
      for image acquisition, 3–26
    acquisition systems, 7–24
    electromagnetic spectrum, 5–7
    nomenclature and abbreviation, 26
  hyperspectral and multispectral imaging,
      27–58
    applications for food and beverage
        products, 54–8
    calibration of spectral imaging
        systems, 41–7
    spectral image acquisition methods,
        28–30
    spectral images and analysis
        techniques, 47–54
    spectral imaging systems
        construction, 30–41
  image processing techniques for
      computer vision, 97–123
    classification, 113–19
    digital image analysis techniques,
        99–113
    relevance, impact and trends,
        119–22
  tomographic techniques for computer
      vision, 64–96
    applications, 89–93
    electrical impedance, 69–86
    image reconstruction, 86–9
    nuclear tomography, 65–9
food industry
  computer vision for automated cutting,
      206–31
    application examples, 213–28
    feature selection, extraction and
        analysis, 211–12
    machine learning algorithms, 212–13
    machine vision and computer vision,
        208–11
  computer vision for automatic sorting,
      150–77
    advanced techniques and their
        application, 160–72
    alternative image modalities, 172–3
    basic techniques and their application,
        152–60
    special real-time hardware for food
        sorting, 173–5

computer vision for foreign body
    detection and removal,
    181–200
  fundamentals of X-ray inspection,
      188–95
  optical inspection, 183–8
  X-ray inspection of food products,
      195–200
food microstructure
  digital imaging quality control
      applications, 234–7
  image analysis, 233–51
  internal structure characterisation,
      237–8
  number and spatial distribution, 243–9
    circular intersections through a
        foamed food produce, 245
    layer thickness measurement, 248
    lipid distribution in custard, 247
    optical section of unmatched features,
        246
  surfaces and fractal dimension,
      249–50
    fractal surface measurement, 251
    two fractal surfaces, 250
  volume, surface and length, 238–43
    bread slice surface, 241
    fibres viewed in transmission light
        microscope, 243
    meat and fat of a beef steak, 239
    threshold and skeletonised cell walls
        in plant tissue, 244
food processing
  application examples, 143–8
  computer vision, 133–48
  image analysis methods selection,
      139–43
  technology selection, 136–9
Food Safety and Inspection Service
    (FSIS), 332, 344–5
foreign body detection and removal
  computer vision in the food industry,
      181–200
  optical inspection, 183–8
    components, 183–4
    computer processing and algorithm
        development, 185–6
    problems encountered, 184–5
    recent research, 186–8

foreign body detection and removal (*Cont.*)
X-ray inspection
attenuated and unattenuated X-ray
images of an apple, 194
components, 189–91
effect of focal spot size on resolution,
192
fundamentals, 188–95
problems with X-ray images, 191–4
X-ray of uniform intensity incident on
curved object, 193
X-ray inspection of food products,
195–200
apples, 199
cherries, 199
digital photographs and X-ray images
of olives with fruit fly infestation,
196
grain inspection, 197–8
granary weevil in wheat kernels, 198
luggage inspection for contraband
food products, 195–6
miscellaneous food products,
199–200
packaged foods, 196–7
poultry inspection, 197
tree nuts, 199
X-ray images of codling moth
infestation in an apple, 199
foreign object detection, 145–6
four-electrode method, 73
Fourier descriptors, 387
Fourier transform, 430, 442
fractal dimension, 249–50
frame grabber, 7
fresh meat, 255–72
computer vision application and
implementation, 260–9
beef quality assessment, 261–2
lamb quality assessment, 263
pork quality assessment, 264
processed meat quality assessment,
265
future trends, 271
image features, 256–60
cut beef carcass, 257
hyperspectral scattering contours, 259
freshness watchdog, 144–5
front half, 215

fruits
computer vision applications in
inspection of external features,
386–8
colour information, 386–7
decay in citrus caused by fungus
infection, 389
external defects detection, 388
morphological parameters, 387
future trends, 392–4
internal quality assessment, 393–4
non-standard computer vision
systems, 392–3
machine vision systems, 381–6
quality evaluation and control using
computer vision, 379–95
real-time automatic inspection system,
388–92
FT-NIR spectroscopy, 466–7
fusarium
wheat damage detection, 461–5
inspector *vs.* predicted grade levels
for CESRW, 462
kernel classification results for a
universal PLS-DA model, 465
spectra of sound kernels, 464
fusarium head blight, 453–4,
461–2

gas chromatography, 463
geometric control points (GCP), 407
GiGE VISION, 403
goat, 288–93
grain products
hyperspectral imaging, 456–9
mildew effect on quality of end-
products, 471–3
multispectral imaging systems
development, 451–78, 473–7
green seeds detection, 475–7
wavelength selection, 473–5
wheat fusarium damage detection,
461–5
wheat mildew damage detection,
459–61
wheat sprout damage, 465–9
grain quality
colour imaging, 402–6
computer vision evaluation, 400–18

hyperspectral imaging, 406–11
overview, 400–2
    line-scan machine vision system, 402
    thermal imaging, 415–18
    X-ray imaging, 411–15
graphics processing unit (GPU), 122, 175
green immature kernels
    determination in cereal grains, 469–71
        green seeds detection in barley and
            wheat, 472
        reflectance spectra for normal and
            green barley, 471
        seed count for green barley kernels,
            470
green seeds
    detection, 475–7
        multispectral system consisting of a
            monochrome GigE camera, 476
        waveband spectral imaging system of
            barley samples, 477
        wavelength measurement of barley
            samples, 476
grey-level co-occurrence matrix (GLCM),
    404, 414, 430
grey-level run length matrix (GLRM), 404
ground-probing radar (GPR), 5

Haar-like features, 212
Haar transform, 431
HACCP-Based Inspection Models Project
    (HIMP), 332, 336–7
Hagberg falling number test, 453
half-value layer (HVL), 283
ham, 270–1
hand-eye calibration, 139
hard field effect, 6
hard field sensing, 65
hard vitreous kernels (HVK), 454
Hazard Analysis and Critical Control Point
    (HACCP), 332, 336–7, 350
hidden units, 115
high-performance CCD cameras, 40–1
high-performance liquid chromatography,
    463
high-power bichromatic emitting diodes,
    463
histogram of oriented gradients (HoG),
    212
Hough transform, 161

hypercube, 34, 53
Hyperspec, 53
hyperspectral imaging, 28–9, 259, 270,
    333, 394, 406–11, 455, 456–9
    applications, 409–11
    features, 408–9
    HSI hypercube, 456
    image acquisition and calibration, 456–8
        system in a light-tight enclosure, 457
    spectral analyses methods, 458–9
    systems, 406–8
        area-scan NIR hyperspectral imaging
            system, 408
        three-dimensional hyperspectral data
            representation in a hypercube, 408

illumination device, 7
image acquisition, 384–6, 456–8
    acquisition systems, 7–24
        essential elements of a typical
            computer vision system, 7
    CCD sensor structure, 385
    computer vision, 7–14
        electronics, 11–14
        frame grabber card general
            structure, 13
        illumination, 8–11
        lighting geometries, 10
        possible CCD architectures, 12
        relative spectral energy distribution
            of daylight, incandescent and
            fluorescent light, 9
        sensor operation in passive and active
            mode, 12
    computer vision and infrared techniques
        in the food and beverage industries,
        3–26
    electromagnetic spectrum, 5–7
        visible and non-visible range, 5
    infrared, 17–24
        bolometer detector schematic
            representation, 23
        cooled IR detectors, 19–21
        decreasing temperature curves of
            different materials, 18
        Dewar schematic diagram, 21
        first-generation pyroelectric tube, 22
        hybrid focal plane architecture for
            HgCdTe-based IR detector, 20

image acquisition (*Cont.*)
  uncooled IR detectors, 21–4
  ultrasound, 14–17
image analysis
  food microstructure, 233–51
    digital imaging quality control
      applications, 234–7
    internal structure characterisation,
      237–8
    number and spatial distribution,
      243–9
    surfaces and fractal dimension,
      249–50
    volume, surface and length, 238–43
image calibration, 456–8
image processing techniques
  classification, 113–19
    application examples, 117–19
    description of techniques, 113–17
    important statistical and neural
      pattern recognition techniques, 116
    multilayer perceptron classifier
      architecture, 118
  computer vision in the food and
    beverage industries, 97–123
  digital image analysis techniques,
    99–113
  digital image processing classes and
    examples of operations within
    them, 99
  feature extraction and selection, 105–13
    application examples, 110–13
    description of techniques, 105–10
    double log spectrum of averaged
      singular values computed from
      CCD images, 111
    MATLAB source code for computing
      the kernel principal components,
      112–13
    MATLAB source code for computing
      the quaternionic singular value
      decomposition, 112
  image segmentation, 99–105
    application examples, 101–4
    hybrid segmentation approach for
      pores/defects, 103
    MATLAB source code for
      segmenting raw beef joint, 104–5
    techniques description, 99–101

relevance, impact and trends, 119–22
  image processing methods used in
    computer vision applications,
    121–2
  impact and practical examples, 120
  introduction and relevance, 119–20
  recent trends in accelerating image
    processing operations, 122
image rectification, 139
image resolution, 6
image segmentation, 99–105, 257, 269
  application examples, 101–4
  hybrid segmentation approach for pores/
    defects, 103
  MATLAB source code for segmenting
    raw beef joint, 104–5
  techniques description, 99–101
independent component analysis (ICA), 107
indium–antimonide (InSb), 407
infrared techniques
  and computer vision for image
    acquisition in the food and
    beverage industries, 3–26
  acquisition systems, 7–24
  electromagnetic spectrum, 5–7
  nomenclature and abbreviation, 26
InGaAs CCD cameras, 39–40
InGaAs-detector-based cameras, 407
input images, 45
intensity reflection, 14
internal microstructure, 237–8, 239
intramuscular fat (IMF), 303–9
  RTU image analysis mathematical
    modelling approaches, 305–9
  cattle lumbar section showing LTL
    muscle, 306
  trials to predict IMF percentage or
    marbling score in cattle and swine,
    307–8
inverse problem, 86, 88

K-means clustering, 211
K-nearest neighbour (k-NN), 213
kernel PCA, 111
Kinect device, 228
Kirchoff current law, 75

LabVIEW, 54
lamb, 269

Lamour effect, 68
laser triangulation systems, 138
lateral chromatic aberration
        (LCA), 407–8
leave-one-out cross-validation (LOOCV),
        355–6, 360
lighting system, 383–4
line-scan method, 29
line-scanning imaging system, 338–9,
        347–8
    set-up on pilot-scale processing line,
        347
linear discriminant analysis (LDA), 109,
        355–6, 406, 417, 458
liquid crystal tunable filter (LCTF), 34,
        37, 407
Lloyd method, 428
long-wave NIR (LWNIR), 411
longissimus thoracis lumborum (LTL),
        291, 299, 309
low pass filtering, 78
lumborum muscle area (LMA), 286–8

machine vision, 208
Machine Vision Association (MVA), 208
machine vision systems, 381–6, 401
    good lighting, 383–4
    image acquisition, 384–6
    singulation, 381–3
magnetic resonance imaging (MRI), 68–9,
        259, 294, 298, 353, 367–8, 380,
        393, 429
mass spectrometry, 463
MATLAB, 54, 104, 111
meat quality
    real-time ultrasound (RTU) imaging for
        quality control, 277–314
    basic imaging principles, 282–5
    carcass composition and meat traits
        evaluation, 278–82
    carcass composition and meat traits
        prediction in large animals, 285–93
    carcass composition and meat traits
        prediction in small animals and
        fish, 293–303
    prediction of intramuscular fat (IMF)
        in vivo, 303–9
    production system and market
        carcass, 310–13

meat traits
    cattle and swine, 285–8
    fish, 300–3
    poultry, 293–8
    rabbit, 298–300
    sheep and goats, 288–93
mechanised automation, 354
mercury–cadmium–telluride (HgCdTe), 407
MIA_Toolbox, 53
mildew, 452–3
    damage detection in wheat, 459–61
        classification image showing normal
            kernel regions, 460
        inspector *vs.* predicted grade levels
            for CESRW, 462
        performance of PLS regression
            models for mildew, 461
        spectral characteristics of normal
            wheat kernels, 460
    effect on quality of end-products, 471–3
        flour colour based on spectral
            measurements of grain, 473
        level grade of grain sample *vs.* flour
            colour, 474
        summary of PLS models to predict
            Agtron and KJ flour colour, 473
minimum bounding rectangle (MBR), 224
muffins
    blueberry muffin with different degrees
        of doneness, 433
    computer vision applications for
        analysis, 432–6
    3-D model of a chocolate muffin, 435
    colour, 432–4
    size and shape, 434–6
multilayer perceptron (MLP) network, 117
multimodal imaging, 93
multiple linear regression (MLR), 458
multispectral imaging, 29–30, 335–6,
        348–9, 360
    line-scan hyperspectral imaging system
        for three birds, 348
    line-scan hyperspectral imaging system
        processing results, 349
    quality evaluation of cereal grains and
        grain products, 451–78
        development, 473–7
        green immature kernels
            determination, 469–71

multispectral imaging (*Cont.*)
  hyperspectral imaging, 456–9
  mildew effect on quality of end-
    products, 471–3
  wheat fusarium damage detection,
    461–5
  wheat mildew damage detection,
    459–61
  wheat sprout damage, 465–9

naval orange worm (NOW), 199
near-infrared HSI, 409, 410
near-infrared (NIR), 401, 418
near-ultraviolet (NUV), 6
nearest neighbour method, 166
neighbouring method, 73
neural networks, 115
nodes, 101
noise, 191
nuclear magnetic resonance (NMR)
    imaging *see* magnetic resonance
    imaging (MRI)
nuts
  computer vision applications in
    inspection of external features,
    386–8
    colour information, 386–7
    external defects detection, 388
    morphological parameters, 387
  future trends, 392–4
    internal quality assessment, 393–4
    non-standard computer vision
      systems, 392–3
  machine vision systems, 381–6
  quality evaluation and control using
    computer vision, 379–95
  real-time automatic inspection system,
    388–92

object number, 243–9
Offner spectrograph, 34
online line-scan safety inspection,
    336–49
  fecal contamination detection, 346–9
  wholesomeness, 337–46
opposite method, 73
optical character recognition (OCR) text
    reading, 143
optical inspection, 182, 183–8

components, 183–4
computer processing and algorithm
    development, 185–6
future developments, 188
problems encountered, 184–5
  accuracy, 184
  image acquisition, 184–5
recent research, 186–7
  almonds, 187
  fresh fruit, 186
  grains, 187
  pistachios, 187
sample optical inspection system,
    183–4

package fill level, 146–7
palatability
  prediction for beef, 266–7
    automatic segmentation of muscle,
      267
parallel-beam geometry, 67
partial least squares discriminant analysis
    (PLS-DA), 458
pattern matching, 140
phase sensitive demodulation, 78
phosphors, 189
Picosecond Pulse Lab (PSPL), 84
piezoelectric effect, 284
pin bones
  pre-rigor removal from Atlantic salmon
    fillet, 364–6
  nomenclature for describing fish
    anatomy, 366
  Trio FDS 2000 PR-PB-4 machine, 366
pixel, 39, 140
pixel co-occurrence algorithm, 258
pixel counting, 140–1
pixel difference histograms (PDH), 258
pixel-pixel operation, 152
pizza base
  computer vision applications for
    analysis, 439–44
    colour, 439
    flowing, poor alignment, poor
      pressing and standard, 440
    size and shape, 439–44
Planck's equation, 6, 68
Planck's law, 18
PLS_Toolbox, 53

point-scan method, 28
polar method, 73
pomegranates, 390
pork, 269–70
poultry, 293–8
poultry industry
    computer vision, 330–51
    future trends, 350
    online line-scan safety inspection,
        336–49
    processing applications, 331–2
    spectral imaging inspection, 333–6
        multispectral imaging, 335–6
        spectroscopy, 334–5
Poultry Product Inspection Act (PPIA),
    332
pre-harvest sprouting, 465
pre-processing, 139
pricking wheel, 187
principal component analysis (PCA), 107,
    458
principal component regression (PCR),
    458
prism-grating-prism (PGP), 33
probabilistic neural network (PNN)
    classifier, 404–5
process tomography, 90
processed meat, 255–72
    computer vision application and
        implementation, 260–9
    beef quality assessment, 261–2
    lamb quality assessment, 263
    pork quality assessment, 264
    processed meat quality assessment,
        265
    computer vision application and
        implementation for lamb and pork,
        269–71
    image features, 256–60
    cut beef carcass, 257
    hyperspectral scattering contours, 259
pulse-forming network (PFN), 84
pushbroom method, 29
pyroelectricity, 22

quadratic discriminant analysis (QDA),
    406, 417
quality control
    digital imaging, 234–7

adjusted count based on image and
    object dimension, 236
bubbles rising in beer, 237
distribution of the length values for
    the rice grains, 236
measured dispersed rice grains
    length, 234
measuring length as a projected
    distance, 235
real-time ultrasound (RTU) imaging of
    meat, 277–314
basic imaging principles, 282–5
carcass composition and meat traits
    evaluation, 278–82
carcass composition and meat traits
    prediction in large animals,
    285–93
carcass composition and meat traits
    prediction in small animals and
    fish, 293–303
future trends, 313–14
prediction of intramuscular fat (IMF)
    in vivo, 303–9
production system and market
    carcass, 310–13
quality grading
whole Atlantic salmon, 355–7
quantum well infrared photodetectors
    (QWIP), 407
quartz tungsten halogen (QTH), 31
quasi-Newton method, 119

rabbit, 298–300
radial basis function (RBF), 443
radiometric calibration, 47
RANSAC, 161
real-time automatic inspection system,
    388–92
automatic sorting of pomegranate arils,
    392
real-time ultrasound (RTU)
basic imaging principles, 282–5
    tissue acoustic properties, 284
    tissue imaging, 284–5
    transducers, 284
carcass composition and meat traits
    evaluation, 278–82
carcass composition and meat traits
    prediction in large animals, 285–93

real-time ultrasound (RTU) (*Cont.*)
  carcass composition and meat traits
      prediction in small animals and
      fish, 293–303
    future trends, 313–14
      novel ultrasound scanning
          development, 314
    prediction of intramuscular fat (IMF)
        in vivo, 303–9
    production system optimization and
        market carcass characteristics,
        310–13
      animal sorting prior to feedlot
          feeding, 311–12
      market strategy optimisation,
          312–13
      slaughter date, 310–11
    quality control of meat, 277–314
receiver operating characteristic
    (ROC), 166
reflection gratings, 34
region of interest (ROI), 140, 257–8, 336
  parameters and wavebands, 341–3
    average spectrum for wholesome and
        unwholesome ROI pixels, 343
    range of the 55-channel spectral
        differences, 342
region segmentation, 100–1
regression and multivariate curve
    resolution (MCR), 458
remote sensing, 65
RGB cameras, 227
ring illuminator, 11
ROI pixels, 341, 343–4
root-mean-squared-error (RMSE), 458

salmonids, 360–2
scab *see* fusarium head blight
scale invariant feature transform (SIFT),
    211, 217
Schottky barrier, 23
scintillators, 66
second and further processing, 214
sensor fusion technique, 93
sheep, 288–93
short-wave NIR (SWNIR), 409–10, 411
shortwave infrared (SWIR), 463, 466–7
Sick LMS 400 laser scanner, 227
silicon CCD cameras, 39

single-shot method, 30
singular value decomposition (SVD), 110
singulation, 381–3
  errors during fruit transport, 382
skeletonisation, 159
slaughter date, 310–11
slaughtering
  salmonids, 360–2
    different steps in the 3-D machine
        vision system, 361
    high throughput and moving un-
        anaesthetized fish, 362
    pilot installation of the slaughter line,
        361
soft field effect, 6
soft-field sensing, 65
soft independent modelling of class
    analogy (SIMCA), 458
spatial calibration, 43–4, 45
  point-scan, line-scan, area-scan
      hyperspectral imaging systems, 45
spatial distribution, 243–9
spatial ratio I (SRI), 441
spatial ratio II (SRII), 441
spatial resolution, 6
special mechanical considerations, 143
spectral angle mapper (SAM), 458
spectral calibration, 41–3
  line-scan hyperspectral imaging system
      using calibration lamps, 42
  Raman imaging system, 43
spectral imaging systems
  applications for food and beverage
      products, 54–8
    hyperspectral imaging applications
        for food and beverage products,
        55–6
    multispectral imaging applications for
        food and beverage products, 57
  area detectors, 39–41
    high-performance CCD cameras,
        40–1
    InGaAs CCD cameras, 39–40
    silicon CCD cameras, 39
    silicon EMCCD detector and an
        InGaAs photodiode detector, 40
  calibration, 41–7
  construction, 30–41
  general configuration, 31

hyperspectral and multispectral imaging in the food and beverage industries, 27–58
image registration, 44–6
light sources, 31–2
  excitation sources, 32
  halogen fiberoptic light and high-power blue LED line light, 32
  illumination sources, 31–2
other calibrations, 47
spatial calibration, 43–4, 45
  point-scan, line-scan, area-scan hyperspectral imaging systems, 45
spectral calibration, 41–3
  line-scan hyperspectral imaging system using calibration lamps, 42
  Raman imaging system, 43
spectral image acquisition methods, 28–30
  hyperspectral imaging, 28–9
  methods for acquiring 3-D hyperspectral cubes, 28
  multispectral imaging, 29–30
spectral image analysis, 49–53
  data pre-processing, 50
  ENVI software for hyperspectral image analysis, 54
  principal component analysis, 53
  software, 53–4
  spectral dimension reduction and optimal band selection, 50–3
  spectral information divergence mapping, 52
spectral images and analysis techniques, 47–54
  main steps for hyperspectral and multispectral image analysis, 49
  representative hyperspectral images acquired by different methods, 48
  spectral image data, 47–9
wavelength dispersive devices, 33–9
  acousto-optic tunable filter (AOTF), 36
  beam-splitting devices, 38–9
  electronically tunable filters, 34–8
  imaging spectrograph-based line-scan hyperspectral fluorescence imaging system, 36
  imaging spectrographs, 33–4
  liquid crystal tunable filter (LCTF), 37
  Offner imaging spectrograph, 35

two-channel plate beam-splitter and three-channel prism beam-splitter, 38
spectral integration process, 9
spectroscopy, 139
spiking neural networks (SNN), 115
sprout
  wheat damage, 465–9
    α-amylase activity, 466–7
    detection, 467–9
    mean reflectance spectra of CWRS from the VNIR system, 468
stereo vision, 138, 227
structure–processing relationship, 237
structure–property relationship, 237
subcutaneous fat depth (SFD), 286–8, 292
support vector machine–recursive feature elimination (SVM-RFE), 109
support vector machine (SVM), 213, 458
surface dimension, 249–50
surface roughness, 249
swine, 285–8

texture features, 257, 258
thermal cameras, 227
thermal imaging, 415–18
  applications, 416–17
  features, 416
  systems, 415–6
    infrared thermal imaging system, 416
thermographic photography, 17
thermography, 380
three-dimensional (3-D) ultrasonography, 314
three-way sorting, 145–6
thresholding, 100
thresholding-based segmentation, 427
TOBEC, 294, 298
tomographic reconstruction, 190
tomographic techniques
  applications, 89–93
    annular-core distributions of wetted granular particles, 92
    conductivity changes in the skimmed milk, 90
    fluidized bed dryer ECT-based instrumentation, 91

tomographic techniques *(Cont.)*
    computer vision in the food and
        beverages industries, 64–96
    electrical impedance, 69–86
        capacitance measuring circuit based
            on AC excitation and analogue
            demodulation, 80
        conversion of incident wavefield
            to reflected and transmitted
            components, 85
        ECT DAS schematic, 82
        ECT system illustration, 79
        electrical capacitance tomography
            (ECT), 78–81, 82
        electrical resistance tomography
            (ERT), 72–8
        ERT instrument, 72
        ERT measuring strategies, 73
        floating load and grounded load
            voltage to current converter, 76
        Fricke's model of food system, 71
        major components of ERT DAS, 75
        ultra wide band tomography
            (UWBT), 81–6
        UWB antenna examples, 83
        UWB DAS schematic, 84
        UWB DAS with VNA, 86
    image reconstruction, 86–9
        sensitivity distributions isometric
            view, 88
    nuclear tomography, 65–9
        computed tomography (CT), 66–7
        different geometries for tomographic
            imaging, 66
        magnetic resonance imaging (MRI),
            68–9, 70
        modern CT with fan beam geometry,
            67
        MRI images showing water
            distribution and freezing
            behaviour, 70
        MRI system block diagram, 69
tomography, 64
transmission coefficients, 14

ultra wide band (UWB), 82
ultrasound, 270
ultrasound transducers, 284
ultraviolet (UV), 6

unfolding, 238
unit conversion, 139
US RS170, 13

value-based marketing system, 312
vector network analysing (VNA), 85
vegetables
    computer vision applications in
        inspection of external features,
        386–8
        colour information, 386–7
        external defects detection, 388
        morphological parameters, 387
    future trends, 392–4
        internal quality assessment, 393–4
        non-standard computer vision
            systems, 392–3
    machine vision systems, 381–6
    quality evaluation and control using
        computer vision, 379–95
    real-time automatic inspection system,
        388–92
vertical sectioning, 242
visible/near-infrared (Vis/NIR)
    spectroscopy, 334–5
visible near-infrared (VNIR), 459, 468
2-D vision, 138
3-D vision, 138
vision system, 137
Visual Basic, 54
voltage controlled current source (VCC), 75
voltage standing wave ratio (VSWR), 83
voltage-to-current converter, 74

water separator
    effect on herring fillet quality, 370–2
        grading scale, 372
        graphical user interface for computer-
            vision-assisted quality grading, 371
        illustration, 371
        traditional *vs.* new, 372
watershed transformation, 437
wavelength selection, 473–5
    coefficients plot of the FDK-PLS model,
        475
weight estimation
    whole herring, 357–8
        error distribution evaluated by
            LOOCV of fresh herring, 359

feature sets, evaluated using LOOCV, 359
geometrically calibrated intensity image, 358
wheat
 fusarium damage detection, 461–5
 mildew damage detection, 459–61
 sprout damage, 465–9
whiskbroom method, 28
wholesomeness inspection, 337–46
 automated inspection algorithm, 343–4
 bird detection algorithm, 339–41
 line-scanning imaging system, 338–9
 online evaluation, 344–6
  online multispectral line-scan inspection of chicken, 345, 346
 overview, 337
 ROI parameters and wavebands, 341–3
window-pixel operation, 152
wrapper methods, 109

X-ray fluoroscopy, 190
X-ray imaging, 411–15
 applications, 413–15
 features, 413
 systems
  illustration, 412
X-ray inspection, 183
 algorithm development, 194–5
 attenuated and unattenuated X-ray images of an apple, 194

components, 189–91
 detection and imaging, 189–91
 X-ray sources, 189
effect of focal spot size on resolution, 192
food products, 195–200
 apples, 199
 cherries, 199
 digital photographs and X-ray images of olives with fruit fly infestation, 196
 grain inspection, 197–8
 granary weevil in wheat kernels, 198
 luggage inspection for contraband food products, 195–6
 miscellaneous food products, 199–200
 packaged foods, 196–7
 poultry inspection, 197
 tree nuts, 199
 X-ray images of codling moth infestation in an apple, 199
fundamentals, 188–95
problems with X-ray images, 191–4
X-ray of uniform intensity incident on curved object, 193
X-ray tomography, 429
X-ray vision, 145

zone sonography, 314